Mathematical Achievements of Pre-modern Indian Mathematicians

Mathematical Achievements of Pre-modern Indian Mathematicians

T.K. Puttaswamy
Professor Emeritus of Mathematical Sciences,
Ball State University, Muncie, IN, USA

ELSEVIER
AMSTERDAM • BOSTON • HEIDELBERG • LONDON • NEW YORK • OXFORD
PARIS • SAN DIEGO • SAN FRANCISCO • SINGAPORE • SYDNEY • TOKYO

Elsevier
32 Jamestown Road, London NW1 7BY
225 Wyman Street, Waltham, MA 02451, USA

First edition 2012

British Library Cataloguing-in-Publication Data
A catalogue record for this book is available from the British Library

Library of Congress Cataloging-in-Publication Data
A catalog record for this book is available from the Library of Congress

ISBN: 978-0-323-28264-2

This book has been manufactured using Print On Demand technology. Each copy is produced to order and is limited to black ink. The online version of this book will show color figures where appropriate.

Transferred to Digital Printing in 2016

Dedication

In the Loving Memory of my parents,
Srimathi Bhagirathamma and Dr. T. Krishnappa

Contents

Preface

On November 8, 1998, I received a letter from Ms. Helaine Selin of Hampshire College that she would be editing a book on the History of Non-Western Mathematics to be published by Kluwer Academic Publishers and whether I would be interested to take on a project in writing an article of about 15 pages on the Ancient Indian Mathematics. I considered it an honor and I humbly accepted her kind offer. When I was collecting material to write this article, the elegance, the ingenuity, and the masterly skill exhibited in the works of ancient Indian mathematicians inspired me to make an attempt to write a book devoted to the comprehensive and detailed presentation of the mathematical achievements of ancient Indians in the fields of Algebra, Geometry, and Hindu Trigonometry, from the Vedic Period (800 BC) to the seventeenth century of the Christian era. The project of writing this book was undertaken at that time. I did mention this intention of mine at the end of my article entitled "Mathematical Accomplishments of Ancient Indian Mathematicians," in the book, "Mathematics Across Cultures," published by Kluwer Academic Publishers in the year 2000.

The plan of this book differs substantially from the books that have been published so far. Most of these books are in the form of "source books," wherein various topics are presented in the light of contributions by various authors at various times. The present book is chronological in its plan, covering all the contributions in Algebra, Geometry, and Hindu Trigonometry of one ancient Indian mathematician to the next from Vedic Period (800 BC) to the seventeenth century of Christian era.

Some of the prominent features of this book are detailed proofs, adequate illustrative examples, mathematical explanations, and critical observations. To make the contributions by various authors authentic, their original Sanskrit verses with English translations are provided whenever it is necessary and useful. This book is supplemented with an exhaustive Bibliography. The highlight of this book is the inclusion of some special features for the first time to make the reading of the book more interesting and challenging.

The main purpose of this book is twofold: to draw the attention of the modern readers of the world to the historical continuity of the development of mathematics in the ancient India starting from the "Sulva sutras" of the Vedic Period to the seventeenth century and to dispel once for all the prevailing misconception that there was utter stagnation of the development of the mathematics in premodern India after Bhaskara II in the twelfth century.

This book can be used as a text for a two-semester course on the History of Premodern Indian Mathematics at the undergraduate or the graduate level. Most of the chapters contain sufficient material, so that a careful selection can be made to get the needs of any such course.

It is always a great pleasure to acknowledge the contribution of those who make a project of this nature possible. In the preparation of this book, I am deeply indebted to the authors of all the published books and research papers, all the expository source works, and manuscripts available in the various Indian Libraries and Archives, which are listed in the Bibliography. I would like to express my profound gratitude to Ms. Helaine Selin of Hampshire College for her tremendous support and continued guidance.

I am pleased to acknowledge my deep appreciation to Mr. Gangadhara C. Yagati for his expert compositing of the manuscript. I am also grateful to Professor Ubiratan D'Ambrosio, Professor M. Rajagopalan, as well as two other anonymous reviewers at Elsevier Publishing Company for their kind review, contributed comments, and criticism to improve my book. I am highy indebted to Professor Ubiratan D'Ambrosio for gracing this book with his invaluable foreword. I would like to express my heartfelt thanks to the superb officers of the Elsevier Publishing Company for their kind support and unfailing assistance. I owe special thanks to Ms. Patricia Osborn, Acquisition Editor, Applied and Computational Mathematics, Elsevier Science for forwarding my book proposal to their new exciting Elsevier Insights program, as she felt my book proposal was best suited there. It was my great privilege to work with Ms. Lisa Tickner, Publishing Director, Elsevier Insights and I do not find adequate words to express my sincere thanks for her help and guidance. I also wish to place on record my deep appreciation to Dr. Erin Hill-Parks, Associate Acquisition Editor, Elsevier Insights, Ms. Elizabeth Mitza, Publishing Assistant, Elsevier Serial, Series, and Insights, and Ms. Tracey Miller, Editorial Project Manager, Insights Science and Technology Books for their excellent coordinating efforts to bring out the book as early as possible. I express my thanks and appreciation to Ms. Sivakami Ramanathan of MPS Limited company in Chennai, India for handling copy-editing and type setting of the manuscript. I am also very grateful to Ms. Sujata T. Sambandam, Elsevier Insights, Chennai, India and to Mr. Stalin Viswanathan, Project Manager, Elsevier Insights, Chennai, India for working with me diligently with me through production and for the nice get up of the book. I am profoundly grateful to the Elsevier Publishing Company for publishing and releasing my book by August 2012, the year in which the entire country of India is celebrating as its national year of mathematics to commemorate the 125th birth anniversary of India's greatest mathematician, Sri Srinivasa Ramanujan with "Panoramic" lectures and conferences throughout the country. Finally, I thank the members of my family, my wife Pramila, my daughters Shashi and Padma, my son-in-law Kamalesh, and my grandchildren Raghavendra and Nidhishree for their patience and encouragement during the time of writing this book.

Foreword

It is a pleasure for me to write a foreword for this book. I have fond memories of enjoyable and fruitful conversations with T.K. Puttaswamy about the global behavior of solutions of ordinary differential equations and ancient Indian Mathematics. I learned much from those conversations, jumping from the results of his important contributions to a research area of central interest in pure and applied mathematics to his personal views of the History of Mathematics focusing on many personalities and topics of ancient India.

This book draws attention to the historical continuity of the development of mathematics in India, from the Sulvasutras of the Vedic period (800 BC), showing the evolution of contributions up to the seventeenth century. It thus dispels the misconception that there was a stagnation of Indian Mathematics after Bhaskara II in the twelfth century. It avoids a common feature of books on the mathematics thinking of other cultures, which is to justify the importance of Indian Mathematics by comparison with Greek Mathematics. Although the objective of this book is not to make comparisons, occasionally the author mentions correspondent developments in Western Mathematics.

Reading this book is a very interesting and challenging experience. It is an original treatment, showing how mathematics was done by ancient mathematicians. This book is organized chronologically. Each author receives a brief presentation and the developments are as close as possible to the original. Since language represents a fundamental role in the original writings, many Sanskrit verses are presented, so that the reader may appreciate the style of the contribution of the ancients to mathematics. Considering that readers most probably will not be knowledgeable in Sanskrit, the author gives English translations, so that it is possible to recognize the spirit of the vernacular comments. The author develops calculations and proofs using modern and current notations and symbols, but without losing the original conceptions and style of the ancients. Throughout this book, proofs are presented step by step, with illustrative examples, mathematical explanations, and critical observations.

This book provides an opportunity to learn about results not easily found in the literature, some of which are new. For example, a derivation of the formulae to find rational solutions to the Altar Garudachayana, as given in the Baudhayana Sulva Sutra and in the Apasthamba Sulva Sutra. Another very interesting topic in this book is the study of the Vargaprakrti equation, known in Western Mathematics as Pell's Equation, using Brahmagupta–Bhaskara II cyclic method. Puttaswamy gives five new examples to find integral solutions of the equation and also makes some original contributions to the Brahmagupta–Bhaskara II cyclic method. He also gives refined proofs of the contributions of Sridhara, Mahavira, Bhaskara II, and Jyesthedeva.

This book gives the reader an overall idea of the contributions of ancient authors to the development of Indian Mathematics. The overall picture may be very helpful for students and even researchers working on specific authors and topics of Indian Mathematics.

Ubiratan D'Ambrosio
São Paulo, April 2012

Introduction

Civilization is spread over faraway different lands and different languages. The history of ancient remotest civilization is often concealed in mystery and surrounded by controversy. When a mathematician writes the History of Mathematics, he first becomes a historian and then a mathematician. The sacred duty of a historian is to set the facts as he finds them or as he understands them and let the modern readers of the world to interpret and assess these facts according to their own understanding of the subject.

The most marvelous accomplishment for which the entire world is beholden to ancient India is the invention of the decimal numerals with place values, zero, and infinity. According to Professor G.B. Halstead [59], the zero symbol was used in metrics ("Chandas") by Pingala before 200 BC in his "Chandas sutra"—a work on prosody. The ancient India's contributions to mathematics are mainly in the fields of Arithmetic, Geometry, Algebra, and Hindu Trigonometry. Indian Arithmetic is what is now used all over the world.

Veda is the scripture of the Hindus. The very word "Veda" has the derivational meaning that it is the fountainhead and illimitable storehouse of all knowledge. The branch of mathematics that got the earliest attention was geometry. The "Sulva sutras" in Vedas is a manual of geometric constructions. None of the nine "Sulva sutras" is later than the fifth century BC and a few of them may be as old as the ninth century BC. The early geometry of "Sulva sutras" and the beginning of Algebra developed out of the necessities of the Altar (fireplace) construction, mathematical series out of Jain Cosmography, and a little later out of the need for better methods of astronomical computations and Hindu Trigonometry.

More or less overlapping the period of Sulvas, we discover a voluminous Jain literature that comprises a significant amount of mathematical information. If the accurate arrangement and dimensions of the sacrifiable Vedis (Altar) was important to the followers of the Vedic religion, corrected computations of their cosmography was equally important to the Jains. The early Jain works that contain mathematical material in them are "Tattravarthadhigama Sutra" and its commentary by Umasvati, "Sthananga Sutra," "Jambudvipa Prajnapti," "Anuvathadhigama sutra," and "Ksetra Samasa." Umasvati lived in the second century BC. Hence, we can safely assume that most of the above-mentioned works belong to a period dating from 500 BC to 200 BC. The ancient Jain contribution to mathematics is really remarkable. The constant π is approximated to $\sqrt{10}$ and computed up to 13 places of decimals.

The enunciation of the theorem of squares of the diagonal (Pythagoras' theorem) is one of the most remarkable achievements of this period. The "Taittiriya

Samahita" of Yajurveda gives the measurement of Mahavedi Altar in the form of a right triangle of sides 15, 36, and hypotenuse 39. "Katyayana Sutra" gives the construction of a right triangle with sides $((n^2-1)/2)a$, na, $((n^2+1)/2)a$. "Apasthamba Sutra" and "Katyayana Sutra" list several other right triangles. In "Baudhayana Sulva Sutra," the oldest of the nine sulva sutras, one finds the earliest explicit statement of Pythagoras theorem.

In AD 1881, a manuscript consisting of 70 birch-bark leaves, now known as Bakshali Manuscript, was discovered and it is believed to belong to the third century to the seventh century AD. We find interesting topics like arithmetic and geometric progressions, simple equations, and the approximation now known as Bakshali formula for the square root of a nonperfect square number in this manuscript.

With the rising ascendancy of the astronomy of Hindus and the cosmography of Jains, the geometry became circle oriented, after the dawn of the Christian era. Geometry along with the rest of the mathematics became a part of astronomical treatises. The original "Suryasiddhanta" (now lost) and the other siddhantas summarized by Varahamihira (sixth century AD) in his "Pancasiddhanta" must have been written during the opening centuries of the Christian era. Mathematical treatises were also composed during this period. Unfortunately, many of these works have been buried in the sands of time. The Bakshali Manuscript may perhaps be a representative remnant of such works.

Aryabhata I (AD 476) is one of the greatest mathematicians and astronomers of ancient India. He dominated the ancient Indian mathematical world as Plato did the Greek philosophical world. Aryabhata I, well known for his innovations of the astronomical methods, was responsible for a new school of astronomy in ancient India—the Aryabhata I School. The whole host of later Kerala astronomers treated him as "acharya" (Master). In his celebrated composition "Aryabhatiyam," he has provided the table for trigonometric ratio, "Hindu Sine" (called "Jya" in Sanskrit) for acute angles at intervals of $3\frac{3^\circ}{4}$. The accuracy of these values is remarkable.

The ancient Hindus were the first to realize the existence of irrational numbers. Simple surd numbers were used quite freely in the "Sulva sutras." Aryabhata I was the first person to give the approximation $\pi = 3.1416$. It is interesting to observe that Aryabhata I uses the word "Asana" (approximate), implying that the rational value of π given by him is only approximate. This accomplishment of Aryabhata I in the fifth century is truly commendable.

Aryabhata I was followed by his exponent, Bhaskara I (seventh century AD), who wrote a celebrated commentary on "Aryabhatiyam." In the Aryabhata School, Bhaskara I's astronomical works, "Mahabhaskariya" and "Laghu Bhaskariya," are very popular as evinced by a large number of commentaries written on them.

Brahmagupta (AD 628) was the first mathematician to provide the formula for the area of a cyclic quadrilateral. His contributions to geometry are significant. He is the first person to discuss the method of finding a cyclic quadrilateral with rational sides. His celebrated theorems on the circumdiameter of a triangle and on obtaining the diagonals of a cyclic quadrilateral are now well known. Bhaskara II (AD 1150) is so much charmed of Brahmagupta's intellect that he respectfully refers to him as "Mahamatiman" (very intelligent person) and even confers the unique "Ganita Chakra Chudamani" (the gem of the circle of mathematicians) title on Brahmagupta.

Brahmagupta was very well acquainted with the mathematics and astronomy of Aryabhata I. He seems to criticize Aryabhata I severely in his work "Brahma Sphuta Siddhanta" but seems to be compelled to bow to Aryabhata's popularity by referring to Aryabhata with tremendous reverence in his later work, "Khandakadyaka."

Between Brahmagupta and Bhaskara II, we find four treatises exclusively on mathematics—the "Patiganita" and "Trisatika" of Sridhara (eighth century), the "Ganita Sara Sangraha" of Mahavira (ninth century), and "Ganita tilaka" of Sripati (eleventh century). Mahavira's excellent work "Ganita Sara Sangraha" is very elaborate and deals with geometry and algebra. Mahavira enunciated the formula ${}_nC_r = n!/((n-r)!r!)$ for the first time in the history of mathematics.

Around AD 860, Pruthudakaswami Caturveda wrote a masterly commentary on Brahmagupta's "Brahma Sphuta Siddhanta," which contains numerous examples to illustrate the mathematical assertions made in the "Brahma Sphuta Siddhanta." Pruthudakaswami Caturveda is said to have proved "Ptolemy's theorem," according to G.A. Kaye [71].

Aryabhata II belongs to the tenth century, who wrote his astronomical text "Mahasiddhanta." The Jain semireligious works "Triloya Panniti" ("Triloka Prajnapti") of Yativrsabhacarya and "Triloka Sara" and "Gommater Sara" of Nemi Candra were also composed during this period. Sripati who also authored "Siddhanta Sekara" and other astronomical texts flourished in the eleventh century.

Bhaskara II, reckoned as the greatest of all the ancient Indian mathematicians, has several original contributions to expound especially in Algebra and Hindu Trigonometry. The correct formula for the volume and surface area of a sphere and an exceedingly close approximation for the length of the arc in terms of its chord and vice versa are some of his geometrical accomplishments. The last of these is a modification of the formula enunciated earlier by Bhaskara I in his work "Mahabhaskariya."

One of the most significant contributions of ancient Indians to mathematics is in providing a systematic method of solution in integers of the indeterminate equation of the first degree, "Kuttaka" (Pulveriser) $by - ax = c$, presently known as "Diophantine equation," after the great Greek mathematician, Diophantus, who lived probably in the latter half of the third century AD, as well as of the indeterminate equation of the second degree, "Vargaprakriti" ("Varga" meaning square and "Prakrti" referring to the coefficient N) $Nx^2 + k = y^2$. The former has been treated for the first time in the history of mathematics by Aryabhata I, while the latter is a phenomenal contribution of Brahmagupta. Brahmagupta solved completely to find the rational solutions of the equation $Nx^2 + k = y^2$. He also exhibited how to find the integral solutions of the equation $Nx^2 + 1 = y^2$, provided an integral solution $(x\ y) = (\alpha\ \beta)$ can be found for the equation $Nx^2 + k = y^2$, when $k = -1,\ \pm 2,\ \pm 4$. The whole modern world would have to admit that Brahmagupta must have been a genius to accomplish this marvelous feat in AD 628, without using continued fractions, which is the method now employed to solve "Vargaprakriti." The equation $Nx^2 + 1 = y^2$ at present bears the name "Pell's equation" after the English mathematician John Pell (AD 1610–1689), although his association with it consists of merely the publication of solutions of it in his edition of Brounker's translation of Rhonius Algebra in AD 1668. It was an accident that the celebrated Swiss mathematician Leonhard Euler (AD 1707–1783) referred to it as "Pell's equation." But this has no historical justification whatsoever. Pell has made no contribution to this topic.

Bhaskara II solved in integers the equation $Nx^2 + 1 = y^2$ by introducing his famous "Chakravala" (cyclic) method. As an illustration of his elegant and simple method, he gave two examples of solving in integers, the equations $67x^2 + 1 = y^2$ and $61x^2 + 1 = y^2$. The latter example is of great historical interest. This example was proposed by the famous French mathematician Pierre De Fermat (AD 1601–1665) to Frenicle in a letter of February AD 1657. Euler solved this in AD 1732.

In AD 1350, Narayana Pandita, a commentator of Bhaskara II's "Bijaganita," solved in integers two more equations $103x^2 + 1 = y^2$ and $97x^2 + 1 = y^2$ in his work "Bijaganita," using the "Chakravala" method of Bhaskara II. To further demonstrate the elegancy, the beauty, and the simplicity of the "chakravala" method, the present author has included integral solutions of five more new equations: $179x^2 + 1 = y^2$, $131x^2 + 1 = y^2$, $231x^2 + 1 = y^2$, $31x^2 + 1 = y^2$, and $71x^2 + 1 = y^2$.

Bhaskara II has exhibited remarkable ingenuity in devising various other problems, some of them being simultaneous equations and some of them involving equations of higher degree by using clever devices to reduce them to the form $Nx^2 + k = y^2$. The British Sanskrit scholar Henry Thomas Colebrooke translated in AD 1817 the entire works of Brahmagupta and Bhaskara II from Sanskrit to English in his work "Algebra with Arithmetic and Mensuration from the Sanskrit of Brahmagupta and Bhaskara II [27]."

It should be regarded as a highly marvelous accomplishment that Bhaskara introduced the concept of differential and integral calculus though not with modern rigor. If x and x^1 are the mean anomalies of the planet at the ends of consecutive intervals, Bhaskara stated that $\sin x^1 - \sin x = (x^1 - x)\cos x$, which is equivalent to $\delta(\sin x) = \cos x \, \delta x$. He further stated that the derivative as instantaneous speed vanishes at the extreme. According to B.B. Dutta, the credit of having conceived and explicitly mentioning this result should go to the earlier ancient Indian mathematician Munjala (AD 950) and his commentator Prashastidara (AD 950). B.B. Dutta pointed out that in his work "Laghumanasam," Munjala mentioned that the tabular differences of the sines for an arc are proportional to the cosines. The English physicist and mathematician Sir Isaac Newton (AD 1642–1727) and the German philosopher and mathematician Gottfried Wilhelm Leibnitz invented modern calculus and are regarded as the founders of differential and integral calculus.

Modern readers will find the beautiful examples from Mahavira's "Ganita Sara Sangraha" and Bhaskara II's "Lilavati" fascinating and even challenging.

Narayana Pandita (AD 1357), the author of "Ganita Kaumudi," is often confused with Narayana the coauthor of "Kriyakramakari" and the commentator of Bhaskara

II's "Lilavati." Some of the Narayana Pandita's theorems, like three and only three diagonals are possible for a cyclic quadrilateral with given sides and the area of a cyclic quadrilateral is the product of the three diagonals divided by twice the diameter, are not found in the earlier works. In his treatment with cyclic quadrilateral, Narayana Pandita displayed some affinity with Aryabhata School.

Sangama grama Madhava, Paramesvara of Dṛgganitam fame, Nilakantha Somyaji, the author of "Tantra sangraha," are astronomers and mathematicians of high caliber. These above-mentioned mathematicians along with the authors of "Karanapaddhati," "Kriyakramakari," an elaborate commentary of Bhaskara II's "Lilavati," Jyesthadeva, the author of "Yukti Bhasa," which professes to be an exposition in Malayalam of Nilakantha Somyaji's "Tantra sangraha," belong to an era of intense mathematical studies in Kerala from fourteenth to seventeenth century.

A method of integration formed on the delicate geometrical analysis and clever methods of summing the series to obtain the infinite series for π, sine, cosine, and the arc are the most significant accomplishments of this period. This procedure was also successfully employed for determining the volume and the surface area of a sphere. Proofs of Brahmagupta's theorems and trigonometric identities are also discovered in the works of this era.

From the references scattered through the works of others, one can observe that the marvelous developments in the geometry of the circle made during this period are the accomplishments of Sangama grama Madhava. Madhava lived during the latter half of the fourteenth century. Paramesvara, who is believed to have carried out continuous observations of the sky over a period of 55 years and recorded the correct astronomical elements in AD 1430, was a prolific writer and authored about 30 works.

Nilakantha Somyaji's (AD 1443–1545) commentary on "Aryabhatiya" is much more elaborate and substantiates his pride in designating his work "Mahabhasya" (great commentary). His "Tantra sangraha," although chiefly astronomical, includes all new mathematical developments. Nilakantha Somyaji's comments, dated AD 1500, about π are breathtaking and of monumental historical value. In his commentary of "Aryabhatiyam," he poses for himself a question, "Why do we assign π an approximate value instead of giving its true value?" He answers it by stating that "for the simple reason, the ratio of circumference to the diameter can never be expressed as the ratio of two integers." This clearly shows that Swiss–German mathematician, astronomer, physicist, and philosopher Johann Heinrich Lambert (AD 1728–1777) is not the first person who enunciated that π is an irrational number.

Puthumana Somayaji's "Karanapaddhati" appears to be in the form of a compendium. On the other hand, "Yuktibhasa" of Jyesthadeva is a unique work providing the rationale as well as the derivations of all the theorems and the formulae, then in use by Kerala astronomers. Its marvelous expositions reappear in the "Kriyakramakari" in part and in a Sanskrit commentary on "Tantra sangraha" preserved in the college library in Trippunithura in Kerala. The authors of the former are most likely Sankaravariyar (AD 1556), a student and a commentator of Nilakantha Somyaji, and Narayana of mahisa mangalam family. The "Sadrantamala," a much later work written by Sankara Varman, contains all the results without proofs.

In 1835, Charles M. Whish [149] referred in his article the four works: "Tantra sangraha," "Yuktibhasa," "Karanapaddhati," and "Sadrantamala." He wrote that "Tantra sangraha" whose date is AD 1500 laid foundations for a complete system of fluxions, while Sadrantamala is abound with fluxional forms and series to be found in the work of no foreign or Indian countries. These four are essentially astronomical texts that provide in accurate form what we now call "Gregory–Leibniz series" and "Euler series." According to Whish, the series that bear the name "Gregory–Leibniz series" and "Euler series" for π had been obtained by Sangama grama Madhava long before the times of Gregory–Leibniz and Euler.

It appears that for nearly 200 years after Bhaskara II, the ancient India which stood forefront of mathematical knowledge sank into a state of mathematical inactivity. The political upheavals due to frequent foreign invasion that began in the Indian subcontinent after Bhaskara II may have contributed to the bareness of the mathematical activity during this time. Much more likely, some mathematical works of very high caliber must have existed between AD 1150 and AD 1500, which were available to Narayana Pandita, Sangama grama Madhava, Nilakantha Somyaji, Jyesthadeva, and others, and are yet to be unearthed.

Aryabhata School of Mathematics appears to have some connection and sympathy for the Jaina mathematics. Bhaskara I quoted "Prakrt" verses of the Jains in his commentary. Another piece of evidence is the voluminous manuscripts of Jaina Mahavira's "Ganita Sara Sangraha" found in Kerala.

In the sixteenth and seventeenth centuries, we discover several commentaries of Bhaskara II's works, "Bijaganita," "Lilavati," and "Siddhanta Siromani." These include "Bijadyaya" of Jnanaraja (AD 1503), Krsna (AD 1580), and "Siddhanta tattva Viveka" of Kamalakara (AD 1658) on "Bijaganita," "Ganitamrita" (AD 1538) and "Suryaprakasa" (AD 1541) of Surydasa, and "Buddhivilasani" of Daivajna Ganesa (AD 1545) on "Lilavati" and "Titichintamani" of Daivjna Ganesa and Ramakrsna (AD 1648) on "Siddhanta Siromani."

Looking back at, when they lived, the general standard of mathematics prevailing then and the elegance, ingenuity, and the masterly skill exhibited in their works, Aryabhata I, Brahmagupta, Mahavira, Bhaskara II, and Sangama grama Madhava rightly deserve to be the members of an elite tier group of the greatest mathematicians of all time.

The Western scholars have not given the deserved credit to Brahmagupta and Bhaskara II, in spite of the work of H.T. Colbrooke in AD 1817. To continue to call the equation $Nx^2 + 1 = y^2$ as Pell's equation seems a misnomer. This equation can be renamed as Brahmagupta–Bhaskara II equation to do fair justice to these two great ancient Indian mathematics stalwarts. Sangama grama Madhava and Nilakantha Somyaji have not received their due credit either, in spite of the research paper of C.M. Whish in AD 1835. In the case of a result being rediscovered, no one can vouchsafe that the person who rediscovered the result later had access to all the extant literature in various languages in all corners of the world. In order to do fair justice to the person who discovered the result first and the person who rediscovered the same result later, the result can be renamed in the name of both. These are only the humble opinions of this author. Until the time, these controversies are satisfactorily resolved, they will continue to remain as a haunting challenge for the future historians of mathematics and mathematicians of this world. Due to modern information technology, the world has become small. There is no longer any more remote civilization.

It is high time that the full and the wonderful story of the continuous mathematics development in Premodern India from the Vedic times (BC 800) to the seventeenth century of the Christian era be well known to open the eyes of the mathematicians and historians of mathematics of this modern world to the abundant legacy of mathematics to which India gave birth.

1 A Glance at Ancient India and the Decimal System of Numeration

If we trace the ancient history of India, one will marvel at the magnificent accomplishments of ancient Indians at a very early period, in the arts as well as in the sciences. The discoveries at Mohenjo Daro and Harappa excavations in Indus Valley reveal advanced and cultured civilization as early as 3000 BC. The ancient Indians dwelled a highly organized life in brick houses in well-planned towns and cities. They were far ahead of any other people on any nation during that period.

"Vedas" are the scriptures of Hindus in India. The derived meaning of the very word "Veda" is that it is the fountain head and illimitable repository of all knowledge. Vedic scripture sets down the wholesome dictum:

युक्तियुक्तं वचो ग्राह्यं बालादपि शुकादपि।
युक्तिहीनं वचस्त्याज्यं वृद्धादपि शुकादपि॥

i.e., whatever is coherent with proper reasoning should be accepted even if it comes from a young boy or a parrot and whatever is incoherent should be rejected even if it comes from an old man or the great learned sage Sri Sukla himself. The Vedas are widely known as four in numbers: *Rigveda*, *Yajurveda*, *Samaveda*, and *Atharvaveda*. "*Vedas*" are also appended by four *upavedas*, namely, "*Ayurveda*" (Medical Science including anatomy, physiology, hygiene, surgery, etc.), "*Dhanurveda*" (Archery and other military Sciences), "*Gandharvaveda*" (the science and art of Music), and "*Sthapatyaveda*" (Engineering and Architecture), and six ancillary branches of knowledge called Sad-vedangas. These six "*Sad-vedangas*" are "*Siksa*" (Phonetics), "*Vyakarana*" (Grammar), "*Chandas*" (metrics), "*Nik-ruta*" (Etymology), "*Jyotisa*" (the science of luminaries like Astronomy, Astrology, etc.), and "*Kalpa*" (ritual). All these subjects are inherent parts of "*Vedas*."

"*Ganita*" literarily means "the science of calculation" and an ancient Hindu name for mathematics. "*Sad-vedanga Jyotisa*" (about 1200 BC) gives mathematics the highest place of honor among all the sciences which form "*Vedas*."

यथा शिखा मयूराणां नागानां मणयो यथा ।
तद्वद्वेदाङ्ग शास्त्राणां गणितं मूर्धनि स्थितम् ॥

(*Vedanga Jyotisa*, 4)

Mathematical Achievements of Pre-modern Indian Mathematicians. DOI: http://dx.doi.org/10.1016/B978-0-12-397913-1.00001-6

i.e., as the crests on the head of peacocks and as the gems on the head of cobras, "*Ganita*" is at the top of all sciences known as "*Vedangas.*"

In the early works, "*Vedas*" (probably much earlier than 3000 BC) and "*Brahmana* literature" (about 2000 BC), we find the beginning of Mathematics (Arithmetic, Geometry, Algebra, etc.) and Astronomy. These two periods were followed by more than 20 centuries of perpetual progress and marvelous accomplishments. It is during this great period, sage Valmiki known as the father of Sanskrit poetry composed the great epic "*Ramayana.*" Panini perfected Sanskrit grammar, Susruta wrote on the sciences of Medicine and Surgery, the immortal Asvaghosa, Bhasa, and Kalidasa dazzled with their poetry, Mahavira and Gautama Buddha preached the unique systems of Religious and Moral philosophy and the doctrine of "Nirvana." The expansion of Jainism and Buddhism sprang the Jaina and Buddhist literatures. Some of the earlier Hindu "*Puranas*" and "Dharma Sastras" were also written during that period.

At the early Vedic period, "*Ganita*" included Astronomy, whereas Geometry known as "*Ksetra ganita*" was included in a different group of Sciences known as "*Kalpasastra.*" It is accepted that some time before Christ, there was a renaissance of Hindu "*Ganita,*" Astronomy (*Jyotisa*) became a separate subject and Geometry was included within its scope. The Hindu "*Ganita*" of the early renaissance period comprised the following: "*Parikrama*" (the four principles of operation, namely, addition, subtraction, multiplication, and division), "*Vyavahara*" (Determinations), "*Rasi*" (Rule of three), "*Rajju*" ("Rope," meaning Geometry), "*Kalasavarna*" (operations with fractions), "*Yavat Tavat*" (as many as) referring to simple equations, "*Varga*" (Square, meaning quadratic equations), "Ghana" (Cube, meaning Cubic equations), "*Varavarga*" (Biquadratic equations), and "*Vikalpa*" (Permutations and combinations). Later on, the section of "*Ganita*" dealing with Algebra was given the name "*Bijaganita.*"

The Decimal System of Numeration

The most important mathematical contribution of ancient India is the invention of the decimal system of numeration, including the number zero. The unique feature of this system is the use of nine digits 1, 2, 3, 4, 5, 6, 7, 8, 9 and a symbol 0 to represent all the integral numbers by assigning a place value to the digits. In 1912, in his book "On the Foundations and Technique of Arithmetic," Professor E.B. Halstead remarks that "The importance of the creation of the zero mark can never be exaggerated. This giving to airy nothing not merely a local habitation and a name, picture, a symbol but helpful power is the characteristic of the Hindu race whence it Sprang. It is like coining the Nirvana into Dynamics. No Single mathematical creation has been more potent for the general on-go of intelligence and power." [58, p. 20]. A few years later (1926), he showed that zero existed in India at the time of Pingala's work "*Chandra Sutra*"—work on Prosody before 200 BC [59].

Pierre Simon Laplace (1749–1827), the celebrated French mathematician and astronomer, regarded as one of the greatest mathematicians of all time wrote that "The idea of expressing all numbers by the ten digits whereby is imparted to them both an absolute and a positional value is so simple that very simplicity is the reason for our not being sufficiently aware how much admiration it deserves. It is India that gave us this ingenious method."

In the Indian Historical Society, vol. 3, pp. 530–540, B.B. Dutta said that "The Hindus adopted decimal system very early. The numerical language of no other notation is so scientific and attained as high a state of perfection as that of Ancient Indians. In symbolism they succeeded with ten digits to express any number most elegantly and simply. It is the beauty of Hindu numerical notation which attracted the attention of all civilized people of the world and charmed them to adopt it." In the same context, Professor J. Ginsburg says in "New light on our numbers" in Bulletin of American Mathematical Society, vol. 23, pp. 366–369, "Hindu notation was carried to Arabia about AD 770 by a Hindu scholar named Kanka who was invited from Ujjain to the famous court of Baghdad by Abbaside Khalif Ali-Mansur. Kanka taught Hindu Astronomy and Mathematics to Arab scholars and with his help, they translated 'Brāhma-Sphuta-Siddhanta' of Brahma Gupta." The recent discovery of French Savant M.F. Nau, Journal Asiatique, vol. 2, 1910, pp. 225–227, proved that Hindu numerals were well known and much appreciated in Syria about the middle of seventeenth century AD. On this point, B.B. Dutta wrote that "From Arabia, the numerals slowly marched towards the west through Egypt and North Arabia and they finally entered Europe in the 11th century." The Europeans called them the Arabic notations because they received from Arabs. But the Arabs themselves, the Eastern as well as the Western, have unanimously called them the Hindu figures (Al-Arqan-Al-Hindu).

The Arab philosopher Djāhiz (who died in AD 869) called the numerals "figures of Hind" and pointed out that huge numbers can be represented with ease with these numerals. The Arab Historian Abul Hassan Al-Masūdi (AD 943) wrote that "A Congress of Sages at the command of creator Brahma invented the nine digits and also their astronomy and other sciences." Al-Maśudi statement is very significant, as he had firsthand knowledge of India. He was in Multan in AD 912 and Cambay in AD 916.

Abū Sahl Ibn Tamin (AD 950), a native of Kairwan, a village in Tunis in North Africa, wrote in his commentary on the "Sefer Yesīrah" that "The Indians have invented the nine digits for marking the units." I have spoken often in a book that I have written on the "Hindu Calculation" under the name of "Hisāb-al-ghobar." Al-Birūni, the famous Persian writer, who lived in India for nearly 13 years from AD 1017 to AD 1030, said in "Tarikh al-Hind" ("Chronical of India") that "The numerical signs which we use are derived from the finest forms of Hindu digits. I have composed a treatise showing how far possibly, the Hindus are ahead of us in this subject."

Rabbi Abraham ibn Meir ibn Ezra (AD 1092–1167) in his work "Sefer ha-Mispar" ("the Book of numbers") gave the Hindu forms of numerals and stated "that is why the wise people of India have designated all their numbers through the nine digits and have built form for the nine ciphers." Leonardo Fibbonacci of Pisa in his work Liber Abaci (1202) often refers to the nine Indian digits. At one place, he wrote that "ubi ex mirabili magisterio in arte per novem figures indorum introductus" etc. In another instance, as a heading to a different division, he stated that "De cognitione novem figurarum Yndorum" etc., "Novem figure indorum he sunt 9 8 7 6 5 4 3 2 1."

The above passages have been given in connection with and in appreciation of ancient India's invention of the zero mark and the nine digits and his contributions of the seventh century AD, and later to the world knowledge.

From the very earliest known time, 10 has become the basis of numeration in India. The Greeks had no terminology for denominations above the "myria" (10^4) and the Romans the "millie" (10^3). The ancient Hindus used at least 18 denominations. Even in modern times, the numeral language of no other nation is as scientific and perfect as that of ancient Indians.

In the "Yajurveda Samhita" XIV 2, the following list of numeral is given:

Ekha	: 1
Dasa	: 10
Sata	: $100 = 10^2$
Sahasra	: $1000 = 10^3$
Ayuta	: $10{,}000 = 10^4$
Niyuta	: $100{,}000 = 10^5$
Prayuta	: $1{,}000{,}000 = 10^6$
Arbuda	: $10{,}000{,}000 = 10^7$
Nyarbuda	: $100{,}000{,}000 = 10^8$
Samudra	: $1{,}000{,}000{,}000 = 10^9$
Madhya	: $10{,}000{,}000{,}000 = 10^{10}$
Anta	: $100{,}000{,}000{,}000 = 10^{11}$
Parardha	: $1{,}000{,}000{,}000{,}000 = 10^{12}$

The same list occurs at two places (IV 40 11.4) and (VII 2 20 1) in "Taittiriya Samhita." The "Maitrayani, ii 8.14" and the "Kathaka Samhita XVII 10" have the same list with slight alterations. The "Pancavimsa Brahmana" has Yajurveda Samhita list up to "nyarbuda" inclusive and then follow "Nikharva," "Vadava," "aksiti," etc. The "Sankhayana Sranta Sutra" contains the series after "nyarbuda" with "nikharva," "Samudra," "Salila," and "Ayuta" (10 billions).

In Valmiki "Ramayana," vi 28, a spy of the evil King Ravana narrates to his King the exact strength of the army of his enemy Sri Rama. Five persons from Ravana's side have left Ravana and joined Sri Rama's Camp. The spy marks the following statement:

शतं शतसहस्त्रा कोटिमाहुर्मनीषि
शतं कोटिसहस्त्राशङ्ख इत्यमिधीयते
शतं शङ्खसहस्त्रा महाशङ्ख इतिस्मृतम्
महाशङ्ख सहस्त्रा शतं वृंद इअतिस्मृतम्

शतं वृंद सहस्त्रा महावृंद इतिस्मृतम्
महावृंद सहस्त्रा शतं पद्मिवोच्यते
शतं पद्मसहस्त्रा महापद्ममितिस्मृतम्
महापद्मसहस्त्रा शतं खर्व मिवोच्यते
शतं खर्व सहस्त्रा महाखर्वमितिस्मृतम्
महाखर्व सहस्त्रा समुद्रमभिधीयते
शतमोघ सहस्त्रा महौघ इति विश्रुतः

In the above stanza, we have the following enumeration:

1 Koti	=	100 Lakhs	=	10^7
10^5 Koti	=	1 Sankha	=	10^{12}
10^5 Sankha	=	1 Mahasanka	=	10^{17}
10^5 Mahasanka	=	1 Vrinda	=	10^{22}
10^5 Vrinda	=	1 Mahavrinda	=	10^{27}
10^5 Mahavrinda	=	1 Padma	=	10^{32}
10^5 Padma	=	1 Mahapadma	=	10^{42}
10^5 Mahapadma	=	1 Kharva	=	10^{42}
10^5 Kharva	=	1 Mahakarva	=	10^{47}
10^3 Mahakarva	=	1 Samudra	=	10^{50}
10^5 Samudra (or Ogha)	=	1 Mahaugha	=	10^{55}

The spy estimated the strength of Sri Rama to be

$$10^{10} + 10^{12} + 10^{20} + 10^{24} + 10^{30} + 10^{34} + 10^{40} + 10^{44} + 10^{52} + 10^{57} + 10^{62} + 5$$

Writers on the Mohenjo Daro and Harappa Civilizations (3000 BC) refer to the decimal system of numeration found in these excavations. The Jain religious works, "Surya Prajnāpti," "Dwipa Prajnāpti," "Sthānāya sūtra," "uttārādhyana sūtra," "Bhagavati Sutra," and "Anuyoga Dwāra Sūtra" dating from 500 to 100 BC, used large numbers in the decimal system. In Buddhist work "Lalita Visara" of the first century BC, Gautama Buddha explained to a mathematician named Arjuna the system of numerals in multiple of 100 starting from 10^7 to 10^{63}.

2 The Sulvasutras

In ancient India, mathematical development became necessary for the religious practices and observances prevailing at that time. A good knowledge of mathematics and astronomy was essential to determine the exact time of the rising and setting of the Sun and Moon as well as the occurrence of Solar and Lunar eclipses. This in turn required a sound knowledge of plane and spherical geometry and trigonometry.

In the ancient Vedic religion, in every house, they had to perform certain types of "Pujas" (acts of worship) every day. For the purpose of "Pujas," they had to preserve three types of "Agnis" (fires) sheltering them in certain Altars (Bhūmikas) of particular design. The three types of "Agnis" were called "Dakshina," "Garhapatya," and "Ahavaneeya." The altar for "Ahavaneeya Agni" was square in shape and the altar for "Dakshna Agni" was semicircular in shape, whereas the altar for "Garhapatya Agni" was of square shape in one system and circular shape in another system. The unit of length used to make these altars was "vyam" (96 in.). Although these "Agnis" were used for daily routine "Pujas," they were also used for some elaborate sacrifices called "Yajnas" and "Yagas" to attain certain desired wishes. These were called "Kamayagnis." Also during the progress of these elaborate sacrifices, transference of the "Agni" from one altar ("Yagna Bhumika") to another altar, either of the same design or different design, was required. So, there was a need to construct the sacrificed altars according to different prescribed shapes and sizes. This meant that they had to have a good knowledge of the properties of the triangle, circle, rectangle, etc.

These "Pujas" or sacrifices can be traced back to very early times. A number of references are mentioned in "Rigveda Samhita." The mode of construction of these altars has been given a special importance in "Taittiriya Samhita," "Taittiriya Brāhmana," and "Ārānyaka." With the demands of the construction of sacrificial "vedis" (altars), it became absolutely necessary for setting out instructions in some written form. In this way, several "Sulvasutras" evolved. These sutras became adjuncts or appendices to the scriptural text known as "Srauta Sutras."

The root meaning of the word "Sulv" is to measure. Since rope was used to measure length and breadth, rope ("rajju"), the word "sulva" or "sulba" in due course came to mean a rope or a cord. Geometry which is at present referred to as "Rekaganita" was originally called "Ksetra ganita" or "Sulva Sastra" or "Rajju Sastra." Only nine of the Sulvasutras are known at the present time. They are known by the names "Āpasthamba," "Baudhayana," "Katyayana," "Laugaksi," "Manava," "Varaha," "Vadhula," "Maitrayana," and "Hiranyakesin," after the

Mathematical Achievements of Pre-modern Indian Mathematicians. DOI: http://dx.doi.org/10.1016/B978-0-12-397913-1.00002-8

rishies or vedic scholars who wrote them. The "Katyayana" sulvasutra opens with the statement "Rajju Samasam Vyakhyamah." Out of these nine sulvasutras, only three of them, namely, "Āpasthamba," "Baudhayana," and "Katyayana," form separate treatises. The remaining ones are either chapters or parts of chapters of the corresponding "Srauta Sutras." All the above nine Sulvasutras belong to "Yajurveda," the Veda particularly designed for the performance of sacrifices. "Katyayana" belongs to the section of the veda called "Sukla Yajurveda," while the remaining eight belong to "Krsna Yajurveda." The "Āpasthamba," "Baudhayana," and "Katyayana" are of immense significance from the point of view of mathematical contribution.

It is not known when exactly the codification of the rule for the construction of the altars began. The existing "Sulvasutras" do not give any information to exact dating. Buhler placed "Āpasthamba Sulvasutra" 150–200 years before Panini, who perfected the Sanskrit grammar. Bürk in "Āpasthamba Sulvasutra" [19] agreed with the view of Buhler. A.B. Keith in his work, A History of Sanskrit Literature, 1948, p. 517, placed "Baudhayana Sulvasutra" before "Āpasthamba Sulvasutra." The language of "Sulvasutras" can alone provide some clue to their age, and from their language, it is definitely pre-Panian. Also keeping in mind the dates assigned to Panini by later researchers, it is generally accepted that 800 BC is the more probable date for the codification of the "Sulvasutras."

The geometry of "Sulvasutras" is predominantly constructive and occasionally formulates some of the geometrical truths implied in the construction. Since the altars used were different geometrical shapes, the shapes, areas, and volumes of these had to be strictly exact, the exactness being as important as the exact pronunciation of vedic mantras. So, the geometrical contents of "Sulvasutras" can, in general, be classified into three categories: (i) theorems correctly stated, (ii) the actual constructions, and (iii) the geometrical truths implied in these constrictions.

"Baudhayana Sulvasutra" is the most ancient of the previously mentioned nine "sulvasutras" and it comprises three chapters. "Āpasthamba Sulvasutra" contains six chapters.

Geometrical Truths Involved in Constructions

Several geometrical theorems have been used for the construction of these altars of desired shapes and sizes. Some of these theorems are the following:

1. The diagonal of a rectangle divides it into two equal parts.
2. The diagonals of a rectangle bisect each other and opposite areas are equal. ("Baudhayana," (iii), 168, 169, 178).
3. The perpendicular through the vertex of an isosceles triangle on the base of the triangle divides the triangle into two equal parts ("Baudhayana" (iii), 256).
4. A rectangle and a parallelogram on the same basis and between the parallels are equal in area.
5. The diagonals of a rhombus bisect each others at right angles.
6. The area of a square by joining the midpoints of the sides of a square is half of that of the original square.

7. A quadrilateral formed by the lines joining the midpoints of the sides of a rectangle is a rhombus whose area is half of that of the rectangle.
8. The tangents to a circle are perpendicular to the radius at the point of contact.
9. A finite line can be divided into any number of equal parts.
10. The perpendicular bisector of a line is the locus of points equidistant from the end points of the line.
11. The line joining the vertex to the middle point of the base of an isosceles triangle is perpendicular to the base.
12. The area of an isosceles triangle is equal to half the area of the rectangle with sides equal to the base and the altitude of the triangle.

There is also the famous theorem named after the Greek mathematician, Pythagoras. The Pythagorean theorem occurs in "Baudhayana Sulvasutra" in the following form:

दीर्घस्याक्ष्णया रज्जुः पाश्र्वमानी तिर्यङ्मानी
च यत् पृथग्भूते कुरुतस्तदुभयं करोति ॥

This means that the diagonal of a rectangle gives an area equal to the sum of areas given by its length and breadth. The Vedic authors of Sulvasutras did not prove the theorem. They stated it and used it. But there is no evidence that Pythagoras proved it either. One has to realize that according to unprejudiced estimates, "Sulvasutras" were written about two centuries before Pythagoras. Pythagoras is believed to have lived from about 572 BC to 501 BC and the Sulvas precede him. Pythagorean theorem could have been proved the following way, according to Professor C.N. Srinivasiengar [132].

Take points E, F, G, and H on the sides of the square $ABCD$ such that

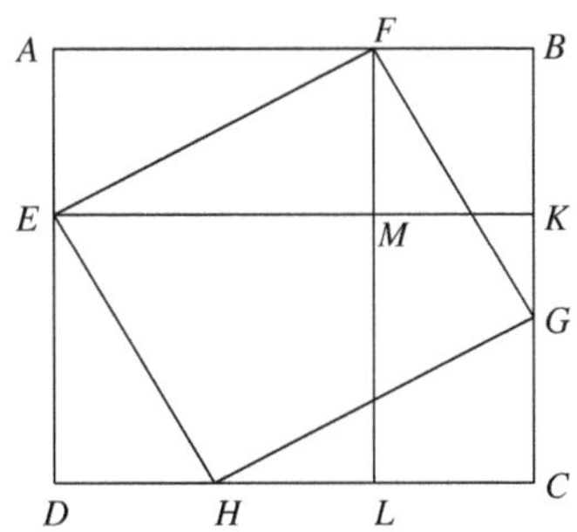

$$BF = BK = CG = CL = DH = AE$$

Join EK and FL. Let them intersect at M.
Now, Area of Square $ABCD$

$$\begin{aligned} &= \text{Area of square } EMLD + \text{Area of square } FBMK + 4 \text{ Area of } \Delta^{\text{le}}\ AEF \\ &= DL^2 + BF^2 + 4 \text{ Area of } \Delta^{\text{le}}\ AEF \\ &= AF^2 + AE^2 + 4 \text{ Area of } \Delta^{\text{le}}\ AEF \end{aligned} \qquad (1)$$

Also, Area of square $ABCD$

$$\begin{aligned} &= \text{Area of square } EFGH + 4 \text{ Area of } \Delta^{\text{le}}\, AEF \\ &= EF^2 + 4 \text{ Area of } \Delta^{\text{le}}\, AEF \end{aligned} \tag{2}$$

From (1) and (2), we see that $EF^2 = AF^2 + AE^2$.

The "Sulvasutras" demonstrate complete mastery with the property of sides and diagonals of figures with right angular corners. The following right triangles have been mentioned in the "Sulvasutras."

15, 36, 39 (Āpasthamba Sl. V. 2; Baudhayana, Sl. I. 49)

3, 4, 5 (Āpasthamba Sl. V. 3; Baudhayana, Sl. I. 49)

$$\left.\begin{array}{lll} 5, & 12, & 13 \\ 7, & 24, & 25 \\ 12, & 35, & 37 \end{array}\right\} \text{(Āpasthamba Sl. V. 4; Baudhayana, Sl. I. 49)}$$

$$\left.\begin{array}{lll} 72, & 96, & 120 \\ 40, & 96, & 104 \\ 2\frac{1}{2}, & 6, & 6\frac{1}{2} \\ 7\frac{1}{2}, & 10, & 12\frac{1}{2} \end{array}\right\} \text{(Manava, Sl. 1)}$$

The "Soutramani" altar uses a right triangle with sides $5\sqrt{3}$, $12\sqrt{3}$, $13\sqrt{3}$ and "Aśwa Medha" altar uses a right triangle with sides $15\sqrt{2}$, $36\sqrt{2}$, $39\sqrt{2}$. These irrational numbers obviously have arisen on account of the need for constructing an altar whose area is a multiple of another altar of same shape. For example, we can observe that the dimensions of "Soutramani" altar give a right triangle whose area is three times of the right triangle altar with dimensions (5, 12, 13). There are also huge altars with dimensions (72, 96, 120) and (40, 96, 104). The use of these irrational numbers also unmistakably indicates an excellent knowledge of the properties of similar triangles.

Pythagoras was therefore not the first person to have discovered the famous theorem named after him. Instead, the full geometrical significance of this theorem was first discovered by the ancient altar constructing Vedic scholars. The famous German Mathematician and Historian Moritz Benedikt Cantor (AD 1829–1920), in his well-known book on History of Mathematics, "Vorlesungen uber die Geschichte der Mathematik," vol. 1, 1880, and German Mathematician Albert Bürk [19] have discussed this question thoroughly and have come to the unmistakable conclusion that Pythagorean theorem was well known in ancient India by at least eighth century BC, which is the date of "Baudhayana Sulvasutra." Moreover, the German Mathematician and Historian Hermann Hankel in "Zur Geschichte der Mathematic" (Leipzig, 1874, p. 98) is of the opinion that the proof of Pythogorean theorem given earlier on pages 9–10 is more Indian in style than Greek.

Professors V. Lakshmikantham, S. Leela, and J. Vasundara Devi [81] in their book "The Origin and History of Mathematics" considered the theory of Pythagoras theorem under two aspects. In aspect I, they point out that the Pythagoras theorem is used in many geometrical constructions like to construct the side of a square whose area is equal to the sum or the difference of areas of two given squares. In aspect II, the Pythagoras theorem is used to calculate the diagonal of a rectangle with given sides. Aspect II comes in handy, when one has a Pythagorean triple to construct a right angle. The old Babylonians ca. 1800 BCE were aware of aspect II and used it. They had no use for aspect I. Both Greeks and Sulvasutras were familiar with aspect I. Sulvasutras knew aspect II, while Greeks and Egyptians had realized aspect II, since they had Pythagorean triples.

Professor David Mumford [88] in his review of Professor Kim Plofker's book "Mathematics in India" stated that "One finds here (Baudhayana Sulvasutra) the earliest explicit statement of the Pythagorean theorem. (So, it might be arguably called "Baudhayana theorem"). It is completely clear that the result was known to Babylonian Circa 1800 BCE, but they did not state as such – like their mathematical results, it is only recorded in examples and problems using it." From what we can see today, both the Egyptians and the Babylonians stuck to specific practical problems; the idea of stating and proving general theorems did not seem to occur in either civilization.

Professor V. Lakshmikantham, S. Leela, and J. Vasundara Devi [81] cite the examples of gnomon and the problem of squaring the circle as common elements of Greek and Indian mathematics not shared by old Babylonian ca. 1800 BCE.

Geometrical Constructions

We will now give some of the constructions contained in the Sulvas:

(1) **To draw the perpendicular bisector of given line**

तदन्तरं रज्वाभ्यस्य, पाशो कृत्वा,
शङ्क्वोः पाशौ प्रतिमुच्य, दक्षिणायम्य मध्ये शंकु निहन्ति।
एवमुत्तरतः सोदीची।

Kātyāyana (I. 3) *Manava* (P. 2)

Method: Here, the method is same as the modern method.

Let *AB* be the given line. Construct identical isosceles triangles *ABC* and *ABD* on either side of the given line *AB*. Join *CD*. Let the line *CD* intersect the given line *AB* at *O*.

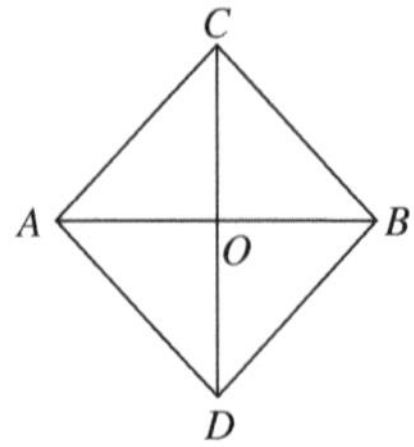

The line COD is the perpendicular bisector of the line *AB*.

(2) **To construct a square whose area is equal to the difference of two squares**

चतुरश्राच्चतुरश्रं निजिहीर्षन् यावन्निचिहीर्षेत्
तस्य करण्या वृध्द्रमुल्लिखेत् ।
वृध्द्रस्य पार्श्वमानीं अक्षणया इतरत्
पार्श्व उपसंहरेत् सा यत्र निपतेत्तदपछिन्द्यात् ।

Baudhayana Sl. 1. 51, (*Āp. Sl.* II. 5) *Katayana Sl.* III. 1

Method: Let a and b be the sides of the given two squares, such that $a > b$. Let *ABCD* be the square with each side $= a$.

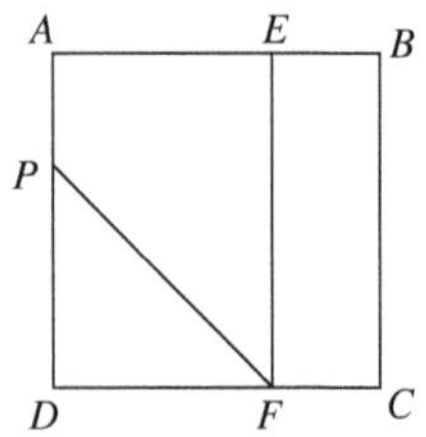

Take points E and F such that $AE = DF = b$. Take a point P on DA such that $PF = EF = a$. Join PF.

Then, the square with side DP is our required square.

Proof: In Δ^{le} *PDF*,

$$PD^2 = PF^2 - DF^2 = a^2 - b^2$$

i.e., the area of square with side $PD = a^2 - b^2 =$ the difference of areas of squares.

(3) **To construct a square of area equal to the sum of two given squares**

ह्रसीयसः करण्या वर्षीयसो वृध्द्रसुल्लिखेत् ।
वृध्द्रस्याक्षणायारज्जुरुभे समस्यति ।

Also *B. Sl.* I. 52 and, (*Āp. Sl.* II. 4) *K. Sl.* II. 22

Method: Let a and b be the sides of the two squares, with $a > b$. Let $ABCD$ be the square with each side $= a$. Take two points P and Q on AB and DC, respectively, such that $AP = DQ = b$. Join AQ.

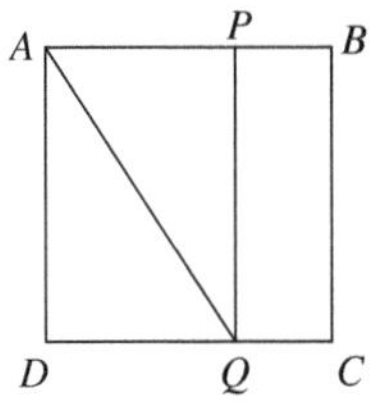

Proof: In Δ^{le} ADQ,

$$AQ^2 = AD^2 + DQ^2 = a^2 + b^2$$

i.e., the area of square with side $AQ = a^2 + b^2$, the sum of areas of two squares.

(4) **Given the shorter parallel side of a trapezium, to construct a trapezium of area equal to that of a given square or rectangle**

चतुरश्रमेकतोऽणिमच्चिकीर्षन् अणिमतः करणीं
तिर्यङ्मानीं कृत्वा शेषमक्षणया विभज्य
विपर्यस्येतरत्रोपदध्यात् ।

(*B. Sl.* I. 55)

Method: Let $ABCD$ be the given rectangle or a square. Let its area be A. Let a be shorter parallel side of the required trapezium.

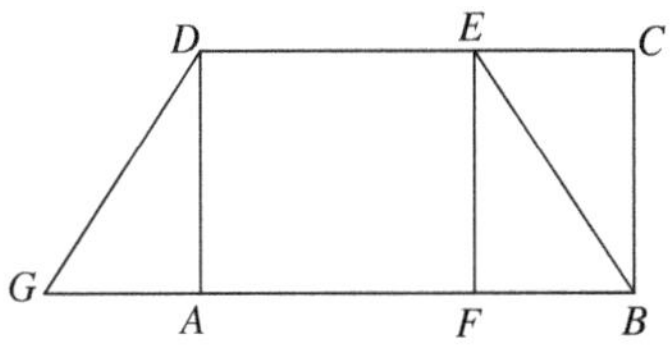

Take points E and F on DC and AB, respectively, such that $AF = DE = a$.

Join EF and BE. Attach the triangle BEC to the side AD of the rectangle $AFED$, in the position of the triangle AGD. Then, $DGBE$ is our required trapezium.

Proof: Since $FBCE$ is a rectangle and EB its diagonal,

$$\text{Area of the triangle } ECB = \frac{1}{2} \text{ the area of the rectangle } FBCE.$$

Now, the area of trapezium $DGBE$ = Area of $DEBA$ + Area of Δ^{le} DGA

$$= \text{Area of } DEBA + \text{Area of } ECB$$
$$= \text{Area of } ABCD$$

(5) To construct a rectangle whose area is equal to that of a given trapezium

दक्षिणसमादंसाद् द्वादशसु दक्षिणस्यां श्रोण्यां निपातयेत्,
छेदं विपर्यस्य उत्तरत उपदध्यात् ।
सा दीर्घा चतुरस्रा । तथायुक्तां संवक्षीत ।

Āpasthamba

Method: Let *ABCD* be the given trapezium. Draw *CP* perpendicular to *AB*. Place the triangle *CPB* in the position triangle *ADE*. Then, *APCE* is our required rectangle.

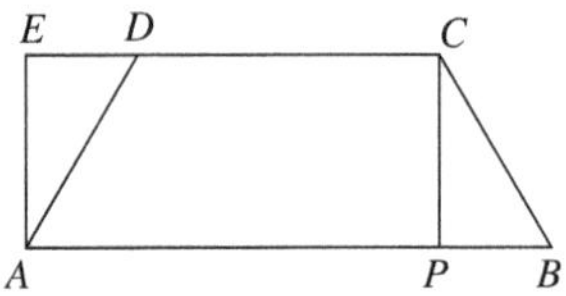

Proof: Area of the rectangle *APCE*

= Area of *APCD* + Area of Δ^{le} *ADE*
= Area of *APCD* + Area of Δ^{le} *BCP*
= Area of the trapezium *ABCD*

(6) To construct a rhombus of given area

तावदेव दीर्घचतुरश्रं विहृत्य पूर्वापरयोः करण्योरर्घात्
तावति दक्षिणोत्तरयोनिपातयेत् सनित्योभयतःप्रउगम् ।

(*Āp. Sl.* XII. 9) Also *B. Sl.* I. 57 and *K. Sl.* IV. 6

Method: Let $A = ab$ be the area of the required rhombus. Let *ABCD* be a rectangle with opposite sides of length $2a$ and b.

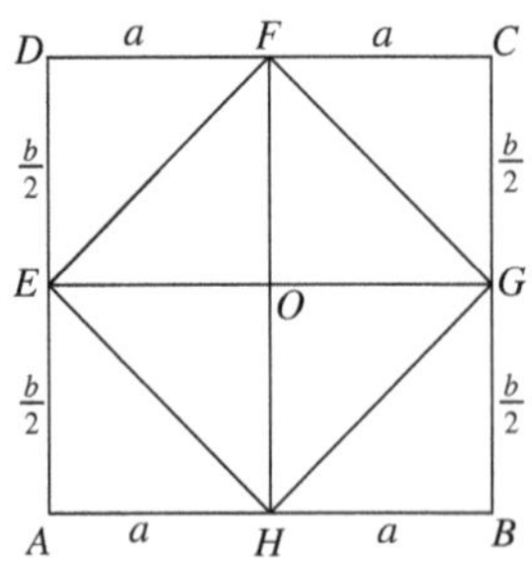

Let *E*, *F*, *G*, and *H* be the midpoints of the four sides of the rectangle.
Join *EF*, *FG*, *GH*, and *HE*. Join *EG* and *FH*. Then, *EFGH* is our required rhombus.

Proof: Now,

$$FG = GH = EH = EF = \sqrt{a^2 + \left(\frac{b}{2}\right)^2}$$

So, *EFGH* is a rhombus.
Area of the rhombus *EFGH*

$$= 4 \text{ Area of the } \Delta^{le}\ FOG$$

$$= 4\left[\frac{1}{2}\text{Area of the rectangle } FOGC\right]$$

$$= 2 \text{ Area of the rectangle } FOGC$$

$$= 2\left[\frac{1}{4}\text{Area of the rectangle } ABCD\right]$$

$$= \frac{1}{2}2ab = ab = A$$

(7) To construct a square *n* times in area to a given square

यावत्प्रमाणानि समचतुरश्राण्येकीकतुं चिकीर्षेत्
एकोनानि तानि भवन्ति तिर्यक् द्विगुणन्येकत एकाधिकानि ।
त्र्यस्रिर्भवति तस्येषुस्तत्करोति ।

(K. Sl. VI. 7)

Method: Let a be the side of a given square. Construct an isosceles Δ^{le} ABC with base $BC = (n-1)a$ and sides $AB = AC = (1/2)(n+1)a$.

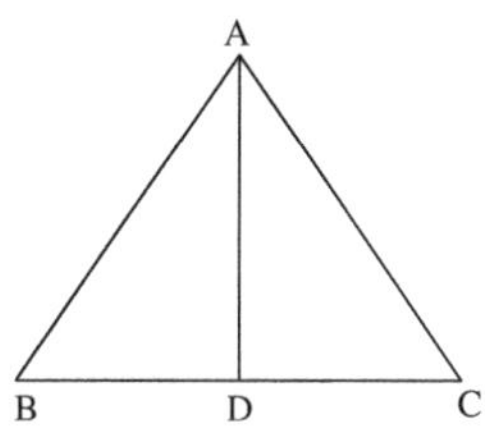

Let AD be the altitude from A on BC. Then, AD bisects BC, since ABC is an isosceles triangle.

So,

$$BD = DC = \frac{1}{2}(n-1)a$$

In the right Δ^{le} ABD,

$$AD^2 = AB^2 - BD^2 = \left\{\frac{1}{2}(n+1)a\right\}^2 - \left\{\frac{1}{2}(n-1)a\right\}^2$$

$$= \frac{1}{4}a^2\left[(n+1)^2 - (n-1)^2\right] = \frac{1}{4}a^2 \cdot 4n = a^2 n$$

So, a square with side $AD = a\sqrt{n}$ will have an area $= a^2 n = n$ times the area of the given square.

(8) To construct an isosceles triangle equal in area to a given square and vice versa

यावानग्निस्सारत्निप्रादेशो द्विस्तावतीं भूमिं चतुरश्रं कृत्वा,
पूर्वस्याः करण्या त्र्यर्धत् श्रोणी प्रत्यालिखेत् ।
स नित्या प्रउगम् ।

B. Sl. 1. 56 and K. Sl. IV. 5 (Ā. Sl. XII. 5)

Method: Let A be the area of the required isosceles triangle. Let $ABCD$ be a square with an area $2A$. Let E be the midpoint of DC. Join AE and BE. Draw EF perpendicular to AB. Now, AEB is our required isosceles triangle.

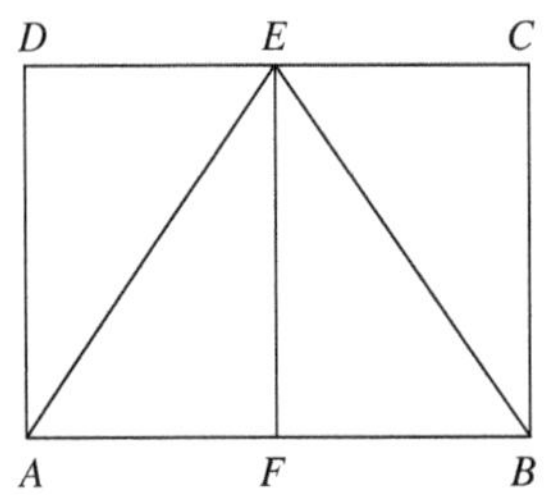

Proof: Now,

$$AF = FB = \frac{1}{2}AB$$

In Δ^{le} AEF,

$$AE^2 = AF^2 + EF^2 = \left(\frac{1}{2}AB\right)^2 + EF^2$$

In Δ^{le} FEB,

$$EB^2 = EF^2 + FB^2 = EF^2 + \left(\frac{1}{2}AB\right)^2$$

So, $AE = EB$, which means AEB is an isosceles triangle.

Now, EF divides the square $ABCD$ into two halves, namely, rectangles $ECBF$ and $EFAD$.

Since *ECBF* is a rectangle and *BE* is its diagonal,

$$\text{Area of } \Delta^{le}\ EFB = \frac{1}{2}\text{Area of the rectangle } ECBF$$

Similarly,

$$\text{Area of } \Delta^{le}\ EFA = \frac{1}{2}\text{Area of the rectangle } EFAD$$

So, Area of Δ^{le} *AEB*

$= \text{Area of } \Delta^{le}\ AEF \text{ and Area of } \Delta^{le}\ EFB$

$= \frac{1}{2}[\text{Area of rectangle } ECBF + \text{Area of rectangle } EFAD]$

$= \frac{1}{2}\text{Area of the square } ABCD = \frac{1}{2} \cdot 2A = A$

Converse Problem:

To construct a rectangle whose area is equal to that of an isosceles triangle

प्रउगं चतुरश्रं चिकीर्षन् मध्ये प्राच्त्रमपच्छिद्य
विपर्यस्येतरत उपधाय तोर्घचतुरश्रसमासेन समस्येत् ।

Katyayana Sulvasutra SL (IV. 7)

Method: Let *ABC* be the given isosceles triangle. Let *CD* be the altitude from *C* on *AB*. Then, *CD* cuts the Δ^{le} *ABC* into two equal halves. Let the part Δ^{le} *DBC* be applied inverted to the side as in the figure, in the position *ACB′*.

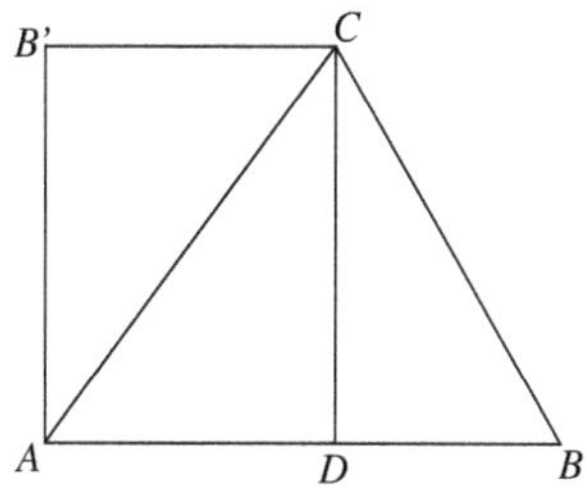

Then, the Area of rectangle *ADCB′*

$= \text{Area of } \Delta^{le}\ ADC + \text{Area of } \Delta^{le}\ CAB'$
$= \text{Area of } \Delta^{le}\ ABC$

(9) To construct a rectangle equal to a given square in area

समचतुरश्रं दीर्घचतुरश्रं चिकीर्षन् मध्येऽक्ष्णयापच्छिद्य
विभच्येतरत्पुरस्तादुत्तारतश्चोपदध्यात् ।
विषमं चेद्यथायोगं उपसंहरेदिति व्यासः ।

(*K. Sl.* III. 4) Also B. SI. I. 52

Let *ABCD* be the given square, with each side $= a$. Join the diagonal *AC*. Thus, the diagonal *AC* divides the square *ABCD* into two equal halves. Then, the area of each part $= a^2/2$.

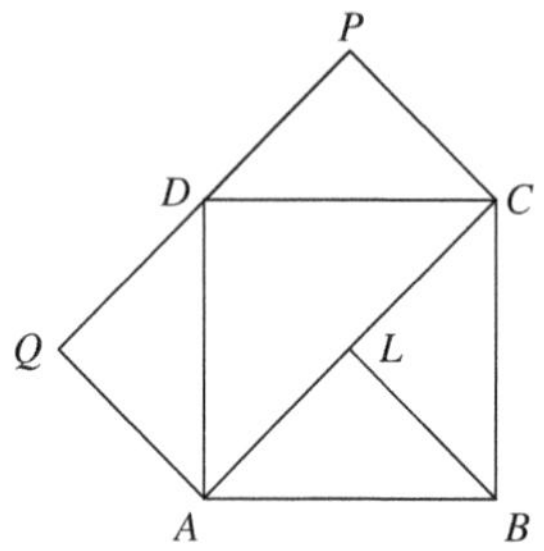

Now, divide the Δ^{le} *ABC* into two equal parts by drawing *BL* perpendicular to *AC* (here, $AL = LC$).

Then, the area of Δ^{le} ALB = the area of Δ^{le} $BLC = a^2/4$.

Then, the two triangles *ALB* and *BLC* are removed to the positions *CPD* and *ADQ*, respectively, as in the figure.

Then, the area of the rectangle *ACPQ*

= Area of Δ^{le} *ACD* + Area of Δ^{le} *CPD* + Area of Δ^{le} *ADQ*

= Area of Δ^{le} *ACD* + Area of Δ^{le} *BLC* + Area of Δ^{le} *ABL*

= Area of the square *ABCD*

The above method is very simple. But, the defect of this method is that the rectangle cannot be given any desired side.

Āpasthamba commentator Sundararaja has given a purely geometrical and exact construction as follows:

यावदिच्छं पाश्वमान्यो प्राच्यौ वधंयित्वा उत्तरपूवां कणंरज्जुमायच्छेत्,
सा दीर्घचतुश्रवध्यस्थायां समचतुरश्रतियंड्मान्यां यत्र निपतति
तत उत्तरं हित्वा दक्षिणांशं तिर्यड्मानीं कुर्यात्,
तद् दीर्घचतुरश्रं भवति ।

Method: Let *ABCD* be the given square with each side $= b$.

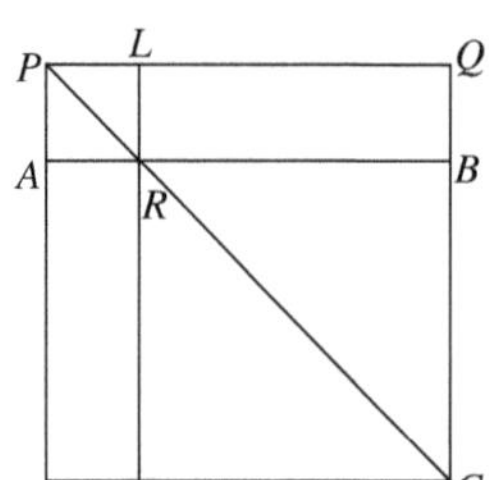

Suppose one of the sides of the required rectangle be a, such that $a > b$.

Then, produce DA to P and CB to Q such that $DP = CQ = a$. Complete the rectangle $DPQC$. Join the diagonal CP intersecting AB at R. Through R, draw a line parallel to AD, intersecting PQ at L and DC at M.

Then, the rectangle $MCQL$ is our required rectangle.

Proof: Now, $AP = DP - DA = a - b$.

Now, Δ^{les} APR and PDC are similar.

So,

$$\frac{AR}{DC} = \frac{AP}{PD} \Rightarrow AR = \frac{AP}{PD} \cdot DC = \frac{(a-b)}{a} \cdot b = \left(1 - \frac{b}{a}\right)b = b - \frac{b^2}{a}$$

Also,

$$RB = AB - AR = b - \left(b - \frac{b^2}{a}\right) = \frac{b^2}{a}$$

But, the area of the rectangle $LQCM$

$$= CQ \cdot LQ = CQ \cdot RB = a\left(\frac{b^2}{a}\right)$$

$$= b^2 = \text{the Area of the given square } ABCD$$

We have to mention that in this proof, we have assumed that the given side a of required rectangle is greater than the side b of the given square. We will now show that with slight change, the method is also applicable when the given side of the required rectangle is less than the side of the given square.

Let $a < b$.

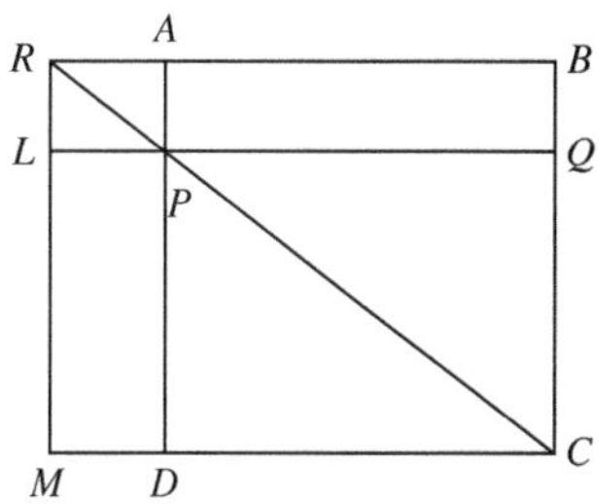

As before, let $ABCD$ be the given square with each side $= b$. Then, take points P and Q, such that $DP = CQ = a$.

Join CP and produce CP to cut the line BA produced at R. Through R, draw a line parallel to AD, intersecting QP produced at L and CD produced at M, respectively.

Then, $LQCM$ is our required rectangle.

Now, Δ^{les} APR and PDC are similar.

$$\frac{AR}{CD} = \frac{AP}{PD} \Rightarrow AR = \frac{AP}{PD} CD = \frac{AD - PD}{PD} \cdot CD$$

$$= \frac{b-a}{a} \cdot b = \frac{b^2}{a} - b$$

Now,

$$RB = RA + AB = \frac{b^2}{a} - b + b = \frac{b^2}{a}$$

Area of rectangle *LQCM*

$$= LQ \cdot QC = RB \cdot QC = a \cdot \frac{b^2}{a} = b^2$$

= the area of the given square *ABCD*.

(10) To construct a rectangle whose area is equal to that of a rhombus

This converse construction occurs in Kātyāyana only.

उभयतःप्रउगं चेन्मध्ये तियंगपच्छिद्य पूवंवत् समस्येत् ।

(K. Sl. Iv. 8)

Method: Let *ABCD* be a rhombus. Join the diagonal *AC* and *BD*. Let *M* be the point of intersection of the diagonal *AC* and *BD*.

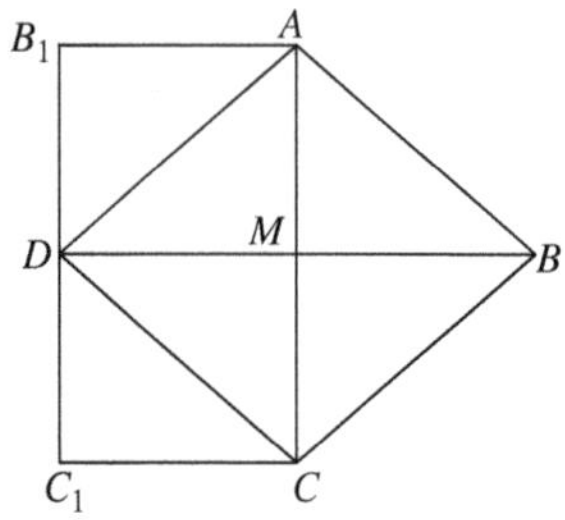

Since the diagonals of a rhombus intersect at right angles, we have four right triangles *AMD*, *AMB*, *BMC*, and *DMC*. The triangles *AMB* and *BMC* are placed as Δ^{les} AB_1D and CDC_1, respectively. Then, AB_1C_1C is our required rectangle.

Area of the rectangle AB_1C_1C

= Area of Δ^{le} *ADC* + Area of $\Delta^{le} AB_1D$ + Area of Δ^{le} C_1DC
= Area of Δ^{le} *ADC* + Area of Δ^{le} *AMB* + Area of Δ^{le} *BMC*
= Area of the rectangle *ABCD*

(11) To construct a square equal to a given rectangle in area

दीर्घचतुरश्रं समचतुरश्रं चिकीर्षन् तिर्यङमान्यापच्छिद्य
शेषं विभज्योभयत उपदध्यात् ।
खण्डमागन्तुनां संपूरयेत् । तस्य निह्रास उक्तः ।

(*Baudhayana* Sl. 1. 53) (Āp. Sl. II. 7)

Method: Let *ABCD* be the given rectangle. Take points *E* on *DC* and *F* on *AB*, such that $DE = AF = AD$. Let *G* be the midpoint of *EC* and let *H* be the midpoint of *FB*. Join *GH*.

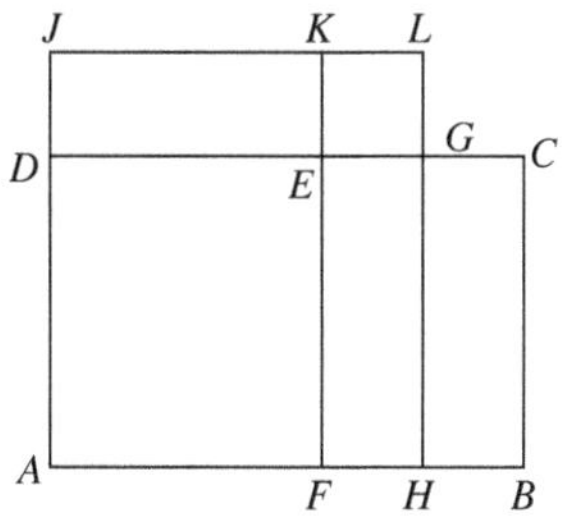

Now, the line *GH* cuts the rectangle *ECBF* into two equal halves, namely, rectangles *EGHF* and *GCBH*.

The rectangle *GCBH* is then cut off and attached to the side of the square *DEFA* with the position *DEKJ* as in the figure. Now, complete the square *AJLH*.

We observe that *EGLK* is a square, since $EG = LG$.

Area of the square *AJLH*

= Area of the rectangle *AHGD* + Area of the rectangle *DEKJ* + Area of the square *KEGL*
= Area of the rectangle *AHGD* + Area of the rectangle *GCBH* + Area of the square *KEGL*
= Area of the rectangle *ABCD* + Area of the square *KEGL*

So, Area of rectangle *ABCD*

= Area of the square *AJLH* − Area of the square *KELG*

Now, we can construct a square, whose area is the difference of the squares *AJLH* and *KEJL*, using the method shown in (2) on page 12.

We will now give a different proof. It is as follows:

Alternate Proof: Let a and b be sides of the given rectangle with $a > b$.

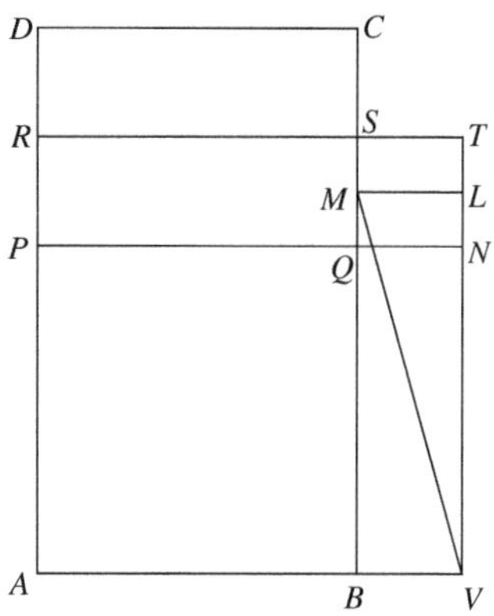

Let $ABCD$ be the given rectangle, with $AD = a$ and $AB = b$. Take a point P on AD such that $AP = AB = b$. Complete the square $APQB$. Let R be the midpoint of PD. Through the point R, draw a line parallel to AB, intersecting BC at S. Complete the square $ARTV$.

Produce PQ to intersect VT at N.

Take a point M on BS, such that $TV = MV$. Through M, draw a line parallel to AB, intersecting VT at L. Then, the square on LV is our required square.

Proof: Now,

$$PD = AD - AP = a - b$$

Since R is the midpoint of PD,

$$PR = RD = \frac{1}{2}PD = \frac{1}{2}(a - b)$$

$$MV = VT = AR = AP + PR = b + \frac{1}{2}(a - b) = \frac{1}{2}(a + b)$$

Also,

$$ML = BV = AV - AB = TV - AB$$

$$= \frac{1}{2}(a + b) - b = \frac{1}{2}(a - b)$$

In the right triangle MVL,

$$VL^2 = MV^2 - ML^2$$

$$= \left\{\frac{1}{2}(a+b)\right\}^2 - \left\{\frac{1}{2}(a-b)\right\}^2$$

$$= \frac{1}{4}\left[(a+b)^2 - (a-b)^2\right] = \frac{1}{4}(4ab) = ab$$

= the area of the given rectangle $ABCD$

(12) **To construct a square of area equal to a fraction of that of another square**

तृतीयकरण्येतेन व्याख्याता ।
प्रमाणविभागस्तु नवधा करणीतृतीयं नवभागः ।
नवभागास्त्रयः तृतीयकरणी ।

(Āpasthamba Sl. I1. 3 and Baudhayana Sl. I. 47 (K. Sl. 15–18)

Method: For example, "Sautramani Vedi" is one-third of "Saumiki Vedi." One-third make is explained as follows:

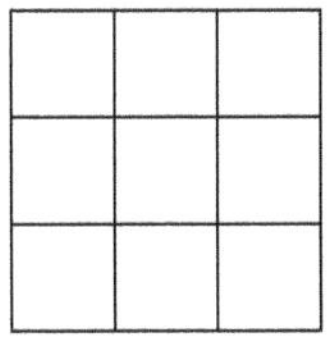

Divide the square into nine equal parts, by dividing the pair of opposite sides by drawing lines parallel to the other pairs of sides. Combine the three of the small squares into a square. The area of this square will be one-third of the original square.

Proof: Let a be the side of the given square. Then, the area of the given square $= a^2$. Then, the area of each of the nine small squares $= a^2/9$.

If we combine three of these small squares, their combined area

$$= 3\left(\frac{a^2}{9}\right) = \frac{a^2}{3}$$

Let x be the side of the square, whose area is equal to the combined area of these three small squares.

Then,

$$x^2 = \frac{a^2}{3}$$

This means that the area of the required square is one-third the area of the original square.

(13) To construct a circle whose area is approximately equal to that of the given square

Method: Let $ABCD$ be the given square that has to be transformed into a circle, whose area will be equal to that of the given square $ABCD$. Let C_1 be the center of the square. Join C_1A.

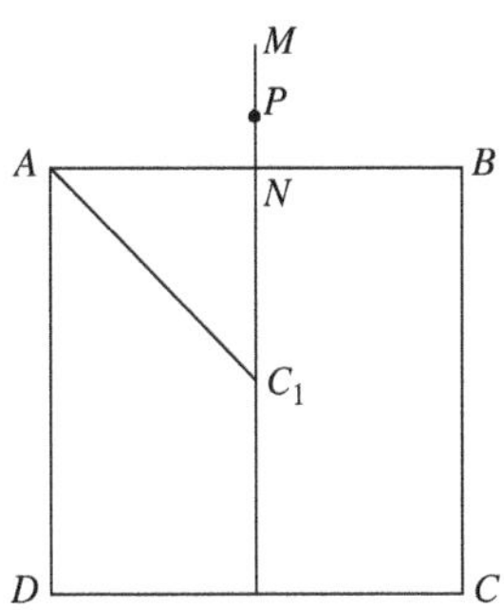

Now with C_1 as center and radius equal to C_1A, draw a circle intersecting the vertical line through C_1 at M, as in the figure. Let N be the midpoint of AB. Take a point P on MN such that $MP = 2PN$.

Then with C_1 as center, draw a circle with C_1P as radius. This circle is the required transformed circle, whose area will be approximately equal to that of the given square $ABCD$.

Proof: Let a be the length of each side of the square.

Then,

$$C_1A = \sqrt{(C_1N)^2 + (AN)^2} = \sqrt{\left(\frac{a}{2}\right)^2 + \left(\frac{a}{2}\right)^2} = \frac{1}{2}a\sqrt{2}$$

Now,

$$C_1P = C_1N + PN = C_1N + \frac{1}{3}MN$$

$$= C_1N + \frac{1}{3}(C_1M - C_1N) = \frac{a}{2} + \frac{1}{3}\left(C_1A - \frac{a}{2}\right)$$

$$= \frac{a}{2} + \frac{1}{3}\left(\frac{1}{2}a\sqrt{2} - \frac{a}{2}\right) = \frac{a}{2} + \frac{1}{6}a\left(\sqrt{2} - 1\right)$$

$$2C_1P = a + \frac{1}{3}a\left(\sqrt{2} - 1\right) = \frac{a[3 + \sqrt{2} - 1]}{3}$$

$$d = \frac{a}{3}\left(2 + \sqrt{2}\right)$$

where d is the diameter of the required circle.

Area of the required circle

$$= \frac{\pi d^2}{4} = \frac{\pi}{4}\left[\left(\frac{2+\sqrt{2}}{3}\right)a\right]^2$$

$$= \frac{\pi}{36}a^2\left(2+\sqrt{2}\right)^2$$

$\approx$ the area of the given square $= a^2$

So,

$$\frac{\pi}{36}\left(2+\sqrt{2}\right)^2 \approx 1$$

$$\pi \approx \frac{36}{(2+\sqrt{2})^2} = \frac{36}{(2+\sqrt{2})^2}\frac{(2-\sqrt{2})^2}{(2-\sqrt{2})^2}$$

$$\pi \approx \frac{36(2-\sqrt{2})^2}{(4-2)^2} = 9\left(2-\sqrt{2}\right)^2$$

$$\pi \approx 9\left(6-4\sqrt{2}\right) = 18\left(3-2\sqrt{2}\right) = 3.088$$

which is not a good approximation for π, since $\pi = 3.145926539$.

The above problem which is commonly known as the "squaring of the circle" had remained unsolved until recently. It has now been proved that this construction of transforming a rectangle into a circle of equal area is impossible. One has to admire the Vedic authors of "Sulvasutras" for tackling such a difficult and complex problem so early in time.

(14) To construct a square whose area is approximately equal to that of a given circle

This problem is the converse of the above (13) complex problem.

Method: Divide the diameter d of the given circle into 15 equal parts. Take the combined length of 13 parts as the side of the required square.

Proof: Let a be the side of the given square.

Then,

$$a = \frac{13}{15}d$$

Area of the square

$$= a^2 = \left(\frac{13}{15}d\right)^2$$

$$\approx \text{Area of the given circle} \approx \frac{\pi d^2}{4}$$

$$\frac{\pi d^2}{4} \approx \left(\frac{13}{15}\right)^2 d^2 \Rightarrow \frac{\pi}{4} \approx \left(\frac{13}{15}\right)^2 = \frac{169}{225}$$

So,

$$\pi \approx \frac{676}{225} = 3.004$$

which is not a good approximation for π.

"Baudhayana Sulvasutra" in Sl. I. 59 gives a little better approximation, by saying the side a of the required square is given by

$$a = d\left[1 - \frac{28}{8.29} - \frac{1}{6.8.29} + \frac{1}{6.8.29.8}\right]$$

which gives the value of π as $= 3.088$, which is also not a good approximation for π.

(15) To construct a square whose area is twice that of a given square

समस्य द्विकरणी।

Āpasthamba (1.5)

Method: Let $ABCD$ be the given square, with each side $= a$.

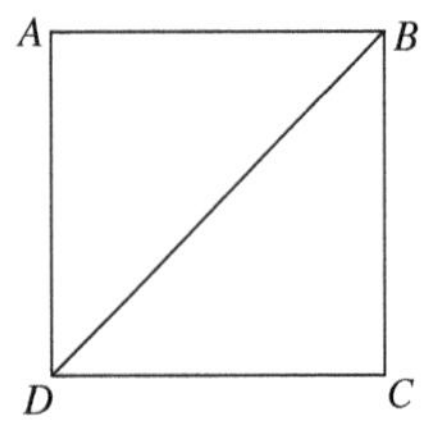

Join the diagonal BD. Then, the square with each side $= BD$ is our required square.

Proof: In the right triangle BCD,

$$BD^2 = BC^2 + DC^2 = a^2 + a^2 = 2a^2$$

So, the square with each side $= BD$ will have an area $2a^2 =$ twice the area of the given square $ABCD$.

Also, the diagonal $BD = a\sqrt{2} = \sqrt{2}$ times the side of the given square. It is noteworthy that a remarkable approximation is given here to $\sqrt{2}$ in the Sulvas:

$$\sqrt{2} = 1 + \frac{1}{3} + \frac{1}{3 \cdot 4} - \frac{1}{3 \cdot 4 \cdot 34}$$

समस्य द्विकर्णि। प्रमाणं तृतीयेन वर्धयेत्
तच्चतुर्थेनात्म चतुस्त्रिंशोनेन सविशेषतः (आपस्तंब)

Kathayana Sl. II. 13 (Āp. Sl. I. 5 and B. Sl. I. 61–62)[1]

This approximation gives $\sqrt{2} = 1.4142156\ldots$, whereas the correct value of $\sqrt{2} = 1.414213\ldots$. So, this approximation is correct to five decimal places and is represented in terms of simple unit fractions. The Sulvas do not give in any information about how this remarkable approximation was derived.

The approximation $\sqrt{Q}=\sqrt{a^2+b}=a+(b/2a)$, where b is small, is known as Heron's formula after the Greek Mathematician, who lived in the second half of first century AD. In Heron's work, it came in connection with a geometric problem according to H. Eves [42]. It is very possible that this approximation was known in ancient India. This formula can be easily derived from the use of Binomial theorem for fractional indices by writing.

$$\sqrt{Q}=\sqrt{a^2+b}=a\sqrt{1+\frac{b}{a^2}}=a\left(1+\frac{b}{a^2}\right)^{\frac{1}{2}}$$

$$=a\left[1+\frac{1}{2}\frac{b}{a^2}\right]$$

neglecting higher powers of b/a^2 since b/a^2 is small.

So,

$$\sqrt{a^2+b}=a+\frac{b}{2a}$$

if b/a^2 is small.

In Pingala's "Chandra Sutra," we find results that require the application of the Binomial theorem. Expansions of expressions like $(a+b)^3$ and $(a+b)^4$ are given in this ancient text. However, it is generally believed that the Binomial theorem was known to ancient Indians only for positive integral index.

It is quite possible that the authors of "Sulvasutra" obtained the approximation, namely, $\sqrt{2}=1+(1/3)+(1/(3\cdot 4))-(1/(3\cdot 4\cdot 34))$, if they were aware of the approximation named after Heron. We will now give a plausible way of obtaining the approximation for $\sqrt{2}$ with the use of Heron's formula.

Now,

$$2=\frac{16}{9}+\frac{2}{9}$$

So,

$$\sqrt{2}=\sqrt{\frac{16}{9}+\frac{2}{9}}$$

$$=\sqrt{\left(\frac{4}{3}\right)^2+\frac{2}{9}}=\frac{4}{3}+\frac{(2/9)}{2\cdot(4/3)}\quad\text{(by Heron's formula)}$$

$$= \frac{4}{3} + \frac{2}{9} \cdot \frac{3}{8} = 1 + \frac{1}{3} + \frac{1}{12} = q_1 \quad \text{(first approximation)}$$

Now,

$$q_1 = 1 + \frac{1}{3} + \frac{1}{12} = \frac{12 + 4 + 1}{12} = \frac{17}{12}$$

So,

$$q_1^2 = \left(\frac{17}{12}\right)^2 = \frac{289}{144} = 2 + \frac{1}{144}$$

Hence,

$$2 = q_1^2 - \frac{1}{144}$$

$$\sqrt{2} = \sqrt{q_1^2 - \frac{1}{144}} = q_1 - \frac{(1/144)}{2q_1} \quad \text{(by Heron's formula)}$$

$$= 1 + \frac{1}{3} + \frac{1}{12} - \frac{(1/144)}{2 \cdot (17/12)} = q_2 \quad \text{(second approximation)}$$

$$= 1 + \frac{1}{3} + \frac{1}{3 \cdot 4} - \frac{1}{144} \cdot \frac{12}{34} = q_2$$

$$= 1 + \frac{1}{3} + \frac{1}{3 \cdot 4} - \frac{1}{12 \cdot 34} = q_2$$

Thus,

$$\sqrt{2} = 1 + \frac{1}{3} + \frac{1}{3 \cdot 4} - \frac{1}{3 \cdot 4 \cdot 34} = q_2$$

Now,

$$q_2^2 = \left(1 + \frac{1}{3} + \frac{1}{3 \cdot 4} - \frac{1}{3 \cdot 4 \cdot 34}\right)^2 = \left(\frac{17}{12} - \frac{1}{408}\right)^2$$

$$= \frac{289}{144} - 2 \cdot \frac{17}{12} \frac{1}{408} + \left(\frac{1}{408}\right)^2$$

So,

$$q_2^2 = 2 + \frac{1}{144} - \frac{1}{144} + \frac{1}{(408)^2} = 2 + \frac{1}{(408)^2}$$

Hence,

$$2 = q_2^2 - \frac{1}{(408)^2}$$

and

$$\sqrt{2} = \sqrt{q_2^2 - \frac{1}{(408)^2}} = q_2 - \frac{(1/408)^2}{2q_2} \quad \text{(by Heron's formula)}$$

$$= q_2 - \frac{(1/408)^2}{2[(17/12) - (1/408)]}$$

$$= q_2 - \frac{1}{2(408)^2[((17)(34) - 1)/408]}$$

So,

$$\sqrt{2} = q_2 - \frac{408}{2(408)^2(578 - 1)} = q_2 - \frac{1}{2(408)577} = q_3 \quad \text{(third approximation)}$$

$$\sqrt{2} = 1 + \frac{1}{3} + \frac{1}{3 \cdot 4} - \frac{1}{3 \cdot 4 \cdot 34} - \frac{1}{3 \cdot 4 \cdot 34 \cdot (1154)}$$

This gives the value of $\sqrt{2}$ as 1.4142135123746899 11..., which is correct to 11 decimal places. We observe that second approximation $q_2 = 1 + (1/3) + (1/(3 \cdot 4)) - (1/(3 \cdot 4 \cdot 34))$ is the approximation that has been given in "Sulvasutras."

A commentator by name Rama, in the middle of the fifteenth century AD, who lived in Naimisha near Lucknow, gave the following approximation:

$$\sqrt{2} = 1 + \frac{1}{3} + \frac{1}{3 \cdot 4} - \frac{1}{3 \cdot 4 \cdot 34} - \frac{1}{3 \cdot 4 \cdot 34 \cdot 33} + \frac{1}{3 \cdot 4 \cdot 34 \cdot 34}$$

Rama's approximation gives $\sqrt{2} = 1.414213502$, which is correct to seven decimal places. So, the approximation we derived on page 29 above, namely,

$$\sqrt{2} = 1 + \frac{1}{3} + \frac{1}{3 \cdot 4} - \frac{1}{3 \cdot 4 \cdot 34} - \frac{1}{3 \cdot 4 \cdot 34 \cdot (1154)}$$

is much superior than Rama's approximation.

But L. Rodét in "Sur une methode d'approximation des racines carrés conne das plnde antirieurement a la conquête d'Alexandre" (Bulletin of Society Math. D France VII (1879), 98–102) gives a totally different approximation for

$$\sqrt{a^2+\lambda} = a + \frac{\lambda}{2a+1} + \frac{(\lambda/(2a+1))(1-(\lambda/(2a+1)))}{2(a+(\lambda/(2a+1)))} + E$$

where

$$E = \frac{\lambda - \{(\lambda/(2a+1)) + ((\lambda/(2a+1))(1-(\lambda/(2a+1)))/2(a+(\lambda/(2a+1))))\} \times \{2a+(\lambda/(2a+1)) + ((\lambda/(2a+1))(1-(\lambda/(2a+1)))/2(a+(\lambda/(2a+1))))\}}{2\{a+(\lambda/(2a+1)) + ((\lambda/(2a+1))(1+(\lambda/(2a+1)))/2(a+(\lambda/(2a+1))))\}}$$

Rodét credits this approximation to the Vedic authors of "Sulvasutras." He believed that this approximation was obtained on geometric devices. B.B. Dutta in "the sciences of the Sulba," University of Calcutta, 1932, thought that this approximation was obtained by the method of continued traditions. But it is not clear even now whether the ancient Indians before Bhaskara II knew anything about continued fractions.

Professor C.N. Srinivasiengar [132] in his book "The History of Ancient Indian Mathematics," wrote that the actual proofs of several formulae still remain a matter for research and provide material for conjecture and speculation. But to this author's knowledge, no where it is mentioned how the approximation mentioned by Rodét can be derived.

We will now give a plausible method that might have been employed by the vedic authors of "Sulvasutras."

$$\sqrt{Q} = \sqrt{a^2+\lambda} = a + \frac{\lambda}{2a} \quad \text{(by Heron's formula)}$$

$$\approx a + \frac{\lambda}{2a+1} = q_1 \quad \text{(first approximation)}$$

Then,

$$q_1^2 = \left(a + \frac{\lambda}{2a+1}\right)^2 = a^2 + \frac{2a\lambda}{2a+1} + \left(\frac{\lambda}{2a+1}\right)^2$$

$$q_1^2 = a^2 + \frac{\lambda(2a+1)-\lambda}{2a+1} + \left(\frac{\lambda}{2a+1}\right)^2$$

$$q_1^2 = a^2 + \lambda - \frac{\lambda}{2a+1} + \frac{\lambda^2}{(2a+1)^2}$$

$$q_1^2 = a^2 + \lambda + \frac{\lambda}{2a+1}\left(\frac{\lambda}{2a+1} - 1\right)$$

$$q_1^2 = Q + \frac{\lambda}{2a+1}\left(\frac{\lambda}{2a+1} - 1\right)$$

So,

$$Q = q_1^2 + E_1, \quad \text{where } E_1 = \frac{\lambda}{2a+1}\left(1 - \frac{\lambda}{2a+1}\right)$$

So,

$$\sqrt{Q} = \sqrt{q_1^2 + E_1} = q_1 + \frac{E_1}{2q_1} \quad \text{(by Heron's formula)}$$

$$\sqrt{Q} = a + \frac{\lambda}{2a+1} + \frac{E_1}{2q_1} = q_2 \quad \text{(second approximation)}$$

So,

$$\sqrt{Q} = a + \frac{\lambda}{2a+1} + \frac{(\lambda/(2a+1))(1-(\lambda/(2a+1)))}{2(a+(\lambda/(2a+1)))} = q_2$$

So,

$$\sqrt{Q} = q_2 = a + \frac{\lambda}{2a+1} + l$$

where

$$l = \frac{\lambda}{2a+1}\frac{(1-(\lambda/(2a+1)))}{2(a+(\lambda/(2a+1)))}$$

Now,

$$q_2^2 = \left(a + \frac{\lambda}{2a+1} + l\right)^2 = a^2 + \left(\frac{\lambda}{2a+1}\right)^2 + l^2 + \frac{2a\lambda}{2a+1} + 2al + \frac{2\lambda l}{2a+1}$$

$$q_2^2 = a^2 + \left(\frac{\lambda}{2a+1}\right)^2 + l^2 + \frac{\lambda(2a+1) - \lambda}{2a+1} + 2al + \frac{2\lambda l}{2a+1}$$

$$q_2^2 = a^2 + \left(\frac{\lambda}{2a+1}\right)^2 + l^2 + \lambda - \frac{\lambda}{2a+1} + 2al + \frac{2\lambda l}{2a+1}$$

$$q_2^2 = Q + \left(\frac{\lambda}{2a+1}\right)^2 + l^2 - \frac{\lambda}{2a+1} + 2al + \frac{2\lambda l}{2a+1}$$

So,

$$q_2^2 = Q + E_2$$

where

$$E_2 = \left(\frac{\lambda}{2a+1}\right)^2 + l^2 - \frac{\lambda}{2a+1} + 2al + \frac{2\lambda l}{2a+1}$$

So,

$$Q = q_2^2 - E_2 \quad \text{and} \quad \sqrt{Q} = \sqrt{q_2^2 - E_2} = q_2 - \frac{E_2}{2q_2} = q_3 \quad \text{(by Heron's formula)}$$

So,

$$\sqrt{Q} = \sqrt{a^2 + \lambda} = a + \frac{\lambda}{2a+1} + l - \frac{E_2}{2q_2}$$

where

$$E_2 = \left(\frac{\lambda}{2a+1}\right)^2 - \frac{\lambda}{2a+1} + l\left(l + 2a + \frac{2\lambda}{2a+1}\right)$$

Now,

$$l = \frac{\lambda}{(2a+1)} \frac{(1 - (\lambda/(2a+1)))}{2(a + (\lambda/(2a+1)))}$$

and

$$q_2 = a + \frac{\lambda}{2a+1} + l$$

Thus,

$$\sqrt{Q} = \sqrt{a^2 + \lambda} = a + \frac{\lambda}{2a+1} + l + E = q_3 \quad (*)$$

where

$$E = \frac{-E_2}{2q_2} = \frac{(\lambda/(2a+1)) - (\lambda/(2a+1))^2 - l(l+2a+(2\lambda/(2a+1)))}{2q_2}$$

$$= \frac{(2a\lambda + \lambda - 2a\lambda/(2a+1)) - (\lambda/(2a+1))^2 - l(l+2a+(2l/(2a+1)))}{2q_2}$$

So,

$$E = \frac{\lambda - (2a\lambda/(2a+1)) - (\lambda^2/(2a+1)^2) - l(l+2a+(2\lambda/(2a+1)))}{2q_2}$$

Hence,

$$E = \frac{\lambda - (2a\lambda/(2a+1)) - (\lambda^2/(2a+1)^2) - (l\lambda/(2a+1)) - l(l+2a+(\lambda/(2a+1)))}{2q_2}$$

So,

$$E = \frac{\lambda - (\lambda/(2a+1))(2a+(\lambda/(2a+1))+l) - l(l+2a+(\lambda/(2a+1)))}{2q_2}$$

$$E = \frac{\lambda - (l+2a+(\lambda/(2a+1)))(l+(\lambda/(2a+1)))}{2a}$$

Thus,

$$\sqrt{Q} = \sqrt{a^2+\lambda} = a + \frac{\lambda}{2a+1} + l + E = q_3$$

where

$$E = \frac{\lambda - (l+2a+(\lambda/(2a+1)))(l+(\lambda/(2a+1)))}{2q_2}$$

$$q_2 = a + \frac{\lambda}{2a+1} + l$$

and,

$$l = \frac{\lambda}{2a+1}\frac{(1-(\lambda/(2a+1)))}{2(a+(\lambda/(2a+1)))}$$

This completes the proof of the approximation, which Rodét credits to the Vedic Scholars of "Sulvasutras." If we put $a = 1$, $\lambda = 1$ in the above approximation, we get $\sqrt{2} = 1 + (1/3) + (1/(3 \cdot 4)) - (1/(3 \cdot 4 \cdot 34))$ which is given in "Sulvasutras," as mentioned before.

Fractions and operations on them are dealt in the "Sulvasutras." We also come across fractions like 3/8 ("thri ashtama"), 2/7 ("dwisaptama"), 3/4 ("chaturbhagona"), etc. Surds of the form $\sqrt{2}$, $\sqrt{3}$, etc. are called "Karanis." For example, $\sqrt{2}$ is "dwi-karani," $\sqrt{3}$ is "trikarni," $\sqrt{1/3}$ is "triteeya karani," $\sqrt{1/7}$ is "saptama karani," $\sqrt{18}$ is "ashtadasa karani," etc.

In the "Sulvasutras," it is given that the shape of the "Aswamedhiki Vedika" is an isosceles trapezium, whose face, base, and altitude are, respectively, $24\sqrt{2}$, $30\sqrt{2}$, and $36\sqrt{2}$ prakramas and its area is 1944 square prakramas.

$$\text{Area of a trapezium} = \text{altitude} \times \frac{1}{2}(\text{face} + \text{base})$$

So, Area of "Aswamedhi Vedika"

$$= 36\sqrt{2} \cdot \frac{1}{2}\left(24\sqrt{2} + 30\sqrt{2}\right)$$

$$= 18\sqrt{2}\left(54\sqrt{2}\right) = 1944 \text{ sq. prakramas}$$

This clearly demonstrates their knowledge of determining the area of a trapezium.

"Manava Sutra" contains the following:

$$40^2 + 40^2 \approx 56^2 \Rightarrow 40\sqrt{2} \approx 56 \Rightarrow \sqrt{2} \Rightarrow \frac{7}{5} = 1.4$$

$$4^2 + 4^2 \approx \left(5\frac{2}{3}\right)^2 \Rightarrow 4\sqrt{2} \approx \frac{17}{3} \Rightarrow \sqrt{2} \approx \frac{17}{12} = 1.4166$$

$$36^2 + 90^2 \approx (97)^2 \Rightarrow \sqrt{9396} \approx 97$$

$$5^2 + 6^2 \approx \left(7\frac{5}{6}\right)^2 \Rightarrow \sqrt{61} \approx \left(7\frac{5}{6}\right) = \frac{47}{6} = 7.833...$$

We also have

$$\sqrt{2} \approx 1 + \frac{1}{3} + \frac{1}{3 \cdot 4} - \frac{1}{3 \cdot 4 \cdot 34} \Rightarrow \sqrt{2} \approx \frac{577}{408}$$

The very fact that three different values of $\sqrt{2}$ were given shows that the ancient Indians knew that $\sqrt{2}$ could not be determined. This in turn leads to the concept of irrational numbers. From this, it is very clear that Indians were the first people to use irrational numbers. Formal mathematical definition and theory were introduced by the celebrated German mathematicians, Julius Wilhelm Richard Dedekind (1831–1916), J.W. George Cantor (1829–1920), and Karl Theodor Wilhelm Weierstrass (1815–1897).

Construction of Altars

Brick construction for the altars has been discussed in detail in the "Sulvasutras." This requires a great skill in the mathematical computation and exhibit problems involving simultaneous indeterminate equations. "Baudhayana Sulvasutra" gives the following two examples.

Example 1

The altar sheltering "Garhapatyagni" has to be constructed so as to have five layers of brick, one over the other. Each layer contains 21 bricks and has an area of one square vyayam. Here, we have to realize that when one layer of brick is placed on the top of another, the cleavage between two bricks of the bottom layer has to be covered by the solid material of the superimposed brick. The shape of the brick may be rectangular or square. The bricks have to be of any required size. The altar itself may be either circular or square.

Let each layer of brick contains x square bricks of one shape and y square bricks of another shape. Let their lengths be $1/a$ and $1/b$, respectively. Then, from the above-mentioned requirements, we have

$$x + y = 21 \tag{i}$$

$$\frac{x}{a^2} + \frac{y}{b^2} = 1 \tag{ii}$$

Multiplying equation (i) by $1/a^2$, we get

$$\frac{x}{a^2} + \frac{y}{a^2} = \frac{21}{a^2} \tag{iii}$$

Subtracting (iii) from (ii), we get

$$y\left(\frac{1}{b^2} - \frac{1}{a^2}\right) = 1 - \frac{21}{a^2}$$

(Continued)

Example 1 (cont'd)

$$y\frac{(a^2-b^2)}{a^2b^2}=\frac{a^2-21}{a^2}\Rightarrow y=\frac{b^2(a^2-21)}{a^2-b^2} \tag{iv}$$

From (i), we have

$$x+\frac{b^2(a^2-21)}{a^2-b^2}=21$$

$$x=21-\frac{b^2(a^2-21)}{a^2-b^2}$$

$$=\frac{21a^2-21b^2-a^2b^2+21b^2}{a^2-b^2}$$

So,

$$x=\frac{a^2(21-b^2)}{a^2-b^2} \tag{iv}$$

Suppose $a>b$ then $b^2<21<a^2$, since $x>0$ and $y>0$.

If $b^2<21$, we must have $b\le 4$.

Substitute $b=1$, 2, 3, and 4 in turn. Then, we will find by trial the values of a that will yield integral values for x and y. Thus, we will obtain the following solutions:

i. $a=6,\ b=3,\ x=16,\ y=5$
ii. $a=6,\ b=4,\ x=9,\ y=12$

This solution explains the construction formulation for the altar given in "Baudhayana Sulvasutras." Suppose three types of bricks of square size, with length 1/6th, 1/4th, and 1/3rd of vyāyam, are used for construction.

Then, if each layer contains x, y, and z of these three different types of bricks, we have

$$x+y+z=21 \tag{1}$$

$$\frac{x}{36}+\frac{y}{16}+\frac{z}{9}=1 \tag{2}$$

Here, x, y, and z are nonnegative integers.

Multiplying (1) by 1/9, we get

$$\frac{x}{9}+\frac{y}{9}+\frac{z}{9}=\frac{21}{9}=\frac{7}{3} \tag{3}$$

Subtracting (2) from (3), we get

$$x\left(\frac{1}{9}-\frac{1}{36}\right)+y\left(\frac{1}{9}-\frac{1}{16}\right)=\frac{7}{3}-1=\frac{4}{3} \tag{4}$$

Multiplying (4) by 144, we get

$$x(16-4)+y(16-9)=192$$

i.e.,

$$12x+7y=192$$

So,

$$12x=192-7y \quad \text{or} \quad x=16-\frac{7}{12}y \tag{5}$$

Since $y \leq x+y+z=21$, taking $y=0, 1, 2, 3, \ldots, 21$, we have integral solutions for x, only when $y=12$ or 0.

If $y=12$, $x=16-7=9$ from (5) and from (1), $z=0$.

If $y=0$, $x=16$ from (5) and from (1), $z=5$.

Hence, $(x, y, z)=(9, 12, 0)$ or $(16, 0, 5)$.

If the first layer has to have second type of brick, then the first layers have to be constructed with nine bricks of first type and 12 of the second type. The second layer can be constructed with 16 bricks of first type and five bricks of third type. Then, the third and the fifth layers can be similar to the first and the fourth similar to the second.

Example 2

A more difficult problem has been given in the construction of the altar for "Garuda Chayana."

As in the previous example, there should be five layers of brick. Each layer is to have 200 bricks to cover an area of $7\frac{1}{2}$ square purushas. Four different sizes of bricks are used for construction. Let $1/m$, $1/n$, $1/p$, and $1/q$ be the area of each brick of these four different sizes. If each layer contains x, y, z, and w of these four different types of bricks, we have

$$x+y+z+w=200 \tag{1}$$

(Continued)

Example 2 (cont'd)

$$\frac{x}{m}+\frac{y}{n}+\frac{z}{p}+\frac{w}{q}=7\frac{1}{2} \quad (2)$$

"Baudhayana Sulvasutra" gives the following solution.
If $(m, n, p, q) = (16, 25, 36, 100)$

$$x = 24,\ y = 120,\ z = 36,\ w = 20 \quad \text{or}$$

$$x = 12,\ y = 125,\ z = 63,\ w = 0$$

This means 1/16, 1/25, 1/36, and 1/100 are the area of these four different square bricks. So, these four different types of square bricks are of length 1/4, 1/5, 1/6, and 1/10. "Baudhayana Sutra" does not give any method of how they arrived at the above solutions. We will now construct a new method of solving the above equations:

$$x+y+z+w=200 \quad (1)$$

$$\frac{x}{m}+\frac{y}{n}+\frac{z}{p}+\frac{w}{q}=\frac{15}{2} \quad (2)$$

Let

$$\frac{x}{m}=k_1 \Rightarrow x = mk_1 \quad \text{and} \quad \frac{y}{n}=k_2 \Rightarrow y = nk_2$$

Then, from (1), we have

$$z+w=200-mk_1-nk_2 \quad (3)$$

From (2), we have

$$\frac{z}{p}+\frac{w}{q}=\frac{15}{2}-k_1-k_2 \quad (4)$$

Multiplying equation (3) by $1/q$, we get

$$\frac{z}{q}+\frac{w}{q}=\frac{200-(mk_1+nk_2)}{q} \quad (5)$$

Subtracting (5) from (4), we get

$$z\left(\frac{q-p}{pq}\right) = \frac{z}{p} - \frac{z}{q} = \frac{15}{2} - (k_1 + k_2) - \frac{200 - (mk_1 - nk_2)}{q}$$

So,

$$z = \frac{p[\{(15/2) - (k_1 + k_2)\}q - \{200 - mk_1 - nk_2\}]}{q - p} \tag{A}$$

Let e_1 be the smallest of the numbers m, n, p, and q.
Then,

$$\frac{x}{e_1} \geq \frac{x}{m}; \quad \frac{y}{e_1} \geq \frac{y}{n}; \quad \frac{z}{e_1} \geq \frac{z}{p}; \quad \text{and} \quad \frac{w}{e_1} \geq \frac{w}{q}$$

Hence,

$$\frac{1}{e}[x + y + z + w] \geq \frac{x}{m} + \frac{y}{n} + \frac{z}{p} + \frac{w}{q} = \frac{15}{2}$$

i.e.,

$$\frac{200}{e_1} \geq \frac{15}{2} \Rightarrow 200 \geq \frac{15}{2} e_1$$

i.e.,

$$\frac{15}{2} e_1 \leq 200 \Rightarrow e_1 \leq \frac{400}{15} = \frac{80}{3} = 26\frac{2}{3}$$

Since e_1 is a square, the possible values of e_1 are 1, 4, 9, 16, and 25.
Similarly, if e_2 is the largest value of m, n, p, and q, we will have $e_2 \geq 26\frac{2}{3}$.
Since e_2 is a square, the possible values of e_2 are 36, 49, 64, 81, 100,. . ..
(a) Suppose we choose $(m, n, p, q) = (16, 25, 36, 100)$
Hence, equation (A) becomes

$$z = \frac{36[\{(15/2) - (k_1 + k_2)\}(100) - (200 - 16k_1 - 25k_2)]}{100 - 36}$$

So,

$$z = \frac{9}{16}[550 - 84k_1 - 75k_2] \tag{B}$$

Now, $x = 16k_1$ and $y = 25k_2$.

We choose

$$k_1 = \frac{3}{2} \quad \text{and} \quad k_2 = \frac{24}{5}$$

Then,

$$x = 16\left(\frac{3}{2}\right) = 24 \quad \text{and} \quad y = 25\left(\frac{24}{5}\right) = 120$$

Now, substituting the values of $k_1 = 3/2$ and $k_2 = 24/5$ in (B), we get

$$z = \frac{9}{16}\left[550 - 84\left(\frac{3}{2}\right) - 75\left(\frac{24}{5}\right)\right]$$

$$= \frac{9}{16}[550 - 126 - 360] = \frac{9}{16} \cdot (64) = 36$$

$$w = 200 - x - y - z = 200 - 24 - 120 - 36 = 20$$

$(x, y, z, w) = (24, 120, 36, 20)$ is a solution of (1) and (2).
(b) Suppose we choose $(m, n, p, q) = (16, 25, 36, 100)$
We choose

$$k_1 = \frac{3}{4} \quad \text{and} \quad k_2 = 5$$

Then,

$$x = mk_1 = (16)\left(\frac{3}{4}\right) = 12 \quad \text{and} \quad y = nk_2 = (25)(5) = 125$$

Then, substituting these values of $k_1 = 3/4$ and $k_2 = 5$ in equation (B), we get

$$z = \frac{9}{16}\left[550 - 84 \cdot \frac{3}{4} - 75(5)\right]$$

$$= \frac{9}{16}(550 - 63 - 375) = \frac{9}{16}(112) = 63$$

Now, $w = 200 - x - y - z - w = 200 - 12 - 125 - 63 = 0$.
So, $(x, y, z, w) = (12, 125, 63, 0)$ is a solution of (1) and (2).

The two solutions (24, 120, 36, 20) and (12, 125, 63, 0) that we have found in (a) and (b) are exactly the two solutions given in "Baudhayana Sutra." For other suitable values of k_1 and k_2, we get other solutions of the above construction, where $(m, n, p, q) = (16, 25, 36, 100)$.

"Baudhayana Sulvasutra" also points out that in the above construction, rectangular bricks can be used instead of square bricks. No solution has been provided in the "Baudhayana Sutra" on the premises that they may have to use large number of these four different types of bricks, because these bricks have to further slit up to have square shape.

We will now give particular solutions of breaking rectangular bricks to have square shape.

Suppose $(m, n, p, q) = (25, 50, 50/3, 100)$.

Substituting in (A), we get

$$z = \frac{50}{3}\frac{[\{(15/2) - (k_1 + k_2)\}100 - (200 - 25k_1 - 50k_2)]}{100 - (50/3)}$$

So,

$$z = \frac{1}{5}[750 - 100k_1 - 100k_2 - 200 + 25k_1 + 50k_2]$$

$$z = \frac{1}{5}[550 - 75k_1 - 50k_2] \tag{C}$$

(a) Let

$$k_1 = \frac{32}{5}, \quad k_2 = \frac{3}{5}$$

Then, $x = mk_1 = 25k_1$, $y = nk_2 = 50k_2$.

So,

$$x = 25\left(\frac{32}{5}\right) = 160, \quad y = 50\left(\frac{3}{5}\right) = 30$$

Also,

$$z = \frac{1}{5}\left[550 - 75\left(\frac{32}{5}\right) - 50\left(\frac{3}{5}\right)\right]$$

So,

$$s = \frac{1}{5}[550 - 480 - 30] = \frac{40}{5} = 8$$

Hence, $w = 200 - x - y - z = 200 - 160 - 30 - 8 = 2$.

$(x, y, z, w) = (160, 30, 8, 2)$ is a solution of (1) and (2).

(b) Let

$$k_1 = \frac{33}{5} \quad \text{and} \quad k_2 = \frac{1}{2}$$

Then,

$$x = mk_1 = (25)\frac{33}{5} = 165$$

$$y = nk_2 = (50)\frac{1}{2} = 25$$

Also from equation (C),

$$z = \frac{1}{5}\left[550 - 75\left(\frac{33}{5}\right) - 50\left(\frac{1}{2}\right)\right]$$

$$z = \frac{1}{5}[550 - 495 - 25] = \frac{30}{5} = 6$$

Hence, $w = 200 - x - y - z = 200 - 165 - 25 - 6 = 4$.

Thus, $(x, y, z, w) = (165, 25, 6, 4)$ is a solution of (1) and (2). For other suitable values of k_1 and k_2, we can obtain other solutions of (1) and (2).

Since the first and the fourth type of bricks are of area 1/25 and 1/100, respectively, these two types of bricks are already square bricks of length 1/5 and 1/10. Also, the second and third types of bricks are rectangular bricks of area 1/50 and 3/50, respectively.

Remembering the fact that we have to use as less bricks as possible, we observe that

$$\frac{1}{50} = \frac{1}{100} + \frac{1}{100} \quad \text{and} \quad \frac{3}{50} = \frac{1}{100} + \frac{1}{100} + \frac{1}{25}$$

This means the second rectangular type brick has to be split up into two square type bricks of length 1/10 and third rectangular type brick has to be split up into types of square bricks of length 1/10 and 1/5.

Now, from (A)

$$z = \frac{p[q\{(15/2) - (k_1 + k_2)\} - \{200 - mk_1 - nk_2\}]}{q - p}$$

Suppose we choose $(m, n, p, q) = (4, n, 9, 36)$, where $n \neq 4$, $n \neq 9$, and $n \neq 36$. Also,

$$\frac{x}{m} = \frac{x}{4} = k_1, \quad \frac{y}{n} = k_2, \quad \frac{z}{p} = \frac{z}{9} = k_3, \quad \text{and} \quad \frac{w}{q} = \frac{w}{36} = k_4$$

So,

$$k_1 + k_2 + k_3 + k_4 = \frac{15}{2}$$

Then, we have

$$z = \frac{9[36\{(15/2) - (k_1 + k_2)\} - \{200 - 4k_1 - nk_2\}]}{36 - 9}$$

Hence,

$$z = \frac{270 - 36(k_1 + k_2) - 200 + 4k_1 + nk_2}{3}$$

i.e.,

$$z = \frac{70 - 32k_1 + k_2(n - 36)}{3} \text{ is an integer} \geq 0$$

Let $k_2 = 1 \Rightarrow y = n$.
Then,

$$z = \frac{70 - 32k_1 + n - 36}{3}$$

i.e.,

$$z = \frac{34 - 32k_1 + n}{3}$$

(a) Let $k_1 = 1 \Rightarrow x = 4$
Then,

$$z = \frac{34 - 32 + n}{3} = \frac{2 + n}{3} \text{ is an integer} \geq 0$$

(1) Let

$$n = 1, \quad \text{then } y = n = 1 \text{ and } z = \frac{2 + 1}{3} = \frac{3}{3} = 1$$

$$w = 200 - x - y - z = 200 - 4 - 1 - 1 = 194$$

Then, $(x, y, z, w) = (4, 1, 1, 194)$ is a solution.

(2) Let

$$n = 16 \Rightarrow y = 16, \quad z = \frac{2+n}{3} = \frac{2+16}{3} = \frac{18}{3} = 6$$

$$w = 200 - x - y - z = 200 - 4 - 16 - 6 = 174$$

Then, $(x, y, z, w) = (4, 16, 6, 174)$ is a solution.

(3) Let

$$n = 25 \Rightarrow y = 25, \quad z = \frac{2+n}{3} = \frac{2+25}{3} = \frac{27}{3} = 9$$

$$w = 200 - x - y - z = 200 - 4 - 25 - 9 = 162$$

Then, $(x, y, z, w) = (4, 25, 9, 162)$ is a solution.

(4) Let

$$n = 49 \Rightarrow y = 49, \quad z = \frac{2+n}{3} = \frac{2+49}{3} = \frac{51}{3} = 17$$

$$w = 200 - x - y - z = 200 - 4 - 49 - 17 = 130$$

Then, $(x, y, z, w) = (4, 49, 17, 130)$ is a solution.

(5) Let

$$n = 64 \Rightarrow y = 64, \quad z = \frac{2+n}{3} = \frac{2+64}{3} = \frac{66}{3} = 22$$

$$w = 200 - x - y - z = 200 - 4 - 64 - 22 = 110$$

Then, $(x, y, z, w) = (4, 64, 22, 110)$ is a solution.

(6) Let

$$n = 100 \Rightarrow y = 100, \quad z = \frac{2+n}{3} = \frac{2+100}{3} = \frac{102}{3} = 34$$

$$w = 200 - x - y - z = 200 - 4 - 100 - 34 = 62$$

Then, $(x, y, z, w) = (4, 100, 34, 62)$ is a solution.

(7) Let

$$n = 121 \Rightarrow y = 121, \quad z = \frac{2+n}{3} = \frac{2+121}{3} = \frac{123}{3} = 41$$

$$w = 200 - x - y - z = 200 - 4 - 121 - 41 = 34$$

Then, $(x, y, z, w) = (4, 121, 41, 34)$ is a solution.
(b) Let

$$k_1 = \frac{5}{4} \Rightarrow x = 5$$

Now,

$$z = \frac{34 - 32k_1 + n}{3} = \frac{34 - 32 \cdot (5/4) + n}{3} = \frac{34 - 40 + n}{3}$$

i.e.,

$$z = -2 + \frac{n}{3} \text{ is an integer} \geq 0$$

(8) Let

$$n = 81 \Rightarrow y = 81 \quad \text{and} \quad z = -2 + \frac{n}{3} = -2 + \frac{81}{3}$$

$$z = -2 + 27 = 25$$

$$w = 200 - x - y - z = 200 - 5 - 81 - 25 = 89$$

Then, $(x, y, z, w) = (5, 81, 25, 89)$ is a solution.
(9) Let

$$n = 144 \Rightarrow y = 144 \quad \text{and} \quad z = -2 + \frac{n}{3} = -2 + \frac{144}{3}$$

$$z = -2 + 48 = 46$$

$$w = 200 - x - y - z = 200 - 5 - 144 - 46 = 5$$

Then, $(x, y, z, w) = (5, 144, 46, 5)$ is a solution.
(c) Let

$$k_1 = \frac{7}{4} \Rightarrow x = 7$$

Now,

$$z = \frac{34 - 32k_1 + n}{3} = \frac{34 - 32 \cdot (7/4) + n}{3}$$

$$z = \frac{34 - 56 + n}{3} = \frac{n - 22}{3} \text{ is an integer} \geq 0$$

(10) Let

$$n = 25 \Rightarrow y = 25 \quad \text{and} \quad z = \frac{n - 22}{3} = \frac{25 - 22}{3} = \frac{3}{3} = 1$$

$$w = 200 - x - y - z = 200 - 7 - 25 - 1 = 167$$

Then, $(x, y, z, w) = (7, 25, 1, 167)$ is a solution.

(11) Let

$$n = 49 \Rightarrow y = 49 \quad \text{and} \quad z = \frac{n - 22}{3} = \frac{49 - 22}{3} = \frac{27}{3} = 9$$

$$w = 200 - x - y - z = 200 - 7 - 49 - 9 = 135$$

Then, $(x, y, z, w) = (7, 49, 9, 135)$ is a solution.

(12) Let

$$n = 64 \Rightarrow y = 64 \quad \text{and} \quad z = \frac{n - 22}{3} = \frac{64 - 22}{3} = \frac{42}{3} = 14$$

$$w = 200 - x - y - z = 200 - 7 - 64 - 14 = 115$$

Then, $(x, y, z, w) = (7, 64, 14, 115)$ is a solution.

(13) Let

$$n = 100 \Rightarrow y = 100 \quad \text{and} \quad z = \frac{n - 22}{3} = \frac{100 - 22}{3} = \frac{78}{3} = 26$$

$$w = 200 - x - y - z = 200 - 7 - 100 - 26 = 67$$

Then, $(x, y, z, w) = (7, 100, 26, 67)$ is a solution.

(14) Let

$$n = 121 \Rightarrow y = 121 \quad \text{and} \quad z = \frac{n - 22}{3} = \frac{121 - 22}{3} = \frac{99}{3} = 33$$

$$w = 200 - x - y - z = 200 - 7 - 121 - 33 = 39$$

Then, $(x, y, z, w) = (7, 121, 33, 39)$ is a solution.

(d) Let $k_1 = 2 \Rightarrow x = 8$
Now,

$$z = \frac{34 - 32k_1 + n}{3} = \frac{34 - 64 + n}{3} = -10 + \frac{n}{3} \text{ is an integer} \geq 0$$

Let

$$n = 81 \Rightarrow y = 81 \quad \text{and} \quad z = -10 + \frac{n}{3} = -10 + \frac{81}{3} = -10 + 27 = 17$$

$$w = 200 - x - y - z = 200 - 8 - 81 - 17 = 94$$

Then, $(x, y, z, w) = (8, 81, 17, 94)$ is a solution.
(16) Let

$$n = 144 \Rightarrow y = 144 \quad \text{and} \quad z = -10 + \frac{n}{3} = -10 + \frac{144}{3} = -10 + 48 = 38$$

$$w = 200 - x - y - z = 200 - 8 - 144 - 38 = 10$$

Then, $(x, y, z, w) = (8, 144, 38, 10)$ is a solution.
(e) Let

$$k_1 = \frac{5}{2} \Rightarrow x = 10$$

$$z = \frac{34 - 32k_1 + n}{3} = \frac{34 - 32 \cdot (5/2) + n}{3}$$

$$z = \frac{34 - 80 + n}{3} = \frac{n - 46}{3}$$

(17) Let

$$n = 49 \Rightarrow y = 49 \quad \text{and} \quad z = \frac{n - 46}{3} = \frac{49 - 46}{3} = \frac{3}{3} = 1$$

$$w = 200 - x - y - z = 200 - 10 - 49 - 1 = 140$$

Then, $(x, y, z, w) = (10, 49, 1, 140)$ is a solution.
(18) Let

$$n = 64 \Rightarrow y = 64 \quad \text{and} \quad z = \frac{n - 46}{3} = \frac{64 - 46}{3} = \frac{18}{3} = 6$$

$$w = 200 - x - y - z = 200 - 10 - 64 - 6 = 120$$

Then, $(x, y, z, w) = (10, 64, 6, 120)$ is a solution.

(19) Let

$$n = 100 \Rightarrow y = 100 \quad \text{and} \quad z = \frac{n - 46}{3} = \frac{100 - 46}{3} = \frac{54}{3} = 18$$

$$w = 200 - x - y - z = 200 - 10 - 100 - 18 = 72$$

Then, $(x, y, z, w) = (10, 100, 18, 72)$ is a solution.

(20) Let

$$n = 121 \Rightarrow y = 121 \quad \text{and} \quad z = \frac{n - 46}{3} = \frac{121 - 46}{3} = \frac{75}{3} = 25$$

$$w = 200 - x - y - z = 200 - 10 - 121 - 25 = 44$$

Then, $(x, y, z, w) = (10, 121, 25, 44)$ is a solution.

(f) Let

$$k_1 = \frac{1}{4} \Rightarrow x = 1$$

Now,

$$z = \frac{34 - 32k_1 + n}{3} = \frac{34 - 32 \cdot (1/4) + n}{3} = \frac{34 - 8 + n}{3}$$

$$z = \frac{26 + n}{3} \text{ is an integer} \geq 0$$

(21) Let

$$n = 1 \Rightarrow y = 1 \quad \text{and} \quad z = \frac{26 + n}{3} = \frac{26 + 1}{3} = \frac{27}{3} = 9$$

$$w = 200 - x - y - z = 200 - 1 - 1 - 9 = 189$$

Then, $(x, y, z, w) = (1, 1, 9, 189)$ is a solution.

(22) Let

$$n = 16 \Rightarrow y = 16 \quad \text{and} \quad z = \frac{26 + 16}{3} = \frac{42}{3} = 14$$

$$w = 200 - x - y - z = 200 - 1 - 16 - 14 = 169$$

Then, $(x, y, z, w) = (1, 16, 14, 169)$ is a solution.

(23) Let

$$n = 25 \Rightarrow y = 25 \quad \text{and} \quad z = \frac{26 + 25}{3} = \frac{51}{3} = 17$$

$$w = 200 - x - y - z = 200 - 1 - 25 - 17 = 157$$

Then, $(x, y, z, w) = (1, 25, 17, 157)$ is a solution.

(24) Let

$$n = 49 \Rightarrow y = 49 \quad \text{and} \quad z = \frac{26 + 49}{3} = \frac{75}{3} = 25$$

$$w = 200 - x - y - z = 200 - 1 - 49 - 25 = 125$$

Then, $(x, y, z, w) = (1, 49, 25, 125)$ is a solution.

(25) Let

$$n = 64 \Rightarrow y = 64 \quad \text{and} \quad z = \frac{26 + 64}{3} = \frac{90}{3} = 30$$

$$w = 200 - x - y - z = 200 - 1 - 64 - 30 = 105$$

Then, $(x, y, z, w) = (1, 64, 30, 105)$ is a solution.

(26) Let

$$n = 100 \Rightarrow y = 100 \quad \text{and} \quad z = \frac{26 + 100}{3} = \frac{126}{3} = 42$$

$$w = 200 - x - y - z = 200 - 1 - 100 - 42 = 57$$

Then, $(x, y, z, w) = (1, 100, 42, 57)$ is a solution.

(27) Let

$$n = 121 \Rightarrow y = 121 \quad \text{and} \quad z = \frac{26 + 121}{3} = \frac{147}{3} = 49$$

$$w = 200 - x - y - z = 200 - 1 - 121 - 49 = 29$$

Then, $(x, y, z, w) = (1, 121, 49, 29)$ is a solution.

(g) Let

$$k_1 = \frac{1}{2} \Rightarrow x = 2$$

Now,

$$z = \frac{34 - 32k_1 + n}{3} = \frac{34 - 32 \cdot (1/2) + n}{3} = \frac{34 - 16 + n}{3}$$

$$z = 6 + \frac{n}{3} \text{ is an integer} \geq 0$$

(28) Let

$$n = 81 \Rightarrow y = 81 \quad \text{and} \quad z = 6 + \frac{n}{3} = 6 + \frac{81}{3} = 6 + 27 = 33$$

$$w = 200 - x - y - z = 200 - 2 - 81 - 33 = 84$$

Then, $(x, y, z, w) = (2, 81, 33, 84)$ is a solution.

(29) Let

$$n = 144 \Rightarrow y = 144 \quad \text{and} \quad z = 6 + \frac{n}{3} = 6 + \frac{144}{3} = 6 + 48 = 54$$

$$w = 200 - x - y - z = 200 - 2 - 144 - 54 = 0$$

Then, $(x, y, z, w) = (2, 144, 54, 0)$ is a solution.

The results obtained above can be summarized as follows:

$$x + y + z = w = 200$$

$$k_1 + k_2 + k_3 + k_4 = \frac{15}{2}$$

where

$$k_1 = \frac{x}{m}; \quad k_2 = \frac{y}{n}; \quad k_3 = \frac{z}{p}; \quad \text{and} \quad k_4 = \frac{w}{q}$$

Here, $m = 4$, $p = 9$, and $q = 36$.

Thus, we have the following solution table:

Solution No.	k_1	k_2	n	k_3	k_4	$k_1 + k_2 + k_3 + k_4$	x	y	z	w
1.	1	1	1	$\frac{1}{9}$	$\frac{97}{18}$	$\frac{15}{2}$	4	1	1	194
2.	1	1	16	$\frac{2}{3}$	$\frac{29}{6}$	$\frac{15}{2}$	4	16	6	174
3.	1	1	25	1	$\frac{9}{2}$	$\frac{15}{2}$	4	25	9	162

(Continued)

(Continued)

Solution No.	k_1	k_2	n	k_3	k_4	$k_1+k_2+k_3+k_4$	x	y	z	w
4.	1	1	49	$\frac{17}{9}$	$\frac{65}{18}$	$\frac{15}{2}$	4	49	17	130
5.	1	1	64	$\frac{22}{9}$	$\frac{55}{18}$	$\frac{15}{2}$	4	64	22	110
6.	1	1	100	$\frac{34}{9}$	$\frac{31}{18}$	$\frac{15}{2}$	4	100	34	62
7.	1	1	121	$\frac{41}{9}$	$\frac{17}{18}$	$\frac{15}{2}$	4	121	41	34
8.	$\frac{5}{4}$	1	81	$\frac{25}{9}$	$\frac{89}{36}$	$\frac{15}{2}$	5	81	25	89
9.	$\frac{5}{4}$	1	144	$\frac{46}{9}$	$\frac{5}{36}$	$\frac{15}{2}$	5	144	46	5
10.	$\frac{7}{4}$	1	25	$\frac{1}{9}$	$\frac{167}{36}$	$\frac{15}{2}$	7	25	1	167
11.	$\frac{7}{4}$	1	49	1	$\frac{15}{4}$	$\frac{15}{2}$	7	49	9	135
12.	$\frac{7}{4}$	1	64	$\frac{14}{9}$	$\frac{5}{12}$	$\frac{15}{2}$	7	64	14	115
13.	$\frac{7}{4}$	1	100	$\frac{26}{9}$	$\frac{67}{36}$	$\frac{15}{2}$	7	100	26	67
14.	$\frac{7}{4}$	1	121	$\frac{11}{3}$	$\frac{13}{12}$	$\frac{15}{2}$	7	121	33	39
15.	2	1	81	$\frac{17}{9}$	$\frac{47}{18}$	$\frac{15}{2}$	8	81	17	94
16.	2	1	144	$\frac{38}{9}$	$\frac{5}{18}$	$\frac{15}{2}$	8	144	38	10
17.	$\frac{5}{2}$	1	49	$\frac{1}{9}$	$\frac{35}{9}$	$\frac{15}{2}$	10	49	1	140
18.	$\frac{5}{2}$	1	64	$\frac{2}{3}$	$\frac{10}{3}$	$\frac{15}{2}$	10	64	6	120
19.	$\frac{5}{2}$	1	100	2	2	$\frac{15}{2}$	10	100	18	72
20.	$\frac{5}{2}$	1	121	$\frac{25}{9}$	$\frac{11}{9}$	$\frac{15}{2}$	10	121	25	44
21.	$\frac{1}{4}$	1	1	1	$\frac{21}{4}$	$\frac{15}{2}$	1	1	9	189
22.	$\frac{1}{4}$	1	16	$\frac{14}{9}$	$\frac{169}{36}$	$\frac{15}{2}$	1	16	14	169
23.	$\frac{1}{4}$	1	25	$\frac{17}{9}$	$\frac{157}{36}$	$\frac{15}{2}$	1	25	17	157
24.	$\frac{1}{4}$	1	49	$\frac{25}{9}$	$\frac{125}{36}$	$\frac{15}{2}$	1	49	25	125

(Continued)

(Continued)

Solution No.	k_1	k_2	n	k_3	k_4	$k_1+k_2+k_3+k_4$	x	y	z	w
25.	$\frac{1}{4}$	1	64	$\frac{10}{3}$	$\frac{35}{12}$	$\frac{15}{2}$	1	64	30	105
26.	$\frac{1}{4}$	1	100	$\frac{14}{3}$	$\frac{19}{12}$	$\frac{15}{2}$	1	100	42	57
27.	$\frac{1}{4}$	1	121	$\frac{49}{9}$	$\frac{29}{36}$	$\frac{15}{2}$	1	121	49	29
28.	$\frac{1}{2}$	1	81	$\frac{11}{3}$	$\frac{7}{3}$	$\frac{15}{2}$	2	81	33	84
29.	$\frac{1}{2}$	1	144	6	0	$\frac{15}{2}$	2	144	54	0

There are four distinct numbers m, n, p, and q. In the list of 29 solutions shown in the above table, we have found, taking fixed values for three of these, namely, $m=4$, $p=9$, and $q=36$ and taking $k_2=1$ and $k_1=1$, 5/4, 7/4, 2, 5/2, 1/4, 1/2, we were able to determine the values of n, x, y, z, and w. Similarly, for other values of k_1 and k_2, such that $k_1+k_2+k_3+k_4=15/2$, other solutions can be found. The above procedure can be used to determine solutions for different values of m, q, and r from the list (1, 4, 9, 16, 25, 36,. . .).

In "Āpasthamba Sulvasutra," the following construction problem has been posed:

Unlike the previous problem, here five types of square bricks are used for construction instead of four. If the areas of these five bricks are $1/m^2$, $1/n^2$, $1/p^2$, $1/q^2$, and $1/r^2$ and their number in each layer is x, y, z, w, and u, we have the following two equations:

$$x+y+z+w+u=200 \qquad \text{(i)}$$

$$\frac{x}{m^2}+\frac{y}{n^2}+\frac{z}{p^2}+\frac{w}{q^2}+\frac{u}{r^2}=7\frac{1}{2} \qquad \text{(ii)}$$

Unfortunately, "Āpasthamba Sulvasutra" does not give clearly its construction method. However, three commentators Karavinda Swamy, Kapardi Swami, and Sundararaja have presented the following solutions. We do not know how they arrived at these solutions.

	Name of the Commentator	m	n	p	q	r	x	y	z	w	u
1.	Karavinda Swamy	16	25	64	100	144	67	58	48	18	9
2.	Karvinda Swamy	16	25	36	64	100	12	157	9	0	22
3.	Kapardi Swami	16	25	36	64	100	10	159	9	8	14
4.	Sundararaja	16	25	64	100	144	74	45	52	20	9
5.	Sundararaja	16	25	36	64	100	70	45	9	56	20
6.	Sundararaja	16	25	64	100	144	77	42	40	32	9

We will show later that the above list is a short list of all possible solutions of (i) and (ii).

Dr. K.N. Kamalamma (Bulletin of Calcutta Mathematical Society, vol. 40 (1948), pp. 140–144) has shown that the equations (i) and (ii) have finite number of solutions. She has also provided a method of solving equations (i) and (ii). However, her method is slightly tedious, since it involves solving indeterminate equations of the form $az + bw = c$. We will now solve equations (i) and (ii) by the method shown to solve the previous problem. As shown in the last example, if e_1^2 is the smallest of the numbers m^2, n^2, p^2, q^2, r^2, the possible values of e_1^2 are 1, 4, 9, 16, 25, and if e_2^2 is the biggest of the numbers m^2, n^2, p^2, q^2, r^2, the possible values of e_2^2 are 36, 49, 64, 81, 100, 144, ...

We have

$$x + y + z + w + u = 200 \tag{i}$$

and

$$\frac{x}{m^2} + \frac{y}{n^2} + \frac{z}{p^2} + \frac{w}{q^2} + \frac{u}{r^2} = \frac{15}{2} \tag{ii}$$

Let

$$\frac{x}{m^2} = k_1 \Rightarrow x = m^2 k_1; \quad \frac{y}{n^2} = k_2 \Rightarrow y = n^2 k_2$$

and

$$\frac{z}{p^2} = k_3 \Rightarrow z = k_3 p^2$$

From (ii), we have

$$k_1 + k_2 + k_3 + \frac{w}{q^2} + \frac{u}{r^2} = \frac{15}{2}$$

$$\Rightarrow k_1 + k_2 + k_3 \le \frac{15}{2}$$

since w and u are nonnegative integers.

Also,

$$\frac{w}{q^2} + \frac{u}{r^2} = \frac{15}{2} - k_1 - k_2 - k_3 \tag{iii}$$

From (1), we have

$$w + u = 200 - x - y - z$$

$$= 200 - m^2k_1 - n^2k_2 - p^2k_3 \qquad \text{(iv)}$$

We get

$$\frac{w}{r^2} + \frac{u}{r^2} = \frac{1}{r^2}(200 - m^2k_1 - n^2k_2 - p^2k_3) \qquad \text{(v)}$$

multiplying (iv) by $1/r^2$. Subtracting (v) from (iii), we get

$$w\left(\frac{1}{q^2} - \frac{1}{r^2}\right) = \left\{\frac{15}{2} - (k_1 + k_2 + k_3) - \frac{1}{r^2}(200 - m^2k_1 - n^2k_2 - p^2k_3)\right\}$$

i.e.,

$$w\left(\frac{r^2 - q^2}{r^2q^2}\right) = \left[\frac{15}{2} - (k_1 + k_2 + k_3)\right] - \frac{1}{r^2}\left[200 - m^2k_1 - n^2k_2 - p^2k_3\right]$$

Then,

$$w = \frac{q^2\{[(15/2) - (k_1 + k_2 + k_3)]r^2 - [200 - m^2k_1 - n^2k_2 - p^2k_3]\}}{(r^2 - q^2)} \qquad \text{(A)}$$

Suppose we take $(m, n, p, q, r) = (1, 5, p, 3, 6)$.
Let $k_1 = k_2 = 1$, here $p \neq 1$, $p \neq 5$, $p \neq 3$, and $p \neq 6$.
Then, $x = m^2k_1 = 1 \cdot 1 = 1$; $y = n^2k_2 = 25 \cdot 1 = 25$.
Thus,

$$w = \frac{9[(15/2) - (1 + 1 + k_3)]36 - [200 - 1 - 25 - p^2k_3]}{(36 - 9)}$$

$$w = \frac{9[((11/2) - k_3)36 - 174 + p^2k_3]}{27}$$

$$= \frac{24 - 36k_3 + p^2k_3}{3}$$

Now,

$$w = 8 - 12k_3 + \frac{p^2k_3}{3} \text{ is an integer} \geq 0$$

Let,

$$\frac{p^2k_3}{3} = s = \text{an integer} \geq 0$$

But

$$z = p^2k_3 \Rightarrow \frac{z}{3} = s \Rightarrow z = 3s$$

Since

$$x = 1, \quad y = 25, \quad z \leq z + w + u = 200 - x - y = 200 - 1 - 25 = 174$$

So, $3s \leq 174 \Rightarrow s \leq 58$.
Then, $o \leq s \leq 58$.
Then, we have

$$w = 8 - 12\left(\frac{3s}{p^2}\right) + s = \text{an integer} \geq 0$$

(**) i.e.,

$$w = 8 + s - \frac{36s}{p^2}$$

where

$$s = 0, 1, 2, 3, \ldots, 58$$

(**) has solutions for $s = 0$, 1, 4, 8, 9, 12, 16, 18, 20, 24, 25, 27, 28, 32, 36, 40, 45, 48, 49 and (**) has no solution for other integral values of s. We will now find the solutions of (**).

$$z = 3s$$

$$w = 8 + s - \frac{36s}{p^2}$$

$$x + y + z + w + u = 200$$

$$1 + 25 + 3s + 8 + s - \frac{36s}{p^2} + u = 200$$

$$u = 200 - 34 - 4s + \frac{36s}{p^2} = 166 - s\left(4 - \frac{36}{p^2}\right)$$

(1) Let $s = 0 \Rightarrow z = 3s = 0$

$$w = 8 + s - \frac{36s}{p^2}, \quad u = 166 - s\left(4 - \frac{36}{p^2}\right)$$

If $s = 0$, $w = 8$, and $u = 166$, $(x, y, z, w, u) = (1, 25, 0, 8, 166)$ is a solution of (i) and (ii) on page 52.

(2) Next Let $s = 1 \Rightarrow z = 3s = 3$

$$w = 8 + s - \frac{36s}{p^2}$$

If $s = 1$,

$$w = 8 + 1 - \frac{36}{p^2} = 9 - \frac{36}{p^2} \text{ is an integer} \geq 0$$

If $p = 2$,

$$w = 9 - \frac{36}{4} = 9 - 9 = 0$$

$$u = 166 - s\left(4 - \frac{36}{p^2}\right)$$

$$= 166 - \left(4 - \frac{36}{4}\right) = 166 + 5 = 171$$

$(x, y, z, w, u) = (1, 25, 3, 0, 171)$ is a solution of (i) and (ii) on page 52.

(3) Let $s = 4 \Rightarrow z = 3s = 12$

$$w = 8 + s - \frac{36s}{p^2} = 12 - \frac{36 \cdot 4}{p^2} \text{ is an integer} \geq 0$$

If $p = 4$,

$$w = 12 - \frac{36 \cdot 4}{4 \cdot 4} = 12 - 9 = 3$$

$$u = 166 - s\left(4 - \frac{36}{p^2}\right) = 166 - 4\left(4 - \frac{36}{16}\right)$$

$$= 166 - 16 + 9 = 159$$

$(x, y, z, w, u) = (1, 25, 12, 3, 159)$ is a solution of (i) and (ii) on page 52.

If $p = 12$,

$$w = 12 - \frac{36 \cdot 4}{12 \cdot 12} = 12 - 1 = 11$$

$$u = 166 - s\left(4 - \frac{36}{p^2}\right) = 166 - 4\left(4 - \frac{36}{144}\right)$$

$$= 166 - 16 + 1 = 151$$

$(x, y, z, w, u) = (1, 25, 12, 11, 151)$ is a solution of (i) and (ii) on page 52.

(4) Let $s = 8 \Rightarrow z = 3s = 24$

$$w = 8 + s - \frac{36s}{p^2} = 8 + 8 - \frac{36 \cdot 8}{p^2}$$

$$w = 16 - \frac{36 \cdot 8}{p^2} \text{ is an integer} \geq 0$$

If $p = 12$,

$$w = 16 - \frac{36 \cdot 8}{12 \cdot 12}$$

$$= 16 - 2 = 14$$

$$u = 166 - s\left(4 - \frac{36}{p^2}\right)$$

$$= 166 - 8\left(4 - \frac{36}{144}\right) = 166 - 32 + 2 = 136$$

$(x, y, z, w, u) = (1, 25, 24, 14, 136)$ is a solution of (i) and (ii) on page 52.

(5) Let $s = 9 \Rightarrow z = 3s = 27$

$$w = s + 8 - \frac{36s}{p^2} = 8 + 9 - \frac{36 \cdot 9}{p^2}$$

$$w = 17 - \frac{36.9}{p^2} \text{ is an integer} \geq 0$$

If $p = 9$,

$$w = 17 - \frac{36 \cdot 9}{9 \cdot 9} = 17 - 4 = 13$$

$$u = 166 - s\left(4 - \frac{36}{p^2}\right)$$

$$= 166 - 9\left(4 - \frac{36}{9 \cdot 9}\right) = 166 - 36 + 4 = 134$$

$(x, y, z, w, u) = (1, 25, 27, 13, 134)$ is a solution of (i) and (ii) on page 52.

(6) Let $s = 12 \Rightarrow z = 3s = 36$

$$w = s + 8 - \frac{36s}{p^2} = 12 + 8 - \frac{36 \cdot 12}{p^2}$$

$$w = 20 - \frac{36 \cdot 12}{p^2} \text{ is an integer} \geq 0$$

If $p = 12$,

$$w = 20 - \frac{36 \cdot 12}{12 \cdot 12} = 20 - 3 = 17$$

$$u = 166 - s\left(4 - \frac{36}{p^2}\right) = 166 - 12\left(4 - \frac{36}{144}\right)$$

$$u = 166 - 48 + 3 = 121$$

$(x, y, z, w, u) = (1, 25, 36, 17, 121)$ is solution of (i) and (ii).

(7) Let $s = 16 \Rightarrow z = 3s = 48$

$$w = 8 + s - \frac{36s}{p^2} = 8 + 16 - \frac{36 \cdot 16}{p^2}$$

$$w = 24 - \frac{36 \cdot 16}{p^2} \text{ is an integer} \geq 0$$

If $p = 8$,

$$w = 24 - \frac{36 \cdot 16}{8 \cdot 8} = 24 - 9 = 15$$

$$u = 166 - s\left(4 - \frac{36}{p^2}\right) = 166 - 16\left(4 - \frac{36}{64}\right)$$

$$= 166 - 64 + 9 = 111$$

$(x, y, z, w, u) = (1, 25, 48, 15, 111)$ is a solution of (i) and (ii).

If $p = 12$,

$$w = 24 - \frac{36 \cdot 16}{12 \cdot 12} = 24 - 4 = 20$$

$$u = 166 - s\left(4 - \frac{36}{p^2}\right) = 166 - 16\left(4 - \frac{36}{144}\right)$$

$$= 166 - 64 + 4 = 106$$

$(x, y, z, w, u) = (1, 25, 48, 20, 106)$ is a solution of (i) and (ii) on page 52.

(8) Let $s = 18 \Rightarrow z = 3s = 54$

$$w = 8 + s - \frac{36s}{p^2} = 8 + 18 - \frac{36 \cdot 18}{p^2}$$

$$w = 26 - \frac{36 \cdot 18}{p^2} \text{ is an integer} \geq 0$$

If $p = 9$,

$$w = 26 - \frac{36 \cdot 18}{9 \cdot 9} = 26 - 4.2 = 18$$

$$u = 166 - s\left(4 - \frac{36}{p^2}\right) = 166 - 18\left(4 - \frac{36}{9 \cdot 9}\right)$$

$$= 166 - 72 + 8 = 102$$

$(x, y, z, w, u) = (1, 25, 54, 18, 102)$ is a solution of (i) and (ii) on page 52.

If $p = 18$,

$$w = 26 - \frac{36 \cdot 18}{18 \cdot 18} = 26 - 2 = 24$$

$$u = 166 - s\left(4 - \frac{36}{p^2}\right) = 166 - 18\left(4 - \frac{36}{18 \cdot 18}\right)$$

$$= 166 - 72 + 2 = 96$$

$(x, y, z, w, u) = (1, 25, 54, 24, 96)$ is a solution of (i) and (ii) on page 52.

(9) Let $s = 20 \Rightarrow z = 3s = 60$

$$w = 8 + s - \frac{36s}{p^2} = 8 + 20 - \frac{36 \cdot 20}{p^2}$$

$$w = 28 - \frac{36 \cdot 20}{p^2} \text{ is an integer} \geq 0$$

If $p = 12$,

$$w = 28 - \frac{36 \cdot 20}{12 \cdot 12} = 28 - 5 = 23$$

$$u = 166 - s\left(4 - \frac{36}{p^2}\right) = 166 - 20\left(4 - \frac{36}{144}\right)$$

$$u = 166 - 80 + 5 = 91$$

$(x, y, z, w, u) = (1, 25, 60, 23, 91)$ is a solution of (i) and (ii) on page 52.

(10) Let $s = 24 \Rightarrow z = 3s = 72$

$$w = 8 + s - \frac{36s}{p^2} = 8 + 24 - \frac{36 \cdot 24}{p^2}$$

$$w = 32 - \frac{36 \cdot 24}{p^2} \text{ is an integer} \geq 0$$

If $p = 12$,

$$w = 32 - \frac{36 \cdot 24}{12 \cdot 12} = 32 - 6 = 26$$

$$u = 166 - s\left(4 - \frac{36}{p^2}\right) = 166 - 24\left(4 - \frac{36}{144}\right)$$

$$= 166 - 96 + 6 = 76$$

$(x, y, z, w, u) = (1, 25, 72, 26, 76)$ is a solution of (i) and (ii).

(11) Let $s = 25 \Rightarrow z = 3s = 75$

$$w = 8 + s - \frac{36s}{p^2} = 8 + 25 - \frac{36 \cdot 25}{p^2}$$

$$w = 33 - \frac{36 \cdot 25}{p^2} \text{ is an integer} \geq 0$$

Let $p = 10$,

$$w = 33 - \frac{36 \cdot 25}{100} = 33 - 9 = 24$$

$$u = 166 - s\left(4 - \frac{36}{p^2}\right) = 166 - 25\left(4 - \frac{36}{100}\right)$$

So,

$$u = 166 - 100 + 9 = 75$$

$(x, y, z, w, u) = (1, 25, 75, 24, 75)$ is a solution of (i) and (ii) on page 52.
Let $p = 15$,

$$w = 33 - \frac{36 \cdot 25}{15 \cdot 15}$$

$$w = 33 - 4 = 29$$

$$u = 166 - s\left(4 - \frac{36}{p^2}\right) = 166 - 25\left(4 - \frac{36}{15 \cdot 15}\right)$$

$$u = 166 - 100 + 4 = 70$$

$(x, y, z, w, u) = (1, 25, 75, 29, 70)$ is a solution of (i) and (ii) on page 52.

(12) Let $s = 27 \Rightarrow z = 3s = 81$

$$w = 8 + s - \frac{36s}{p^2} = 8 + 27 - \frac{36 \cdot 27}{p^2}$$

$$w = 35 - \frac{36 \cdot 27}{p^2} \text{ is an integer} \geq 0$$

Let $p = 9$,

$$w = 35 - \frac{36 \cdot 27}{9 \cdot 9} = 35 - 12 = 23$$

$$u = 166 - s\left(4 - \frac{36}{p^2}\right) = 166 - 27\left(4 - \frac{36}{9 \cdot 9}\right)$$

$$u = 166 - 108 + 12 = 70$$

$(x, y, z, w, u) = (1, 25, 81, 23, 70)$ is a solution of (i) and (ii) on page 52.
Let $p = 18$,

$$w = 35 - \frac{36 \cdot 27}{18 \cdot 18} = 35 - 3 = 32$$

$$u = 166 - s\left(4 - \frac{36}{p^2}\right) = 166 - 27\left(4 - \frac{36}{18 \cdot 18}\right)$$

$$u = 166 - 108 + 3 = 61$$

$(x, y, z, w, u) = (1, 25, 81, 32, 61)$ is a solution of (i) and (ii) on page 52.

(13) Let $s = 28 \Rightarrow z = 3s = 84$

$$w = 8 + s - \frac{36s}{p^2} = 8 + 28 - \frac{36 \cdot 28}{p^2}$$

$$w = 36 - \frac{36 \cdot 28}{p^2} \text{ is an integer} \geq 0$$

Let $p = 12$,

$$w = 36 - \frac{36 \cdot 28}{12 \cdot 12}$$

$$w = 36 - 7 = 29$$

$$u = 166 - s\left(4 - \frac{36}{p^2}\right) = 166 - 28\left(4 - \frac{36}{12 \cdot 12}\right)$$

$$u = 166 - 112 + 7 = 61$$

$(x, y, z, w, u) = (1, 25, 84, 29, 61)$ is a solution of (i) and (ii) on page 52.

(14) Let $s = 32 \Rightarrow z = 3s = 96$

$$w = 8 + s - \frac{36s}{p^2} = 8 + 32 - \frac{36 \cdot 32}{p^2}$$

$$w = 40 - \frac{36 \cdot 32}{p^2} \text{ is an integer} \geq 0$$

Let $p = 8$,

$$w = 40 - \frac{36 \cdot 32}{8 \cdot 8} = 40 - 18 = 22$$

$$u = 166 - s\left(4 - \frac{36}{p^2}\right) = 166 - 32\left(4 - \frac{36}{8 \cdot 8}\right)$$

$$u = 166 - 128 + 18 = 56$$

$(x, y, z, w, u) = (1, 25, 96, 22, 56)$ is a solution of (i) and (ii) on page 52.

Let $p = 12$,

$$w = 40 - \frac{36 \cdot 32}{12 \cdot 12} = 40 - 8 = 32$$

$$u = 166 - s\left(4 - \frac{36}{p^2}\right) = 166 - 32\left(4 - \frac{36}{12 \cdot 12}\right)$$

$$u = 166 - 128 + 8 = 46$$

$(x, y, z, w, u) = (1, 25, 96, 32, 46)$ is a solution of (i) and (ii) on page 52.
Let $p = 24$,

$$w = 40 - \frac{36 \cdot 32}{24 \cdot 24} = 40 - 2 = 38$$

$$u = 166 - s\left(4 - \frac{36}{p^2}\right) = 166 - 32\left(4 - \frac{36}{24 \cdot 24}\right)$$

$$u = 166 - 128 + 2 = 40$$

$(x, y, z, w, u) = (1, 25, 96, 38, 40)$ is a solution of (i) and (ii) on page 52.
(15) Let $s = 36 \Rightarrow z = 3s = 108$

$$w = 8 + s - \frac{36s}{p^2} = 8 + 36 - \frac{36 \cdot 36}{p^2}$$

$$w = 44 - \frac{36 \cdot 36}{p^2} \text{ is an integer} \geq 0$$

If $p = 9$,

$$w = 44 - \frac{36 \cdot 36}{9 \cdot 9} = 44 - 16 = 28$$

$$s = 166 - s\left(4 - \frac{36}{p^2}\right) = 166 - 36\left(4 - \frac{36}{9 \cdot 9}\right)$$

$$= 166 - 144 + 16 = 38$$

$(x, y, z, w, u) = (1, 25, 108, 28, 38)$ is a solution of (i) and (ii) on page 52.
If $p = 12$,

$$w = 44 - \frac{36 \cdot 36}{12 \cdot 12} = 44 - 9 = 35$$

$$s = 166 - s\left(4 - \frac{36}{p^2}\right) = 166 - 36\left(4 - \frac{36}{144}\right)$$

$$s = 166 - 144 + 9 = 31$$

$(x, y, z, w, u) = (1, 25, 108, 35, 31)$ is a solution of (i) and (ii) on page 52.
If $p = 18$,

$$w = 44 - \frac{36 \cdot 36}{18 \cdot 18} = 44 - 4 = 40$$

$$s = 166 - s\left(4 - \frac{36}{p^2}\right) = 166 - 36\left(4 - \frac{36}{18 \cdot 18}\right)$$

$$= 166 - 144 + 4 = 26$$

$(x, y, z, w, u) = (1, 25, 108, 40, 26)$ is a solution of (i) and (ii) on page 52.
If $p = 36$,

$$w = 44 - \frac{36 \cdot 36}{36 \cdot 36} = 44 - 1 = 43$$

$$u = 166 - s\left(4 - \frac{36}{p^2}\right) = 166 - 36\left(4 - \frac{36}{36 \cdot 36}\right)$$

$$= 166 - 144 + 1 = 23$$

$(x, y, z, w, u) = (1, 25, 108, 43, 23)$ is a solution of (i) and (ii) on page 52.
(16) Let $s = 40 \Rightarrow z = 3s = 120$

$$w = 8 + s - \frac{36s}{p^2} = 8 + 40 - \frac{36 \cdot 40}{p^2}$$

$$w = 48 - \frac{36 \cdot 40}{p^2} \text{ is an integer} \geq 0$$

If $p = 12$,

$$w = 48 - \frac{36 \cdot 40}{12 \cdot 12} = 48 - 10 = 38$$

$$u = 166 - s\left(4 - \frac{36}{p^2}\right) = 166 - 40\left(4 - \frac{36}{144}\right)$$

$$= 166 - 160 + 10 = 16$$

$(x, y, z, w, u) = (1, 25, 120, 38, 16)$ is a solution of (i) and (ii) on page 52.

(17) Let $s = 44 \Rightarrow z = 3s = 132$

$$w = 8 + s - \frac{36s}{p^2} = 8 + 44 - \frac{36 \cdot 44}{p^2}$$

$$w = 52 - \frac{36 \cdot 44}{p^2} \text{ is an integer} \geq 0$$

If $p = 12$,

$$w = 52 - \frac{36 \cdot 44}{12 \cdot 12}$$

$$w = 52 - 11 = 41$$

$$u = 166 - s\left(4 - \frac{36}{p^2}\right) = 166 - 44\left(4 - \frac{36}{144}\right)$$

$$= 166 - 176 + 11 = 1$$

$(x, y, z, w, u) = (1, 25, 132, 41, 1)$ is a solution of (i) and (ii) on page 52.

(18) Let $s = 45 \Rightarrow z = 3s = 135$

$$w = s + 8 - \frac{36s}{p^2} = 45 + 8 - \frac{36 \cdot 45}{p^2}$$

$$w = 53 - \frac{36 \cdot 45}{p^2} \text{ is an integer} \geq 0$$

If $p = 9$,

$$w = 53 - \frac{36 \cdot 45}{9 \cdot 9} = 53 - 20 = 33$$

$$u = 166 - s\left(4 - \frac{36}{p^2}\right) = 166 - 45\left(4 - \frac{36}{9 \cdot 9}\right)$$

$$u = 166 - 180 + 20 = 6$$

$(x, y, z, w, u) = (1, 25, 135, 33, 6)$ is a solution of (i) and (ii) on page 52.

(19) Let $s = 48 \Rightarrow z = 3s = 144$

$$w = s + 8 - \frac{36s}{p^2} = 48 + 8 - \frac{36 \cdot 48}{p^2}$$

$$w = 56 - \frac{36 \cdot 48}{p^2} \text{ is an integer} \geq 0$$

If $p = 8$,

$$w = 56 - \frac{36 \cdot 48}{8 \cdot 8}$$

$$w = 56 - 27 = 29$$

$$u = 166 - s\left(4 - \frac{36}{p^2}\right) = 166 - 48\left(4 - \frac{36}{8 \cdot 8}\right)$$

$$= 166 - 192 + 27 = 1$$

$(x, y, z, w, u) = (1, 25, 144, 29, 1)$ is a solution of (i) and (ii) on page 52.

(20) Let $s = 49 \Rightarrow z = 3s = 147$

$$w = s + 8 - \frac{36s}{p^2} = 49 + 8 - \frac{36 \cdot 49}{p^2}$$

$$w = 57 - \frac{36 \cdot 49}{p^2} \text{ is an integer} \geq 0$$

If $p = 7$,

$$w = 57 - \frac{36 \cdot 49}{7 \cdot 7} = 57 - 36 = 21$$

$$u = 166 - s\left(4 - \frac{36}{p^2}\right) = 166 - 49\left(4 - \frac{36}{49}\right)$$

$$= 166 - 196 + 36 = 6$$

$(x, y, z, w, u) = (1, 25, 147, 21, 6)$ is a solution of (i) and (ii) on page 52.

Results obtained above can be summarized as follows:

$$x + y + z + w + u = 200$$

$$\frac{x}{m^2} + \frac{y}{n^2} + \frac{z}{p^2} + \frac{w}{q^2} + \frac{u}{r^2} = \frac{15}{2}$$

$$\frac{x}{m^2} = k_1; \quad \frac{y}{n^2} = k_2; \quad \frac{z}{p^2} = k_3; \quad \frac{w}{q^2} = k_4; \quad \text{and} \quad \frac{u}{r^2} = k_5$$

So,

$$k_1 + k_2 + k_3 + k_4 + k_5 = \frac{15}{2}$$

$$(m\, n\, p\, q\, r) = (1\ 5\ p\ 3\ 6)$$

If $k_1 = 1$, $x = k_1 m^2 = 1 \cdot 1^2 = 1$ and $k_2 = 1$, $y = k_2 n^2 = 1 \cdot 5^2 = 25$.
So, we have the following solution table:

Solution Number	p	k_3	k_4	k_5	$k_3+k_4+k_5$	x	y	z	w	u
1.	p	0	$\frac{8}{9}$	$\frac{83}{18}$	$\frac{11}{2}$	1	25	0	8	166
2.	2	$\frac{3}{4}$	0	$\frac{19}{4}$	$\frac{11}{2}$	1	25	3	0	171
3.	4	$\frac{3}{4}$	$\frac{1}{3}$	$\frac{159}{36}$	$\frac{11}{2}$	1	25	12	3	159
4.	12	$\frac{1}{12}$	$\frac{1}{3}$	$\frac{151}{36}$	$\frac{11}{2}$	1	25	12	11	151
5.	12	$\frac{1}{6}$	$\frac{14}{9}$	$\frac{34}{9}$	$\frac{11}{2}$	1	25	24	14	136
6.	9	$\frac{1}{3}$	$\frac{13}{9}$	$\frac{67}{18}$	$\frac{11}{2}$	1	25	27	13	134
7.	12	$\frac{1}{4}$	$\frac{17}{9}$	$\frac{121}{36}$	$\frac{11}{2}$	1	25	36	17	121
8.	8	$\frac{3}{4}$	$\frac{5}{3}$	$\frac{37}{12}$	$\frac{11}{2}$	1	25	48	15	111
9.	12	$\frac{1}{3}$	$\frac{20}{9}$	$\frac{53}{18}$	$\frac{11}{2}$	1	25	48	20	106
10.	9	$\frac{2}{3}$	2	$\frac{17}{6}$	$\frac{11}{2}$	1	25	54	18	102
11.	18	$\frac{1}{6}$	$\frac{8}{3}$	$\frac{8}{3}$	$\frac{11}{2}$	1	25	54	24	96
12.	12	$\frac{5}{12}$	$\frac{23}{9}$	$\frac{91}{6}$	$\frac{11}{2}$	1	25	60	23	91
13.	12	$\frac{1}{2}$	$\frac{26}{9}$	$\frac{91}{9}$	$\frac{11}{2}$	1	25	72	26	76

(Continued)

(Continued)

Solution Number	p	k_3	k_4	k_5	$k_3+k_4+k_5$	x	y	z	w	u
14.	10	$\frac{3}{4}$	$\frac{8}{3}$	$\frac{25}{2}$	$\frac{11}{2}$	1	25	75	24	75
15.	15	$\frac{1}{3}$	$\frac{29}{9}$	$\frac{35}{8}$	$\frac{11}{2}$	1	25	75	29	70
16.	9	1	$\frac{23}{3}$	$\frac{35}{18}$	$\frac{11}{2}$	1	25	81	23	70
17.	18	$\frac{1}{4}$	$\frac{32}{9}$	$\frac{61}{36}$	$\frac{11}{2}$	1	25	81	32	61
18.	12	$\frac{7}{12}$	$\frac{29}{9}$	$\frac{61}{36}$	$\frac{11}{2}$	1	25	84	29	61
19.	8	$\frac{3}{2}$	$\frac{22}{9}$	$\frac{14}{9}$	$\frac{11}{2}$	1	25	96	22	56
20.	12	$\frac{2}{3}$	$\frac{32}{9}$	$\frac{23}{18}$	$\frac{11}{2}$	1	25	96	32	46
21.	24	$\frac{1}{6}$	$\frac{38}{9}$	$\frac{10}{9}$	$\frac{11}{2}$	1	25	96	38	40
22.	9	$\frac{4}{3}$	$\frac{28}{9}$	$\frac{19}{18}$	$\frac{11}{2}$	1	25	108	28	38
23.	12	$\frac{3}{4}$	$\frac{35}{9}$	$\frac{31}{36}$	$\frac{11}{2}$	1	25	108	35	31
24.	18	$\frac{1}{3}$	$\frac{40}{9}$	$\frac{31}{18}$	$\frac{11}{2}$	1	25	108	40	26
25.	36	$\frac{1}{12}$	$\frac{43}{9}$	$\frac{23}{36}$	$\frac{11}{2}$	1	25	108	43	23
26.	12	$\frac{5}{6}$	$\frac{38}{9}$	$\frac{4}{9}$	$\frac{11}{2}$	1	25	120	38	16
27.	12	$\frac{11}{12}$	$\frac{41}{9}$	$\frac{1}{36}$	$\frac{11}{2}$	1	25	132	41	1
28.	9	$\frac{5}{3}$	$\frac{11}{3}$	$\frac{1}{6}$	$\frac{11}{2}$	1	25	135	33	6
29.	8	$\frac{9}{4}$	$\frac{29}{9}$	$\frac{1}{36}$	$\frac{11}{2}$	1	25	144	29	1
30.	7	3	$\frac{7}{3}$	$\frac{1}{6}$	$\frac{11}{2}$	1	25	147	21	6

There are five distinct numbers m, n, p, q, and r. In the list of 30 solutions shown in the above table, we have found, taking fixed values for four of these, namely, $m=1$, $n=5$, $q=3$ and $r=6$ and taking $k_1=k_2=1$, which means that $x=1$ and $y=25$, we were able to determine the values of p, z, w, and u. Similarly, for other values of k_1 and k_2 such that $k_1+k_2+k_3+k_4+k_5=15/2$, other solutions can be found. The above precedence can be used to determine solutions for other different values of m, n, p, q, and r from the list (1, 2, 3, 4, 5, 6,. . .).

Next, we will show how to find the solutions (1), (2), (3), (4), (5), and (6) as given by commentators Karavinda Swamy, Kapardi Swami, and Sundararaja.

Let $(m, n, p, q, r) = (4, 5, 8, 10, 12)$

Let

$$\frac{x}{m^2} = k_1, \quad \frac{y}{n^2} = k_2, \quad \text{and} \quad \frac{z}{P^2} = k_3$$

Since $m = 4$, $n = 5$, and $p = 8$, we have $x = 16k_1$, $y = 25k_2$, and $z = 64k_3$.

Then, from (A), we have

$$w = \frac{q^2[\{(15/2) - (k_1 + k_2 + k_3)\}r^2 - (200 - 16k_1 - 25k_2 - 64k_3)]}{r^2 - q^2}$$

$$= \frac{100[\{(15/2) - (k_1 + k_2 + k_3)\}144 - 200 + (16k_1 + 25k_2 + 64k_3)]}{144 - 100}$$

$$w = \frac{100[1080 - 144(k_1 + k_2 + k_3) - 200 + (16k_1 + 25k_2 + 64k_3)]}{44}$$

$$w = \frac{25}{11}[880 - 128k_1 - 119k_2 - 80k_3] \quad (*)$$

(1) Let

$$k_1 = \frac{37}{8}, \quad k_2 = \frac{9}{5}, \quad \text{and} \quad k_3 = \frac{13}{16}$$

$$\Rightarrow x = \frac{37}{8} \cdot 16 = 74; \quad y = 25k_2 = (25)\left(\frac{9}{5}\right) = 45; \quad \text{and} \quad z = \frac{13}{16} \cdot 64 = 52$$

Also from (*),

$$w = \frac{25}{11}\left[880 - 128\left(\frac{37}{8}\right) - 119\left(\frac{9}{5}\right) - 80\left(\frac{13}{16}\right)\right]$$

$$= \frac{25}{11}\left[880 - 592 - \frac{1071}{5} - 65\right]$$

$$= \frac{25}{11}\left[223 - \frac{1071}{5}\right] = \frac{25}{11}\left[\frac{1115 - 1071}{5}\right] = \frac{25}{11} \cdot \frac{44}{5} = 20$$

So, $u = 200 - x - y - z - w = 200 - 74 - 45 - 52 - 20 = 9$.

Hence, $(x, y, z, w, u) = (74, 45, 52, 20, 9)$ is a solution of (i) and (ii).

We observe that this solution is exactly the solution (4) of Sundararaja on page 52.
(2) Let

$$k_1 = \frac{77}{16}, \quad k_2 = \frac{42}{25}, \quad k_3 = \frac{5}{8}$$

$$\Rightarrow x = \frac{77}{16}(16) = 77, \quad y = \frac{42}{25}(25) = 42, \quad \text{and} \quad z = \frac{5}{8}(64) = 40$$

Also from (*), we get

$$w = \frac{25}{11}\left[880 - 128 \cdot \left(\frac{77}{16}\right) - 119\left(\frac{42}{25}\right) - 80\left(\frac{5}{8}\right)\right]$$

$$= \frac{25}{11}\left[880 - 616 - \frac{4998}{25} - 50\right]$$

$$w = \frac{25}{11}\left[214 - \frac{4998}{25}\right] = \frac{25}{11}\left[\frac{5350 - 4998}{25}\right]$$

$$w = \frac{25}{11} \cdot \frac{352}{25} = 32$$

So, $u = 200 - x - y - z - w = 200 - 77 - 42 - 40 - 32 = 9$.

Thus, $(x, y, z, w, u) = (77, 42, 40, 32, 9)$ is a solution of (i) and (ii). We observe that this solution is the same as solution (6) given by Sundararaja on page 52.
(3) Let

$$k_1 = \frac{67}{16}, \quad k_2 = \frac{58}{25}, \quad \text{and} \quad k_3 = \frac{3}{4}$$

$$\Rightarrow x = \left(\frac{67}{16}\right)(16) = 67, \quad y = \left(\frac{58}{25}\right)(25) = 58, \quad \text{and} \quad z = \left(\frac{3}{4}\right)(64) = 48$$

Also from (*), we get

$$w = \frac{25}{11}\left[880 - (128)\left(\frac{67}{16}\right) - 119\left(\frac{58}{25}\right) - 80\left(\frac{3}{4}\right)\right]$$

$$= \frac{25}{11}\left[880 - 536 - \frac{6902}{25} - 60\right]$$

$$= \frac{25}{11}\left[284 - \frac{6902}{25}\right] = \frac{25}{11}\left[\frac{7100 - 6902}{25}\right]$$

So,

$$w = \frac{25}{11} \cdot \frac{198}{25} = 18$$

Hence, $u = 200 - x - y - z - w = 200 - 67 - 58 - 48 - 18 = 9$.

So, $(x, y, z, w, u) = (67, 58, 48, 18, 9)$ is a solution of (i) and (ii).

We observe that this solution is exactly the same solution (1) as that of Karavinda Swamy on page 52.

Let $(m, n, p, q, r) = (4, 5, 6, 8, 10)$. Then, from (A), we have

$$w = \frac{8^2[\{(15/2) - (k_1 + k_2 + k_3)\}100 - [200 - 16k_1 - 25k_2 - 36k_3]]}{100 - 64}$$

$$w = \frac{64[750 - 100k_1 - 100k_2 - 100k_3 - 200 + 16k_1 + 25k_2 + 36k_3]}{36}$$

$$w = \frac{64[550 - 84k_1 - 75k_2 - 64k_3]}{36} = \frac{16}{9}[550 - 84k_1 - 75k_2 - 64k_3] \quad (*)$$

(1) Let

$$k_1 = \frac{5}{8}, \quad k_2 = \frac{159}{25}, \quad k_3 = \frac{9}{36} = \frac{1}{4}$$

$$x = m^2k_1 = 16k_1 = 16\left(\frac{5}{8}\right) = 10$$

$$y = n^2k_2 = 25k_2 = 25\left(\frac{159}{25}\right) = 159$$

$$z = p^2k_3 = 36k_3 = 36\left(\frac{1}{4}\right) = 9$$

From (*), we have

$$w = \frac{16}{9}\left[550 - 84 \cdot \frac{5}{8} - 75 \cdot \frac{159}{25} - 64 \cdot \frac{1}{4}\right]$$

$$= \frac{16}{9}\left[550 - \frac{105}{2} - 477 - 16\right] = \frac{16}{9}\left[57 - \frac{105}{2}\right]$$

$$= \frac{16}{9}\left(\frac{114 - 105}{2}\right) = \frac{16}{9} \cdot \frac{9}{2} = 8$$

So, $u = 200 - x - y - z - w = 200 - 10 - 159 - 9 - 8 = 14$.

Thus, $(x, y, z, w, u) = (10, 159, 9, 8, 14)$, which is exactly the solution (3) of Kapardi Swami on page 52.

(2) Let

$$k_1 = \frac{35}{8}, \quad k_2 = \frac{9}{5}, \quad k_3 = \frac{1}{4}$$

From (*), we have

$$w = \frac{16}{9}\left[550 - 84 \cdot \frac{35}{8} - 75 \cdot \frac{9}{5} - 64 \cdot \frac{1}{4}\right]$$

$$= \frac{16}{9}\left[550 - \frac{(21)(35)}{2} - 135 - 16\right] = \frac{16}{9}\left[399 - \frac{735}{2}\right]$$

$$= \frac{16}{9}\left(\frac{798 - 735}{2}\right) = \frac{16}{9} \cdot \frac{63}{2} = 8.7 = 56$$

Now,

$$x = m^2 k_1 = 16 \cdot \frac{35}{8} = 70; \quad y = n^2 k_2 = 25 \cdot \frac{9}{5} = 45; \quad z = p^2 k_3 = 36 \cdot \frac{1}{4} = 9$$

So, $u = 200 - x - y - z - w = 200 - 70 - 45 - 9 - 56 = 20$.

Thus, $(x, y, z, w, u) = (70, 45, 9, 56, 20)$, which is exactly the solution (5) of Sundararaja.

(3) Let

$$k_1 = \frac{3}{4}, \quad k_2 = \frac{157}{25}, \quad k_3 = \frac{1}{4}$$

$$x = m^2 k_1 = 16 \cdot \frac{3}{4} = 12; \quad y = n^2 k_2 = 25 \cdot \frac{157}{25} = 157$$

$$z = p^2 k_3 = \frac{1}{4} \cdot 36 = 9$$

Then from (*),

$$w = \frac{16}{9}\left[550 - 84 \cdot \frac{3}{4} - 75\left(\frac{157}{25}\right) - 64\frac{1}{4}\right]$$

$$w = \frac{16}{9}[550 - 63 - 471 - 16] = \frac{16}{9}(0) = 0$$

Then, $u = 200 - x - y - z - w = 200 - 12 - 157 - 9 - 0 = 22$.

Thus, $(x, y, z, w, u) = (12, 157, 9, 0, 22)$, which is exactly the solution (3) of Karavinda Swamy.

Quadratic Equations

Constructions of Vedic altars needed the complete solution of quadratic equation $ax^2 + bx + c = 0$. The quadrate equation that had to be solved was

$$7x^2 + \frac{1}{2}x = 7\frac{1}{2} + c$$

i.e., $14x^2 + x - (15 + 2c) = 0$.
So,

$$x = \frac{-1 \pm \sqrt{1 + 14(15 + 2c)4}}{28}$$

$$x = \frac{1}{28}\left[-1 \pm \sqrt{841 + 112c}\right]$$

Since,

$$x > 0, \quad x = \frac{1}{28}\left[-1 + \sqrt{841 + 112c}\right]$$

So,

$$x^2 = \frac{1}{784}\left[841 + 112c + 1 - 2\sqrt{841 + 112c}\right]$$

$$x^2 = 1 + \frac{58 + 112c - 2\sqrt{841 + 112c}}{784}$$

i.e.,

$$x^2 = 1 + \frac{29 + 56c - \sqrt{841 + 112c}}{392}$$

$$x^2 = 1 + \frac{29 + 56c - 29(1 + (112c/841))^{\frac{1}{2}}}{392}$$

i.e.,

$$x^2 \approx 1 + \frac{29 + 56c - 29(1 + (1/2)(112c/841))}{392}$$

neglecting higher powers of $112c/841$,

$$x^2 \approx 1 + \frac{56c - (56/29)c}{392} = 1 + \frac{c - (1/29)c}{7}$$

i.e.,

$$x^2 \approx 1 + \frac{28c}{7(29)} = 1 + \frac{4c}{29}$$

Katyayana gives the value of x^2 as approximately equal to $1 + (c/7)$.

3 Mathematics of Jains

Mathematics played an important role in the Jain religion. In fact, the Jains went to the extent of regarding mathematics as a part of their religion. The knowledge of mathematics contained in the Jain religious writings has been more or less parallel to that in the Vedic literature. A part of their religious literature is called "Ganitanuyogi" (the system of calculations). The Jains were devoted to mathematical calculation by an abstract love of precision. The trapezium, the shape of "Mahavedi" of Vedic religion, shows up again in Jain cosmography to portray the shapes of the universe, mountains, and the continents. Jains believed that the earth and the heavenly bodies were circular in shape and so the study of circle became an area of significant importance. As a consequence, the mensuration of circle and trapezium forms the geometry of Jains.

The original mathematical works of Jains have not yet been found. Our present knowledge about Jaina mathematics is mostly based on the existing commentaries. Among the religious works of Jains, those that are mathematics texts are "Suryaprajnapti," "Jamboo Dwipa Prajnapti," "Candra Prajnapti," "Sthananga Sutra," "Uttarayana Sutra," "Bhagavati Sutra," and "Anuyoga Dwara Sutra." The approximate date of the first two is about 500 BC and the rest of them may be about 300 BC.

Another mathematical text of great importance is "Tattwarthadhiga Sutra Bhashya" of Umaswati. It is believed that Umswati was born in a place called Nyagrodhika and lived around 150 BC. His name seems to be a combination of his parents, mother Uma and father Swati. He lived in Kusumapura, which was later called Pataliputra near Patna. There is strong historical evidence for the existence of school of mathematics at Kusumapura.

Umaswati was a prolific writer on Jain metaphysics, but he was not known as a mathematician. So the mathematical formulae and results quoted in his work must have been taken from some mathematical treatise available in his time.

In "Suryaprajnapti," the ratio π of the circumference of the circle to its diameter is taken as $\sqrt{10}$. The following geometric figures are mentioned in "Suryaprajnapti."

"Samacaturaśra" (equal four sided figure, i.e., "Rhombus")
"Visamacaturaśra" (unequal four sided figure, i.e., the quadrilateral)
"Samacatuskona" (the equiangular quadrilateral, i.e., rectangle)
"Visamacatushkona" (the quadrilateral with unequal angles)
"Samacakravāla" (circle), "Visama cakravāla" (ellipse)
"Cakrārdha cakravala" (semicircle).

Mathematical Achievements of Pre-modern Indian Mathematicians. DOI: http://dx.doi.org/10.1016/B978-0-12-397913-1.00003-X

Among the mathematical results mentioned in "Tattwadhigama Sutra Bashya" are the following mensuration formulae:

i. Circumference of a circle = $\sqrt{10}$ diameter.
ii. Area of a circle $= \frac{1}{4}$ circumference $\times$ diameter.
iii. Chord = $\sqrt{4 \text{ shara(diameter} - \text{shara)}}$ (shara means the height or arrow of the segment.
iv. Arc length of a segment (less than a semicircle):

$$= \sqrt{6(\text{shara})^2 + (\text{chord})^2}$$

v. Shara $= \frac{1}{2}\left[\text{diameter} - \sqrt{(\text{diameter})^2 - (\text{chord})^2}\right]$.

vi. Diameter $= \dfrac{(\text{shara})^2 + (1/4)(\text{chord})^2}{\text{shara}}$.

Let c, d, and h be the chord, diameter, and the height of the segment, respectively.

To prove (iii), we have to show $c = \sqrt{4h(d-h)}$.

Proof: Let $AB = c$ be a chord of the circle with centre O. Draw OC perpendicular to AB and produce both ways to meet the circle at D and E, respectively. Join OA, AD, and AE.

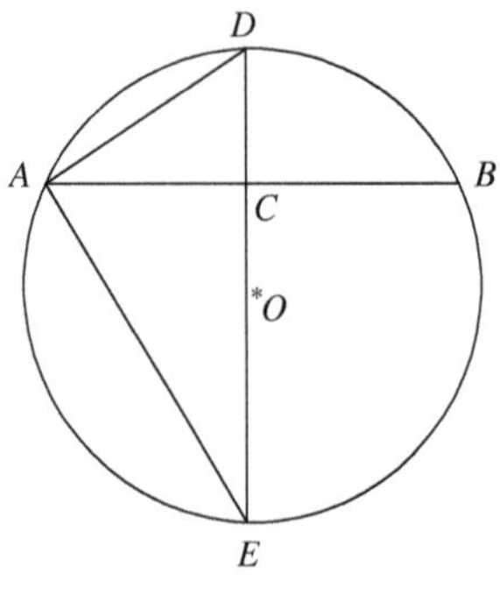

Then, $DE = d$ and $CD = h$.
Since OC bisects AB, we have $AC = c/2$.
Now in triangles ACD and ACE,

$$A\hat{C}D = A\hat{C}E = 90^\circ$$

$$D\hat{A}C = 90^\circ - A\hat{D}C$$

$$= 90^\circ - A\hat{D}E = A\hat{E}D = A\hat{E}C$$

Hence the triangles ACD and ACE are similar.

So,

$$\frac{CD}{AC} = \frac{AC}{CE} \Rightarrow \frac{h}{c/2} = \frac{c/2}{DE - CD}$$

i.e., $\dfrac{h}{c/2} = \dfrac{c/2}{d-h}$

So, $\dfrac{c^2}{4} = h(d-h); \quad c^2 = 4h(d-h) \Rightarrow c = \sqrt{4h(d-h)}$

To prove (v), we have to show

$$h = \frac{1}{2}\left[d - \sqrt{d^2 - c^2}\right]$$

Proof of (v),
$c^2 = 4hd - 4h^2$.
i.e., $4h^2 - 4hd + c^2 = 0$.

$$h = \frac{4d \pm \sqrt{16d^2 - 16c^2}}{8} = \frac{4d \pm 4\sqrt{d^2 - c^2}}{8} = \frac{d \pm \sqrt{d^2 - c^2}}{2}$$

Since $h < d$, we discard

$$h = \frac{d + \sqrt{d^2 - c^2}}{2}$$

Hence,

$$h = \frac{d - \sqrt{d^2 - c^2}}{2}$$

To prove (vi), we have to show

$$d = \frac{h^2 + (1/4)c^2}{h}$$

Proof of (vi)
We have $c^2 = 4hd - 4h^2$.
So, $4hd = c^2 + 4h^2$.
Then,

$$d = \frac{c^2 + 4h^2}{4h} = \frac{(c^2/4) + h^2}{h}$$

To prove (iv), we have to show arc length of segment $= \sqrt{6h^2 + d^2}$.

Proof: If d is the diameter of the circle

$$\text{Semiperimeter of the circle} = \pi\frac{d}{2} = \sqrt{10}\frac{d}{2}$$

$$= \sqrt{10\left(\frac{d^2}{4}\right)} = \sqrt{\left\{\frac{6d^2 + 4d^2}{4}\right\}}$$

For a semicircle, $h = d/2$ and $c = d$.
Hence, semiperimeter of the circle

$$= \sqrt{\frac{6(4h^2) + 4c^2}{4}} = \sqrt{6h^2 + c^2}$$

This formula for the semiperimeter of the circle was perhaps accepted as true for all arc length of segments.

Actually, arc length of a segment less than a semicircle will be less than $\sqrt{6h^2 + c^2}$.

All the above formulae are restarted in "Jambūdwipasāmasa" written by Umaswati himself.

The "Jambudwipa Prajnapati" mentioned about the arc ("dhanuh prstha") and the chord ("Jivā"). Since the arc is often calculated from chord, we have to believe that the approximate formulae present in other Jain works were known to the author of "Jambudwipa prajnapti." Again, the authors of "Suryaprajñapti" and "Chandra prajñapti" must have known these formulae, since the two texts have the same cosmographical descriptions in the background.

In fact, the "Jyotiṣkarandaka" which professes to expound the mathematical information contained in "Surya prajnapti" gives the following approximate formulae:

a. Chord $= c = \sqrt{4h(d-h)}$.
b. Arc $= a = \sqrt{6h^2 + c^2}$.
c. Height of the segment $= h = \sqrt{\dfrac{a^2 - c^2}{6}}$.
d. Chord $= c = \sqrt{a^2 - 6h^2}$.
e. Circumference of the circle $= \sqrt{10d^2} = \pi d$, where $d =$ diameter of the circle.
f. Area of the circle $= \dfrac{\text{circumference} \times d}{4}$.

These formulae also occur in "Laghusānghāyani" of Haribhadra Sūri.

The next three are rough approximations:

For, when the segment is a semicircle,

$$c = d, \quad h = \frac{d}{2}$$

So,

$$a = \frac{\sqrt{10d^2}}{2}$$

can be written and

$$a = \frac{\sqrt{4d^2 + 6d^2}}{2} = \sqrt{d^2 + 6\left(\frac{d}{2}\right)^2} = \sqrt{d^2 + h^2}$$

German Mathematician and Historian, Moritz Benedict Cantor [21], pp. 647–649, gave the following explanation why Jains have taken π to be $\sqrt{10}$:

The arrow (height) of the segment formed by the side of regular inscribed hexagon is given by

$$h_6 = \frac{1}{2}\left(d - \sqrt{d^2 - \frac{d^2}{4}}\right) = \frac{1}{4}d(2 - \sqrt{3}) = \frac{d}{12}$$

taking $\sqrt{3} = 5/3$ approximately.

Now the side of the regular inscribed polygon of 12 sides is given by $s_{12}^2 = h_6^2 + (1/4)s_6^2$, where s_6 and s_{12} are the sides of the hexagon and the 12 sided figure, respectively.

Now,

$$s_{12}^2 = \left(\frac{d}{12}\right)^2 + \frac{1}{4}\left(\frac{d}{2}\right)^2 = \frac{d^2}{144} + \frac{d^2}{16} = \frac{10d^2}{144}$$

Hence, its perimeter

$$= 12\sqrt{\frac{10d^2}{144}} = \sqrt{10d^2}$$

So, $(\text{perimeter})^2 = 10d^2$.

If this is equated to the square of the circumference, then we have

$$\pi^2 d^2 = 10d^2 \Rightarrow \pi^2 = 10 \Rightarrow \pi = \sqrt{10}$$

Jains have used this approximation of $\pi = \sqrt{10}$ from 500 BC till fifteenth century AD. In Jaina cosmography system, Jambudwipa (earth) is taken as a circle of diameter 100000 yojanas and is divided into seven parts by six mountain ranges from east to west. The circumference of the earth is stated in "Jambudwipa prajñapti" and other join works as 316227 yojanas, 3 graviyutis, 128 dhanus, and a little over 13½ angulas, while the area of the earth is given as 7905694150 yojanas 1 graviyuti, 1515 dhanus, and 60 angulas approximately.

According to the square root method, $\sqrt{10} = 3.162776601683$.

This explains their table of measurement, namely

1 yojana = 4 graviyuti
1 graviyuti = 2000 dhanus
1 dhanu = 100 angulas.

Among the seven divided parts of the earth, the southern most part is "Bharatavarsha" (India).

According to "Jambudwipa prajnapti," we have the following dimensions for this part:

The shara or height or breadth $= 526\frac{6}{19}$ yojanas.

The length of the chord is a little more than $14,471\frac{6}{19}$ yojanas.

The length of the southern boundary = arc $= 14,528\frac{11}{18}$ yojanas.

We will verify one of the formulae here.
Using the formula,

$$\text{Chord} = \sqrt{4\ \text{shara}(\text{diameter} - \text{shara})}$$

$$= \sqrt{4 \times \left(526\frac{6}{19}\right)\left[100,000 - 526\frac{6}{19}\right]}$$

$$= \sqrt{4 \times \frac{10,000}{19}\left[100,000 - \frac{10,000}{19}\right]}$$

$$= \sqrt{4 \times \frac{10,000}{19}\left[\frac{1,900,000 - 10,000}{19}\right]}$$

$$= \sqrt{4 \times \frac{10,000}{19} \times \frac{1,890,000}{19}}$$

$$= \frac{20,000}{19}\sqrt{189} = \frac{20,000}{19}(13.74772)$$

$$= \frac{274,955}{19} = 14,471\frac{6}{19} \text{ approximately.}$$

The Jains used huge numbers for the measurement of space and time. No people in any country have used such huge numbers as Jains and Buddhists in ancient India.

The Buddhist work "Lalita Visthara" of the first century mentioned numbers up to 10^{53}.

In "Kachachayana Pali" grammar, the number 10^{140} is called "Asankeya" (uncountable).

According to some measurement of time:

One purvi = 75600,000,000,000 years
One Shirsha Prahelika = $(8400000)^{28}$ years.

This number contains 194 digits.

The introduction of such huge numbers led the Jains to the concept of ∞. The numbers were classified into three categories, namely enumerable, unnumerable, and infinite.

The "Anuyogadwara Sutra," we find the enumeration of powers and roots of numbers, like $2^2, 2^4, 2^6, \ldots$ and $2^{\frac{1}{2}}, 2^{\frac{1}{4}}, 2^{\frac{1}{8}}, \ldots$

We also see statements like $a^{\frac{1}{2}} \times a^{\frac{1}{4}} = (a^{\frac{1}{4}})^3$ and $2^{\frac{1}{4}} \times 2^{\frac{1}{8}} = (2^{\frac{1}{8}})^3$, etc.

In the "Anuyogadwora Sutra," we will see a statement that the population of the world is $2^{64} \cdot 2^{32} = 2^{96}$, which has 29 digits.

In "Uttaradhyayana Sutra" (about 300 BC or earlier), we come across other powers like "varga" (square), "ghana" (4th power), "ghana-varga" (6th powers), "varga-varga" (12th power), "third varga-mula-ghana" ($\frac{3}{8}$th power).

Topics for Study in Mathematics

In "Sthananga Sutra" (around 300 BC), it is pointed out that there are 10 topics for study in mathematics. They are "parikarma," "vyavahara," "rajju," "rasi," "kalasavarma," "yavat-tawat," "varga," "ghana," "varga-varga," and "vikalpa." "Parikarma" means the four operations of arithmetic, namely addition, subtraction, multiplication, and division. "Vyavahara" means applied arithmetic (application of arithmetic to solve particular problems). "Kalasa varma" refers to fraction. "Rajju" is the ancient name for geometry. "Rasi" means heap and it may refer to mensuration of planes and solid figures. "Yavat-tavat" means the unknown quantity or the unknown numbers. "Vikalpa" is the Jaina name for permutations and combinations. The topic "vikalpa" has been discussed in some detail for the first time in the ancient Jaina works. The following formulae have been correctly stated, which in the modern notation are

$$nC_1 = n, \quad nC_2 = \frac{n(n-1)}{1 \cdot 2}, \quad nC_3 = \frac{n(n-1)(n-2)}{1 \cdot 2 \cdot 3}$$

$$nP_1 = n, \quad nP_2 = n(n-1), \quad nP_3 = n(n-1)(n-2)$$

From a reading of "Anuyoga dwara Sutra" and its commentary by Hemachandra Suri (AD 1087), one can easily come to the conclusion that the value of ${}_nP_r$ and $r!$ and their proof was familiar to the ancient Jains.

Pingala and His "Chandas-Sutra"

Pingala, a writer on prosody in the third century BC exhibited in his "Chandas-Sutra," the method of obtaining the number of combinations by choosing one letter ("ekaka samyoga"), two letters ("dwika samyoga"), etc. out of given number of letters. Halayudha, a commentator of the tenth century AD explained the meaning of the above rule as follows:

First draw a square (see the following figure). Below it, starting from the middle of the lower side, draw two squares. Similarly draw three squares below this.

1
1 1
1 2 1
1 3 3 1
1 4 6 4 1
1 5 10 10 5 1
1 6 15 20 15 6 1
1 7 21 35 35 21 7 1

Proceed this way writing squares so that the row below has one square more than the row above, write 1 in the top square and inside the first square and the last square of each row. Inside every square, write a number which is the sum of numbers in the two squares above it and overlapping it. The diagram we get this way is called "Meru Prastara" and the numbers in the rth row are ${}_rC_1$, ${}_rC_2$, ${}_rC_3$, etc.

The diagram demonstrates the truth of the formula

$$ {}_{n+1}C_r = {}_nC_r + {}_nC_{r-1} $$

If we compare this diagram with the well-known Pascal's triangle, we observe that this "Meru Prastara" rule is simpler than that of Pascal. Most important fact one has to remember is that this rule was formulated more than six centuries earlier than that of Pascal.

In "Sushruta Samahita" (Chapter 63, "Rasabheda vikalpadyaya") written in about sixth century BC, it was mentioned that 63 combinations can be made of six different "rasas" or tastes, namely bitter, saltish, sour, sweet, hot, and astringent by taking "rasas" one at a time, two at a time, and so on, until all six at a time, we would obtain 6, 15, 20, 15, 6, and 1 combinations, which would give a total of 63. So the word "vikalpa" for combinations has been used prior to Jainism. As pointed out earlier, it was Jains who dealt this subject first systematically in detail.

Jains believed that the universe is of the form of three trapezoidal solids piled one upon another. As a consequence, trapezium was a very familiar geometrical figure in their writings. In "Jyotikarandaka," a simple rule of computing the diameter at any height of a mountain in the form of a frustum of a cone has been given.

The rate r of increase of diameter is given by

$$r = \frac{\text{diameter at base} - \text{diameter at top}}{\text{height}}$$

Then, the required diameter d at a height h is to be calculated from $d = hr + \text{diameter at the top}$.

An important branch of Jaina religio-computational literature "Karana-gathas" is either entirely lost or yet to be unearthed.

4 The Bakshali Manuscript

In the year AD 1881, a farmer who was doing some excavation found a manuscript rich with mathematical content of a high order, at a village called Bakshali, about 70 miles away from the historically well-known town of Taxila near Peshawar. The manuscript consists of only about 70 birch bark leaves, some of which are mere scraps. Major portion of the manuscript has been lost. Nothing is known about the author of this manuscript. We also do not know what else had been discussed in the lost pages of this manuscript.

Bakshali manuscript is in the Gatha language, a sort of refinement of Prakrit and is written in Sharada script. This manuscript is estimated to belong to third to seventh century AD. It deals mostly with arithmetic and algebra problems and with just a few problems in geometry and mensuration. The topic in arithmetic contains fractions, square root, and profit and loss, where as topics in algebra contains simple and simultaneous equations, quadratic equations, and arithmetic and geometric progressions.

Circumstantial indication is present to point out that the Bakshali manuscript is perhaps a commentary of an earlier work. A close examination of this manuscript reveals that the manuscript displays the writings of more than one scribe possibly as many as four to five. In a certain place, there is a remark in between two lines, that a certain rule is incorrect. No author would skip over such a mistake in his work, telling that it is a mistake. The remark can only obviously be that of the scribe who copied the work. We also find a statement in the manuscript that it was written by a Brahmana, son of Chajaka for the education of Vasishta and his descendants. If the son of Chajaka had written this manuscript, he would have preferred to use the word "kritam" or "vrachitam" instead of "likhitam." There appears to be a few technical terms particular to this work, which are not used in the later ancient Indian writings.

For example, the word commonly used for the reduction of fractions to a common denomination is "savarnana," but in the Bakshali work, it is called by the word "Sadrisi-Karana" or "Hara-samaya-karana," which means making them similar. In later ancient Indian writings, the statement of a problem is called "Nyasa," while in Bakshali work, the word "Stapana" is used most of the time and the words "Nyasa" or "Nyasa Stapana" are used now and then. The usual word for series is "Sredi." But Bakshali work uses the terms like "varga" (group), "partha" (derived from), and "ruguna-karana" (summation).

Mathematical Achievements of Pre-modern Indian Mathematicians. DOI: http://dx.doi.org/10.1016/B978-0-12-397913-1.00004-1

The author of the Bakshali work knew how to reduce the number of fractions to a common denomination to find their sum. We will now give a few such examples that are found in this work:

1. Find the sum of 2, $1\frac{1}{2}, 1\frac{1}{3}, 1\frac{1}{4}$, and $1\frac{1}{5}$.

First, they are reduced to have a common denominator ("sadrisam kriyate") to become.

$\frac{120}{60}, \frac{90}{60}, \frac{80}{60}, \frac{75}{60}, \frac{72}{60}$

and the sum is

$$\frac{120+90+80+75+72}{60} = \frac{437}{60}$$

2. Similarly the sum of $\frac{1}{2}, \frac{1}{3}, \frac{3}{4}, \frac{3}{5}$ is found to be $\frac{131}{60}$.

(3) $\dfrac{13\frac{1}{3}}{3\frac{1}{8}} + \dfrac{13-\frac{1}{4}}{8\frac{1}{2}} + \dfrac{1\frac{1}{3}}{3\frac{1}{5}} + \dfrac{\frac{1}{2}}{1\frac{1}{2}} + \dfrac{1}{5\frac{1}{3}} + \dfrac{2\frac{1}{4}}{5} + \dfrac{12\frac{1}{2}}{33\frac{1}{3}} = \dfrac{1807}{240}.$

Some indeterminate equations, which are not trivial, are solved in the Bakshali work. We will discuss two such problems here:

(a) To solve in integers:

$$\sqrt{x+a} = l \text{ and } \sqrt{x-b} = m$$

The following is the solution:

$$l^2 - m^2 = (x+a) - (x-b) = a+b$$

i.e.,

$$(l+m)(l-m) = a+b$$

Let

$$l - m = z \quad \text{(i)}$$

where z is an arbitrary number $\neq 0$.

Then,

$$l + m = \frac{a+b}{z} \quad \text{(ii)}$$

Adding (i) and (ii), we get

$$2l = z + \frac{a+b}{z} \Rightarrow l = \frac{1}{2}\left[z + \frac{a+b}{z}\right]$$

$$x = l^2 - a = \frac{1}{4}\left[z + \frac{a+b}{z}\right]^2 - a$$

(b) Solve the indeterminate equation

$$xy - bx - cy = d$$

Following is the solution:
Adding bc to both sides, we get

$$xy - bx + bc - cy = d + bc$$

$$x(y - b) - c(y - b) = d + bc$$

$$(y - b)(x - c) = d + bc$$

Let $y - b = z$, where z is a nonzero arbitrary number.
Then,

$$x - c = \frac{d + bc}{z}$$

Hence,

$$x = c + \frac{d + bc}{z} \text{ and } y = z + b$$

Approximate Values for Surds

A mathematical contribution of extreme historical importance in the Bakshali work is the formula for computing the square root of a nonsquare number.

The following is the correct version of the sutra:

अकृते श्लिष्टकृत्युनात् शेषच्छेदो द्विसंगुणः ।
तद्वर्गदल संश्लिष्टहृति शुद्धिकृति क्षयः ॥

This correct version of the above "sutra" has been provided by Dr. M.N. Channabasappa [25]. He gave the correct translation of this "sutra" as follows:

> *"In the case of a number whose square root is to be found, divide it by the approximate root (the root of the nearest square numbers); multiply the denominator of the resulting 'śesa' (the ratio of the remainder to the devisor) by two; square it (the fraction just obtained); halve it; divide it by the composite fraction (the first approximation); subtract (from the composite fraction); the result is the refined root."*

So,

$$\sqrt{Q} = \sqrt{a^2 + b} \approx a + \frac{b}{2a} - \frac{(b/2a)^2}{2(a + (b/2a))}$$

There is no mention in the Bakshali work of how this formula was derived. Dr. M.N. Channabasappa, in the above-mentioned paper, has given a plausible method, which could have been employed by the author of Bakshali manuscript. His solution is based on the principle of iteration. Taking Heron's formula

$$\sqrt{Q} = \sqrt{a^2 + b} = a + \frac{b}{2a}$$

if $b/2a^2$ is small, that was mentioned earlier on page 27 as the first approximation, he proves that the second approximation is Bakshali formula. His proof is as follows:

Let

$$\sqrt{Q} = \sqrt{a^2 + b} = a + \frac{b}{2a} = q_1 \qquad \text{(i)}$$

$$= \text{the first approximation.}$$

Let q_2 be the second approximation of $\sqrt{Q}$.

Thus,

$$q_2 = q_1 + \frac{b_1}{2q_1}$$

where $Q = q_1^2 + b_1$.

So,

$$q_2 = q_1 + \frac{Q - q_1^2}{2q_1} \qquad \text{(ii)}$$

Substituting the value of q_1 from (i) in (ii), we get

$$q_2 = a + \frac{b}{2a} + \frac{a^2 + b - (a+(b/2a))^2}{2(a+(b/2a))}$$

$$q_2 = a + \frac{b}{2a} + \frac{a^2 + b - (a^2 + b + (b^2/4a^2))}{2(a+(b/2a))}$$

Thus,

$$q_2 = a + \frac{b}{2a} - \frac{(b/2a)^2}{2(a+(b/2a))}$$

We will now compute an approximate value of $\sqrt{2}$ using Bakshali formula. Now

$$\begin{aligned} 2 &= \frac{288}{144} = \frac{289-1}{144} = \frac{289}{144} - \frac{1}{144} \\ &= \left(\frac{17}{12}\right)^2 - \frac{1}{144} \end{aligned}$$

So,

$$\sqrt{2} = \sqrt{\left(\frac{17}{12}\right)^2 - \frac{1}{144}}$$

$$= \frac{17}{12} + \frac{-(1/144)}{2(17/12)} - \frac{(-(1/408))^2}{2((17/12) + (-1/144)/(2\cdot(17/12)))}$$

$$\sqrt{2} = \frac{17}{12} - \frac{1}{408} - \frac{1}{408}\frac{1}{1154} = q_1 \quad \text{(first approximation)}$$

So,

$$q_1 = 1 + \frac{1}{3} + \frac{1}{3\cdot 4} - \frac{1}{3\cdot 4\cdot 34} - \frac{1}{3\cdot 4\cdot 34}\ \frac{1}{1154}$$

$$q_1^2 = \left(\frac{17}{12} - \frac{1}{408} - \frac{1}{(408)(1154)}\right)^2$$

$$= \left(\frac{17}{12} - \frac{1155}{(408)(1154)}\right)^2$$

$$= \left(\frac{17}{12}\right)^2 - 2\cdot\frac{17}{12}\frac{1155}{(408)(1154)} + \left(\frac{1155}{(408)(1154)}\right)^2$$

$$= \frac{289}{144} - \frac{1155}{(144)1154} + \left\{\frac{1155}{(408)(1154)}\right\}^2$$

$$= 2 + \left[\frac{1}{144} - \frac{1155}{(144)(1154)}\right] + \left\{\frac{1155}{(408)(1154)}\right\}^2$$

So,

$$q_1^2 = 2 - \frac{1}{(144)(1154)} + \frac{(1155)^2}{144(34)^2(1154)^2}$$

$$= 2 + \left[\frac{(1155)^2 - (1154)(1156)}{(144)(34)^2(1154)^2}\right]$$

$$q_1^2 = 2 + \left[\frac{(1155)^2 - \{(1155-1)(1155+1)\}}{(144)(1156)(1154)^2}\right]$$

$$= 2 + \frac{\{(1155)^2 - [(1155)^2 - 1]\}}{(144)(1156)(1154)^2}$$

$$= 2 + \frac{1}{(144)(1156)(1154)^2}$$

$$q_1^2 = 2 + \frac{1}{(408)^2(1154)^2} = 2 + E_1$$

where

$$E_1 = \frac{1}{(408)^2(1154)^2}$$

$$\sqrt{2} = \sqrt{q_1^2 - E_1} = q_1 - \frac{E_1}{2q_1} - \frac{(-E_1/2q_1)^2}{2(q_1 + (-E_1/2q_1))} = q_2$$

where q_2 is the second approximation.

So,

$$q_2 = q_1 - \frac{E_1}{2q_1} - \frac{(E_1/2q_1)^2}{2(q_1 - (E_1/2q_1))}$$

Now,

$$\frac{E_1}{2q_1} = \frac{1/(408)^2(1154)^2}{2((17/12) - (1155/(408)(1154)))}$$

$$= \frac{1}{2(408)^2(1154)^2} \frac{(408)1154}{[(17)(34)(1154) - 1155]}$$

$$= \frac{1}{2(408)(1154)[(578)(1154) - 1155]}$$

So, the second approximation q_2 for $\sqrt{2}$ is given by

$$q_2 = q_1 - \frac{E}{2q_1} - \frac{(E_1/2q_1)^2}{2(q_1 - (E/2q_1))}$$

where

$$q_1 = 1 + \frac{1}{3} + \frac{1}{3 \cdot 4} - \frac{1}{3 \cdot 4 \cdot 34} - \frac{1}{3 \cdot 4 \cdot 34(1154)}$$

= the first approximation of $\sqrt{2} = 1.4142135623746899106262934555$, which is correct to 11 decimal places:

$$\frac{E_1}{2q_1} = \frac{1}{2(408)(1154)[(578)(1154) - 1155]}$$
$$= 0.00000000000159486182395 2$$

and

$$\left(\frac{E_1}{2q_1}\right)^2 < 0.00000000000000000000000003$$

we find q_2 = the second approximation of $\sqrt{2}$.

$$= 1.4142135623730950488802341$$

which is correct to 20 decimal places. It is really remarkable that the author of the Bakshali manuscript had found this excellent approximation formula to determine the square root of a nonsquare number at such ancient time.

We will now give a few examples found in the Bakshali work, which are applications of the Bakshali formula:

$$\sqrt{41} = \sqrt{6^2 + 5} \approx 6 + \frac{5}{12} - \frac{(5/12)^2}{2(6+(5/12))}$$

$$\sqrt{105} = \sqrt{10^2 + 5} \approx 10 + \frac{5}{20} - \frac{(5/20)^2}{2(10+(5/20))}$$

Similar approximations are proposed for $\sqrt{481}$, $\sqrt{889}$, and $\sqrt{339009}$.

The formation of terms in arithmetic progression (AP) and its sum to n terms has been given in the Bakshali work. As an illustration of this, we will present an example found in the Bakshali work.

Example

"A person travels a 'yojanas' in the first day and b 'yojanas' more on each successive day. Another person who travels at the uniform rate of c 'yojanas' per day has a head start of t days. When will the first man overtake the second?"

Solution

Let x be the number of days after which the first man overtakes the second man.

Then we have

$$(t+x)c = a + (a+b) + (a+2b) + \cdots x \text{ terms}$$

So, $(t+x)c = x\left[a + \frac{x-1}{2}b\right]$

i.e., $ct + cx = ax + \frac{b(x^2 - x)}{2}$

So, $2ct + 2cx = 2ax + bx^2 - bx$

or $bx^2 + x(2a - b - 2c) - 2ct = 0$

Thus,

$$x = \frac{2(c-a) + b + \sqrt{(2c-2a+b)^2 + 8bct}}{2b}, \quad \text{since } x > 0$$

This value of x agrees exactly with the solution presented in the Bakshali work. This clearly demonstrates that the author of the Bakshali work not only knew how to solve quadratic equations but also had an excellent knowledge of AP.

The above approximation is used in the Bakshali work to determine e_{rr} in certain types of problems. For example, if the first term of an AP is a, common difference is d, number of terms t, S the sum of t terms in that AP, then

$$S = \left\{a + \frac{1}{2}(t-1)d\right\}t \tag{1}$$

i.e.,

$$2S = 2at + dt^2 - dt \text{ or } dt^2 + t(2a-d) - 2S = 0$$

If S is given,

$$(^*)\ t = \frac{-(2a+d) + \sqrt{(2a-d)^2 + 8dS}}{2d}, \quad \text{since } t > 0$$

Let

$2a - d = P$ and $(2a-d)^2 + 8dS = q$

Then,

$$t = \frac{-P + \sqrt{q}}{2d} \text{ from } (^*).$$

Suppose q is not a perfect square. Let $q_1, q_2, \ldots$ be the successive approximations to the value of $\sqrt{q}$ and $t_1, t_2, \ldots$ be the corresponding values of t. It is readily perceived that when these values are substituted in (1), they will not give S precisely. Let the different values of S corresponding to $t_1, t_2, \ldots$ be $S_1, S_2, \ldots$.

Then from (1), we have, $2S_1 = t_1^2 d + Pt_1$.

Multiplying by $4d$, we get

$$8S_1 d = 4t_1^2 d^2 + 4dPt_1 = (2t_1 d + P)^2 - P^2$$

So,

$$8S_1 d + P^2 = (2t_1 d + P)^2 = q_1^2 \tag{2}$$

Also,

$$8Sd = 4t^2 d^2 + 4Pdt = (2td + P)^2 - P^2$$

(Continued)

Example (cont'd)

So,

$$8Sd + P^2 = (2td+P)^2 = q \tag{3}$$

Subtracting (3) from (2), we get

$$8d(S_1 - S) = q_1^2 - q \tag{4}$$

Using the first approximation of

$$\sqrt{q} = \sqrt{a^2 + b} = a + \frac{b}{2a}$$

we have

$$q_1 = a + \frac{b}{2a} \text{ (first approximation)}$$

So,

$$q_1^2 = \left(a + \frac{b}{2a}\right)^2 = a^2 + b + \left(\frac{b}{2a}\right)^2 = q + \left(\frac{b}{2a}\right)^2$$

So,

$$q_1^2 - q = \left(\frac{b}{2a}\right)^2 = E_1 \text{ (say)}$$

Then, E_1 is the error for the first approximation.
Then from (4),

$$S_1 - S = \frac{q_1^2 - q}{8d} = \frac{E_1}{8d}$$

If q_2 is the second approximation for $\sqrt{q}$,

$$q_2 = a + \frac{b}{2a} - \frac{(b/2a)^2}{2(a+(b/2a))}$$

$$\text{So, } q_2^2 = \left\{(a+(b/2a)) - \frac{(b/2a)^2}{2(a+(b/2a))}\right\}^2$$

$$= \left(a + \frac{b}{2a}\right)^2 - 2\left(a + \frac{b}{2a}\right)\frac{(b/2a)^2}{2(a+(b/2a))} + \left\{\frac{(b/2a)^2}{2(a+(b/2a))}\right\}^2$$

$$= \left(a + \frac{b}{2a}\right)^2 - \left(\frac{b}{2a}\right)^2 + \left\{\frac{(b/2a)^2}{2(a+(b/2a))}\right\}^2$$

$$= a^2 + b + \left(\frac{b}{2a}\right)^2 - \left(\frac{b}{2a}\right)^2 + E_2$$

where

$$E_2 = \left\{\frac{(b/2a)^2}{2(a+(b/2a))}\right\}^2$$

$$q_2^2 = a^2 + b + E_2 = q + E_2$$

So,

$$q = q_2^2 - E_2$$

So, for the second approximation, the error i.e,

$$E_2 = \left\{\frac{(b/2a)^2}{2(a+(b/2a))}\right\}^2$$

As before,

$$S_2 - S = \frac{E_2}{8d}$$

We will now examine a particular case of the above approximation:
Suppose $a = 1$, $d = 1$, $S = 60$.
Then, $8dS = 8(1)(60) = 480$.

$$\sqrt{q} = \sqrt{(2a-d)^2 + 8dS} = \sqrt{(2-1)^2 + 8 \cdot 1 \cdot 60} = \sqrt{1 + 480}$$

$$\sqrt{481} = \sqrt{441 + 40} = \sqrt{(21)^2 + 40}$$

So,

$$\sqrt{481} = 21 + \frac{40}{2(21)} = 21 + \frac{20}{21} \quad (\text{here } a = 21,\ b = 40)$$

(*Continued*)

Example (cont'd)

Now, $P = 2a - d = 2 - 1 = 1$.
Then,

$$t_1 = \frac{-P + \sqrt{q}}{2d} = \frac{-1 + \sqrt{481}}{2}$$

$$t_1 = \frac{-1 + 21 + (20/21)}{2} = \frac{20 + (20/21)}{2} = \frac{220}{21}$$

But, $2S_1 = t_1^2 d + Pt_1 = t_1^2 + t_1 = t_1(t_1 + 1)$, since $d = P = 1$.

$$S_1 = \frac{t_1(t_1 + 1)}{2} = \frac{(220/21)((220/21) + 1)}{2} = \frac{(220)(241)}{2(21)^2}$$

$$S_1 = \frac{(110)(241)}{441} = \frac{26,510}{441}$$

e_{rr} in the first approximation

$$= E_1 = \left(\frac{b}{2a}\right)^2 = \left\{\frac{40}{2(21)}\right\}^2 = \left(\frac{20}{21}\right)^2 = \frac{400}{441}$$

$$S_1 - S = \frac{E_1}{8d} = \frac{(400/441)}{8} = \frac{50}{441}$$

Thus,

$$S = S_1 - \frac{50}{441} = \frac{26,510}{441} - \frac{50}{441} = \frac{26,460}{441} = 60$$

Now, for the second approximation,

$$\sqrt{481} = \sqrt{441 + 40} = \sqrt{(21)^2 + 40}$$

$$= 21 + \frac{40}{2(21)} - \frac{(40/2(21))^2}{2(21 + (20/21))} = 21 \cdot \frac{20}{21} - \frac{(20/21)^2}{2(21 + (20/21))}$$

$$= \frac{461}{21} - \frac{200}{(21)(461)} = \frac{(461)^2 - 200}{(21)(461)} = \frac{212,321}{9681}$$

Now,

$$t_2 = \frac{-P+\sqrt{q}}{2d} = \frac{-1+\sqrt{481}}{2} = \frac{-1+(212,321/9681)}{2}$$

$$t_2 = \frac{-9681+212,321}{19,362} = \frac{202,640}{19,362} = \frac{101,320}{9681}$$

$$S_2 = \frac{t_2(t_2+1)}{2} = \frac{(101,320/9681)((101,320/9681)+1)}{2}$$

$$= \frac{101,320(111,001)}{2(9681)^2}$$

$$S = S_2 - \frac{E_2}{8d} = \frac{(101,320)(111,001)}{2(9681)^2} - \frac{E_2}{8}$$

where

$$E_2 = \left\{\frac{(b/2a)^2}{2(a+(b/2a))}\right\}^2 = \left\{\frac{(40/42)^2}{2(21+(40/42))}\right\}^2 = \left\{\frac{200}{21(461)}\right\}^2$$

Hence,

$$S = \frac{(101,320)(111,001)}{2(9681)^2} - \frac{40,000}{8(9681)^2}$$

$$= \frac{(50,660)(111,001) - 5000}{(9681)^2}$$

$$= \frac{5,623,310,660 - 5000}{(9681)^2} = \frac{5,623,305,660}{(9681)^2} = 60$$

We will now give the following two examples that have been presented in the Bakshali work for the solution of linear equations with several variables.

Example 1

Five merchants want to purchase a jewel. Its cost is equal to half of the money the first merchant has together with the money of others or one-third of the

(*Continued*)

Example 1 (cont'd)

money the second merchant has together with the money of others or one-fourth of the money the third merchant has together with the money of others or one-fifth of the money the fourth merchant has together with the money or others or one-sixth of the money the fifth merchant has together with the money of others. Find the price of the jewel as well as the money which each merchant has, if the money possessed by each merchant is an integral number.

Let the five merchants have money $x_1,\ x_2,\ x_3,\ x_4,\ x_5$, respectively. Let y be the cost of the jewel.

Then,

$$\frac{x_1}{2} + x_2 + x_3 + x_4 + x_5 = y \tag{1}$$

$$x_1 + \frac{1}{3}x_2 + x_3 + x_4 + x_5 = y \tag{2}$$

$$x + x_2 + \frac{1}{4}x_3 + x_4 + x_5 = y \tag{3}$$

$$x + x_2 + x_3 + \frac{1}{5}x_4 + x_5 = y \tag{4}$$

$$x_1 + x_2 + x_3 + x_4 + \frac{1}{6}x_5 = y \tag{5}$$

From (1) and (2),

$$\frac{1}{2}x_1 = \frac{2}{3}x_2$$

From (1) and (3),

$$\frac{1}{2}x_1 = \frac{3}{4}x_3$$

From (1) and (4),

$$\frac{1}{2}x_1 = \frac{4}{5}x_4$$

From (1) and (5),

$$\frac{1}{2}x_1 = \frac{5}{6}x_5$$

Thus,

$$\frac{1}{2}x_1 = \frac{2}{3}x_2 = \frac{3}{4}x_3 = \frac{4}{5}x_4 = \frac{5}{6}x_5 = k$$

So, if x_1, x_2, x_3, x_4, and x_5 are integral numbers, then k must be divisible by 2, 3, 4, 5.

So, the least value of k is LCM of these numbers, which is 60 for,

$$\begin{array}{r|l} 2 & 2, 3, 4, 5 \\ \hline 3 & 1, 3, 2, 5 \\ \hline & 1, 1, 2, 5 \end{array}$$

Since $60 = 2 \times 3 \times 1 \times 1 \times 2 \times 5 \times 1 = 60$.

If $k = 60$, $x_1 = 120$, $x_2 = 90$, $x_3 = 80$, $x_4 = 75$, and $x_5 = 72$.

Then price of the jewel

$$= \frac{1}{2}(120) + 90 + 80 + 75 + 72$$

$$= 60 + 90 + 80 + 75 + 72 = 377$$

Example 2

Three persons own 7 aswas, 9 hayas, and 10 camels, respectively. Each gives one animal to the other. Then, they are equally rich. Find the price of each animal in integral numbers.

Let x_1, x_2, and x_3 be the price of aswa, haya, and camel, respectively.

Then, $5x_1 + x_2 + x_3 = x_1 + 7x_2 + x_3 = x_1 + x_2 + 8x_3 = k$.

From these equations, we have

$$4x_1 = 6x_2$$

$$4x_1 = 7x_3$$

So, $4x_1 = 6x_2 = 7x_3 = k$ (say).

If k is to be a whole number, the least value of k is the LCM of 4, 6, 7.

$$\begin{array}{r|l} 2 & 4, 6, 7 \\ \hline & 2, 3, 7 \end{array}$$

So, $\text{LCM} = 2 \times 2 \times 3 \times 7 = 84 = k$.

If $k = 84$, $x_1 = 21$, $x_2 = 14$, $x_3 = 12$.

Example on AP

"Two persons start with different initial velocities a_1 and a_2, travel on successive days increasing at different rates b_1 and b_2. But they cover the same distance after the same period of time. What is the period?"

Let x be the period.

Then,

$$a_1 + (a_1 + b_1) + (a_1 + 2b_1) + (a_1 + 3b_1) + \cdots x \text{ terms}$$

$$= a_2 + (a_2 + b_2) + (a_2 + 2b_2) + (a_3 + 2b_2) + \cdots x \text{ terms}$$

So,

$$x\left[a_1 + \frac{x-1}{2}b_1\right] = x\left[a_2 + \frac{x-1}{2}b_2\right]$$

Hence,

$$a_1 - a_2 = \frac{(x-1)}{2}(b_2 - b_1)$$

Hence,

$$x - 1 = \frac{2(a_1 - a_2)}{b_2 - b_1}$$

$$\Rightarrow x = \frac{2(a_1 - a_2)}{b_2 - b_1} + 1$$

which is the solution given in the Bakshali work, which states the following:

> *"Twice the difference of initial terms divided by the difference of the common differences, is increased by unity."*

Examples of Linear Equations with Several Variables

Example 1

Three persons possess a certain amount of riches each. The riches of the first and the second taken together amount to 13; the riches of the second and the third together are 14; and the riches of the first and third mixed are known to be 15. Tell me the riches of each.

Let x_1, x_2, and x_3 be the riches of the three persons.

Then,

$$x_1 + x_2 = 13 \quad (1)$$

$$x_2 + x_3 = 14 \quad (2)$$

$$x_1 + x_3 = 15 \quad (3)$$

Adding (1), (2), and (3), we get

$$2(x_1 + x_2 + x_3) = 13 + 14 + 15 = 42$$

So,

$$x_1 + x_2 + x_3 = \frac{42}{2} = 21 \quad (4)$$

Subtracting (1) from (4), we get, $x_3 = 21 - 13 = 8$.
Subtracting (2) from (4), we get, $x_1 = 21 - 14 = 7$.
Subtracting (3) from (4), we get, $x_2 = 21 - 15 = 6$.
The solution given in the Bakshali work is as follows.
Let $x_1 = P$, $x_2 = 13 - P$, and $x_3 = 14 - x_2$.
So, $x_3 = 14 - (13 - P) = 1 + P$.
But

$$x_1 + x_3 = 15 \Rightarrow P + 1 + P = 15 \Rightarrow 2P + 1 = 15 \Rightarrow P = \frac{15 - 1}{2} = 7$$

So, $x_1 = P = 7$, $x_2 = 13 - P = 13 - 7 = 6$, and $x_3 = 1 + P = 1 + 7 = 8$.

Example 2

Solve $x_1 + x_2 = a_1$; $x_2 + x_3 = a_2$, $x_3 + x_4 = a_3, \ldots$, and $x_{n-1} + x_n = a_{n-1}$, and $x_n + x_1 = a_n$.

Now,

$$\begin{aligned} a_2 - a_1 &= x_3 - x_1 \\ a_4 - a_3 &= x_5 - x_3 \\ a_6 - a_5 &= x_7 - x_5 \\ &\vdots \\ a_{n-1} - a_{n-2} &= x_n - x_{n-2} \end{aligned}$$

(Continued)

Example 2 (cont'd)

Adding we get, $(a_2 - a_1) + (a_4 - a_3) + \cdots + (a_{n-1} - a_{n-2}) = x_n - x_1)$.
Also we have $x_1 + x_n = a_n$.
So, $(a_2 - a_1) + (a_4 - a_1) + \cdots + (a_{n-1} - a_{n-2}) = (a_n - x_1) - x_1$.
Thus, $(a_2 - a_1) + (a_4 - a_1) + \cdots + (a_{n-1} - a_{n-2}) + 2x_1 = a_n$.
Assume $x_1 = P$, where P is arbitrary.
Then, $(a_2 - a_1) + (a_4 - a_1) + \cdots + (a_{n-1} - a_{n-2}) + 2P = b_n$ say.
Then subtracting, we get $2(x_1 - P) = a_n - b_n$.

$$x_1 - P = \frac{a_n - b_n}{2} \Rightarrow x_1 = P + \frac{a_n - b_n}{2}$$

$$x_2 = a_1 - x_1 = a_1 - \left(P + \frac{a_n - b_n}{2}\right)$$

$x_3 = a_2 - x_2$, and so on.

Example 3

Five persons possess a certain amount of riches each. The riches of the first and the second mixed together amount to 16; the riches of the second and the third taken together are known to be 17; the riches of the third and the fourth together are known to be 18; the riches of the fourth and the fifth mixed together are 19; and the riches of the first and the fifth amount to 20. Tell me what is the amount each have.

Let x_1, x_2, x_3, x_4, and x_5 be the amount that the five persons have.
Then,

$$x_1 + x_2 = 16 \quad (1)$$

$$x_2 + x_3 = 17 \quad (2)$$

$$x_3 + x_4 = 18 \quad (3)$$

$$x_4 + x_5 = 19 \quad (4)$$

$$x_1 + x_5 = 20 \quad (5)$$

Adding (1), (2), (3), (4), and (5), we get

$$2(x_1 + x_2 + x_3 + x_4 + x_5) = 16 + 17 + 18 + 19 + 20 = 90$$

So, $x_1 + x_2 + x_3 + x_4 + x_5 = 45$.
From (1) and (3), we have $16 + 18 + x_5 = 45$.
So, $x_5 = 45 - 34 = 11$.

Then from (5), $x_1 + 11 = 20 \Rightarrow x_1 = 9$.
From (1), $9 + x_2 = 16 \Rightarrow x_2 = 7$.
Then from (2), $7 + x_3 = 17 \Rightarrow x_3 = 10$.
Finally from (3), $10 + x_4 = 18 \Rightarrow x_4 = 8$.
From (4), $9 + x_5 = 20 \Rightarrow x_5 = 11$.

5 Aryabhata I

Date of Aryabhata I and the Composition of Āryabhateeya

Aryabhata I wrote his celebrated work "Aryabhateeya." W.E. Clark translated this work into English in AD 1930, "The Āryabhateeya" translated with notes, Chicago, University of Chicago Press. "Ārya Bhateeya" was lost for a long time. In 1864, an Indian scholar named Bahudaji found a copy of this work. In this work, Aryabhata I mentioned the year of his birth as well as the year in which he wrote his work, "Aryabhateeya." According to the ancient Indian astronomers, the present period of time is called "Kaliyuga" and it started at midnight between 17th and 18th February, 3102 BC. Aryabhata I stated:

षष्ट्यब्दानां षष्टिर्यदा व्यतीतास्त्रयश्च युगापदाः ।
त्र्यधिका विंशतिरब्दास्तदेह मंम जन्मनोऽतीताः ।

This means, Aryabhata composed his work, "Aryabhateeya" when he was 23 years old, when 60 multiplied by 60 (3600) years have passed by, since the beginning of "Kaliyuga." According to K.S. Shukla in "Arya Bhateeya of Aryabhata I," edited and translated with notes by K.S. Shukla and K.V. Sarma, Aryabhata I was born on March 21, AD 476 and wrote "Arya Bhateya" in AD 499. Aryabhata I did not give any information about the place of his birth in his work. However, he mentioned that he was from Kusumapura School. Kusama Pura had been identified by his commentator, Bhaskara I (AD 629) with Patiliputra near the city of Patna in Bihar state. There was a very strong belief that Aryabhata I was a native of "Asmaka country." Its exact location was not clearly known. It was generally thought that "Asmaka country" is some region either in northwest part of India or a region between Godavari and Narmada rivers. But K. Sambasiva Sastry in "Nilkantha Somasutran's commentary, edited and published (in three parts), Trivandrum, 1977 (reprint)" had a firm conviction that "Asmaka country" was a part of the present Kerala state. His main argument was that "Aryabhateeya" had been extremely popular in Kerala than any other parts of India. Moreover, almost all of the commentators of Aryabhata I had been from the state of Kerala. He further suggested that it is highly probable that Aryabhata I was a native of Kerala state and subsequently might have migrated to Kusumapura, the imperial capital of Gupta Dynasty for patronage.

"Arya Bhateeya" is a very small and concise work. Except for three slokas which form the introduction and conclusion, the work contains 10 slokas in "Geetika" form and additional 108 slokas in "Aryavritta" meter. Brahmagupta (AD 628)

Mathematical Achievements of Pre-modern Indian Mathematicians. DOI: http://dx.doi.org/10.1016/B978-0-12-397913-1.00005-3

splits "Aryabhateeya" into two parts and calls them "Dasageethika" and "Aryashta Shatam," respectively. The "Aryashta Shatam" consists of three padams (parts), called "Ganita," "Kalakriya," and "Golam."

"Dasa Geetika" contains explanation of basic definitions and significant parameters and tables of astronomy. In this, we are introduced to

i. very large units of time, namely Kalpa, Manu, and Yuga;
ii. circular measure of arc, namely degree and minutes;
iii. distance units such as yojanas, hasta, and angula.

The part "Ganita Pada" contains only 33 slokas pertaining to mathematics. It contains very significant and useful topics of pure mathematics like determination of square and cube roots of numbers, geometrical figures with their properties and mensuration, arithmetic progression problems on the shadow of gnomon, simple and compound interests, simple, quadratic, and simultaneous equations, and linear interdeterminate equations.

The part "Kalakriyapada" comprises 25 slokas deserving the various units of time, finding the position of planets for any particular day and determining the speeds of planetary motions, etc.

The fourth part, "Goladyaya," contains 50 slokas where significant geometrical and trigonometric aspects of celestial sphere have been discussed.

Representation of Numbers by Alphabets

Aryabhata I introduced for the first time in "Dasageetika Pada" a totally original method of representing numbers using the alphabets. In Devanagari script, there are 25 "Vargiya Vyajanas" (classified consonants) from *k* to *m* which constitute "five vargas" (classes) namely k-varga, ca-varga, ṭa-varga, ta-varga, and pa-varga to denote numbers from 1 to 25, eight "Avargeeya" (unclassified consonants), namely ya, ra, la, va, sha, śha, sa, ha to denote numbers 30, 40, 50, 60, 70, 80, 90, and 100, respectively, along with 10 vowels indicating powers of 10 in order. These are shown in the following table.

Varga (Class)	Alphabets and Represented Numbers
Ka-varga	k 1 kh 2 g 3 gh 4 ṅ 5
ca-varga	c 6 ch 7 j 8 jh 9 n 10
ṭa-varga	ṭ 11 ṭh 12 ḍ 18 ḍh 14 ṇ 15
ta-varga	t 16 th 17 d 18 dh 19 n 20
pa-varga	p 21 ph 22 b 23 bh 24 m 25

The following examples given in "Aryabhateeya" will demonstrate how any number can be expressed in terms of alphabets:

$$khya = 2 + 30 = 32; \quad khyu = (2 + 30)10^3 = 32,000$$
$$ghr = 4 \times 10^6; \quad khy\bar{u}ghr = 43,20000$$

and *cha ya gi ngu shu chchlr*

$$\begin{aligned}&= 6 + 30 + 300 + 30 \times 100 + 5 \times 10000 + 70 \times 10,000 \\ &\quad + (7 + 50)(1,000,000) \\ &= 57,753,336\end{aligned}$$

$makhi = 225$.

Approximate Value of π

The ratio of circumference of a circle to its diameter $= c/d =$ a constant, which is denoted by π. The value of π is given by Aryabhata in the following sloka:

चतुरधिकं शतमष्टगुणं द्वाषष्टिस्तथा सहस्राणाम्
अयुतद्वयविष्कम्भस्यासन्नो वृत्तपरिणाहः ॥

i.e., "Add 4 to 100 and multiply by 8 and add 62,000.

This number gives approximately the circumference of a circle whose diameter is 20,000."

This means that the circumference of a circle of diameter $d = 20,000$ units is approximately equal to

$$(100 + 4)8 + 62,000 = 62,832$$

Hence,

$$\pi = \frac{c}{d} = \frac{62,832}{20,000} = 3.1416$$

which is correct to four decimal places. It is interesting to note that Aryabhata I says that the above value of π is "Asana" (approximate). It is very likely that Aryabhata I knew that π is incommensurable. This is evidently an enormous achievement, since only after 1300 years later that in 1761, Lambert proved that π is irrational ($\pi \neq p/q$, where p and q are integers and $q \neq 0$) and Lindemann in AD 1882 proved that π is transcendental.

Aryabhata I did not believe in the concept of creation or destruction of the universe. He was firm in his conviction that time is a perpetual existence without beginning or end, which means that time is "Anadi" (no beginning) and "Ananta" (no ending).

Theory of Yuga—Division of Time

According to ancient "smritis" and "puranas," as well as "Surya siddhanta," the notion of yuga for division of time is as follows:

1 kalpa = 14 Manus and 15 Sandhyās (period of intermediate twilight)
1 Manu = 71 Mahayugas

1 Mahayuga = 43,20,000 years
1 Sandhyā = 2/5 Mahayuga

$$\text{Thus 1 kalpa} = 14 \times 71 \times 43{,}20{,}000 + 15 \cdot \frac{2}{5}\,(43{,}20{,}000) \text{ years}$$

$$= 43,20,000(994 + 6) \text{ years}$$

$$= 43,20,000(1000) \text{ years}$$

$$= 4.32 \times 10^9 \text{ years.}$$

One Mahayuga consists of four yugas called "Ḳṛta," "Treta," "Dwapara," and "Kalki," with their periods, in the ratio 4:3:2:1.

Thus 1 "Kaliyuga" = 4,32,000 years
1 "Dwaparayuga" = 8,64,000 years
1 "Treta yuga" = 12,96,000 years
1 "Krta yuga" = 17,28,000 years.

We are at present, as mentioned earlier, in "Kaliyuga," which started according to Indian astronomers at the midnight between 17th and 18th February 3102 BC.

However, Aryabhata I discarded the ancient yuga theory and advanced his own a simpler and more astronomically acceptable periods of time as follows:

1 kalpa = 14 Manus
1 Manu = 72 Mahayugas
1 Mahayuga = 43,20,000 years.

Thus, 1 kalpa (according to Aryabhata I) = 1008 Mahayugas.

Since 1008 is divisible by 7, each kalpa starts on the same week day.

Moreover, 1 Mahayuga = 4 equal "yugas," each of period 10,80,000 years. This equal division of time is astronomically acceptable, since at the beginning of each "yuga," all planets would be in conjunction at the commencement of the zodiac. Aryabhata I also discarded "Sandhya" period of time.

Hindu Trigonometry

Hindu Sine and Hindu Cosine

In ancient Indian astronomy, trigonometry is called "Jyotipatti ganita," which means mathematics of generating sines. In modern India, trigonometry is called "trikonamiti," which means mensuration of a triangle.

We will now take up the meaning of sine. In the following figure let $\overset{\frown}{PQ}$ be an arc of a circle with center O. Let M be the middle point of arc $\overset{\frown}{PQ}$. Because of its shape, PMQ is called "chapa" or "Dhanus" (bow) and the chord PBQ is called the bow string or rope, "jya" in Sanskrit.

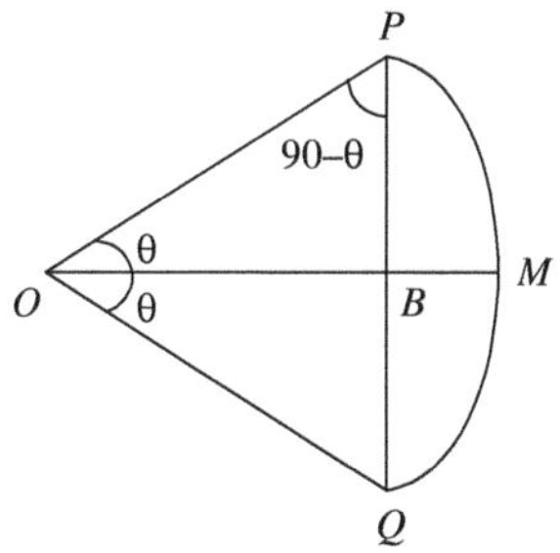

As time passed, PB which is half this "*jya*," itself came to be called "*jya*." This "*jya*" is the Hindusine, Hsine. If $M\hat{O}P = \theta$, "jya" $\theta = PB = R \sin\theta = H \sin\theta$, where R is the radius of the circle. This means that Hindu sine is the modern sine multiplied by the radius R. It was common practice to choose a specific value of R.

According to Prof. C.N. Srinivasiengar [132] (p. 46), an alternate word to "jya," "jiva" became "jiba" in the hands of Arabs and later it became "jiab." There is an Arabic word with similar pronunciation which means heart. In the hands of Romans, these two words were interchanged by mistake and "jya" became "sinus" (heart). In this way, the word "jya" was transformed into sine.

Ancient Indian astronomers made use of three trigonometric functions "jya" $\theta = R \sin\theta$, "Kotījyā" $\theta = H \cos\theta = OB$, and "utkrama jya"$\theta$ = versin $\theta = R(1 - \cos\theta) = OM - OB = BM$. Also $R \sin(90 - \theta) = OB = R\cos\theta$ and $R\cos(90 - \theta) = PB$ from the figure.

Thus,

$$H\cos(90 - \theta) = R\cos(90 - \theta) = R\sin\theta = \text{jya}\,\theta = H\sin\theta$$

Also,

$$H\sin(90 - \theta) = R\sin(90 - \theta) = R\cos\theta = H\cos\theta$$

In the above figure, if the coordinates of P are $(x,\ y)$, then from triangle OBP,

$$r\sin\theta = OP\sin\theta = PB = y$$

$$r\cos\theta = OP\cos\theta = OB = x$$

In triangle OBQ,

$$r\sin(-\theta) = BQ = -y$$

$$r\cos(-\theta) = OB = x$$

Hence,

$$r\cos(-\theta) = x = r\cos\theta$$

$$r\sin(-\theta) = -y = -r\sin\theta$$

Āryabhaṭa I's selection of R and sine table.

To select the value of R, the circumference of the circle was divided into $360 \times 60 = 21{,}600$ equal parts. Taking the length of arc of each one of these equal part as one unit length, the circumference of the circle will be 21,600 and the radius $R = \text{circumference}/2\pi = 21{,}600/2\pi$. Aryabhata I's value for π was 3.1416. This gives the value of R as 3437.7 or 3438 as the nearest integer. Aryabhata I took $R = 3438$ for the construction of sine table. It has remained a matter of speculation whether he took from "Suryasiddhanta" which was available in his time. Since 96 of the above-mentioned 21,600 equal arcs subtend an angle of $3°45'$ at the centers, the table gives sines of multiples of $3°45'$. The sines of other angles between $0°$ and $90°$ were found by interpolation by employing the method of proportion to the differences.

Since $3°45' = 225'$ is a very small angle, the arc which subtends this small angle is taken as the chord subtended by this angle. So "jya" $3°45' = 225'$. "Surya siddhanta" in the following table gives "jya" θ of multiples of $3°45'$. For the construction of the "jya" table, Aryabhata I gave the following in the 10th sloka of "Dasageetika" as follows:

मखि भखि फखि धखि णखि ञखि
ङखि हस्झ स्ककि किष्ग श्घकि किघ्व ।
घ्लकि किग्र हक्य धकि किच
स्ग श्झ ङ्व क्ल प्त फ छ कलार्धज्या ॥

i.e., the "jya" differences (R sine differences) at intervals of 225 min of arc in minutes of arc 225, 224, 222, 219, 215, 210, 205, 199, 191, 183, 174, 164, 154, 143, 131, 119, 106, 93, 79, 65, 51, 37, 22, and 7.

The above sloka (stanza) gives the value of "jya" $\theta = R \sin \theta$ of 24 angles at equal intervals of $3°45' = 225'$. The first value of jya$225' = R \sin 225' = 225'$. The successive values for the subsequent angles at equal intervals if $225'$ are obtained by successively adding jya (R sine) differences given in the stanza, namely 225, 224, 222,...

The values of "jya" $\theta = R \sin \theta = 3438 \sin \theta$ and "utkrama" $\theta = r - \cos \theta = 3438 - \cos \theta$ as given in "Surya siddhanta" are provided in this table. The values of "utkrama" $\theta = R(1 - \cos \theta) = 3438(1 - \cos \theta)$ are obtained by subtracting from 3438 the jya (sine) values taken from the bottom, in the reverse order. This clearly demonstrates that the author of "Suryasiddhanta" knew the formula $\cos \theta = \sin(90 - \theta)$.

For the sake of comparison, jya $\theta = 3438 \sin \theta$ has been listed in the third column, as computed from modern sine table. It is noteworthy to point out that the difference between the modern sine θ and 1/3438 jya θ is very close.

The last column provides the differences between consecutive jya $\theta = (R \sin \theta)$ listed in the second column. As stated earlier, these are sine differences provided by Aryabhata I in the 10th sloka of the "Dasageethika."

Angle θ	Jya or Hindu Sine	$(\sin\theta) \times 3438$	Hindu Versine $r - jya(90^\circ - \theta)$	Hindu Sine Differences
3°45′	225	224.85	7	
				224
7°20′	449	448.95	29	
				222
11°15′	671	670.72	66	
				219
15°0	890	889.82	117	
				215
18°45′	1105	1105.01	182	
				210
22°30′	1315	1315.01	261	
				205
26°15′	1520	1520.58	354	
				199
30°0′	1719	1719.00	460	
				191
33°45′	1910	1910.05	579	
				183
37°30′	2093	2092.09	710	
				174
41°15′	2267	2266.08	853	
				164
45°0	2431	2431.01	1007	
				154
48°45′	2585	2584.08	1171	
				143
52°30′	2728	2727.55	1345	
				131
56°15′	2859	2858.55	1528	
				119
60°0′	2976	2977.04	1719	
				106
63°45′	3084	3038.45	1918	
				93
67°30′	3177	3176.06	2123	
				79
71°15′	3256	3255.75	2333	
				65
75°0′	3321	3320.85	2548	
				51
78°45′	3372	3371.95	2767	
				37
82°30′	3409	3408.75	2989	
				22
86°30′	3431	3430.85	3213	
				7
90°0′	3438	3438	3438	

The next big question is as to how this table was formulated. We find the following rule in "Suryasiddhanta" (chapter 2, Sloka 15–16):

राशि लिप्ताष्टमो भागः प्रथमं ज्याऽर्घसुच्यते
तत्तद्विभक्तलब्धोन मिश्रितं तद्वितीयकं
त्र्याद्येनैवं क्रमात् पिंडान् भक्त्वा लब्धोन संयुताः
खंडकाः स्युश्चतुर्विंश ज्जगर्धपिडाः क्रमादसी

"Rasi" literally means 30°. One eighth part of $30° = 3(3°/4) = 225'$. This is the first "jya," namely jya $225' = 225$. This is the English translation of the first line. Divide the first jya by itself and subtract the quotient 1 from the same (the first jya) and add the result to it (first jya). We thus get the second jya, which means jya $7°30' = 225 + (225 - 225/225) = 225 + 224 = 449$. In this manner, divide the successive "jyas" by 225, subtract from 225. The sum of the quotients and add the remainder to the previous "jya." We thus get the next "jya." Let S_1, S_2,... be the successive values of jya θ and let q_1, q_2,...,q_{24} be the quotient, when those are divided by 225. Thus, we have the following:

$$
\begin{aligned}
S_1 &= 225 \\
S_2 &= S_1 + 225 - q_1 \\
S_3 &= S_2 + 225 - (q_1 + q_2) \\
&\vdots \\
S_r &= S_{r-1} + 225 - (q_1 + q_2 + \cdots + q_{r-1}) \quad r = 1, 2, \ldots, 24
\end{aligned}
$$

Aryabhata I gave the following rule:

प्रथमाच्यापज्यार्घाद्यैरूनं खडितं द्वितीयार्घे
तत्प्रथमज्यर्घां शैस्त्यैस्त्यै रूनानि शोषारिग

(Ganita Pāda, sl. 12)

S.N. Narharayya (Journal of Indian Mathematical Society, Vol. 15, 1924-25, Second part, Notes and Questions, page 113) describes that the meaning of this sloka is essentially the same as that is given in "Surya siddhanta." He translates the sloka the following way:

"The first 'jya' θ being divided by itself and the quotient found, the first 'jya' θ together with the same (the first 'jya' θ) minus the quotient is the second 'jya' θ. Other 'jya' θ are found by successively subtracting the sum of all the quotients from the first 'jya' θ and adding results in succession to the last of already found 'jya' θ."

This means that

$$S_r = S_{r-1} + S_1 - \sum_{i=1}^{r-1} q_i$$

We will now give Naraharyya's method of calculating these successive "jya" θ differences.

Divide the arc AB of a quadrant of a circle of radius 3438 units into 24 equal parts at the points P_1, P_2, P_3,...,P_{24}. Draw P_1R_1, P_2R_2, P_3R_3,...,$P_{24}R_{24}$. Draw P_1R_1, P_2R_2, P_3R_3,..., $P_{24}R_{24}$ perpendicular to x-axis. These, then, are in order the first, second, third,... "jyas." Draw tangents at P_1, P_2, P_3,... to meet the lines OX, OP_1, OP_2,... at $X^1, P_1^1, P_2^1, \ldots$ Draw P_1K_1, P_2K_2,... perpendiculars to P_2R_2, P_3R_3, ... Then, P_2K_1, P_3K_2,... are the successive differences of the jyas(sines). Draw P_2M_1, P_3M_2,... (not shown in the figure) parallel to $PX^1, P_2, P_1^1, \ldots$

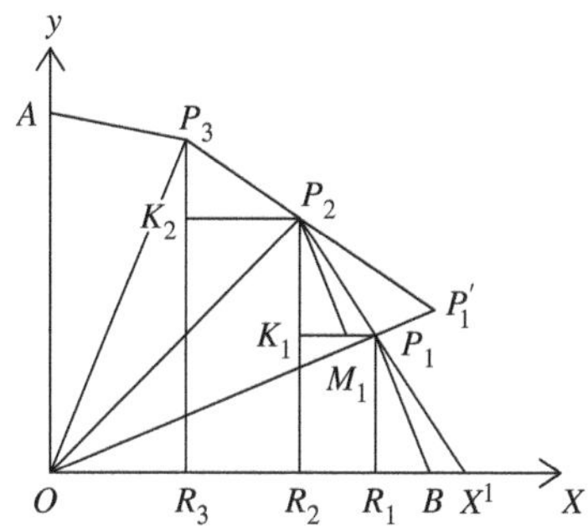

Then, $R_1B = OB - OR_1$

$$= OB - \sqrt{OP_1^2 - P_1R_1^2} \qquad \text{(i)}$$

But, $P_1R_1 \approx P_1B = 225$

$$\therefore \quad R_1B = 3438 - \sqrt{(3438)^2 - (225)^2} \qquad \text{from (i)}$$

So, $R, \ B \approx 7$, to the nearest integer.
Now, B is approximately the middle point of R_1X^1

$$R_1X^1 = 14$$

Now triangle $P_2K_1M_1$ may be treated as the displaced position of the triangle $P_1R_1X^1$.

Hence $K_1M_1 = R_1X^1 = 14$ and $M_1P_1 = 7$ being approximately taken as the displaced position of BX^1.

So, $K_1P_1 = K_1M_1 + M_1P_1 = 14 + 7 = 21$.

Thus, $P_2K_1 = \sqrt{P_1P_2^2 - K_1P_1^2} = \sqrt{(225)^2 - (21)^2} \approx 224$ to the nearest integer.

In a similar manner, $P_2K_2 = 35$ and $P_3K_2 = \sqrt{P_2P_3^2 - K_2P_2^2} = \sqrt{225^2 - 35^2} \approx 222$, to the nearest integer.

If Δ_1, Δ_2,... are the successive "jya" (sine) differences,
We have,

$$\Delta_1 = 224 = 225 - 1 = S_1 - q_1$$

$$\Delta_2 = 222 = 225 - (1 + 2) = S_1 - (q_1 + q_2)$$

and so on.

A.A. Krishnsvamy Ayyangar (Journal of Indian Mathematics Society, Vol. 15 (1924–25), First part, page 121–126) gives a different interpretation for Aryabhata I's sloka 12, in Ganitapāda:

> *"Each of the 'jya' differences is less than the first 'jya' by two quantities, the difference between the first 'jya' and the preceding 'jya' differences as well as the quotient found by dividing the sum of all the preceding 'jya' differences by the first jya."*

Let Δ_1 = Aryabhata I's jya 3(3°/4) = the first jya and let Δ_2, Δ_3,... be the successive "jya" differences. Then the meaning of Aryabhata I's rule is equivalent to the formula

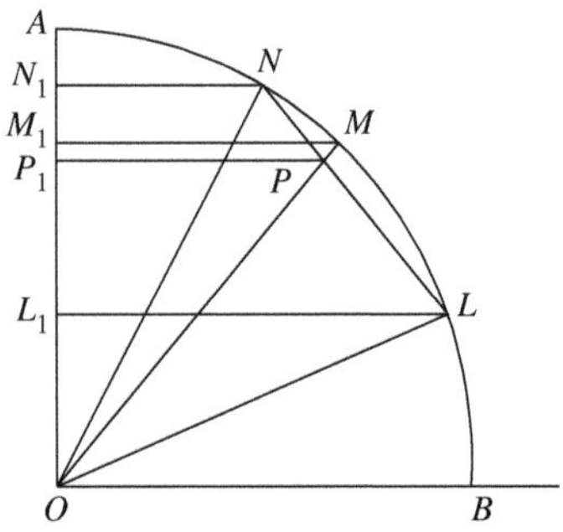

$$\Delta_{r+1} = \Delta_1 - (\Delta_1 - \Delta_r) - \frac{\Delta_1 + \Delta_2 + \cdots + \Delta_r}{\Delta_1}$$

$$= \Delta_r - \frac{\Delta_1 + \Delta_2 + \cdots + \Delta_r}{\Delta_1}$$

Hence,

$$\Delta_r - \Delta_{r+1} = \frac{\Delta_1 + \Delta_2 + \cdots + \Delta_r}{\Delta_1}$$

i.e.,

$$\Delta_r - \Delta_{r+1} = \frac{\sum_{i=1}^{r} \Delta_i}{\Delta_1}$$

which can be written as

$$jya\,(r+1)\alpha - jya\; r\alpha = jya\; r\alpha - jya\,(r-1)\alpha - \frac{jya\; r\alpha}{225}$$

Proof: Let α denote the angle $3°45'$. On the quadrant AB take points L, M, N corresponding to the angles $B\hat{O}L = (r-1)\alpha, B\hat{O}M = r\alpha$, and $B\hat{O}N = (r+1)\alpha$. Let NL cut OM at P. Draw LL_1, MM_1, NN_1, and PP_1 perpendicular to OA.

Now $L_1P_1 = P_1N_1$, since $LP = PN$.

Then,

$$\sum_{i=1}^{r} \Delta_i = OM_1, \quad \Delta_{r+1} = M_1N_1, \quad \Delta_r = L_1M_1$$

Hence,

$$\Delta_r - \Delta_{r+1} = L_1M_1 - M_1N_1 = (L_1P_1 + P_1M_1) - (P_1N_1 - M_1P_1)$$

$$= 2P_1M_1 \;\; \text{since} \;\; L_1P_1 = P_1N_1$$

Since Δ^{les} OMM_1 and OPP_1 are similar, we have

$$\frac{OP_1}{OM_1} = \frac{OP}{OM}$$

So,

$$1 - \frac{OP_1}{OM_1} = 1 - \frac{OP}{OM} \Rightarrow \frac{OM_1 - OP_1}{OM_1} = \frac{OM - OP}{OM}$$

i.e.,

$$\frac{M_1P_1}{OM_1} = \frac{PM}{OM}$$

So,

$$P_1M_1 = OM_1 \cdot \frac{PM}{OM}$$

Hence,

$$\Delta_r - \Delta_{r-1} = 2P_1M_1 = 2\frac{PM}{OM}OM_1 \quad (*)$$

If the tangent at N meets OM in M', $2PM = PM'$ approximately.
Now, from the right Δ^{le} ONM',

$$PM' \cdot OM' = NM'^2$$

$$2PM \cdot OM' = NM'^2$$

$$2PM = \frac{NM'^2}{OM'} \approx \frac{MN^2}{OM}$$

$$\frac{2PM}{OM} \approx \frac{MN^2}{OM^2} = \left(\frac{225}{3428}\right)^2 \approx \frac{1}{225}$$

Thus

$$\Delta_r - \Delta_{r-1} \cong \frac{1}{225} \cdot \sum_{i=1}^{r} \Delta_i \text{ from (*)}$$

The values of "jyas" computed by the use of the above formula do not completely coincide with those values provided in the table. In order to make the computed values of "jyas" completely agree with those in the table, special, subsidiary, rules have been adopted for the computation for some of these q's:

ऽकविशाञ्च विंशञ्च षष्टाप्तं च दशादपि
सप्तमात् द्वादशात्साप्तदशानर्घो त्तरंमतम्

Brahmasphuta siddhanta, cited by Ranganatha

i.e., if the fractional part of the q's is greater than half, it should be discarded and only $||q||$ (integral part of q) is considered, when q = q_6, q_{12}, q_{15}, q_{17}, q_{20}, and q_{21}.

Moreover reduce $\sum_{i=1}^{r} q_i$ by

i. 1, 2, 3, 4, 5 respectively before subtracting from S_1, in the case of $r = 7, 8, 9, 10$, and 11;
ii. 7, 9, 11 respectively in the case of $r = 12$, 13, and 14;
iii. 14, 17, 20 respectively in the case of $r = 15$, 16, and 17;
iv. 36, 43, 50, 58 respectively in the case of $r = 20$, 21, 22, and 23.

It is hard to understand the basis of the above-mentioned subsidiary rules. It leaves some doubt as to the actual method that Aryabhata I or Suryasiddanta might have employed for the construction of their sine table.

Extraction of Square Roots

We know for certain that Aryabhata I is not the first one who has given a method of extracting square roots, since in Jain mathematics, we find the extraction of square roots of huge numbers. But the actual procedure of extracting square roots is not described in Jain mathematics. Aryabhata I gave this in a neat form, which is exactly the present method.

We now explain Aryabhata I's method of extracting the square root:

भाग हरे दवगीन्नित्यं द्विगुणेन वर्गमूलेन
वर्गाद्विगें शुध्दे लब्धं स्थानान्तरे मूलम्।

Given the numbers, indicate the even spaces by horizontal lines and odd space by a small circle, O, starting from right to left.

For example, $7\overset{0}{4}\overset{-}{8}\overset{0}{5}\overset{-}{6}\overset{0}{7}\overset{-}{9}\overset{0}{2}$

i. Subtract the greatest possible square from the last odd place.
ii. Divide the next even space, – by twice the square root up to the preceding odd place.
iii. Subtract from the odd place (standing on the right) the square of quotient found in (ii).
iv. Continue the process as long as there are still digits on the right.

Example 1

Find the square root of 80604484.

Now $8^2 = 64 < 80 < 81 = 9^2$.

$$8)\overset{0}{8}0\overset{0}{6}0\overset{0}{4}4\overset{0}{8}4(8$$
$$\underline{64}$$
$$2\times 8 = 16)166(9$$
$$\underline{144}$$
$$220$$
$$9^2 = 81 \quad \underline{81}$$
$$2\times 89 = 178)1394(7$$
$$\underline{1246}$$
$$1484$$
$$7^2 = 49 \quad \underline{49}$$
$$2\times 897 = 1794)14358(8$$
$$\underline{14352}$$
$$64$$
$$8^2 = 64 \quad \underline{64}$$
$$\cdots$$

So the required square root is 8978.

Example 2

Find the square root of 1522756.

Now $1 = 1^2$.

```
                0 - 0 - 0 - 0
               1)1522756(1
                 1
       1 × 2 = 2)05(2
                 4
                 12
     2² = 4      4
    2 × 12 = 24)82(3
                 72
                  107
     3² = 9         9
   2 × 123 = 246)985(4
                  984
                    16
     4² = 16        16
                    ...
```

Hence, the required square root is 1234.

Extraction of Cube Roots

As far as cube roots are concerned, there is no historical evidence that any method was known before Aryabhata I. Aryabhata I described the following rule for the extraction of cube roots of a given number: He called "Ghana" (cube) and "aghana" (noncube) places and denoted it by 0 and –, respectively. So, the unit place is "Ghana," the 10th place is first "aghana" and 10^2 places is second "aghana," and so on. He took in groups of three digits from the right.

Second "Aghana"	First "Aghana"	"Ghana"
10^2 places	10th place	Unit place
10^5 places	10^4 place	10^3 place
10^8 places	and so on	10^6 place

His method is as follows:

- **i.** Subtract the greatest cube root from the last ghana place in the left.
- **ii.** Divide the second "aghana" place (next in right to the last "ghana" place in the left) by thrice the square of the cube root found in (i).
- **iii.** Subtract from the first "aghana" place (next on right of the second "aghana" place) the square of the quotient obtained in (ii) multiplied by three times the previous cube root.
- **iv.** Subtract the cube of the quotient from the "ghana" place (standing on the right on the first aghana place).
- **v.** Continue the process as long as there are digits left.

Example 1

Find the cube root of 1,860,867.

$$\begin{array}{lr}
 & 0 - - 0 - - 0 \\
 & 1)1860867(1 \\
 & \underline{1} \\
 & 3 \times 1^2 = 3)08(2 \\
 & \underline{6} \\
 & 26 \\
\text{Subtract } 2^2 \times 3 \times 1 = 12 & \underline{12} \\
 & 140 \\
\text{Subtract } 2^3 = 8 & \underline{8} \\
 & 3 \times 12^2 = 432)1328(3 \\
 & \underline{1296} \\
 & 326 \\
\text{Subtract } 3^2 \times 3 \times 12 = 324 & \underline{324} \\
 & 27 \\
\text{Subtract } 3^3 = 27 & \underline{27} \\
 & \ldots
\end{array}$$

So, the required cube root is 123.

Example 2

Find the cube root of 1,879,080,904

$$\begin{array}{lr}
 & 0 - - 0 - - 0 - - 0 \\
 & 1)1879080904(1 \\
 & \underline{1} \\
 & 3 \times 1^2 = 3)08(2 \\
 & \underline{6} \\
 & 27 \\
\text{Subtract } 3 \times 2^2 \times 1 = 12 & \underline{12} \\
 & 159 \\
\text{Subtract } 2^3 = 8 & \underline{8} \\
3 \times 12^2 = & 432)1510(3 \\
 & \underline{1296} \\
 & 2148 \\
\text{Subtract } 3 \times 3^2 \times 12 = 324 & \underline{324} \\
 & 18240 \\
\text{Subtract } 3^3 = 27 & \underline{27} \\
 & 18213 \\
 & 3(123)^2 = 45387)182139(4 \\
 & \underline{181548} \\
 & 5910 \\
\text{Subtract } 3 \times 4^2(123) & \underline{5904} \\
 & 64 \\
\text{Subtract } 4^3 = 64 & \underline{64} \\
 & \ldots
\end{array}$$

The required cube root is 1234.

Example 3

Find the cube root of 927,715,200,777.

$$9^3 = 729 < 927 < 1000 = 10^3$$

$$\begin{array}{lr}
 & \overset{0\;-\;-\;0\;-\;-\;0\;-\;-\;0}{9)927715200777(9} \\
 & \underline{729} \\
3 \times 9^2 = 243)1987(7 & \\
 & \underline{1701} \\
 & 2861 \\
\text{Subtract } 3 \times 7^2 \times 9 = 1323 & \underline{1323} \\
 & 15385 \\
\text{Subtract } 7^3 = 343 & \underline{343} \\
3 \times 97^2 = 28227)150422(5 & \\
 & \underline{141135} \\
 & 92870 \\
\text{Subtract } 3 \times 5^2 \times 97 = 7275 & \underline{7275} \\
 & 855950 \\
\text{Subtract } 5^3 = 125 & \underline{125} \\
3 \times (975)^2 = 2851875)8558257(3 & \\
 & \underline{8555625} \\
 & 26327 \\
\text{Subtract } 3 \times 3^2 \times 975 = 26325 & \underline{26325} \\
 & 27 \\
\text{Subtract } 3^3 = 27 & \underline{27} \\
 & \ldots
\end{array}$$

So, the required cube root is 9753.

Area of a Triangle

The earliest mention of a triangle in ancient Indian literature is in the vedas. The authors of "Sulvasutra" were familiar with a special type of triangle, namely right triangle. It was Aryabhata I who enunciated a formula for the area of any triangle. He said

त्रिभुजस्य फलशरीरं सभदलकोटीभुजार्धसंवर्गः

(A.B. Ganitapada, 6)

i.e., the area of a triangle is equal to the product of half of its base and the altitude.

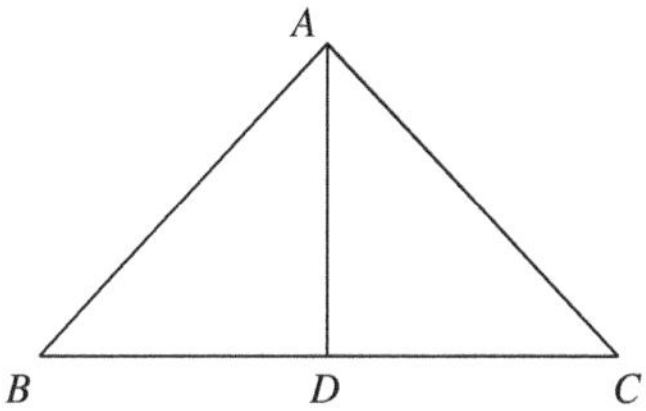

Let *ABC* be a triangle. Let *AD* be the perpendicular from *A* on the base *BC*.

Then the area of the triangle $ABC = \frac{1}{2}BC \cdot AD$.

Circle

Aryabhata I enunciated the following theorem:

परिधेः षड्भागज्या विष्कम्भार्घे न सा तुल्या ।

(A.B. Ganitapada, 9)

i.e., the chord of one-sixth of circumference is equal to half its diameter.

Let *O* be the center of the circle. Let the arc *APB* in the figure be of length equal to one-sixth of the circumference.

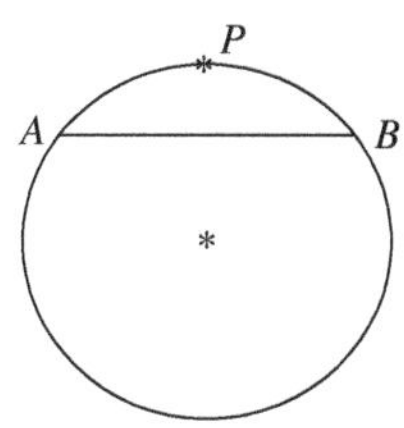

Then $AB = d/2$, where *d* is the diameter of the circle.

Proof of this theorem has been given by Nilakantha Somayaji (AD 1444–1) and Jyesthadeva (AD 1500–1610), the author of yuktibhāsā which is given on page 635 in this book. The same formula is given by Brahmagupta, Mahavira, Sripati, and Narayana.

Aryabhata I stated the following theorem:

वृत्त शरसंवर्गो ऽर्धज्यावर्गः स खलु धनुषोः

(A.B. Gaṇitapāda, 17)

i.e., "In a circle, the product of the arrows is the square of half the chord of the arcs."

Brahamagupta also gives this and derives it.

Let *AB* be a chord of the circle with center *O*. Let $AB = c$. Draw *OD* perpendicular to *AB*. Let the line *OD* meet the circumference at *E* and *F*.

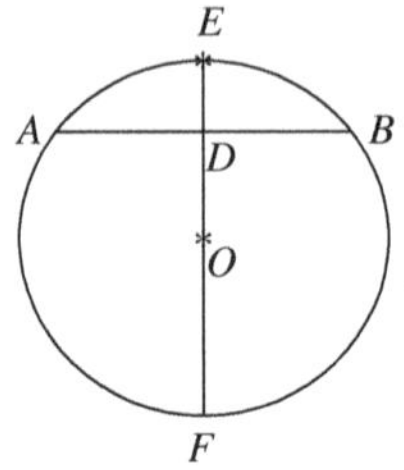

Then, two arrows are *DE* and *DF*.

Now let $DE = h$, so $DF = d - h$, if d is the diameter of the circle.

Then,

$$h(d - h) = \left(\frac{c}{2}\right)^2$$

For the proof, see page 76 in mathematics of Jainas.

The above theorem can also be stated as follows:

If in a circle, a chord *AB* and the diameter *EF* intersect at right angles, then $ED \cdot DF = AD^2$.

Shadow Problems

Aryabhata I devoted more than three slokas to describe the problems dealing with the gnomon and the shadow:

शंको प्रमाणवर्गं छायावर्गेण संयुतं कृत्वा ।
यत् तस्य योगमूलं विष्कम्भार्घ स्ववृत्तस्य ॥

(A.B. Gaṇitapāda, 14)

i.e., the square root of the sum of the squares of the gnomon and the shadow is the radius of "svavrtta" (i.e., the circle of the gnomon and shadow).

In the figure, OA = radius of "svavrtaa" $= \sqrt{AB^2 + OB^2}$

$$= \sqrt{\text{square of gnomon} + \text{square of the shadow}}$$

शङ्कुगुणं शङ्कुभुजाविवरं शङ्कुभुजयोर्विशेषहृतम्।
यल्लब्धं सा छाया शङ्को स्वमूलाद्धि ॥

(A.B. Gaṇitapāda, 15)

i.e., the length of the shadow of the gnomon from its foot is given by the product of the distance between the gnomon and the "bhuja" (sine chord) and the length of the gnomon and then divided by the difference between the gnomon and the "bhuja."

Let a source of light S be on a circle concentric to "svavrtta," which has the tip of the shadow. Let O be the center of both circles. Let the line joining the center O to the top point A of the gnomon AB be the radius of the circle "svavrtta."

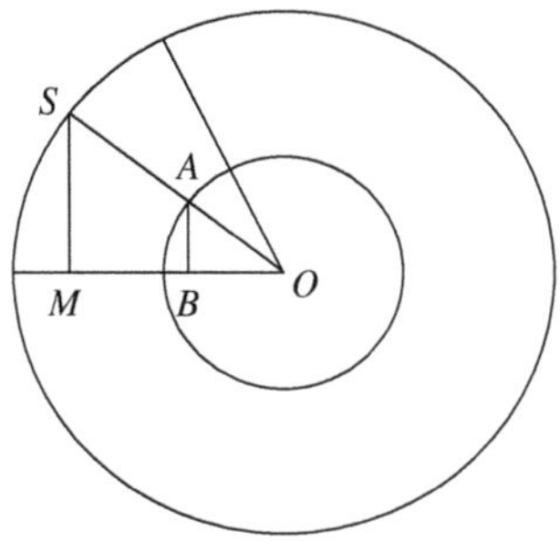

Then, SM, the vertical line through the source of light S is the "bhuja" (sine chord) and MO is the "koti" (cosine chord).

Then the shadow

$$OB = \frac{AB \cdot MB}{SM - AB}$$

Proof: In the similar triangles OBA and OMS,

$$\frac{SM}{AB} = \frac{OM}{OB}$$

Then

$$\frac{SM}{AB} - 1 = \frac{OM}{OB} - 1$$

i.e.,

$$\frac{SM - AB}{AB} = \frac{OM - OB}{OB}$$

i.e.,

$$\frac{SM - AB}{AB} = \frac{MB}{OB} \Rightarrow \frac{MB \cdot AB}{SM - AB} = OB$$

छांयागणितं छायाग्रविवरमूनेन भाजिता कोटी ।
शंकुगुणा कोटी सा छायाभक्ता भुजा भवति ॥

(A.B. Gaṇitapāda, 16)

The "koti" (cosine chord) is the product of the distance between the tips of shadows and the length of the shadow and divided by the difference between the shadows. When this "*koti*" is multiplied by the length of the gnomon and then divided by the length of the shadow, we get the length of the shadow.

This third sloka tells us how to find the height of the source of light and its horizontal distance from the observer in the case of two shadow throwing gnomons. Let S be the source of light. Let AB and CD be two equal gnomons. Let BE and DF be their corresponding shadows.

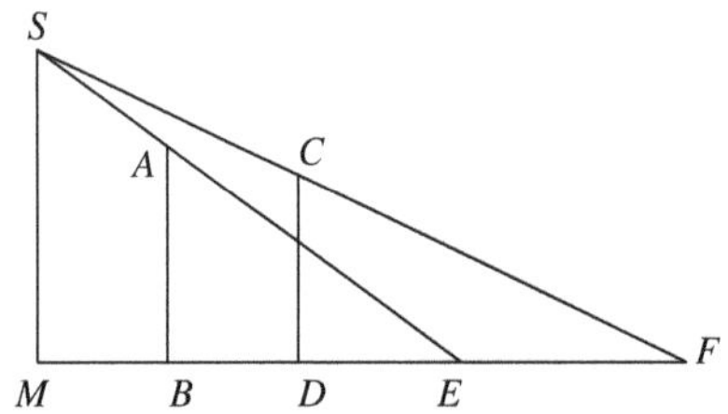

Then, the "koti," either ME or MF

$$= \frac{\text{distance between the tips of two shadows} \times \text{length of the shadow}}{\text{difference between the length of two shadows}}$$

$$= \frac{EF \cdot BE}{DF - BE} \text{ or } \frac{EF \cdot DF}{DF - BE}$$

and the "bhuja" or

$$SM = \frac{\text{"koti"} \times \text{length of the gnomon}}{\text{length of the shadow}}$$

Proof: Δ^{les} SEM and ABE are similar.

$$\frac{ME}{BE} = \frac{SM}{AB}$$

Similarly, Δ^{les} SMF and CDF are similar.

$$\frac{MF}{DF} = \frac{SM}{CD}$$

Since $AB = CD$, we have

$$\frac{ME}{BE} = \frac{SM}{AB} = \frac{SM}{CD} = \frac{MF}{DF} \tag{1}$$

So,

$$\frac{ME}{BE} = \frac{MF}{DF} = K \text{ (say)} \tag{2}$$

Then $ME = KBE$ and $MF = KDF$.
Hence,

$$\frac{MF - ME}{DF - BE} = \frac{KDF - KBE}{DF - BE} = K$$

So,

$$K = \frac{ME}{BE} = \frac{MF - ME}{DF - BE} = \frac{EF}{DF - BE} \tag{3}$$

So,

$$ME = \frac{EF}{DF - BE} \cdot BE$$

So from (2) and (3), we have

$$\frac{MF}{DF} = K = \frac{EF}{DF - BE}$$

So,

$$MF = \frac{EF \cdot DF}{DF - BE}$$

Also, from (1),

$$MS = \frac{ME \cdot AB}{BE} \text{ or } \frac{CD \cdot MF}{DF}$$

Arithmetic Progression

The sequence $a,\ a + d,\ a + 2d,\ a + 3d, \ldots$ is called an arithmetic progression or AP. Let the sum of the first n terms $= a + a + d + \cdots + a + (n - 1)d$ be denoted by S_n.

Then,

$$S_n = \frac{n}{2}\{2a + (n - 1)d\} \text{ or } n\left\{a + \frac{(n - 1)}{2}d\right\}$$

If S_n is sum of a finite number of n terms in an AP, where the first term is a and the common difference is d, Aryabhata I gave the following rule to find the number n:

> *"Multiply* S_n *by* 8d *and then add to the square of the difference between twice the first term* a *and the common difference* d. *Take the square root of this results and diminish it by twice the first term* a *and then divide by the common difference* d; *half of this previous quotient plus unity is the number of terms."*

i.e.,

$$n = \frac{1}{2}\left\{\frac{\sqrt{8dS_n + (2a-d)^2} - 2a}{d} + 1\right\}$$

Thus,

$$n = \frac{\sqrt{8dS_n + (2a-d)^2} - (2a-d)}{2d}$$

In an example, where the solution of a particular problem on interest, Aryabhata I provided the solution of the quadratic equation $tx^2 + ax - Aa = 0$ as

$$x = \frac{\sqrt{Aat + (a/2)^2} - (a/2)}{t}$$

i.e.,

$$x = \frac{\sqrt{(4At + a^2)/4} - (a/2)}{t} = \frac{\sqrt{4At + a^2} - a}{2t}$$

The above solutions clearly demonstrate that Aryabhata had a clear knowledge of determining the solution of a quadratic equation.

The following identities are found out for the first time, in "Aryabhatteya":

$$1^2 + 2^2 + 3^2 + \cdots + n^2 = \frac{1}{6}n(n+1)(2n+1)$$

$$1^3 + 2^3 + 3^3 + \cdots + n^3 = [1+2+3+\cdots+n]^2 = \frac{1}{4}n^2(n+1)^2$$

$$1 + (1+2) + (1+2+3) + (1+2+3+4) + \cdots n \text{ terms} = \frac{1}{6}n(n+1)(n+2)$$

$$= \frac{1}{6}\{(n+1)^3 - (n+1)\}$$

For the area of a circle, Aryabhata states

समपरिणाहस्यार्धं विष्कम्भार्धहतमेव वृत्तफलम्

(A. B. Ganitapāda)

i.e., half the circumference multiplied by half the diameter is the area of the circle.

Let O be the center of the circle. Let c be the circumference and let d be the diameter.

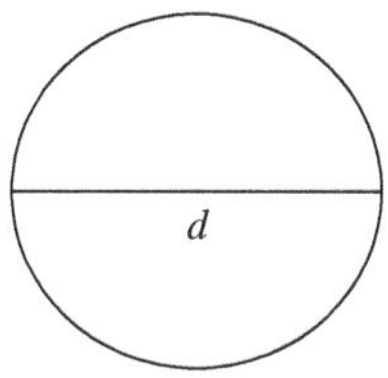

Then the area of the circle $= \frac{1}{2}c \cdot \frac{1}{2}d = \frac{1}{4}cd.$

Brahmagupta also gives this formula and derives the formula.

Aryabhata I enunciated the following theorem:

The common chord and its height of two intersecting circles:

ग्रासेने द्वे वृत्ते ग्रासगुणे भाजयेत् पृथक्त्वेन ।
ग्रासोनयोगलब्धौ सम्पातशरौ परस्परतः ॥

(Ganitapāda, 18)

i.e., "If two circles intersect, then each of two diameters minus the 'grāsa' (the largest breadth of the common position) is multiplied by the 'grāsa' and then divided by the sum of the two diameters minus the 'grasa,' the result is the arrow (the height of the arcs of the circles in the common portion, each of the other)."

Let the two circles with centers O_1 and O_2 intersect at the points B and C. Let the line O_1O_2 meet the circles at A, F, E, and D. Let the common chord BC intersect O_1O_2 at G. Then GE and GF are the arrows and FE is the "grasa." Let them be h_1 and h_2, respectively. If d_1 and d_2 are two diameters, then

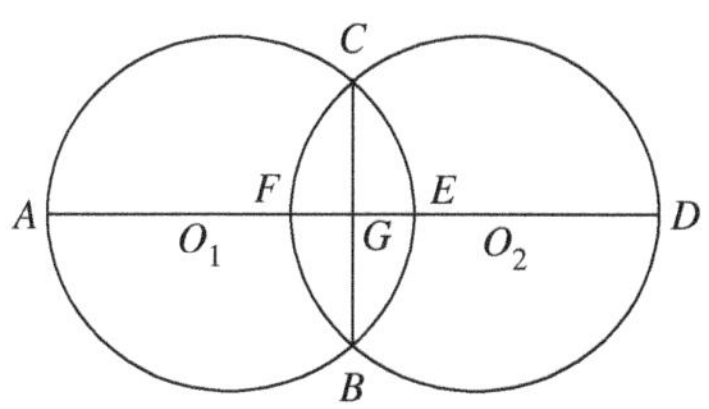

$$h_1 = \frac{\{d_2 - [h_1 + h_2]\}\{h_1 + h_2\}}{d_1 + d_2 - 2(h_1 + h_2)}$$

and

$$h_2 = \frac{\{d_1 - (h_1 + h_2)\}\{h_1 + h_2\}}{d_1 + d_2 - 2(h_1 + h_2)}$$

For proof, see Nilakantha Somayaji on page 604.

Theorem: If in a circle, a chord PQ and a diameter AB intersect each other at right angle at R, then

$$AR \cdot RB = PR^2$$

Proof:

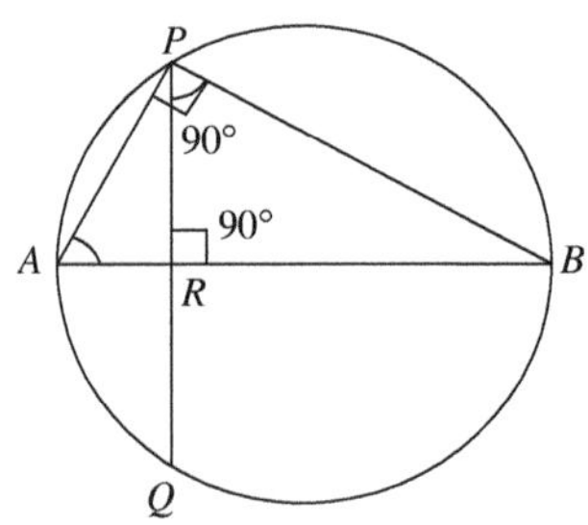

Join AP and PB.
Now $A\hat{P}B = 90^\circ$ (angle in semicircle).

$$R\hat{P}B = 90^\circ - P\hat{B}R = 90^\circ - P\hat{B}A = P\hat{A}B$$

So, Δ^{les} APR and BPR are similar.
Hence,

$$\frac{PB}{AP} = \frac{RB}{RP} = \frac{RP}{AR}$$

So, $PR^2 = RB \cdot AR$.

Indeterminate Equation of the First Degree

The earliest Hindu mathematician to provide a method of computing general solutions of the indeterminate equation of first degree, namely $by = ax + c$, where a, b, c are integers has been Aryabhata I. In ancient Indian Mathematics, this equation is called "Kuttaka," which literally means pulvarizer. This name has been given due to the process of continued division necessary to solve the equation. In the west, this equation has been given the name "Diophantine" equation after the Greek mathematician, Diaphantus, who was supposed to have lived during the later part of third century.

Aryabhata I came across this problem in the following way:

> *"Find an integer which when divided by a leaves a remainder* r_1 *and when divided by* b *leaves a remainder* r_2*. Then, what is the integer?"*

Thus, if N is the required integer, we have

$$N = ax + r_1 \text{ and } N = by + r_2$$

Hence, we get $by - ax = c$, where $c = r_1 - r_2$.

It must be noted that if a and b have a common factor d, then c must also be divisible by d, in order that the equation $by = ax + c$ has a solution. In case a, b, and c are divisible by d, then we have to divide equation $by = ax + c$ by d, to get a new equation, $b_1y = a_1x + c_1$, where $b = db_1$, $a = da_1$, and $c = dc_1$ such that a_1 and b_1 are prime to each other. Hence, we shall presume that a and b are prime to each other in the equation $by = ax + c$. Aryabhata I discovered a method of determining the general solution of the equation $by = ax + c$ and this may be Aryabhata I's best and the most remarkable accomplishment in the branch of pure mathematics. He has described his entire method in just two slokas, sloka 32 and sloka 33 of his "Ganitapada" as follows:

त्र्रधिकाग्र भागहारं छिंद्यादूनाग्र भागहरिए
शेषपरस्पर भक्तं मतिगुएामग्रांतरे न्दिप्तं
त्र्रधउपरि गुएितमंत्ययुगूनाग्र च्धेद भाजितेशेषं
त्र्रधिकाग्र चछेदाग्रमधिकाग्रयुतं

Due to the difficulty and ambiguity involved in the translation of the above two slokas, Dr. B.B. Dutta in "Elder Aryabhata's rule for the solution of indeterminate equations of first degree," Bulletin of Calcutta Mathematics Society, Vol. 24, pages 35–53, 1932, gives the following translation of the above two slokas on the basis of the interpretation of Aryabhata I's most famous commentator Bhaskara I:

> *"Divide the devisor corresponding to the greater remainder by the divisor corresponding to the smaller remainder. The residue and the divisor corresponding to the smaller remainder must be mutually divided; the last residue should be multiplied by an integer that has to be chosen such that the product being either added on to (if the number of quotients in the continued division process is even) or subtracted (if the number of questions in the continued process is odd) by the difference of the remainders, will be divisible by the penultimate remainder. Put the successive quotients of the continued division one below the other in a column; place the found optional number below them; below it place the quotient found with the optional number; the penultimate is multiplied by the one above it and added to the one just below it. Repeat this process; divide the last number obtained by the division corresponding to the smaller remainder and then multiply the residue by the divisor corresponding to the greater remainder; add to this the greater remainder. The result will be number, which corresponds to the two divisors."*

We now perform the repeated division, as given by the above translation:

$$
\begin{array}{l}
b)a\ (q \\
\underline{bq} \\
r_1)b\ (q_1 \\
\underline{r_1q_1} \\
r_2)r_1\ (q_2 \\
\underline{r_2q_2} \\
r_3)r_2(q_3 \\
\underline{r_3q_3} \\
::::: \\
r_{m-1})r_{m-2}(q_{m-1} \\
\underline{q_{m-1}r_{m-1}} \\
r_m)r_{m-1}(q_m \\
\underline{q_mr_m} \\
r_{m+1} \text{ and so on}
\end{array}
$$

Thus we have,

$$a = bq + r_1$$

$$b = rq_1 + r_2$$

$$r_1 = r_2q_2 + r_3$$

$$\ldots\ldots\ldots$$

$$r_{m-1} = r_mq_m + r_{m+1}$$

and so on

Substituting $a = bq + r_1$ in the equation $by = ax + c$, we get $by = (bq + r_1)x + c = bqx + r_1x + c = bqx + by_1$, where $by_1 = r_1x + c$.

So we have, $y = qx + y_1$.

Also, since $a = bq + r_1$ and $y = qx + y_1$, the equation $by = ax + c$ becomes $b(qx + y_1) = (bq + r_1)x + c$, which gives $by_1 = r_1x + c$.

Again, in view of $b = rq_1 + r_2$ and taking $x = q_1y_1 + x_1$, the equation $by_1 = r_1x + c$ becomes

$$(rq_1 + r_2)y_1 = r_1(q_1y_1 + x_1) + c$$

i.e., $rq_1y_1 + r_2y_1 = r_1q_1y_1 + r_1x_1 + c$

i.e., $r_2y_1 = r_1x_1 + c \Rightarrow r_1x_1 = r_2y_1 - c$

and so on.

If we write the successive values and the above reduced equations, we get the following:

I	II
$y = qx + y_1 \quad (1)$	$by_1 = r_1x + c \quad (\bar{1})$
$x = q_1y_1 + x_1 \quad (2)$	$r_1x_1 = r_2y_1 - c \quad (\bar{2})$
$y_1 = q_2x_1 + y_2 \quad (3)$	$r_2y_2 = r_3x_1 + c \quad (\bar{3})$
$x_1 = q_3y_2 + x_2 \quad (4)$	$r_3x_2 = r_4y_2 - c \quad (\bar{4})$
$y_2 = q_4x_2 + y_3 \quad (5)$	$r_4y_4 = r_5x_2 + c \quad (\bar{5})$
$x_2 = q_2y_3 + x_3 \quad (6)$	...
...	
$y_{n-1} = q_{2n-2}x_{n-1} + y_n \quad (2n-1)$	$r_{2n-2}y_n = r_{2n-1}x_{n-1} + c \quad (\overline{2n-1})$
$x_{n-1} = q_{2n-1}y_n + x_n \quad (2n)$	$r_{2n-2}x_n = r_{2n}y_n - c \quad (\overline{2n})$
$y_n = q_{2n}x_n + y_{n+1} \quad (2n+1)$	$r_{2n}y_{n+1} = r_{2n+1}x_n + c \quad (\overline{2n+1})$

We now have the option of either to continue the mutual division to the finish or to get a certain number of quotient and then stop.

Case (i)

Suppose the mutual division is continued until we obtain the remainder to be zero. Since a and b are prime to each other, the last but one remainder must be 1. Also, the quotient, q_1, q_2, ... (excluding q) may be even or odd. So two subcases arise here.

Subcase (1)

Let the number of quotients q_1, q_2,... be even.

Then, we have the quotients to be $q_1, q_2, \ldots, q_{2n}$.

Also, we have $r_{2n} = 1$ and $r_{2n+1} = 0$, $q_{2n} = r_{2n-1}$.

Thus from equation $(\overline{2n+1})$ in (II), we get

$$y_{n+1} = c$$

And the from equation $(2n + 1)$ in (I), we get

$$y_n = q_{2n}x_n + c$$

Giving an arbitrary integral value t for x_n, we get an integral value for y_n, namely $y_n = q_{2n}t + c$. Then we obtain the value of x_{n-1} from the equation $(2n)$ of (I). Proceeding backward, step by step, we ultimately obtain the values of x and y in positive integers.

Subcase (2)

Let the number of quotients be odd. Then, we have quotients $q_1, q_2, \ldots, q_{2n-1}$. We have $r_{2n-1} = 1$ and $r_{2n} = 0$, and $q_{2n-1} = r_{2n-2}$.

Then from equations $(2n)$ of (I) and $(\overline{2n})$ of (II), we get

$$x_n = -c \text{ and } x_{n-1} = q_{2n-1}y_n - c$$

Giving arbitrary integral value t_1 for y_n, we get

$$x_{n-1} = r_{2n-2}t_1 - c$$

Proceeding backward as before, we obtain the values of x and y in positive integers.

Case (ii) Next suppose that the mutual division has to be stopped at some stage. Here, two subcases arise, depending whether the number of quotients obtained are even or odd.

Subcase (1) Suppose the number of quotients are even.

Then from equation $(\overline{2n+1})$ of (II), we get

$$r_{2n}y_{n+1} = r_{2n+1}x_n + c$$

or

$$y_{n+1} = \frac{r_{2n+1}x_n + c}{r_{2n}}$$

If we choose a suitable integer t for x_n, we have

$$y_{n+1} = \frac{r_{2n+1}x_n + c}{r_{2n}} = \frac{r_{2n+1}t + c}{r_{2n}} = \text{an integer } m \text{ (say)}$$

Then from the equation $(2n + 1)$ of (I), we get

$$y_n = q_{2n}x_n + y_{n+1} = q_{2n}t + m = \text{ an integer}$$

By proceeding backward, integral values of x and y can be obtained as before.

Subcase (2) Let the number of quotients be odd.

Then from the equation $(\overline{2n+1})$ of (II), we get

$$r_{2n-1}x_n = r_{2n}y_n - c$$

So,

$$x_n = \frac{r_{2n}y_n - c}{r_{2n-1}}$$

Take $y_n = t_1$, an integer, we get

$$x_n = \frac{r_{2n}t - c}{r_{2n-1}} = \text{an integer } m \text{ (say)}$$

Now, from the equation $(2n)$ of (I), we get $x_{n-1} = q_{2n-1}y_n + x_n = q_{2n-1}t_1 + m =$ an integer.

The integral values of x and y can be found as before by proceeding backward in the following two tables.

We now prove the following lemma.

If $x = \alpha$, $y = \beta$ be the least integral solution of the equation $ax + c = by$, then $x = bm + \alpha$, $y = am + \beta$ is the general solution of $by = ax + c$ where m is an arbitrary integer and the minimum value α of x is equal to the remainder left on dividing its calculated value (of x), $q_1y_1 + x_1$ by b, i.e., $q_1y_1 + x \equiv \alpha(\text{mod } b)$.

Proof: Let $(\alpha\beta)$ be a solution of $ax + c = by$.

Then,

$$a\alpha + c = b\beta$$

Adding abm to both sides, we get $a\alpha + abm + c = b\beta + abm$, where m is an arbitrary integer.

So, we get $a(\alpha + \beta m) + c = b(am + \beta)$.

Thus, $x = bm + \alpha$ and $y = am + \beta$ is the general solution of $ay + c = by$.

As seen before, $x = q_1y_1 + x_1 = bm + \alpha$.

Thus, $q_1y_1 + x_1 = bm + \alpha \equiv \alpha(\text{mod } b)$.

We observe that the above result self-explains the rationale behind the operations described in the two slokas 32 and 33 of "Ganithapada."

We will now give some examples to illustrate Aryabhata I's method of solving the equation $ax + c = by$ in integers.

Example 1

Find the integral solutions of $137x + 10 = 60y$.

$$60)137(2 = q$$
$$\underline{120}$$
$$17)60(3 = q_1$$
$$\underline{51}$$
$$9)17(1 = q_2$$
$$\underline{9}$$
$$8)9(1 = q_3$$
$$\underline{8}$$
$$1$$

Thus, we see that number of quotients leaving the first one $q = 2$, we have three quotients q_1, q_2, and q_3. So, the number of quotients is 3 and odd.

Now, we have to choose an integer m_1 such that when m_1 is multiplied by the last residue (remainder) 1 and 10 is subtracted from the product, the resulting number is divisible by the penultimate remainder 8.

Using modern notation, this means

$$m_1 \cdot 1 - 10 \equiv 0 (\text{mod } 8)$$

The least value of m_1 will be 18, so

$$8)18 - 10 = 8(1 = q^* = \text{quotient}$$
$$\underline{8}$$
$$\ldots$$

We now form the table proposed by Aryabhata I's rule

	2	2	2	2	297	$= Y$
	3	3	3	130	130	$= X$
	1	1	37	37	37	
	1	19	19	19	19	
m_1 = the multiplier	18	18	18	18	18	
The quotient q^*	1	1	1	1	1	

We observe that the above table is formed by the operations described as follows:

The number 18 in the first column and the number above it namely 1 is multiplied and added to 1 (the number below 18) gives the number 19 in the second column. The same process applied in the second column, namely $19(1) + 18 = 37$.

Similarly, $37 \times 3 + 19 = 130$ and $130 \times 2 + 37 = 297$.

Then,

$$X = 130 \text{ and } Y = 297$$

Now,

$$Y = 297 \equiv 23(\text{mod } 137)$$

$$X = 130 \equiv 10(\text{mod } 60)$$

So, $\alpha = 10$, $\beta = 23$.

Then $(\alpha\ \beta) = (10\ 23)$ is the minimum solution of $(x\ y)$.

Then, the general solution is

$$x = 60m + 10$$

$$y = 137m + 23$$

In the above process, the mutual division is continued to the finish.

Next, suppose, the mutual division is stopped with the remainder 8, in the above process of mutual division. Then, the number of quotients leaving the first namely $q = 2$, we have two quotients q_1 and q_2. Hence the number of quotients is 2 and even.

Now, we have to choose an integer m_2 such that which when multiplied by the last remainder 8 and added 10 to the product, the resulting number should be divisible by the penultimate remainder 9.

i.e.,

$$8m_2 + 10 \equiv 0(\text{mod } 9)$$

The smallest integral value of m_2 is 1, since

$$8 \cdot 1 + 10 = 18 \equiv 0(\text{mod } 9)$$

Also,

$$9)18(2 = q^*$$
$$\underline{18}$$
$$\ldots$$

(*Continued*)

Example 1 (cont'd)

Now we form the table:

	2	2	2	23	$= Y$
	3	3	10	10	$= X$
	1	3	3	3	
$m_2 = 1$ the multiplier	1	1	1	1	
The quotient $q^* = 2$	2	2	2	2	

So, $X = 10$; $Y = 23$.

$$X = 10 \equiv 10(\text{mod } 60)$$

$$Y = 23 \equiv 23(\text{mod } 137)$$

So, $(\alpha\ \beta) = (10\ 23)$ is the minimum solution general solution. General solution is $X = 60m + 10$; $Y = 137m + 23$.

Example 2

Solve in positive integers, $179x + 21 = 29y$.

$$29)179(6 = q$$
$$\underline{174}$$
$$5)29(5 = q_1$$
$$\underline{25}$$
$$4)5(1 = q_2$$
$$\underline{4}$$
$$1 = \text{last remainder}$$

So, the quotients leaving $q = 6$, are q_1 and q_2. So the number of quotient is 2 and even.

We have to find an integer m_1 such that

$$m_1 \cdot 1 + 21 \equiv 0(\text{mod } 4)$$

The smallest integer value of m_1 is 3.

$$3 \cdot 1 + 21 = 24 \equiv 0(\text{mod } 4)$$

Also

$$4)24(6 = q^*$$
$$\underline{24}$$
$$\cdots$$

	6	6	6	297	$= Y$
	5	5	48	48	$= X$
	1	9	9	9	
$m_1 = 3$ the multiplier	3	3	3	3	
The quotient q^*	6	6	6	6	

$X = 48;\ Y = 297$

$297 \equiv 118 (\text{mod } 179) = \beta$

$48 \equiv 19 (\text{mod } 29) = \alpha$

$(\alpha\ \beta) = (19\ 118)$ is the minimum solution

So, $x = 29m + 19$; $y = 179m + 118$ is the general solution.

Example 3

Find the integral solution of $79x + 5 = 31y$.

$$31)79(2$$
$$\underline{62}$$
$$17)31(1$$
$$\underline{17}$$
$$14)17(1$$
$$\underline{14}$$
$$3)14(4$$
$$\underline{12}$$
$$2)3(1$$
$$\text{penultimate} \quad \underline{2}$$
$$\text{remainder} \quad = 1$$

The number of quotients leaving the first is 4 and so is even.

(Continued)

Example 3 (cont'd)

Now, we choose an integer m_1 such that $m_1 \cdot 1 + 5 = 0 \pmod 2$.
The least integral value of $m_1 = 1$.
Also $1 \cdot 1 + 5 = 6$.

$$2)6(3 = q^*$$
$$\underline{6}$$
$$0$$

We now form the table

	2	2	2	2	2	97	$= Y$
	1	1	1	1	38	38	$= X$
	1	1	1	21	21	21	
	4	4	17	17	17	17	
	1	4	4	4	4	4	
$m_1 = 1$ the multiplier	1	1	1	1	1	1	
The quotient $q^* = 3$	3	3	3	3	3	3	

$X = 38$; $Y = 97$

Now,

$$Y = 97 \equiv 18 \pmod{79}$$

$$X = 38 \equiv 7 \pmod{31}$$

So, $\alpha = 7$; $\beta = 18$.
$(\alpha\ \beta) = (7\ 18)$ is the minimum solution.
Hence $x = 31m + 7$; $y = 79m + 18$ is the general solution.

Astronomy

Revolutions of Heavenly Bodies

Aryabhata I was the first well-known ancient Indian astronomer. His statements about the heavenly bodies have been based on actual visual observations as well as careful verifications conducted over long periods of time. He strongly believed that the earth is spherical. He enunciated the following:

भूगोलः सर्वतो वृत्तः ।

(Golapāda, 6)

i.e., "Halves of the earth and other planets are dark due to their own shadows, whereas the other halves which face the sun are bright."

Aryabhata I was the first person to state that diurnal motion of the heavenly bodies is due to the rotation of the earth on its own axis. He states the following:

> *i.e., "it looks as through the entire celestial sphere together with other planets are moving towards the west causing their rising and setting" and further states,*
>
> ...अचलानि भानि समपश्चिमगानि...
>
> (Golapāda, 6)

i.e., "A man rowing forward in a boat sees the stationary objects on either side of the river moving backward. Just the same way, the people of Sri Lanka (on the equator) see the stationary stars moving toward west."

He also recognized that the earth and the other planets receive and reflect sun's light and they themselves are not self-illuminating planets. He points out.

"The conjunction of the earth and the sun determines the sun, the conjunction of the moon and the sun determines the moon and conjunction of all other planets and the moon determines the planets."

The accuracy of the period of one sidereal rotation of the earth with reference to stationary stars in the sky as given by Aryabhata I is truly phenomenal.

His time is 23 h 56 min, and 4.1 s, whereas the actual modern time is 23 h, 56 min, and 4.091 s.

Aryabhata I gave the following table for the number of revolutions of different heavenly bodies in the course of a yuga = 4.32×10^6 years:

Number of Revolutions in a Yuga

Name of the Heavenly Bodies		**Number of Revolutions**
Indian Name	**Western Name**	
Ravi	Sun	43,20,000
Candra	Moon	5,77,53,336
Bhumi	Earth	158,223,75,00
Kuja	Mars	22,96,824
Guru	Jupiter	3,64,224
Sani	Saturn	1,46,564

6 Varahamihira and Bhaskara I

Varahamihira

Amaraja, a commentator of Brahmagupta in his commentary on Brahmagupta's "Khandakhadyaka," quoted the following:

> *"Varahamihira passed away in Saka 509 (i.e., ad 587). Varahamihira has taken the saka year 427 (i.e., ad 505) as the epoch for the computations of the planets in his famous astronomical treatise 'Pañca Siddhāntikā.' "*

Dr. S.B. Dikshit believed that Varahamihira must have been around 20 years old when he composed "Pañca Siddhāntikā." It was also believed that Varahamihira was in the court of King Vikramaditya along with the immortal poet Kalidasa.

Major works of Varahamihira are "Pañca Siddhāntikā," "Bṛhat Samhitā," and "Bṛhat Jatakam." "Pañca Siddhāntikā" was a collection of five astronomical treatises that were existing at his time. "Bṛhat Samhitā" was a sort of encyclopedia containing extracts from different branches of knowledge prevailing at his time, and "Bṛhat Jatakam" was an astronomical text. In his work, "Bṛhat Jatakam," Varahamihira mentioned that he is the son of Ādityadasa. Ādityadasa lived in the city of Avanti and a native of Kāpittaka and taught his knowledge of Jyotisa to his son. Kāpittaka was believed to be the birthplace of Varahamihira and this place is now called Kayatha, 12 miles away from the city of Ujjain. The invocatory slokas (stanzas) in his works clearly demonstrated that Varahamihira was a great worshipper of the sun-god.

Bhattotpala, a commentator of Varahamihira, mentioned that Varahamihira was a Magadha Brahmin. The word "Maga" was believed to mean "the worshipper of sun-god." Hence, it is quite possible that Magadha may represent the region where the sun's worshippers resided. Since Avanti was not in Magadha, Varahamihira's ancestors might have migrated to Avanti from Magadha.

Varahamihira has also composed two more works, "Yogayatra" and "Vivahapataka." His famous work "Pancasiddhantika" as mentioned earlier was a collection of five astronomical texts, "Paulisa," "Romaka," "Vasistha," "Sūrya," and "Pitamaha." "Paulisa" and "Romaka" siddhantas were attributed to foreign sources—Greek and Babilonian. "Vasistha Siddhanta" appeared to have different versions and existed prior to the third century AD. Ranganatha quoted half a "sloka" from "Laghu Vasistha Siddhanta" and quoted another sloka from "Vṛddha Vasistha Siddhanta" regarding the eclipses. The word "Pitamaha" means "grandfather" and refers here to Lord Brahama, the creator of the Universe. This astronomical system

Mathematical Achievements of Pre-modern Indian Mathematicians. DOI: http://dx.doi.org/10.1016/B978-0-12-397913-1.00006-5

is also called "Brahma Siddhanta." Varahamihira has given the epoch for this text as January 11, 80 AD. There appears to be another "Brahma Siddhānta" given in "Vishnu dharmotttara Purāna." This siddhanta has been mentioned by Aryabhata I and Brahmagupta. But Varahamihira seemed to be unaware of the second "Brahma Siddhanta." Dr. S.B. Dikshit pointed out that "Suryasiddhānta" given by Varahamihira has taken the commencement of Kaliyuga at midnight and this assumption seemed to be correct, since the revolutions of the planet completely agree with the planetary positions computed on the assumption that Kaliyuga begins at midnight.

The present version of "Suryasiddhanta" available at present to us is different from the original version known to Varahamihira and his commentator, Bhattotpala. Dr. K.S. Shukla believed that "Suryasiddhanta" must have been revised after AD 628 and before AD 966. Dr. Shukla's statement was based on the knowledge that the present version of "Suryasiddhanta" has an unmistakable influence of "Brahma Sphuta Siddhanta" written by Brahmagupta in AD 628 and before AD 966 because Bhattotpala wrote a commentary on it in the year AD 966. To complete the astronomical constants provided by the above-mentioned two "Suryasiddhantas," we list below the following table given by Dr. K.S. Shukla: "Introduction to Suryasiddhānta."

Revolutions of the Planets in a Yuga (a Period of 43,20,000 Years)

Planet	Varahamihira's Suryasiddhanta	Present Suryasiddhanta
Sun	43,20,000	43,20,000
Mars	22,96,824	22,96,832
Jupiter	3,64,220	3,64,220
Venus	70,22,288	70,22,376
Saturn	1,46,564	1,46,568
Moon	5,77,53,336	5,77,53,336
Moon's apogee	4,88,219	4,88,203*
		4,88,211**
Moon's ascending Node	2,32,226	2,32,238*
		2,32,234**

*According to versions of Mallikārjuna Sūri, Yallaya, Tamma Yajva, and Ranganatha.
**According to the versions of Parameśvara, Rāmakṛsna, and Bhūdhara.

Bhaskara I

Nowhere in his works, Bhaskara I mentioned about himself or his family. It is well known that Bhaskara I was the greatest earlier exponent of Aryabhata I's mathematics and astronomy. He composed three works in the chronological order, "Mahabhaskarīyam," "Aryabhatiya Bhasya," and "Lagubhaskariam." Evidently, he must have authored only these three works, since his commentator Śankaranārayana and other later astronomers mentioned only these three works.

In "Mahabhaskarīyam," Bhaskara I elucidated the astronomy of Aryabhata I. He himself has renamed this work "Aryabhata Karma Nibandana" and firmly believed that this work contained simple methods of computation with clean expositions, so that even a lesser intelligent person can understand.

His second work, "Aryabhatiya bhasya," was a commentary on Aryabhata I's "Aryabhatiam." It comprised of two parts called "Dasagitikā Sūtra Vyākhya" and "Aryabhata tantra bhasya."

The word "Maha" means big and the word "Laghu" means small. The "Laghubhaskariyam" is an abridged version of "Mahabhaskariyam."

Based on the statements that Bhaskara I himself has made in "Aryabhatiya Bhasya" about the time that has elapsed since the beginning of Kaliyuga, K.S. Shukla made the conclusion that Bhaskara I composed the said commentary when 3750 years had elapsed since the beginning of Kaliyuga. This date corresponded to AD 629 (3730−3101). It is interesting to note that this date is exactly 1 year after Brahmagupta authored his well-known famous treatise, "Brahma Sphuta Siddānta" (AD 628). If this is accurate, the two great scholars were contemporaries, but it is unlikely that they knew each other.

As pointed out earlier, nowhere in his works, Bhaskara I has mentioned his place, birth, or stay. But he has referred several times to Asmaka country in his works. In fact, he called the followers of Aryabhata I School of astronomy as Asmakiyas and the treatise of Aryabhata I as "Asmaka tantra."

With regard to the place, where Bhaskara I composed his works, he made occasional references here and there about three places called "Valabhi," "Harukaccha," and "Śivabhāgapura." All these three places are close to each other and are situated in Katiwar. There are several references to indicate that Bhaskara I wrote his famous commentary "Aryabhatiya Bhasya" on the "Aryabhatiam" as a resident of "Valabhi."

During Bhaskara's time of seventh century AD, "Valabhi" was the capital of "Surastra" kingdom and a highly reputed seat of Buddhist learning. K.S. Shukla thought that Bhaskara I might have been an astronomer in the court of Dhruva Bhatta, the ruler of "Surāstra."

We will proceed to give some of the mathematical contributions of Bhaskara I.

Solution to the Problem of Remainders

Bhaskara I provided the solution of two such examples.

Example 1

"Find the number which when divided by 8 leaves 5 as the remainder, when divided by 9 leaves 4 as the remainder, and when divided by 7 leaves 1 as a remainder."

(*Continued*)

Example 1 (cont'd)

Solution

Let N be that number. Let x, y, z be the respective quotients. Then, we have

$$N = 8x + 5 = 9y + 4 = 7z + 1 \quad (^*)$$

If $8x + 5 = 9y + 4$, we get

$$9y = 8x + 1$$

We now solve this indeterminate equation of first degree.

$$\begin{array}{l} 9)8(0 = q \\ \underline{0} \\ 8)9(1 = q_1 \\ \underline{8} \\ 1 \end{array}$$

The number of quotients leaving $q = 0$ is one and an odd number.

Hence we have to choose a multiplier m_1^* such that on multiplication by the last residue 1 and subtracting 1 from the product, the result is divisible by the penultimate residue 8.

We have $m_1^* 1 - 1 \equiv 0 (\text{mod } 8)$.

So, $m_1^* = 9$.

So, $9 - 1 = 8 \equiv 0 (\text{mod } 9)$.

$$\begin{array}{l} 8)8(1 = q_1^* \\ \underline{8} \\ 0 \end{array}$$

We now form the following table:

	0	0	9
	1	10	10
$m_1^* =$	9	9	9
$q_1^* =$	1	1	

Then, $Y = 9 \equiv 1 (\text{mod } 8)$; $X = 10 \equiv 1 (\text{mod } 9)$.

Then, $(\alpha \;\; \beta) = (1 \;\; 1)$ is the least solution of $9y = 8x + 1$.

Now general solution of x for $9y = 8x + 1$ is $x = 9m_1 + 1$, where m is an arbitrary integer.

Now $N = 8x + 5 = 8(9m_1 + 1) + 5 = 72m_1 + 13$ is the general solution of N.

Now $N = 7z + 1$.
So, $72m_1 + 13 = 7z + 1$.
So, $7z = 72m_1 + 12$.
Now we will solve this indeterminate equation just like before.

$$7)72(10 = q$$
$$\underline{70}$$
$$2)7(3 = q_1$$
$$\underline{6}$$
$$1$$

Now the number of quotients leaving q is one and an odd number.

As before, we have to choose a multiplier m_2^* such that on multiplying by the last residue 1 and subtracting 12 from the product, the result is divisible by the penultimate remainder 2.

$$m_2^* \cdot 1 - 12 \equiv 0 (\text{mod } 2)$$

Then, $m_2^* = 14$.
Then, $14 - 12 = 2 \equiv 0 (\text{mod } 2)$.

$$2)2(1 = q_2^*$$
$$\underline{2}$$
$$0$$

We now form the following table:

	10	10	444
	3	43	43
$m_2^* =$	14	14	14
$q_2^* =$	1	1	

$$Y = 444 \equiv 12 (\text{mod } 72); \quad X = 43 \equiv 1 (\text{mod } 7)$$

(1 12) is the least value of (m_1 z).
General value of $m_1 = 7m_2 + 1$, where m_2 is an arbitrary integer.
General value of $z = 72m_2 + 12$.
General value of $N = 7z + 1 = 7(72m_2 + 12) + 1 = 504m_2 + 84 + 1$ or

$$N = 72m_1 + 13 = 72(7m_2 + 1) + 13 = 504m_2 + 85$$

where m_2 is an arbitrary integer.
Thus the least value of N is 85 (when $m_2 = 0$).

Example 2

"Tell me immediately, 0! Mathematician, that number which leaves one when divided by any one of the number 2, 3, 4, 5, and 6 but is exactly divisible by 7."

Solution

Let N be the number. Let X_1, X_2, X_3, X_4, X_5, and y be the quotients.

Then,

$$N = 2X_1 + 1 = 3X_2 + 1 = 4X_3 + 1 = 5X_4 + 1 = 6X_5 + 1 = 7y$$

Now

$$2X_1 + 1 = 3X_2 + 1 \Rightarrow 2X_1 = 3X_2$$

$$3X_2 + 1 = 4X_3 + 1 \Rightarrow 3X_2 = 4X_3$$

$$4X_3 + 1 = 5X_4 + 1 \Rightarrow 4X_3 = 5X_4$$

$$5X_4 + 1 = 6X_5 + 1 \Rightarrow 5X_4 = 6X_5$$

Now solve $2X_1 + 1 = 7y$, which is an indeterminate equation of first degree.

$$7)2(0 = q$$
$$\underline{0}$$
$$2)7(3 = q_1$$
$$\underline{6}$$
$$1$$

Number of quotients leaving q is one and an odd number.

So, we have to choose a multiplier m_1^* such that on multiplication by the residue 1 and subtracting 1 from the product, the result is divisible by the penultimate residue 2.

So $m_1^* \cdot 1 - 1 \equiv 0 (\text{mod } 2)$.

Then, $m_1^* = 3$ and $3 \cdot 1 - 1 = 2 \equiv 0(\text{mod } 2)$.

$$2)2(1 = q_1^*$$
$$\underline{2}$$
$$0$$

Now we form the following table:

0	0	3	
3	10	10	$Y = 3 \equiv 1(\text{mod } 2)$
3	3		$X = 10 \equiv 3(\text{mod } 7)$
1	1		

$(X_1, y) = (\alpha \;\; \beta) = (3 \;\; 1)$ is the least solution of $2X_1 + 1 = 7y$.
General solution:

$$X_1 = (7m_1 + 3)$$

$$y = 2m_1 + 1$$

where m_1 is an arbitrary integer.
Now $N = 7y = 7(2m_1 + 1) = 14m_1 + 7$.
But $N = 3X_2 + 1 \Rightarrow 14m_1 + 7 = 3X_2 + 1$.
So, we have $3X_2 = 14m_1 + 6$.
Now we solve this indeterminate equation:

$$3)14(4 = q$$
$$\underline{12}$$
$$2)3(1 = q_1$$
$$\underline{2}$$
$$1$$

Number of quotients leaving q is one and hence an odd number.
We have to choose a multiplier m_2^*, such that on multiplication by the last residue 1 and subtraction of 6 from the product, the result is divisible by the penultimate remainder 2.
So, we have $m_2^* \cdot 1 - 6 \equiv 0 (\text{mod } 2)$.

$$m_2^* = 8, \quad 8 \cdot 1 - 6 = 2 \equiv 0 (\text{mod } 2)$$

$$2)2(1 = q_2^*$$
$$\underline{2}$$
$$0$$

Now we form the following table:

	4	4	44
	1	9	9
$m_2^* =$	8	8	8
$q_2^* =$	1	1	

$$Y = 44 \equiv 2 (\text{mod} 14), \quad X = 9 \equiv 0 (\text{mod } 3)$$

Least solution of $3X_2 = 14m_1 + 6$ is $(m_1 \;\; X_2) = (0 \;\; 2)$.
General value of $X_2 = 14m_2 + 2$.

(*Continued*)

Example 2 (cont'd)

$$m_1 = 3m_2$$

where m_2 is an arbitrary integer.

General value of $N = 3X_2 + 1 = 3(14m_2 + 2) + 1$.

General value of $N = 42m_2 + 7$.

But $N = 4X_3 + 1$.

So, $42m_2 + 7 = 4X_3 + 1$.

So, $4X_3 = 42m_2 + 6$.

Dividing by 2, we get $2X_3 = 21m_2 + 3$.

We now solve this indeterminate equation.

$$\begin{array}{l} 2)21(10 \\ \underline{20} \\ 1 \end{array}$$

The number of quotients leaving 10 is zero and so even.

Now we have to choose a multiplier m_3^* such that on multiplication by the last residue 1 and adding 3 for the product, the result is divisible by the penultimate remainder 2.

So, $m_3^* \cdot 1 + 3 \equiv 0 \pmod 2$.

Then, $m_3^* = 1$ and $1 \cdot 1 + 3 = 4 \equiv 0 \pmod 2$.

Also,

$$\begin{array}{l} 2)4(2 = q_3^* \\ \underline{4} \\ 0 \end{array}$$

We now form the following table:

	10	12	$= Y$
$m_2^* =$	1	1	$= X$
$q_3^* =$	2	2	

$$Y = 12 \equiv 12 \pmod{21} \quad X = 1 \equiv 1 \pmod 2$$

Least solution is $(X_3 \;\; m_2) = (12 \;\; 1)$ for the equation $2X_3 = 21m_2 + 3$.

General solution:

$$X_3 = 21m_3 + 12$$

$$m_2 = 2m_3 + 1$$

where m_3 is an arbitrary integer.

General value of $N = 4X_3 + 1 = 4(21m_3 + 12) + 1$.
General value of $N = 84m_3 + 49$.
But $N = 5X_4 + 1$.
So, $5X_4 + 1 = 84m_3 + 49$, i.e., $5X_4 = 84m_3 + 48$.
We now solve this indeterminate equation.

$$\begin{array}{l} 5)84(16 = q \\ \underline{80} \\ \quad 4)5(1 = q_1 \\ \qquad \underline{4} \\ \qquad 1 \end{array}$$

Number of quotients leaving q is one and so is an odd number.

As before, we have to choose a multiplier m_4^* such that on multiplication by the last residue 1 and subtracting 48 from the product, the result is divisible by the penultimate residue 4.

So, $m_4^* \cdot 1 - 48 \equiv 0 (\text{mod } 4)$.
Hence, $m_4^* = 52$. Also, $52 \cdot 1 - 48 = 4 \equiv 0 (\text{mod } 4)$.

$$\begin{array}{l} 4)4(1 = q_4^* \\ \quad \underline{4} \\ \quad 0 \end{array}$$

We now form the following table:

16	16	900
1	53	53
52	52	52
1	1	

$$Y = 900 \equiv 60 (\text{mod } 84)$$

$$X = 53 \equiv 3 (\text{mod } 5)$$

Least solution of $5X_4 = 84m_3 + 48$ is $m_3 = 3$ and $X_4 = 60$.
General solution:

$$m_3 = 5m_4 + 3$$

$$X_4 = 84m_4 + 60$$

where m_4 is an arbitrary integer.

(Continued)

Example 2 (cont'd)

Now $N = 5X_4 + 1 = 5(84m_4 + 60) + 1$.

$$N = 420m_4 + 301$$

So the least value of $N = 301$, which when divided by 6 leaves 1 as a remainder. Hence this value of N satisfies the last condition.

Thus, the least value of N is 301.

It is worthy to note this very example was later treated by Ibn-al-Haitam in (AD 1000) and Leonardo Fibonacci of Pisa in AD 1202, according to L.E. Dickon, History of Theory of Numbers.

Bhaskara I gave the formulae for the two segments of the base called "abadhas" and the attitude to the base in terms of the sides of a triangle.

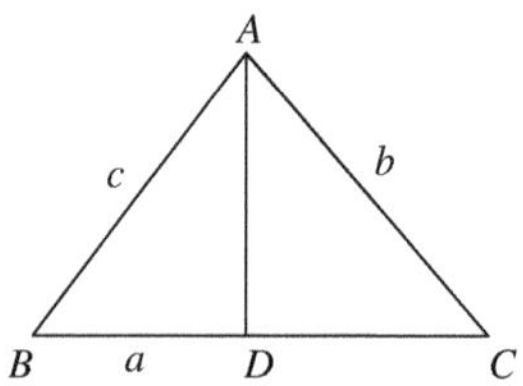

Let ABC be a triangle. Let AD be the attitude from A on the base BC. Then, the segments BD and DC are called "abadhas" of the sides AB and AC, respectively.

$$BD = \frac{1}{2}\left[a - \frac{b^2 - c^2}{a}\right],$$

$$DC = \frac{1}{2}\left[a + \frac{b^2 - c^2}{a}\right]$$

and

$$AD = \left\{\frac{(a + b + c)(c + a - b)(b + c - a)(b - c + a)}{4a^2}\right\}^{\frac{1}{2}}$$

where $a = BC$, $b = AC$, and $c = AB$.

Proof: In the right triangle ABD,

$$AB^2 = BD^2 + AD^2 = (BC-DC)^2 + AD^2$$
$$AB^2 = BC^2 - 2BC\cdot DC + AD^2 + DC^2$$
$$AB^2 = BC^2 - 2BC\cdot DC + (DC^2 + AD^2)$$
$$AB^2 = BC^2 - 2BC\cdot DC + AC^2$$

i.e., $c^2 = a^2 - 2a\cdot DC + b^2$.
Hence, $2a\cdot DC = a^2 + b^2 - c^2$.

$$DC = \frac{a^2+b^2-c^2}{2a} = \frac{1}{2}\left(a + \frac{b^2-c^2}{a}\right)$$

Now

$$BD = BC - DC = a - \frac{1}{2}\left(a + \frac{b^2-c^2}{a}\right)$$

$$BD = \frac{1}{2}\left(a - \frac{b^2-c^2}{a}\right)$$

Now

$$AD^2 = AB^2 - BD^2 = c^2 - \left[\frac{1}{2}\left(a - \frac{b^2-c^2}{a}\right)\right]^2$$

i.e.,

$$AD^2 = c^2 - \frac{1}{4}\left(\frac{a^2-b^2+c^2}{a}\right)^2$$

$$AD^2 = \frac{4a^2c^2 - (a^2+c^2-b^2)^2}{4a^2} = \frac{(2ac)^2 - (a^2+c^2-b^2)^2}{4a^2}$$

i.e.,

$$AD^2 = \frac{(2ac + a^2 + c^2 - b^2)[2ac - (a^2+c^2-b^2)]}{4a^2}$$

$$= \left[(a+c)^2 - b^2\right]\frac{[b^2 - (a^2 + c^2 - 2ac)]}{4a^2}$$

$$= \frac{[(a+c)^2 - b^2][b^2 - (a-c)^2]}{4a^2}$$

i.e.,

$$AD^2 = \frac{(a+c+b)(a+c-b)(b+a-c)[b-(a-c)]}{4a^2}$$

$$AD^2 = \frac{(a+b+c)(a+c-b)(a+b-c)(b+c-a)}{4a^2}$$

$$AD = \left\{\frac{(a+b+c)(b+c-a)(a+c-b)(a+b-c)}{4a^2}\right\}^{\frac{1}{2}}$$

Corollary

If $2s = a+b+c$, then the area of the triangle ABC is given by $A = \sqrt{s(s-a)(s-b)(s-c)}$.

Proof:

$$b+c-a = a+b+c-2a = 2s-2a$$

$$a+c-b = a+b+c-2b = 2s-2b$$

$$a+b-c = a+b+c-2c = 2s-2c$$

Then the altitude

$$AD = \left\{\frac{2s(2s-2a)(2s-2b)(2s-2c)}{4a^2}\right\}^{\frac{1}{2}}$$

i.e.,

$$AD = \frac{\{4s(s-a)(s-b)(s-c)\}^{\frac{1}{2}}}{a}$$

Area of the triangle ABC

$$= \frac{1}{2}BC \cdot AD = \frac{1}{2}a \cdot AD$$

$$= \frac{1}{2}a \cdot \frac{2\{s(s-a)(s-b)(s-c)\}^{\frac{1}{2}}}{a}$$

$$= \{s(s-a)(s-b)(s-c)\}^{\frac{1}{2}}$$

$$A = \sqrt{s(s-a)(s-b)(s-c)}$$

To illustrate this formula, Bhaskara I provided the following example.

Example

"What is the area of the scalene triangle (triangle with unequal sides), if the base is 14 and the two sides are 13 and 15?"

Solution

Let the base be a. Then, $a = 14$.

Let $b = 13$ and $c = 15$.

Then, $2s = a + b + c = 14 + 13 + 15 = 42$, so $s = 21$.

Area

$$A = \sqrt{s(s-a)(s-b)(s-c)}$$

The area

$$A = \sqrt{21(21-14)(21-13)(21-15)}$$

$$= \sqrt{(21)(7)(8)(6)} = \sqrt{(21)(21)(16)}$$

$$A = 21.4 = 84 \text{ square units}$$

We will now give a few examples given by Bhaskara I.

Example 1

"Two parallel faces of a figure looking like a 'Panava' (a musical instrument in the shape of a drum) are each 8 units, the central width 2 units, and breadth between the faces is 16 units. What is the area of this above figure?"

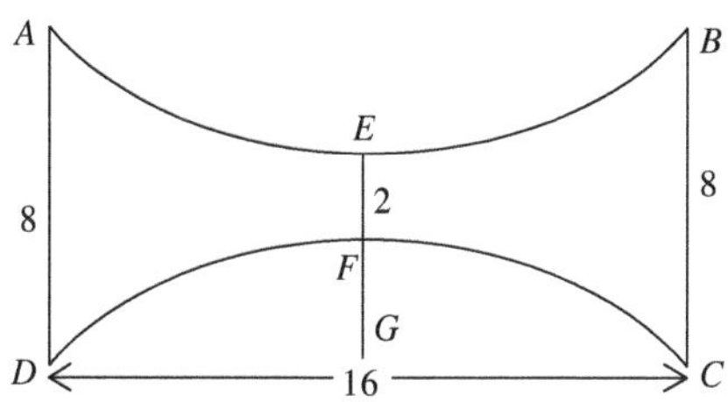

(*Continued*)

Example 1 (cont'd)

Area of $ABCD$

$$= \text{Area of the trapezium } ADFE + \text{area of the trapezium } BCFE$$

$$= \frac{AD+EF}{2} \cdot DG + \frac{BC+EF}{2} \cdot GC$$

$$= \frac{8+2}{2} \cdot 8 + \frac{8+2}{2} \cdot 8 = 5 \cdot 8 + 5 \cdot 8 = 80 \text{ square units}$$

Example 2

"In a trapezium, face and base are a and b, respectively. If the flanks are c and d, find the area of the trapezium."

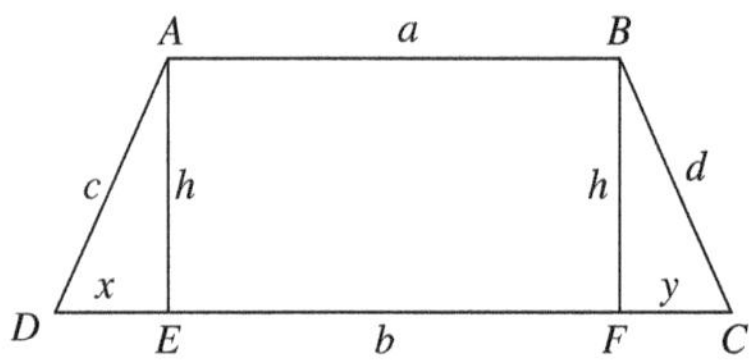

Let $ABCD$ be a trapezium with $AB = a$, $DC = b$, $AD = c$, and $BC = d$.

Let AE and BF be altitudes from A and B respectively on DC.
Let $DE = x$ and $FC = y$ and let $AE = BF = h$.
In the right triangle ADE,

$$AD^2 = AE^2 + DE^2$$

i.e.,

$$c^2 = x^2 + h^2 \tag{1}$$

Also in the right triangle BFC,

$$BC^2 = BF^2 + FC^2$$

i.e.,

$$d^2 = h^2 + y^2 \tag{2}$$

Subtracting (2) from (1), we get

$$x^2 - y^2 = c^2 - d^2 \tag{3}$$

Also, $DE + EF + FC = DC = b$.

$$x + AB + y = DC = b$$

i.e., $x + a + y = b$.
Hence,

$$x + y = b - a \tag{4}$$

Dividing (3) by (4), we get,

$$x - y = \frac{c^2 - d^2}{b - a} \tag{5}$$

Adding (4) and (5), we get

$$2x = (b - a) + \frac{c^2 - d^2}{b - a}$$

Thus,

$$x = \frac{1}{2}\left[(b - a) + \frac{c^2 - d^2}{b - a}\right]$$

From (1),

$$h^2 = c^2 - x^2 = c^2 - \frac{1}{4}\left[(b - a) + \frac{c^2 - d^2}{b - a}\right]^2$$

$$h = \sqrt{c^2 - \left(\frac{1}{4}\left[(b - a) + \frac{c^2 - d^2}{b - a}\right]\right)^2}$$

(*Continued*)

Example 2 (cont'd)

Area of the trapezium

$$= \frac{1}{2}(a+b)h = \frac{1}{2}(a+b)\sqrt{c^2 - \frac{1}{4}\left[(b-a) + \frac{c^2-d^2}{b-a}\right]^2}$$

Example 3

"A hawk is sitting on a pole of height 18 cubits. A rat has left its dwelling at the foot of the pole to a distance of 81 cubits and when returning to its dwelling is pounced by the hawk. Find the distance of the hole, from where the rat was pounced by the hawk. Find also the distance traveled by the hawk, assuming the speeds of the rat and the hawk are the same."

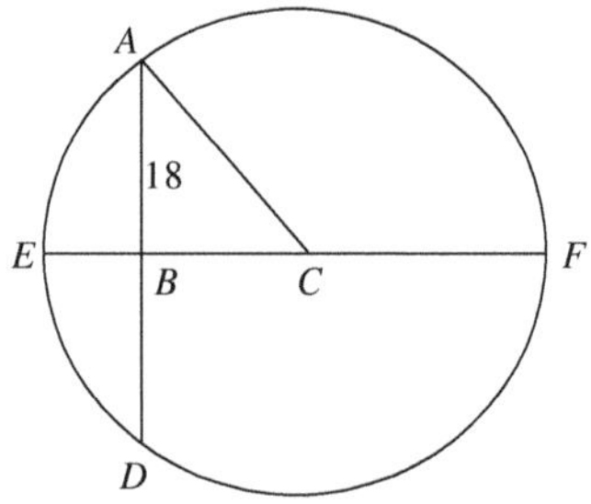

Let AB be the pole, at the top of which the hawk is sitting. We have to assume that the hawk sees the rat, after the rat had left its dwelling at B and had gone to F, which is 81 cubits from B. Let C be the place where the hawk pounces on the rat on its return.

Since the speeds of rat and hawk are the same, the distance traveled by the rat from the position when the hawk sees it and when it is pounced by the hawk $= CF$.

$$= \text{distance traversed by the hawk} = AC$$

So A and F lie on the same circle with center C. Let AB produced meet the circumference of the circle at D. Then, $AB = BD$. Now the two chords AD and EF intersect at B.

Hence $EB \cdot BF = AB \cdot BD = AB^2$.

i.e., $EB \cdot 81 = 18^2$.

So, $EB = (18 \times 18)/81 = 4$ cubits.

Now $BF = BC + CF = BC + CE = BC + BC + EB$.

So, $BF = 2BC + EB$.

Hence, $2BC = BF - EB = 81 - 4 = 77$.

$$BC = \frac{77}{2} \text{ cubits} = 38\frac{1}{2} \text{cubits}$$

Also,

$$AC = CF = CE = EB + BC = 4 + 38\frac{1}{2} = 42\frac{1}{2} \text{ cubits}$$

Example 4

"At the north-east corner of water reservoir of dimensions 6×12, there is a fish. At the north-west corner of the reservoir, there is a crane. The fish starts swimming in an oblique southern direction. The crane observing the swim of the fish comes along the sides of the reservoir and kills the fish. Find the distance traveled by them, assuming that their speeds are the same."

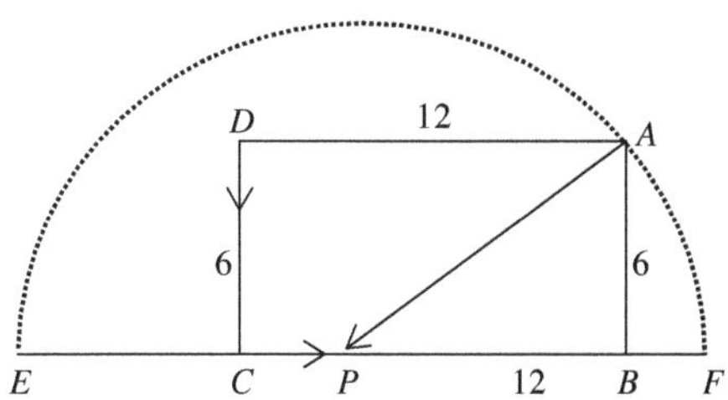

Let $ABCD$ be the reservoir. Let A and D be the initial positions of the fish and crane. Let P be the point where the fish swimming in an oblique southern direction is killed by the crane coming on the side DC and then CP of the reservoir.

Now $DC = AB = 6$ and $AD = CB = 12$.

Since the speeds of the fish and the crane are the same, the distance swam by the fish in the oblique southern direction $= AP =$ the distance traveled by the crane on the side DC and then CP of the reservoir $= DC + CP = 6 + CP$.

Now produce BC to E, such that $AP = PE$.

Then, A and E lie on the same semicircle with center P.

Let PB produced meet the semicircle at F.

Now $BE = BC + EC = BC + EP - CP = BC + AP - CP$.

i.e., $BE = BC + DC + CP - CP = BC + DC = 12 + 6 = 18$ cubits.

Also, $BF \cdot BE = (PF - PB)(EP + PB)$.

(*Continued*)

Example 4 (cont'd)

$$= (EP - PB)(EP + PB) = EP^2 - PB^2$$

So, $BF \cdot BE = AP^2 - PB^2 = AB^2 = 6^2 = 36.$
i.e.,

$$BF = \frac{36}{BE} = \frac{36}{18} = 2 \text{ cubic units}$$

Now

$$AP = PF = \frac{1}{2}EF = \frac{1}{2}(BE + BF) = \frac{1}{2}(18 + 2) = 10$$

Thus, the crane (travels) and the fish swims the same distance of 10 cubits.

Trigonometry: In his "Mahabhaskariam," Bhaskara I has provided a remarkable approximate formula for calculating $H \sin \theta = R \sin \theta$ of an angle $\theta \leq \pi/2$ without using the table. His formula is as follows:

$$H \sin \theta = R \sin \theta = \frac{4R(180^\circ - \theta)\theta}{40{,}500^\circ - \theta(180^\circ - \theta)}$$

where θ is in degrees.

If θ is in radians, the above formula becomes

$$\sin \theta = \frac{16\theta(\pi - \theta)}{5\pi^2 - 4\theta(\pi - \theta)}$$

For example,

$$\text{If } \theta = \frac{\pi}{2}, \sin \frac{\pi}{2} = \frac{16(\pi/2)(\pi - (\pi/2))}{5\pi^2 - 4(\pi/2)(\pi - (\pi/2))} = \frac{16(\pi/2) \cdot (\pi/2)}{5\pi^2 - 4(\pi/2) \cdot (\pi/2)}$$

$$= \frac{4\pi^2}{5\pi^2 - \pi^2} = \frac{4\pi^2}{4\pi^2} = 1$$

$$\text{If } \theta = \frac{\pi}{4}, \sin \frac{\pi}{4} = \frac{16(\pi/4)(\pi - (\pi/4))}{5\pi^2 - 4(\pi/4)(\pi - (\pi/4))}$$

$$= \frac{4\pi \cdot (3\pi/4)}{5\pi^2 - \pi \cdot (3\pi/4)} = \frac{12}{20 - 3} = \frac{12}{17} = 0.70588$$

$$\text{If } \theta = \frac{\pi}{3}, \ \sin\frac{\pi}{3} = \frac{16(\pi/3)(\pi-(\pi/3))}{5\pi^2 - 4(\pi/3)(\pi-(\pi/3))}$$

$$= \frac{32(\pi^2/9)}{5\pi^2-(8\pi^2/9)} = \frac{32\pi^2}{37\pi^2} = \frac{32}{37} = 0.8648$$

$$\text{If } \theta = \frac{\pi}{6}, \ \sin\frac{\pi}{6} = \frac{16(\pi/6)(\pi-(\pi/6))}{5\pi^2 - 4(\pi/6)(\pi-(\pi/6))} = \frac{(8/3)\cdot(5/6)}{5-(2/3)\cdot(5/6)} = \frac{(40/18)}{(80/18)} = \frac{1}{2}$$

These values are remarkably close to the actual values. This formula has been employed extensively by later mathematicians and astronomers.

Bhaskara I gave the trigonometric results:

i. $R\sin(90^\circ+\theta) = R\sin 90^\circ - R \text{ versin } \theta$
ii. $R\sin(180^\circ+\theta) = R\sin 90^\circ - R \text{ versin } 90^\circ - R\sin\theta = -R\sin\theta$
iii. $R\sin(270^\circ+\theta) = R\sin 90^\circ - R \text{ versin } 90^\circ - R + R \text{ versin } \theta$
$= -R\sin 90^\circ + R \text{ versin } \theta$

where $R \text{ versin } \theta = R(1-\cos\theta)$.

Proof:

1. $\text{RHS} = R\sin 90^\circ - R \text{ versin } \theta = R - R(1-\cos\theta) = R\cos\theta$

$= R\cos(-\theta) = R\sin[90-(-\theta)] = R\sin(90+\theta) = \text{LHS}$

2. $R\sin 90^\circ - R \text{ versin } 90^\circ - R\sin\theta = R - R(1-\cos 90^\circ) - R\sin\theta$

$= R - R(1-0) - R\sin\theta = -R\sin\theta$

$= R\sin(-\theta) = R\cos[90-(-\theta)]$

$= R\cos(90+\theta) = R\cos(-90-\theta)$

$= R\sin(90-(-90-\theta)) = R\sin(180+\theta)$

3. $R\sin(270+\theta) = R\sin[90-(-\theta-180)] = R\cos(-\theta-180)$

$= R\cos(180+\theta) = R\cos[90-(-\theta-90)]$

$= R\sin(-\theta-90^\circ) = -R\sin(90+\theta)$

$= -R\sin 90 + R \text{ versin } \theta \quad \text{from (1)}$

Also, $R \sin 90 - R \text{ versin } 90 - R + R \text{ versin } \theta$

$= R - R(1 - \cos 90^\circ) - R \sin 90^\circ + R \text{ versin } \theta$

$= R - R(1 - 0) - R \sin 90 + R \text{ versin } \theta$

$= -R \sin 90^\circ + R \text{ versin } \theta = R \sin(270 + \theta)$

It is extremely strange that Bhaskara I has disappeared into oblivion in the north India for several centuries. It is mostly in Kerala, the remarkable treatise of Bhaskara I have been brought into limelight through the commentaries of Govindaswami, Śankaranarāyana Parameśvara, and all residents of Kerala.

Bhaskara I pointed out that Pandurangaswami, Lātadeva, and Niśanka as direct disciples of Aryabhata I. He also mentioned the names of four more mathematicians with very unusual names, Maskari, Purana, Mudgala, and Putana. He gave their names in the following Sloka (verse):

पतदेकैकस्य ग्रन्थलक्षया न
मस्करि–पुरण–मुद्गलप्रभृतिभिः ऋादार्यैर्निबद्धं कृतं,
स कथं ऋनेनाचार्येण ऋल्पेन ग्रन्थेन शक्यते

"This work (his work) has not been written as a text by 'acaryas' like Maskari, Purana, and Mudgola with their one lakh (hundred thousand) verses each topic. How can this 'acaryas' (him) accomplish it with a short treatise."

गणितविदो मस्करीपूरण–पूतनादयः सर्वेषां क्षे त्त्राणां
फलम् ऋायतचतुरश्रक्षेत्त्रे प्रत्याययन्ति ।

i.e., "mathematicians Maskari, Purana, and Putana demonstrate the rationale of the areas of figures." It is very sad to note that mathematicians of the caliber of Aryabhata I according to Bhaskara I should simply remain as mere names.

7 Brahmagupta

Brahmagupta is the most distinguished mathematician of Ujjain school. He was born in AD 598. Brahmagupta mentioned in his famous work "Brahma Sphuta Siddhanta" that he wrote that work in Śaka 550 (i.e., AD 628) when he was 30 years old, during the reign of King Vyāghramukha of Capa dynasty. Before he wrote his work, "Brahma Sphuta Siddhanta," there existed two more "Brahma Siddhantas" one written by Sakalya and other incorporated in "Vishnudharmottara Purana." It is believed that Brahmagupta corrected and modified one of these old Brahma Siddhantas and called his work "Brahma Sphuta Siddhanta." Brahmagupta was a resident of Bhilamala in north Gujarat. Bhilamala was the capital of north Gujarat in the seventh century AD, and it is a small village now.

The "Brahma Sphuta Siddhanta" consists of 24 chapters and contains 1008 verses. This work is predominantly on astronomy and only 4½ chapters deal with pure mathematics. The first 10 chapters are devoted to the corrected and modified version of topics from the "Brahma Siddhanta." The remaining 14 chapters are devoted to new topics not discussed in the earlier "Brahma Siddhanta." Brahmagupta has named his twelfth chapter as "Ganita" and eighteenth chapter as "Kuttaka." "Ganita" has 56 verses and deals with arithmetic, arithmetic and geometric progressions, and a few topics in geometry. "Kuttaka" has 102 verses dealing with topics in algebra. The literal meaning of "Kuttaka" is Pulvariser. "Kuttaka" is essentially devoted to the determination of solutions of first-order indeterminate equation of the form $ax \pm c = by$ and of second-order indeterminate equation "Vargaprakrti," $Nx^2 + K = y^2$. The word "Prakrti" means coefficient and refers to the coefficient N in this equation.

When Brahmagupta was 67 years old, he wrote another expository book on astronomy called "Khandakhadyaka" in AD 665. The literal meaning of "Khandakhadyaka" is some special type of sweet. This book contains nine chapters with 194 slokas. The second part of this book is called "uttara-khandakhadyaka" and has five chapters with 71 slokas. Amaraja, Bhattotpala, Prthudakasvami, and Varuna are some of the well-known commentators of Brahmagupta.

As mentioned earlier on page 3 of this book, it was through "Brahma Sphuta Siddhanta" of Brahmagupta, the Arabs became familiar with Indian Astronomy. Professor Sachau in his translation of Al Biruni's India (see Ref. [27]) remarks that Arabs learnt their astronomy from Brahma Gupta's work before they heard the name of Ptolemy.

Mathematical Achievements of Pre-modern Indian Mathematicians. DOI: http://dx.doi.org/10.1016/B978-0-12-397913-1.00007-7

In Brahma Sphuta Siddhanta XII, Sloka 58, we observe the following rule:

$$\frac{a}{b} = \frac{a}{b+h}\frac{b+h}{b} = \frac{a}{b+h}\left[1 + \frac{h}{b}\right]$$

$$= \frac{a}{b+h} + \left(\frac{a}{b+h}\right)\frac{h}{b}$$

Example 1

Evaluate 1920/93.

Take $h = 3$, $a = 1920$, $b = 93$.

So,

$$\frac{1920}{93} = \frac{1920}{93+3} + \left(\frac{1920}{93+3}\right)\frac{3}{93}$$

$$= \frac{1920}{96} + \left(\frac{1920}{96}\right)\frac{3}{93} = 20 + 20 \cdot \frac{3}{93} = 20 + \frac{60}{93} = 20\frac{20}{31}$$

Example 2

Evaluate 9999/97.

Take $h = 2$, $a = 9999$, $b = 97$.

So,

$$\frac{9999}{97} = \frac{9999}{97+2} + \left(\frac{9999}{97+2}\right)\frac{2}{97}$$

$$= 101 + (101)\frac{2}{97} = 101 + \frac{202}{97}$$

$$\frac{202}{97} = \frac{202}{97+4} + \left(\frac{202}{97+4}\right)\frac{4}{97} \quad \text{(by taking } h = 4\text{)}$$

$$= 2 + 2 \cdot \frac{4}{97} = 2 + \frac{8}{97}$$

So,

$$\frac{9999}{97} = 101 + 2 + \frac{8}{97} = 103\frac{8}{97}$$

In "Brahma Sphuta Siddhanta" XII, Sloka 63, we observe the following rule: $x^2 = (x - y)(x + y) + y^2$.

Example 1

$$\begin{aligned} 77^2 &= (77 - 3)(77 + 3) + 9, \text{ taking } y = 3 \\ &= (74)(80) + 9 = 5920 + 9 = 5929 \end{aligned}$$

Example 2

$$\begin{aligned} 138^2 &= (138 - 2)(138 + 2) + 2^2, \text{ taking } y = 2 \\ &= (136)(140) + 4 = 19{,}040 + 4 = 19{,}044 \end{aligned}$$

Example 3

$$\begin{aligned} 2997^2 &= (2997 - 3)(2997 + 3) + 3^2 \\ &= (2994)(3000) + 9 \\ &= 8{,}982{,}000 + 9 = 8{,}982{,}009 \end{aligned}$$

Example 4

$$\begin{aligned} 599{,}996^2 &= (599{,}996 - 4)(599{,}996 + 4) + 4^2 \\ &= (599{,}992)(600{,}000) + 16 \\ &= 359{,}995{,}200{,}000 + 16 = 359{,}995{,}200{,}016 \end{aligned}$$

Vargaprakrti

The indeterminate equation $Nx^2 + k = y^2$, where N is a positive integer and not a perfect square, has been called by the ancient Indian mathematicians as "varga-prakrti" or "krti-prakrti." "Varga" means square and "prakrti" means nature. So, "vargaprakrti" means square nature. However, ancient Indian algebraists employed the term "prakrti" to denote N, the coefficient of the square of the unknown.

It is Brahmagupta, in AD 628, who treated this equation first. He showed that the integral solution for $Nx^2 + 1 = y^2$ can be found provided we can have a solution $(\alpha\ \beta)$ for the equation $Nx^2 + k = y^2$, where $k = \pm 1, \pm 2, \pm 4$. This is indeed a remarkable achievement, if we consider that Brahmagupta demonstrated this in AD 628, without the use of continued fractions. Four centuries later, in AD 1050, Bhaskara II gave the famous "Chakravala" (cyclic) method to find the integral solutions of $Nx^2 + 1 = y^2$.

Pierre de Fermat (AD 1601–1665), the French mathematician in one of his well-known challenges to the fellow British mathematicians of his time, proposed to find the integral solution of the equation $Nx^2 + 1 = y^2$, where N is a positive integer and not a perfect square.

The equation $Nx^2 + 1 = y^2$ even now bears incorrectly the name of John Pell (AD 1610–1685), an English mathematician, although his connection with it consists of simply the publications of the solutions of it in his edition of Brounker's Translation of Rhonius's algebra (AD 1668). It was an accident that Leonhard Euler (AD 1707–1783), the famous Swiss mathematician, referred to this equation as the "Pell Equation." Since there is no contribution by John Pell to this topic, to call this equation "Pell's Equation" even now has no historical justification. To continue to call this equation "Pell's Equation" is a misnomer. It is fitting and justified that this "vargaprakrti" equation should be renamed as Brahmagupta Bhaskara II equation.

Brahmagupta's Lemma 1

If $(\alpha\ \beta)$ is a solution of the equation $Nx^2 + K = y^2$ and if (α^1, β^1) is a solution of the equation $Nx^2 + K^1 = y^2$, then $(\alpha\beta^1 \pm \beta\alpha^1\ \ \beta\beta^1 \pm N\alpha\alpha^1)$ is a solution of the equation $Nx^2 + KK^1 = y^2$.

This result is called the "Samasa" ("समास") or the principle of composition. It appears that Brahmagupta never gave a proof of this lemma. The ancient Indians never paid attention to the proofs.

The proof was given by Krsna (AD 1580), which has been given in this book.

Remark: If $(\alpha\ \beta)$ is a solution of the equation $Nx^2 + K = y^2$, we symbolically write this as $(\alpha\ \beta\ K)$. Then Brahmagupta's lemma 1 tells that if we do "Samasa" of $(\alpha\ \beta\ K)$ on $(\alpha^1\ \beta^1\ K^1)$, we get $(\alpha\beta^1 \pm \alpha^1\beta\ \ \beta\beta^1 \pm N\alpha\alpha^1\ \ KK^1)$. We write this symbolically as

$$(\alpha\ \beta\ K) \circ (\alpha^1\ \beta^1\ K^1) \Rightarrow (\alpha\beta^1 \pm \beta^1\alpha\ \ \beta\beta^1 \pm N\alpha\alpha^1\ \ KK^1)$$

where the symbol $\circ$ denotes the "Samasa" of one over the other.

Brahmagupta's Corollary

If $(\alpha\ \beta)$ is a solution of $Nx^2+K=y^2$, then $(2\alpha\beta\ \beta^2+N\alpha^2)$ is a solution of $Nx^2+K^2=y^2$.

Proof: Let $\alpha^1=\alpha$, $\beta^1=\beta$, and $K^1=K$ in the Brahmagupta's lemma 1. Then, we have

$$(\alpha\ \beta\ K)\circ(\alpha\ \beta\ K)\Rightarrow(\alpha\beta+\alpha\beta\ \beta\beta+N\alpha\alpha\ K^2)$$

i.e., $(\alpha\ \beta\ K)\circ(\alpha\ \beta\ K)\Rightarrow(2\alpha\beta\ \beta^2+N\alpha^2\ K^2)$, which means that $(2\alpha\beta\ \beta^2+N\alpha^2)$ is a solution of the equation $Nx^2+K^2=y^2$.

Brahmagupta's Lemma 2

If $(\alpha\ \beta)$ is a solution of $Nx^2\pm p^2d=y^2$, then $(\alpha/p\ \beta/p)$ is a solution of $Nx^2\pm d=y^2$.

Proof: If $(\alpha\ \beta)$ is a solution of the equation $Nx^2\pm p^2d=y^2$, then, we have, $N\alpha^2\pm p^2d=\beta^2$.

Dividing by p^2, we get

$$N\left(\frac{\alpha}{p}\right)^2\pm d=\left(\frac{\beta}{p}\right)^2$$

which means that $(\alpha/p\ \beta/p)$ is a solution of $Nx^2\pm d=y^2$.

Remark: Symbolically, this means

$$(\alpha\ \beta\pm p^2d)\Rightarrow\left(\frac{\alpha}{p}\frac{\beta}{p}\pm d\right)$$

Rational solutions of the equation $Nx^2+1=y^2$.
Brahmagupta's theorem on rational solution
If $(\alpha\ \beta)$ is an integral solution of $Nx^2+K=y^2$, then

$$\left(\frac{2\alpha\beta}{K}\ \frac{\beta^2+N\alpha^2}{K}\right)$$

is a rational solution of the equation $Nx^2+1=y^2$.

Proof: If $(\alpha \ \beta)$ is a solution of $Nx^2 + K = y^2$, then by Brahmagupta's corollary, $(2\alpha\beta \ \beta^2 + N\alpha^2)$ is a solution of $Nx^2 + K^2 = y^2$. Then, by Brahmagupta's lemma 2, we have

$$\left(\frac{2\alpha\beta}{K} \ \frac{\beta^2 + N\alpha^2}{K}\right)$$

is a solution of $Nx^2 + 1 = y^2$.

We note that $(2\alpha\beta)/K$ and $(\beta^2 + N\alpha^2)/K$ are rational numbers.

Remark: Symbolically, this theorem tells us that

$$(\alpha \ \beta \ K) \Rightarrow \left(\frac{2\alpha\beta}{K} \frac{\beta^2 + N\alpha^2}{K} \ 1\right)$$

Integral solutions of $Nx^2 + 1 = y^2$.

Brahmagupta's theorems for integral solutions of the equation $Nx^2 + 1 = y^2$ are as follows.

Theorem 1: If $(\alpha \ \beta)$ is an integral solution of $Nx^2 - 1 = y^2$, then $(2\alpha\beta \ \beta^2 + N\alpha^2)$ is an integral solution of $Nx^2 + 1 = y^2$.

Proof: Let $k = -1$ in Brahmagupta's corollary on page 165, we get $(a \ b) = (2\alpha\beta \ \beta^2 + N\alpha^2)$ is a solution of $Nx^2 + 1 = y^2$.

Since α, β are integers, so are a and b.

Remark: Symbolically, this theorem tells us that

$$(\alpha \ \beta \ -1) \Rightarrow (2\alpha\beta \ \beta^2 + N\alpha^2 \ 1)$$

Theorem 2: If $(\alpha \ \beta)$ is an integral solution of $Nx^2 \pm 2 = y^2$, then

$$\left(\alpha\beta \ \frac{\beta^2 + N\alpha^2}{2}\right)$$

is an integral solution of $Nx^2 + 1 = y^2$.

Proof: Let $K = \pm 2$ in Brahmagupta's theorem on rational solutions on page 165.

We get

$$(a \ b) = \left(\alpha\beta \ \frac{\beta^2 + N\alpha^2}{2}\right)$$

is a solution of $Nx^2 + 1 = y^2$.

Since α and β are integers, so is $a = \alpha\beta$.

We have to show that $b = (\beta^2 + N\alpha^2)/2$ is an integer.

Since $(\alpha\ \beta)$ is an integral solution of $Nx^2 \pm 2 = y^2$, we have $N\alpha^2 \pm 2 = \beta^2$ $N\alpha^2 = \beta^2 \pm 2$. So,

$$b = \frac{\beta^2 + N\alpha^2}{2} = \frac{\beta^2 + \beta^2 \pm 2}{2} = \beta^2 \pm 1$$

Since β is an integer, so is $b = \beta^2 \pm 1$.

Remark:

$$(\alpha\ \beta \pm 2) \Rightarrow \left(\alpha\beta\ \frac{\beta^2 + N\alpha^2}{2}\ 1\right)$$

Theorem 3: If $(\alpha\ \beta)$ is an integral solution of $Nx^2 + 4 = y^2$, then

$$(a\ b) = \begin{cases} \left(\dfrac{\alpha\beta}{2}\ \dfrac{\beta^2 - 2}{2}\right) & \text{if } \beta \text{ is even} \\ \left(\dfrac{1}{2}\alpha(\beta^2 - 1)\ \dfrac{1}{2}\beta\,(\beta^2 - 3)\right) & \text{if } \beta \text{ is odd} \end{cases}$$

is an integral solution of $Nx^2 + 1 = y^2$.

Proof: We let $K = 4$ in Brahmagupta's corollary on page 165. Since $(\alpha\ \beta)$ is a solution of the equation $Nx^2 + 4 = y^2$, we see $(2\alpha\beta\ \beta^2 + N\alpha^2)$ is a solution of the equation $Nx^2 + 16 = y^2$.

Hence, $N(2\alpha\beta)^2 + 16 = (\beta^2 + N\alpha^2)^2$.

Dividing by 16, we get

$$N\frac{4\alpha^2\beta^2}{16} + 1 = \left(\frac{\beta^2 + N\alpha^2}{4}\right)^2$$

i.e.,

$$N\left(\frac{\alpha\beta}{2}\right)^2 + 1 = \left(\frac{\beta^2 + N\alpha^2}{4}\right)^2$$

Thus,

$$(a\ b) = \left(\frac{\alpha\beta}{2}\ \frac{\beta^2 + N\alpha^2}{4}\right)$$

is a solution of $Nx^2 + 1 = y^2$.

Since $(\alpha\ \beta)$ is a solution of $Nx^2 + 4 = y^2$, we have $N\alpha^2 + 4 = \beta^2$ (*).
So, $N\alpha^2 = \beta^2 - 4$.
Now,

$$b = \frac{\beta^2 + N\alpha^2}{4} = \frac{\beta^2 + \beta^2 - 4}{4} = \frac{\beta^2 - 2}{2}$$

So,

$$(a\ b) = \left(\frac{\alpha\beta}{2}\ \frac{\beta^2 - 2}{2}\right)$$

is a solution of $Nx^2 + 1 = y^2$.

We have to show a and b are integers.

Case 1: First, let β be an even integer $= 2m$, where m is an integer.
Then,

$$a = \frac{\alpha\beta}{2} = \frac{\alpha \cdot 2m}{2} = m\alpha = \text{an integer}$$

$$b = \frac{\beta^2 - 2}{2} = \frac{(2m)^2 - 2}{2} = \frac{4m^2 - 2}{2} = 2m^2 - 1 = \text{an integer}$$

Case 2: Let β be an odd integer $= 2m + 1$, where m is an integer. Since $(\alpha\ \beta)$ is an integral solution of $Nx^2 + 4 = y^2$, then by Brahmagupta's lemma 2 on page 165, we have $(\alpha/2\ \beta/2)$ is a solution of $Nx^2 + 1 = y^2$.
Now,

$$\left(\frac{\alpha\beta}{2}\ \frac{\beta^2 - 2}{2}\ 1\right) \circ \left(\frac{\alpha}{2}\ \frac{\beta}{2}\ 1\right)$$

$$\Rightarrow \left(\frac{\alpha\beta}{2} \cdot \frac{\beta}{2} + \frac{\beta^2 - 2}{2} \cdot \frac{\alpha}{2}\ \frac{\beta}{2} \cdot \frac{\beta^2 - 2}{2} + N\frac{\alpha\beta}{2} \cdot \frac{\alpha}{2}\ 1\right)$$

$$\Rightarrow \left(\frac{\alpha\beta^2}{4} + \frac{\alpha\beta^2 - 2\alpha}{4}\ \frac{\beta(\beta^2 - 2) + N\alpha^2\beta}{4}\ 1\right)$$

which means that

$$(a_1\ b_1) = \left(\frac{\alpha\beta^2}{4} + \frac{\alpha\beta^2 - 2\alpha}{4}\ \frac{\beta(\beta^2 - 2) + N\alpha^2\beta}{4}\right)$$

is a solution of $Nx^2 + 1 = y^2$, where

$$a_1 = \frac{\alpha\beta^2}{4} + \frac{\alpha\beta^2 - 2\alpha}{4} = \frac{\alpha}{4}(\beta^2 + \beta^2 - 2) = \frac{\alpha}{2}(\beta^2 - 1)$$

$$b_1 = \frac{\beta(\beta^2 - 2) + N\alpha^2\beta}{4}$$

$$= \frac{\beta}{4}\left[\beta^2 - 2 + N\alpha^2\right] = \frac{\beta}{4}\left[\beta^2 - 2 + \beta^2 - 4\right] \quad \text{from (*)}$$

$$= \frac{\beta}{4}(2\beta^2 - 6) = \frac{\beta}{2}(\beta^2 - 3)$$

Now, we have to show a_1 and b_1 are integers.
Since

$$\beta = 2m + 1, \quad a_1 = \frac{\alpha}{2}(\beta^2 - 1) = \frac{\alpha}{2}\left[(2m+1)^2 - 1\right]$$

$$= \frac{\alpha}{2}\left[4m^2 + 2m + 1 - 1\right] = \frac{\alpha}{2}(4m^2 + 2m)$$

$$= \alpha(2m^2 + m) = \text{an integer}$$

$$b_1 = \frac{\beta^2}{2}(\beta^2 - 3) = \frac{\beta^2}{2}((2m+1)^2 - 3)$$

$$= \frac{\beta^2}{2}(4m^2 + 4m + 1 - 3) = \frac{\beta^2}{2}(4m^2 + 4m - 2)$$

$$= \beta_2(2m^2 + 2m - 1) = \text{an integer}$$

since β and m are integers.

Remark:

$$(\alpha\ \beta\ 4) = (a\ b\ 1)$$

where

$$(a, b) = \begin{cases} \left(\dfrac{\alpha\beta}{2}\ \dfrac{\beta^2 - 2}{2}\right) & \text{if } \beta \text{ is even} \\ \left[\dfrac{1}{2}\alpha(\beta^2 - 1)\ \dfrac{1}{2}\beta(\beta^2 - 3)\right] & \text{if } \beta \text{ is odd} \end{cases}$$

Theorem 4: If $(\alpha\ \beta)$ is a solution of $Nx^2 - 4 = y^2$, then

$$(a\ b) = \left[\frac{\alpha\beta}{2}(\beta^2 + 1)(\beta^2 + 3)\ \frac{\beta^2 + 2}{2}\{(\beta^2 + 1)(\beta^2 + 3) - 2\}\right]$$

is an integral solution of $Nx^2 + 1 = y^2$.

Proof: $(\alpha\ \beta)$ is a solution of $Nx^2 - 4 = y^2$,

$$N\alpha^2 - 4 = \beta^2 \tag{1}$$

Then, by Brahmagupta's corollary on page 165, we get $(2\alpha\beta\ \beta^2 + N\alpha^2)$ is a solution of $Nx^2 + 16 = y^2$.

So, $N(2\alpha\beta)^2 + 16 = (\beta^2 + N\alpha^2)^2$.

Dividing by 16, we get

$$N\left(\frac{\alpha\beta}{2}\right)^2 + 1 = \left(\frac{\beta^2 + N\alpha^2}{4}\right)^2 = \left(\frac{\beta^2 + \beta^2 + 4}{4}\right)^2 \quad \text{from (1)}$$

i.e.,

$$N\left(\frac{\alpha\beta}{2}\right)^2 + 1 = \left(\frac{\beta^2 + 2}{2}\right)^2$$

which means that

$$\left(\frac{\alpha\beta}{2}\ \frac{\beta^2 + 2}{2}\right)$$

is a solution of $Nx^2 + 1 = y^2$.

$$\left(\frac{\alpha\beta}{2}\ \frac{\beta^2 + 2}{2}\ 1\right) \circ \left(\frac{\alpha\beta}{2}\ \frac{\beta^2 + 2}{2}\ 1\right)$$

$$\Rightarrow \left(\frac{\alpha\beta}{2}\left(\frac{\beta^2 + 2}{2}\right) + \frac{\alpha\beta}{2}\left(\frac{\beta^2 + 2}{2}\right)\left(\frac{\beta^2 + 2}{2}\right)\left(\frac{\beta^2 + 2}{2}\right) + N\left(\frac{\alpha\beta}{2}\right)\left(\frac{\alpha\beta}{2}\right)\ 1\right)$$

$$= \left(\frac{\alpha\beta(\beta^2 + 2)}{2}\left(\frac{\beta^2 + 2}{2}\right)^2 + N\left(\frac{\alpha\beta}{2}\right)^2\ 1\right)$$

which means that

$(a_0\ b_0)$ is a solution of $Nx^2 + 1 = y^2$, where

$$a_0 = \frac{\alpha\beta(\beta^2+2)}{2}$$

$$b_0 = \left(\frac{\beta^2+2}{2}\right)^2 + N\left(\frac{\alpha\beta}{2}\right)^2 = \left(\frac{\beta^2+2}{2}\right)^2 + \frac{\beta^2}{4}(N\alpha^2)$$

$$b_0 = \left(\frac{\beta^2+2}{2}\right)^2 + \frac{\beta^2}{4}(\beta^2+4) \quad \text{from (1)}$$

So,

$$b_0 = \frac{\beta^4+4\beta^2+4+\beta^4+4\beta^2}{4} = \frac{2\beta^4+8\beta^2+4}{4} = \frac{\beta^4+4\beta^2+2}{2}$$

Now,

$$\left(\frac{\alpha\beta}{2}\ \frac{\beta^2+2}{2}\ 1\right)\circ\left(\frac{\alpha\beta(\beta^2+2)}{2}\ \frac{\beta^4+4\beta^2+2}{2}\ 1\right)$$

$$= (a_1\ b_1\ 1)$$

where

$$a_1 = \frac{\alpha\beta}{2}\left(\frac{\beta^4+4\beta^2+2}{2}\right) + \left(\frac{\beta^2+2}{2}\right)\left(\frac{\alpha\beta(\beta^2+2)}{2}\right)$$

$$b_1 = \left(\frac{\beta^2+2}{2}\right)\left(\frac{\beta^4+4\beta^2+2}{2}\right) + \frac{\alpha\beta}{2}\left(\frac{\alpha\beta(\beta^2+2)}{2}\right)N$$

So,

$$a_1 = \frac{\alpha\beta}{4}\left[\beta^4+4\beta^2+2+(\beta^2+2)^2\right]$$

$$= \frac{\alpha\beta}{4}\left[\beta^4+4\beta^2+2+\beta^4+4\beta^2+4\right]$$

$$= \frac{\alpha\beta}{4}(2\beta^4+8\beta^2+6)$$

$$= \frac{\alpha\beta}{2}(\beta^4+4\beta^2+3) = \frac{\alpha\beta}{2}(\beta^3+3)(\beta^2+1) \tag{2}$$

$$b_1 = \frac{(\beta^2+2)}{4}\left[\beta^4 + 4\beta^2 + 2 + N\alpha^2\beta^2\right]$$

$$= \frac{\beta^2+2}{4}\left[\beta^4 + 4\beta^2 + 2 + \beta^2(\beta^2+4)\right] \quad \text{from (1)}$$

i.e.,

$$b_1 = \frac{(\beta^2+2)}{4}(2\beta^4 + 8\beta^2 + 2) = \frac{(\beta^2+2)}{2}(\beta^2 + 4\beta^2 + 1)$$

$$= \frac{(\beta^2+2)}{2}\left[(\beta^2+1)(\beta^2+3) - 2\right]$$

$$= (\beta^2+2)\left[\frac{1}{2}(\beta^2+1)(\beta^2+3) - 1\right] \tag{3}$$

which means that $(a_1\ b_1)$ is a solution of $Nx^2 + 1 = y^2$, where $(a_1\ b_1)$ are given from (2) and (3).

We now have to show that a_1 and b_1 are integers.

Case 1: First, let β be even, say $\beta = 2m$, where m is an integer.

$$a_1 = \frac{\alpha\beta}{2}(\beta^2+1)(\beta^2+3) = \alpha \cdot \frac{2m}{2}(\beta^2+1)(\beta^2+3)$$

$$= \alpha m(\beta^2+1)(\beta^2-1) = \text{an integer}$$

$$b_1 = (\beta^2+2)\left[\frac{1}{2}(\beta^2+1)(\beta^2+3) - 1\right]$$

$$= \frac{(\beta^2+2)}{2}\left[(\beta^2+1)(\beta^2+3) - 2\right]$$

$$= \left(\frac{4m^2+2}{2}\right)\left[(\beta^2+1)(\beta^2+3) - 2\right]$$

$$= (2m^2+1)[(\beta^2+1)(\beta^2+3) - 2] = \text{an integer}$$

since m and β are integers.

Case 2: Next, let β be an odd integer = $(2m + 1)$, where m is an integer.

$$a_1 = \frac{\alpha\beta}{2}(\beta^2 + 1)(\beta^2 + 3)$$

$$= \frac{\alpha\beta}{2}\left[\left\{(2m+1)^2 + 1\right\}\left\{(2m+1)^2 + 3\right\}\right]$$

$$= \alpha\beta\left[\frac{(4m^2 + 4m + 2)}{2}\right]\left[4m^2 + 4m + 4\right]$$

$$= \alpha\beta[(2m^2 + 2m + 1)][4m^2 + 4m + 4] = \text{an integer}$$

since m, α, and β are integers.

$$b_1 = (\beta^2 + 2)\left[\frac{1}{2}(\beta^2 + 3)(\beta^2 + 1) - 1\right]$$

$$= (\beta^2 + 2)\left[\frac{1}{2}\left\{(2m+1)^2 + 3\right\}\left\{(2m+1)^2 + 1\right\} - 1\right]$$

$$= (\beta^2 + 2)\left[\frac{1}{2}\left\{(4m^2 + 4m + 4)(4m^2 + 4m + 2)\right\} - 1\right]$$

$$= (\beta^2 + 2)\{(2m^2 + 2m + 2)(4m^2 + 4m + 2) - 1\} = \text{an integer}$$

since β and m are integers.

Remark:

$$(\alpha\ \beta - 4) \Rightarrow (a_1\ b_1\ 1)$$

where

$$a_1 = \frac{\alpha\beta}{2}(\beta^2 + 1)(\beta^2 + 3)$$

$$b_1 = \frac{(\beta^2 + 2)}{2}\left\{(\beta^2 + 1)(\beta^2 + 3) - 2\right\}$$

We will now provide examples to show the power of Brahmagupta's above four theorems.

Example 1

Solve in integers

$$197x^2 + 1 = y^2 \tag{1}$$

(Continued)

Example 1 (cont'd)

We have the identity (1) $197 - 1 = 196 = (14)^2$.

So, $(\alpha\ \beta) = (1\ 14)$ is a solution of $197x^2 - 1 = y^2$.

Then, by Brahmagupta's theorem 1, $(a\ b)$ is a solution of $197x^2 + 1 = y^2$, where

$$a = 2\alpha\beta = 2(1)(14) = 28$$

$$b = \beta^2 + N\alpha^2 = (14)^2 + (197) \cdot 1 = 196 + 197 = 393$$

Thus, (28 393) is an integral solution of $197x^2 + 1 = y^2$.

Remark: We can get infinite solutions of (1) by doing "Samasa" of (28 393 1) on itself repeatedly.

Example 2

Find the integral solutions of

$$14,639x^2 + 1 = y^2 \tag{2}$$

We have the identity $14{,}639 \cdot 1^2 + 2 = 14{,}641 = (121)^2$.

So, (1 121) is an integral solution of $14{,}639x^2 + 2 = y^2$.

Then, by Brahmagupta's theorem 2, $(a\ b)$ is a solution of

$$14,639x^2 + 1 = y^2$$

where

$$a = \alpha\beta = (1)(121) = 121$$

$$b = \frac{\beta^2 + N\alpha^2}{2} = \frac{(121)^2 + (14,639) \cdot 1}{2} = \frac{14,641 + 14,639}{2} = 14,640$$

Thus, (121 14,640) is an integral solution of $14{,}639x^2 + 1 = y^2$.

As we remarked earlier, a repeated application of "Samasa" of (121 14,640 1) on itself gives the infinite solutions of (2).

Example 3

Solve in integers $2021x^2 + 1 = y^2$.

We have the identity, $2021 \cdot 1^2 + 4 = 2025 = (45)^2$.

So, $(\alpha\ \beta) = (1\ 45)$ is an integral solution of $2021x^2 + 4 = y^2$.
So, by Brahmagupta's theorem 3, since $\beta = 45$ is odd,

$$(a\ b) = \left\{\frac{1}{2}\alpha(\beta^2 - 1)\ \frac{1}{2}\beta(\beta^2 - 3)\right\}$$

is an integral solution of $2021x^2 + 1 = y^2$.
So,

$$a = \frac{1}{2}\alpha(\beta^2 - 1) = \frac{1}{2}(1)\{(45)^2 - 1\} = \left(\frac{2025 - 1}{2}\right) = 1012$$

$$b = \frac{1}{2}\beta(\beta^2 - 3) = \frac{1}{2}(45)\left[(45)^2 - 3\right]$$

$$= \frac{45}{2}(2025 - 3) = \left(\frac{45}{2}\right)(2022) = 45(1011) = 45,495$$

Thus, (1 45, 495) is an integral solution of $2021x^2 + 1 = y^2$.
As we remarked before, repeated application of Samasa of (1 45495 1) on itself will be the infinite solutions of the equation (3).

Example 4

Find integral solutions of

$$1765x^2 + 1 = y^2 \tag{1}$$

We have the identity,

$$1765(2)^2 - 4 = (1765)4 - 4$$

$$= 7060 - 4 = 7056 = (84)^2$$

So, $(\alpha\ \beta) = (2\ 84)$ is an integral solution of $1765x^2 - 4 = 1$.
Then, by Brahmagupta's theorem 4, $(a\ b)$ is an integral solution of $1765x^2 + 1 = y^2$, where

$$a = \frac{\alpha\beta}{2}(\beta^2 + 1)(\beta^2 + 3) = \frac{2(84)}{2}\left[(84)^2 + 1\right]\left[(84)^2 + 3\right]$$

$$= (84)[7057][7059] = 84(49,815,363) = 199,261,452$$

(Continued)

Example 4 (cont'd)

$$b = (\beta^2 + 2)\left[\frac{1}{2}(\beta^2 + 1)(\beta^2 + 3) - 1\right]$$

$$= \{(84)^2 + 2\}\left\{\frac{1}{2}(7057)(7059) - 1\right\}$$

$$= \left\{\frac{(84)^2 + 2}{2}\right\}\{49, 815, 363 - 2\}$$

$$= \frac{7058}{2}(49, 815, 361) = (3529)(49, 815, 361)$$

$$= 173, 798, 408, 969$$

Hence, (199,261,452 173,798,408,969) is an integral solution of $1765x^2 + 1 = y^2$.

As remarked before, repeated application of "Samasa" of (199,261,452 1,737,984,089,691) gives the infinite solutions of (4).

Example 5

Find the integral solutions of the equation

$$291x^2 + 1 = y^2 \tag{1}$$

We have the identity $291x^2 - 2 = 289 = (17)^2$.

So, (1 17) is an integral solution of the equation $291(1)^2 - 2 = y^2$.

Then, by Brahmagupta's theorem 2, (a b) is an integral solution of $291x^2 + 1 = y^2$, where

$$\alpha = \alpha\beta = 1(17) = 17$$

$$b = \frac{\beta^2 + N\alpha}{2} = \frac{(17)^2 + 291 \cdot 1}{2} = \frac{289 + 291}{2} = 290$$

Thus, (17 290) is an integral solution of $291x^2 + 1 = y^2$.

As remarked before, repeated application of "Samasa" of (17 290 1) over itself will give the infinite solutions of (1).

Example 6

Find an integral solution of the equation $2499x^2 + 1 = y^2$.

We have the identity $2499 \cdot 2^2 + 4 = 9996 + 4 = 10{,}000 = (100)^2$.

So, $(\alpha\ \beta) = (2\ 100)$ is an integral solution of $2499x^2 + 4 = y^2$.

So, by Brahmagupta's theorem 3, since $\beta = 100$ is even, $(a\ b)$ is an integral solution of $2499x^2 + 1 = y^2$, where

$$a = \frac{\alpha\beta}{2} = \frac{2 \cdot 100}{2} = 100$$

$$b = \frac{\beta^2 - 2}{2} = \frac{10{,}000 - 2}{2} = 4999$$

Thus, $(100\ 4999)$ is an integral solution of $2499x^2 + 1 = y^2$.

Repeated applications of Samasa of $(100\ 4999\ 1)$ gives infinite solutions of $2499x^2 + 1 = y^2$.

Linear Equations with Several Unknowns

Brahmagupta has treated the following system of equations:

$$\sum_{r=1}^{n} x_r \pm cx_1 = a_1 \quad (1)$$

$$\sum_{r=1}^{n} x_r \pm cx_2 = a_2 \quad (2)$$

$$\vdots$$

$$\sum_{r=1}^{n} x_r \pm cx_r = a_r \quad (r)$$

$$\sum_{r=1}^{n} x_r \pm cx_n = a_n \quad (n)$$

Adding (1), (2),. . ., (n), we get

$$n\sum_{r=1}^{n} x_r \pm c\sum_{r=1}^{n} x_r = \sum_{r=1}^{n} a_r$$

i.e.,

$$\sum_{r=1}^{n} x_r[n \pm c] = \sum_{r=1}^{n} a_r$$

So,

$$\sum_{r=1}^{n} x_r = \frac{\sum_{r=1}^{n} a_r}{n \pm c}$$

Now from equation (r),

$$\pm cx_r = a_r - \sum_{r=l}^{n} x_r = a_r - \frac{\sum_{r=1}^{n} a_r}{n \pm c}$$

Hence,

$$x_r = \frac{1}{c}\left[\pm a_r \mp \frac{\sum_{r=1}^{n} a_r}{n \pm c}\right], \quad r = 1, 2, \ldots, n$$

Quadratic Equations

Brahmagupta has given two rules for the solution of the quadratic equation $ax^2 + bx - c = 0$.

In the first, he gives

$$x = \frac{\sqrt{b^2 + 4ac} - b}{2a}$$

In the second, he gives

$$x = \frac{\sqrt{ac + (b/2)^2} - (b/2)}{a}$$

We find the above rules are identically the same as that of Aryabhata I.

The root of the quadratic equation for the number of terms n of an arithmetic progression has been given by Brahmagupta as

$$n = \frac{\sqrt{8ds + (2a-d)^2} - (2a - d)}{2d}$$

where a is the first term, d the common difference, and s the sum of n terms. We observe that is exactly the same as given by Bakshali work.

Brahmagupta gave the following example.

Example (Brahmagupta)

A certain sum A is lent out for a period t_1. The interest accrued x is lent out again at the same rate as before for another period t_2. If total interest accrued is B. Find x.

Let $r\%$ be the rate of interest.

Since x is the interest accrued on amount A for a period t_1 at the interest rate of $r\%$, we have

$$x = Art_1 \Rightarrow r = \frac{x}{At_1} \tag{1}$$

Let x_1 be the interest accrued on amount x for a period t_2 at the same rate of interest $r\%$. Then,

$$x_1 = xrt_2 = xt_2\left(\frac{x}{At_1}\right) \quad \text{from (1)}$$

So,

$$x_1 = \frac{x^2 t_2}{At_1}$$

Now,

$$B = x + x_1 = x + \frac{x^2 t_2}{At_1}$$

i.e.,

$$\frac{x^2 t_2}{At_1} + x - B = 0$$

i.e.,

$$x^2 t_2 + xAt_1 - BAt_1 = 0$$

So,

$$x = \frac{-At_1 + \sqrt{(At_1)^2 + 4t_2(BAt_1)}}{2t_2}, \quad \text{since } x > 0$$

(*Continued*)

Example (Brahmagupta) (cont'd)

i.e.,

$$x = \frac{-At_1}{2t_2} + \sqrt{\left(\frac{At_1}{2t_2}\right)^2 + \frac{BAt_1}{t_2}}$$

which is exactly in the form given by Brahmagupta.

Astronomical Problem

Brahmagupta has also tackled a certain astronomical problem, where the following quadratic equation occurs:

$$(72 + a^2)x^2 \pm 24abx = 144\left(\frac{R^2}{2} - b^2\right) \quad (*)$$

where a = "agra" (the sine of the amplitude of the sun), b = "Palabha" (the equinoxial shadow of gnomon, which is 12 anguli long), R = radius of the sun, and x = "Konasanku" (the sine of the altitude of the sun, when his altitude is 45°).

Dividing (*) by $72 + a^2$, we get

$$x^2 \pm \frac{24ab}{72 + a^2}x = \frac{144((R^2/2) - b^2)}{72 + a^2}$$

or

$$x^2 \pm 2mx = n$$

where

$$m = \frac{12ab}{72 + a^2}$$

and

$$n = \frac{144((R^2/2) - b^2)}{72 + a^2}$$

So, we have $x^2 \pm 2mx - n = 0$.

Hence,

$$x = \frac{-2m + \sqrt{4m^2 + 4n}}{2}, \quad \text{since } x > 0$$

$$= -m + \sqrt{m^2 + n}, \quad \text{since } x > 0$$

as stated by Brahmagupta. The same result has also been provided by Suryasiddhanta (AD 300) and Sripati (AD 1039).

Brahmagupta gave the following two examples.

Example 1 (Brahmagupta)

"The square root of the residue of the revolution of the sun less 2 is diminished by 1, the resulting number is multiplied by 10 and added by 2. When will this be equal to the residue of the revolution of the sun minus 1?"
(Br. Sp. Si. VIII. 49)

Brahmagupta gave the residue of the revolution of the sun as 83. The proof has been provided by Brahmagupta's commentator Prthudakaswami in this book.

Example 2 (Brahmagupta)

"When will the square of one fourth the residue of the exceeding months minus three, be equal to the residue of the exceeding months?"
(Br. Sp. Si. XVIII. 50)

Brahmagupta gave the residue of the exceeding months as 4. Prthudakaswami has given a proof of this example in this book.

Brahmagupta gave the solution of $x-y=a$ and $xy=b$ as:

$$x=\frac{\sqrt{(x-y)^2+2^2xy}+x-y}{2}=\frac{\sqrt{a^2+4b}+a}{2}$$

$$y=\frac{\sqrt{(x-y)^2+2^2xy}-(x-y)}{2}=\frac{\sqrt{a^2+4b}-a}{2}$$

Brahmagupta gave the solution of $x^2+y^2=c$ and $x+y=d$ as:

$$x=\frac{x+y+\sqrt{2(x^2+y^2)-(x+y)^2}}{2}=\frac{d+\sqrt{2c-d^2}}{2}$$

$$y=x+y-\frac{\sqrt{2(x^2+y^2)-(x+y)^2}}{2}=\frac{d-\sqrt{2c-d^2}}{2}$$

(*Continued*)

Example 2 (Brahmagupta) (cont'd)

Brahmagupta gave the solution of $x^2 - y^2 = m$ and $x - y = n$ as:

$$x = \frac{1}{2}\left[\frac{x^2 - y^2}{x - y} + x - y\right] = \frac{1}{2}\left[\frac{m}{n} + n\right]$$

$$y = \frac{1}{2}\left[\frac{x^2 - y^2}{x - y} - (x - y)\right] = \frac{1}{2}\left[\frac{m}{n} - n\right]$$

Brahmagupta has proposed the following astronomical problem of a linear equation:

(*) $197x - 1644y - z = 6302$ (Brahma – Sphuta Siddhanta xviii 55)

Let $z = 3x + 8y$.

Then (*) becomes $197x - 1644y - (3x + 8y) = 6302$.

i.e.,

$194x - 1652y = 6302$ or $194x - 6302 = 1652y$.

Dividing by 2, this becomes

$$97x - 3151 = 826y$$

We now solve this first-order indeterminate equation

```
826) 97 (0
      0
     ----
     97) 826 (8
         776
         ----
          50) 97 (1
              50
              ----
              47) 50 (1
                  47
                  ----
                   3) 47 (15
                      45
                      ----
                       2) 3 (1
                          2
                          --
                          1
```

So, the number of quotients is 6 and so even.

To continue to solve the equation $97x - 3151 = 826y$, we now form the following table:

0	0	0	0	0	0	103,983
8	8	8	8	8	885,431	885,431
1	1	1	1	103,983	103,983	103,983
1	1	1	53,567	53,567		
15	15	50,416	50,416			
1	3151	3151				
3151	3151					
0	0					

So,

$$y = 103,983 \equiv 96 (\text{mod } 97)$$

$$x = 885,431 \equiv 785 (\text{mod } 826)$$

So, $(x, \ y) = (785 \ 96)$ is the least solution of the equation $97x + 3151 = 826y$. Thus, $(x \ y) = (826 - 785 \ 97 - 96)$ is the least solution of $97x - 3151 = 826y$. i.e., $(x \ \ y) = (41 \ 1)$ is the least solution of $97x - 3151 = 826y$.

General solution of $97x - 3151 = 826y$ is given by $x = 826m + 41$ and $y = 97m + 1$, where m is an arbitrary integer.

Now,

$$z = 3x + 8y = 3(826m + 41) + 8(97m + 1)$$

$$= 2478m + 123 + 776m + 8$$
$$= 3254m + 131$$

Thus, general solution of (*) is

$$x = 826m + 41$$
$$y = 97m + 1$$

and

$$z = 3254m + 131$$

where m is an arbitrary integer. The least solution of (*) is given by

$$x = 41, \quad y = 1, \quad \text{and} \quad z = 131$$

Quadrilateral

Brahmagupta treated triangle in detail, but oddly enough, to him, the triangle is a quadrilateral with one side equal to zero.

स्थूलफल त्रिचतुर्भुजबाहुर्प्रतिबाहुयोगदलधातः ।
भुजयोगार्धचतुष्टयभुजोनधातात् पदं सूक्ष्मम् ॥

(Br. Sp. Si. XII. 21)

This means that the "Sthula" (gross) area of three- or four-sided figure is equal to the product of half the sum of the opposite sides. The second part of the verse says that the exact area is equal to the square root of the product of four factors of half the sum of sides each respectively decreased by one side.

So, if a, b, c, and d are the four sides of a quadrilateral, the "Sthula" (gross) A^* area of that quadrilateral

$$= \frac{a+c}{2} \frac{b+d}{2}$$

The exact area A of the quadrilateral is

$$= \sqrt{(s-a)(s-b)(s-c)(s-d)}$$

where $2s = a + b + c + d$.

In the case of a triangle, one side $d = 0$.

Hence, the "Sthula" (gross) area of the triangle is b/2 (a + c)/2 and the exact area of the triangle $= \sqrt{s(s-a)(s-b)(s-c)}$, a formula presently credited to Heron (between second century BC and third century AD) of Alexandria. Brahmagupta does not give proof of his formulae or theorems.

Theorem on "Abadhas" (Brahmagupta)

भुजकृत्यन्तरहृतहीनयुता भूर्द्विभाजितबाधे।
स्वाबाघावर्गोनात् भुजवर्गान्मूलमवलम्बः॥

(Br. Sp. Si. XII. 21)

i.e., "The base decreased by and added with the difference of the squares of the sides divided by the base and the result is divided by two gives the two 'abadhas' (the segments of the base made by the altitude). Also, the square root of the side decreased by the square of its 'abadha' gives the altitude."

Let AD be the altitude from A on the base BC of the triangle ABC. Then, BD and DC are the two "abadhas." Let $BC = a$, $AB = c$, and $AC = b$

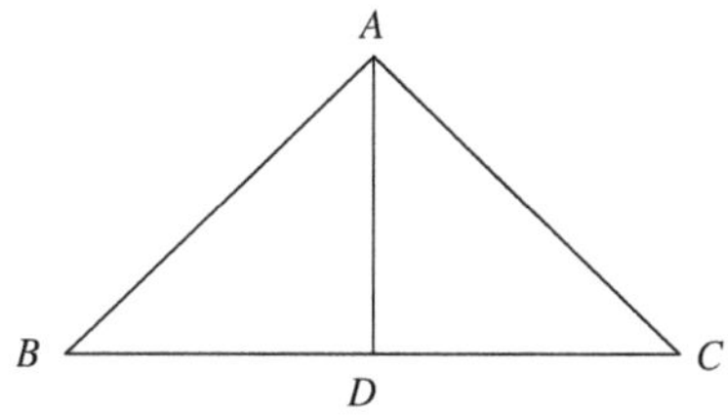

$$BD \text{ or } DC = \frac{1}{2}\left(a \pm \frac{b^2 \sim c^2}{a}\right) \qquad \text{(i)}$$

$$\text{Altitude } AD = \sqrt{AB^2 - BD^2} \text{ or } \sqrt{AC^2 - DC^2} \qquad \text{(ii)}$$

For proof of (i), see Bhaskara I in this book.

Theorem on the Circum Radius of a Triangle (Brahmagupta)

त्रिभुजस्य वधो भुजयोर्द्विगुणितलम्बोद्धृ तो हृदय-रज्जुः।
सा द्विगुणा त्रिचतुर्भुजकोणस्पृगवृत्तविष्कम्भः॥

(Br. Sp. Si. XII. 27)

i.e., "In a triangle, the product of the sides divided by twice the altitude gives the circum radius and twice that is the circum diameter of the circum circle."

If r is the circum radius of the triangle ABC and AD is its altitude, then $r = (AB \cdot AC)/2AD$. Let AE be a diameter through the center O of the circum circle proof by present methods:

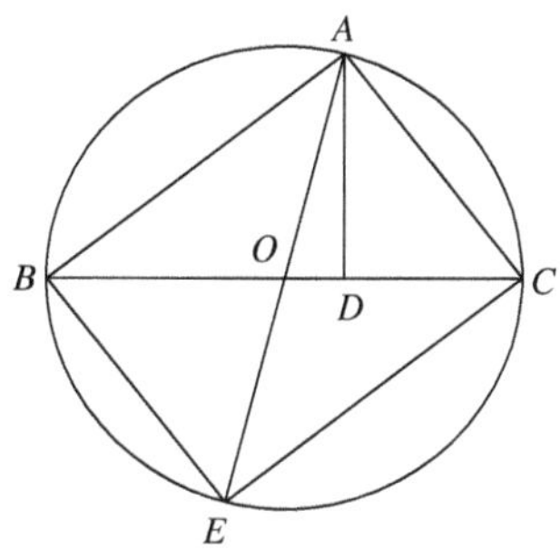

Join BE.

Now in triangles ABE and ADC

$$A\hat{D}C = 90^\circ = A\hat{B}E \quad \text{(angle in a semicircle)}$$

$$A\hat{E}B = A\hat{C}B \quad \text{(angles in the same segment)} = A\hat{C}D$$

Hence, the ΔABE and ADC are similar.
So,

$$\frac{AB}{AE} = \frac{AD}{AC}$$

So,

$$AE = \frac{AB \cdot AC}{AD} = 2r$$

since AE is the diameter of the circumcircle.

It may be observed here that Brahmagupta might have used the method given later in the book under Jyesthadeva.

Brahmagupta's Solution for the Rational Isosceles Triangle

कृतियुतिरसदृशराश्येर्बाहुर्घातो द्विसंगुणो लम्बः ।
कृत्यन्तरमसदृशयोर्द्विगुणं द्विसमत्त्रिभुजभूमिः ॥

(Br. Sp. Si. XII. 33)

i.e., "In an isosceles triangle, the sum of the squares of two unequal rational numbers is the side, twice the difference of the squares of the two unequal numbers is the base and twice the product of the two unequal numbers is the altitude."

If m and n are two unequal rational numbers, the triangle is obtained by the juxtaposition of two equal rational right triangles of sides $m^2 - n^2$ and $2mn$ with hypotenuse $m^2 + n^2$ and the sides $2mn$ being made to coincide with each other.

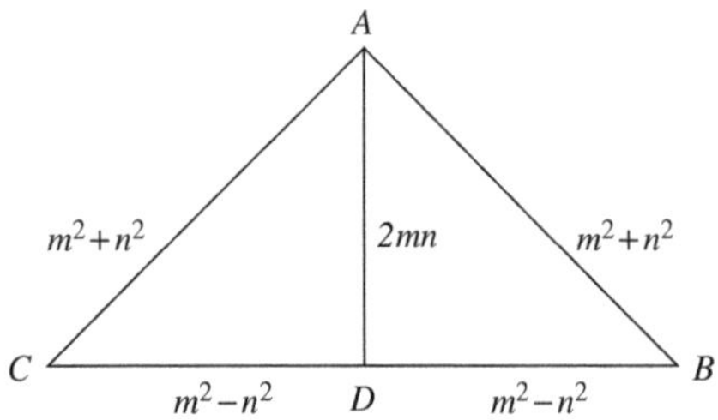

To Construct a Rational Scalene Triangle (Brahmagupta)

इष्टद्वयेन भक्तो द्विधेष्टवर्गः फलेष्टयोगार्धे ।
विषमत्त्रिभुजस्य भूजाविष्टोनफलाधंयोगो भूः ॥

(Br. Sp. Si. XII. 34)

i.e., "Divide the square of an optional number by two other optional numbers separately. Halves of the sum of the quotients and respective divisor optional number will be the sides of the scalene triangle. Half the sum of the quotients decreased by respective divisor optional number will be the base."

Let m, p, and q be three unequal rational numbers.

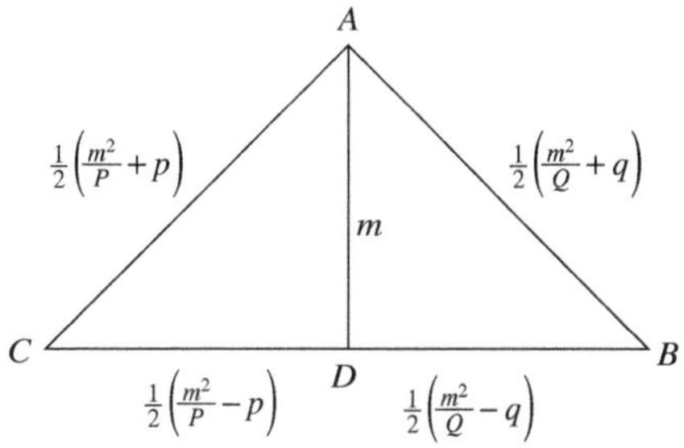

Then the sides of scalene rational triangle are

$$\frac{1}{2}\left(\frac{m^2}{p}+p\right)$$

and

$$\frac{1}{2}\left(\frac{m^2}{q}+q\right)$$

and the base is

$$\frac{1}{2}\left(\frac{m^2}{p}-p\right)+\frac{1}{2}\left(\frac{m^2}{q}-q\right)$$

So, the rational scalene triangle is obtained by the juxtaposition of two right triangles with common perpendicular side m, which becomes the altitude of the required scalene triangle.

Then it becomes a problem of obtaining a general solution for the construction of the right triangle, when one side about the right angle is given. This is given in the following verse:

इष्टस्य भुजस्य कृतिर्भक्तोनेष्टेन तद् दलं कोटिः ।
त्र्यायतचतुरश्रस्य क्षे त्रस्येष्टाधिकः कर्णः ॥

(Br. Sp. Si. XII. 35)

i.e., "The square of the given number divided by an optional number and added by the same optional number and halved is the hypotenuse and the same quotient diminished by same optional number and halved is the base."

If a is the given side (perpendicular side) and if m^* is the optional rational number $\neq 0$,

$$\frac{1}{2}\left(\frac{a^2}{m^*}+m^*\right)$$

is the hypotenuse and

$$\frac{1}{2}\left(\frac{a^2}{m^*}-m^*\right)$$

is the base.

Proof: Suppose b is the base and c be the hypotenuse.
Then $a^2 = c^2 - b^2 = (c+b)(c-b)$.
Let $c - b = m^*$.
Then,

$$c+b=\frac{a^2}{m^*}$$

Adding, we get

$$2c = m^* + \frac{a^2}{m^*} \Rightarrow c = \frac{1}{2}\left(m^* + \frac{a^2}{m^*}\right)$$

Subtracting the first equation from the second, we get

$$2b = \frac{a^2}{m^*} - m^* \Rightarrow b = \frac{1}{2}\left(\frac{a^2}{m^*} - m^*\right)$$

We juxtapose right triangle with $a = m$ and $m^* = p$ with another right triangle with $a = m$ and $m^* = q$ and get the scalene triangle with sides

$$\frac{1}{2}\left(p+\frac{m^2}{p}\right)$$

and

$$\frac{1}{2}\left(q+\frac{m^2}{q}\right)$$

and base

$$\frac{1}{2}\left(\frac{m^2}{p} - p\right) + \frac{1}{2}\left(\frac{m^2}{q} - q\right)$$

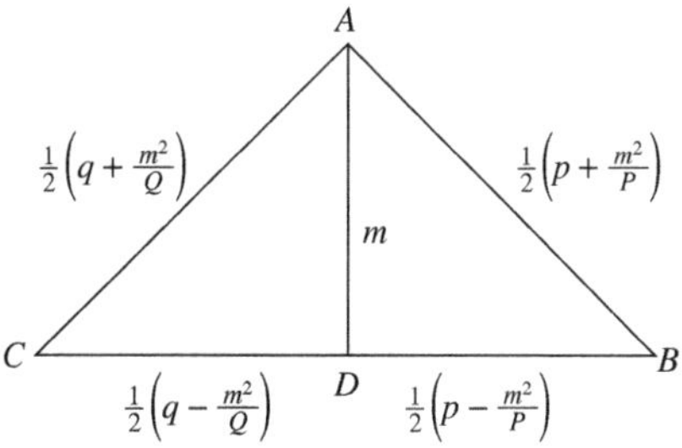

This triangle will have m as the altitude.

Construction of Cyclic Quadrilateral with Rational Sides

Suppose (a, b, c) and (p, q, r) be the sides of two right triangles such that $c^2 = a^2 + b^2$ and $r^2 = p^2 + q^2$.

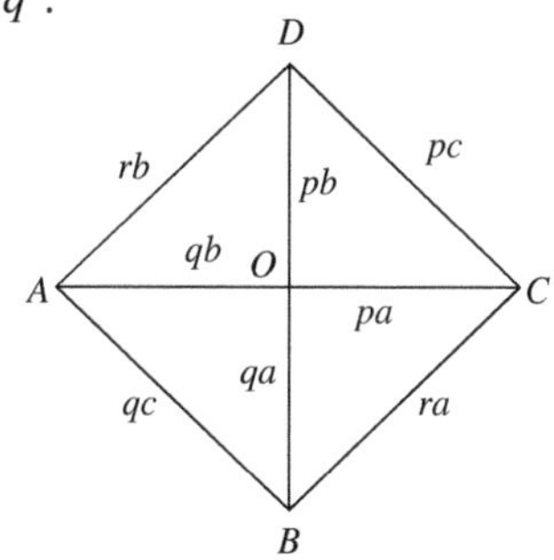

Construct the triangle BOC and COD with sides (pa, qa, ra) and (pa, pb, pc), respectively. On the other side of BD, construct triangles DOA and AOB with sides (pb, qb, rb) and (qa, qb, qc), respectively. We thus have a quadrilateral $ABCD$ with sides qc, ra, pc, rb in order and the diagonals BD and AC intersect at right angles.

We have to show $ABCD$ is a cyclic quadrilateral.

Proof:

$$R_1 = \text{Circum radius of the triangle } ABD = \frac{AD \cdot AB}{2OA}$$

$$= \frac{rb \cdot qc}{2qb} = \frac{1}{2} rc$$

Similarly,

$$R_2 = \text{Circum radius of the triangle } BCD = \frac{CD \cdot BC}{2OC}$$

$$= \frac{pc \cdot ra}{2pa} = \frac{rc}{2}$$

So, $R_1 = R_2$.

So, a circle with circumcenter, say O of the Δ^{le} ABD and radius $= rc/2$ will pass through the points A, B, C, and D, which means $ABCD$ is a cyclic quadrilateral.

To construct a trapezium with three equal sides.

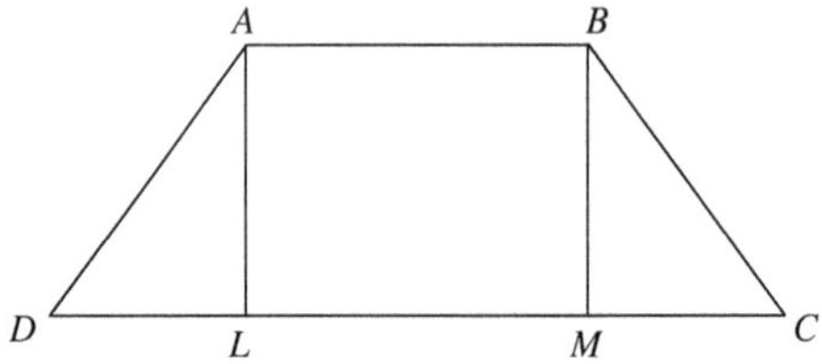

Let $ABML$ be a rectangle with $AB = a^2 + b^2$ and $AL = 2ab$. Produce LM both ways to C and D, respectively, such that $DL = MC = a^2 - b^2$.

Join AD and BC.

Then $ABCD$ is our required trapezium.

Proof: In right Δ^{le} ADL,

$$AD^2 = AL^2 + DL^2$$

$$= (2ab)^2 + (a^2 - b^2)^2$$
$$= 4a^2b^2 + a^4 - 2a^2b^2 + b^4$$
$$= a^4 + 2a^2b^2 + b^4 = (a^2 + b^2)^2$$

So, $AD = a^2 + b^2$.

Again, in right Δ^{le} BMC,

$$BC^2 = BM^2 + CM^2$$

$$= (2ab)^2 + (a^2 - b^2)^2$$
$$= (a^2 + b^2)^2$$

So, $BC = a^2 + b^2$.

So, since $LM \,||^{le}\, AB$, $ABCD$ is a trapezium.

Also, $AB = BC = AD = a^2 + b^2$.

Now base $DC = DL + LM + MC = a^2 - b^2 + a^2 + b^2 + a^2 - b^2 = 3a^2 - b^2$.

Thus, $ABCD$ is our required trapezium with three equal sides.

To Construct a Rational Isosceles Trapezium

त्र्यायतकर्णौ बाहू भुजकृतिरिष्टेन भाजितेष्टोना ।
द्विहृता कोट्यधिका भूर्मुखमूना द्विसमचतुरश्रे ॥

(Br. Sp. Si. XII. 36)

i.e., "The diagonals of the rectangle (generated) are the flank sides of an isosceles trapezium. The square of its 'bhuja' (base) is divided by an optional number and then lessened by that number and divided by two. The result increased by the 'Koti' (upright or the perpendicular side of the rectangle) is the base and lessened by it is the face."

Here, the "bhuja" is chosen to be the altitude *DE* of the isosceles trapezium *ABCD*.

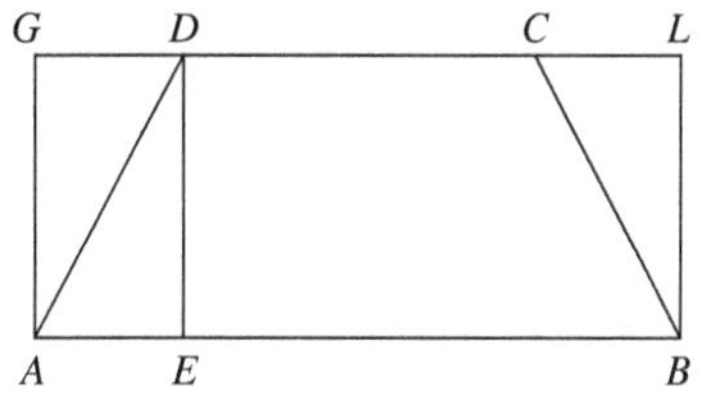

Let $DE = 2mn$.
"Koti"(upright) $= AE = m^2 - n^2$.
Flank side $= AD = BC = m^2 + n^2$.
Let P be the optional rational number $\neq 0$,

$$\frac{1}{2}\left[\frac{\text{Square of bhuja}}{P} - P\right] + \text{Koti} = \text{base } AB \text{ of the trapezium}$$

$$\frac{1}{2}\left[\frac{(DE)^2}{P} - P\right] + m^2 - n^2 = AB$$

Also,

$$\frac{1}{2}\left[\frac{(DE)^2}{P} - P\right] - (m^2 - n^2) = \text{face } CD$$

$$\frac{1}{2}\left[\frac{4m^2n^2}{P} - P\right] + m^2 - n^2 = AB$$

$$\frac{1}{2}\left[\frac{4m^2n^2}{P} - P\right] - (m^2 - n^2) = CD$$

$$EB = AB - AE = \frac{1}{2}\left[\frac{4m^2n^2}{P} - P\right] + (m^2 - n^2) - (m^2 - n^2)$$

$$= \frac{1}{2}\left[\frac{4m^2n^2}{P} - P\right]$$

$$AC = BD = \sqrt{EB^2 + DE^2}$$

$$= \sqrt{\frac{1}{4}\left(\frac{4m^2n^2}{P} - P\right)^2 + (2mn)^2}$$

$$= \frac{1}{2}\sqrt{\left(\frac{4m^2n^2}{P} - P\right)^2 + 16m^2n^2}$$

$$= \frac{1}{2}\sqrt{\frac{16m^4n^4}{P^2} - \frac{8m^2n^2}{P} \cdot P + 16m^2n^2 + P^2}$$

$$= \frac{1}{2}\sqrt{\frac{16m^4n^4}{P} + 8m^2n^2 + P^2}$$

$$= \frac{1}{2}\sqrt{\left(\frac{4m^2n^2}{P} + P\right)^2} = \frac{1}{2}\left(\frac{4m^2n^2}{P} + P\right)$$

Area of the isosceles trapezium $ABCD$

$$= \text{Area of the triangle } ABC + \text{Area of the triangle } CAD$$

$$= \frac{1}{2}AB \cdot DE + \frac{1}{2}CD \cdot DE$$

$$= \frac{1}{2}DE(AB + CD)$$

$$= \frac{1}{2}(2mn)\left[\frac{4m^2n^2}{P} - P\right]$$

$$= mn\left[\frac{4m^2n^2}{P} - P\right]$$

The two rectangles $DEBL$ and $AEDG$ are juxtaposed and then the right triangles AGD and BCL are cut off from either end of the resultant rectangle $ABLG$.

Inscribed Quadrilaterals

Brahmagupta described an ingenious method to obtain all quadrilaterals that can be inscribed within circles having sides, diagonals, perpendiculars, segments areas, and the diameters of circumscribed circles as integers. Such quadrilaterals are called Brahmagupta quadrilaterals.

Brahmagupta gave the following solution for the formation of quadrilaterals that can be inscribed in circles.

Let the sides of the two right Δ^{les} be

$$m^2 - n^2,\ 2mn,\ m^2 + n^2 \text{ and } a^2 - b^2,\ 2ab,\ a^2 + b^2$$

where m, n, a, and b are all integers.

Then, the sides of Brahmagupta quadrilaterals are $(m^2 - n^2)(a^2 + b^2)$, $(a^2 - b^2)(m^2 + n^2)$, $2mn(a^2 + b^2)$, and $2ab(m^2 + n^2)$.

Proof: Let $(a', b', c') = (m^2 - n^2, 2mn, m^2 + n^2)$ and $(p, q, r) = (a^2 - b^2, 2ab, a^2 + b^2)$.

Since m, n, a, and b are integers, so are a', b', c', p, q, and r. Hence, a', b', c', p, q, and r are also rational numbers.

Then, construct the cyclic quadrilateral with rational sides as on page 189.

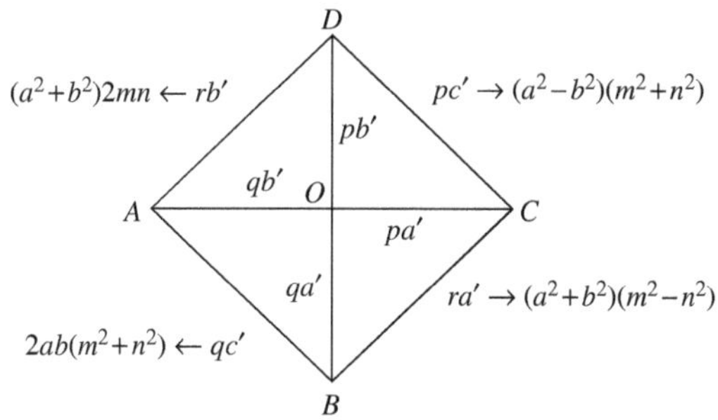

Then, the sides BC, CD, AD, and AB of the rational cyclic quadrilateral are, respectively, ra', pc', rb', and qc', which are integers. The diagonals AC and BD are, respectively, $qb' + pa'$ and $pb' + qa'$, which are integers. The perpendicular DO, OB, AO, and OC are, respectively, pb', qa', qb', and pa', which are integers.

The circum diameter $= rc-$, which is an integer.

Segment areas are the following:

Area of the triangle AOD

$$= \frac{1}{2} AO \cdot OD = \frac{1}{2} qb' \cdot pb' = \text{an integer}$$

Area of the triangle *DOC*

$$= \frac{1}{2} OC \cdot OB = \frac{1}{2} pa' \cdot pb' = \text{an integer}$$

Area of the triangle *AOB*

$$= \frac{1}{2} AO \cdot OB = \frac{1}{2} qb' \cdot pa' = \text{an integer}$$

Area of the triangle *BOC*

$$= \frac{1}{2} OB \cdot OC = \frac{1}{2} qa' \cdot pa' = \text{an integer}$$

The Formula for the Volume of a Pyramid (Brahmagupta)

क्षेत्त्रफलं वेधगु णं समखातफलं हृतं त्त्रिभिः सूच्याः ।
भुखतलतुल्यभुजैक्यान्येकाग्रहृतानि समरज्जुः ॥

(Br. Sp. Si. XII. 44)

i.e., "the volume of pit of uniform depth is equal to the depth multiplied by the area of the section. When this product is divided by 3 gives the volume of 'Suci,' a figure tapering to a point."

The first part of the formula gives the volume of any regular solid of uniform height. The second part of the formula gives the volume when the regular solid tapers uniformly to a point as in the case of a pyramid or cone. Brahmagupta has not provided any method of derivation of this formula.

Frustum

Brahmagupta was the first person to give a formula for finding the volume of a frustum.

मुखतलयुतिदलगुणितं वेधगु णं व्यावहारिकं गणितम् ।
मुखतलन्गणितैक्यार्घं वेधगुणं स्याद् गणितमौत्त्रम् ॥
त्त्रौत्त्रगणिताद्विशोध्य व्यवहारफलं भजेत्त्रिभिः शेषम् ।
लब्धं व्यवहारफले प्रक्षिप्य भवति फलं सूक्ष्मम् ॥

(Br. Sp. Si. XII. 45, 46)

i.e., "If the product of the sum of the sides in the face and the base is multiplied by the depth, it gives 'Vyavaharika' (business measure) volume. If half the sum of the areas of the face and base is multiplied by the depth, it gives 'autra' volume. If 'autra' volume minus the 'vyavarika' volume is divided by 3 and the quotient is added to the 'Vyavaharika' volume, it gives the exact volume of the frustrum."

Here, the one rough value of the volume obtained by multiplying the depth and the area of section got from the average length of the sides is called "Vyavaharika volume." Then, the other rough value of the volume obtained by multiplying the depth and the average of the areas of the face and the base is called "autra" volume. If these are v_v and v_a, respectively, the exact volume v is given by

$$v = v_v + \frac{v_a - v_v}{3}$$

Now v_v (for a pyramid frustum)

$$= \left(\frac{a+b}{2}\right)^2 \times h$$

where a and b are the sides of the base and face and h is the height. But v_a (for a pyramid frustum)

$$= \left(\frac{a^2 + b^2}{2}\right) h$$

So,

$$v = v_v + \frac{v_a - v_v}{3} = \left(\frac{a+b}{2}\right)^2 h + \frac{((a^2 + b^2)/2)h - ((a+b)/2)^2 h}{3}$$

Hence,

$$v = h\left[\frac{a^2 + b^2 + 2ab}{4} + \frac{2a^2 + 2b^2 - (a^2 + b^2 + 2ab)}{12}\right]$$

$$= h\left[\frac{3a^2 + 3b^2 + 6ab + a^2 + b^2 - 2ab}{12}\right]$$

$$= h\left[\frac{4a^2 + 4b^2 + 4ab}{12}\right] = \frac{h}{3}\left[a^2 + b^2 + ab\right]$$

which is the correct formula for the frustrum.

Brahmagupta does not give a formula for the area of a trapezium. He points out some other properties.

अविषमचतुरस्रभुजप्रतिभुजवधयोर्युतेः पदं कर्णः ।
कर्णकृतिर्भूमुखयुतिदलवर्गोना पदं लम्ब ॥

(Br. Sp. Si. XII. 23)

i.e., “In ‘avisma’ quadrilaterals, the square root of the sum of the products of the opposite sides is the diagonal. The square root of square of the diagonal minus the square of half the sum of the base and face is the altitude.”

“Avisma” quadrilateral includes the square, the rectangle, and the isosceles trapezium. The proof of the above result is given by Brahmagupta’s commentator Prthudakaswami.

Circum Radius of an Isosceles Trapezium

Brahmagupta gave the following:

त्र्यविषमपर्श्वभुजगुणः कर्णौ द्विगुणावलमबकविभक्तः हृदयं . . . (ड्लृल्. २६)

i.e., “The product of the diagonal of an isosceles trapezium and the lateral side divided by twice the altitude is equal to circum radius of the isosceles trapezium.”

In the figure, let *ABCD* be an isosceles trapezium. Let *DE* be the altitude. From the triangle *ADB*, $R = (BD \cdot AD)/2DE$, where *R* is the circum radius of the triangle *ADB*.

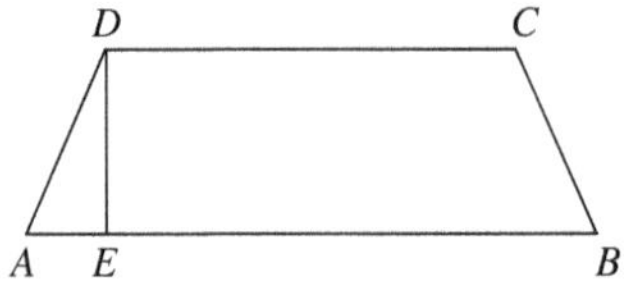

R is also the circum radius of the isosceles trapezium *ABCD*.

Some of the rules given by Brahmagupta for cyclic quadrilateral are also applicable for any quadrilateral.

(१) विषमचतुरश्रमध्ये विषमत्त्रिभूजद्वयं प्रकल्प्य पृथक् ।
कर्णद्वयेन पूर्ववदाबाधे लम्बकौ च पृथक् ॥

(Br. Sp. Si. XII. 29)

i.e., “Considering two scalene triangles in a scalene quadrilateral separately with two diagonals, one should be able to find the altitude and ‘abadhas.’ If *ABCD* is a scalene quadrilateral, let *DE* and *CF* be the altitudes from *D* and *C*, respectively. Let *a*, *b*, *c*, *d* be the four sides of *AB*, *BC*, *CD*, and *DA*, respectively.”

In Δ^{le} *ADB*,

$$DE = \frac{AD \cdot DB}{2R_1}$$

where R_1 is the circum radius of the ΔADB

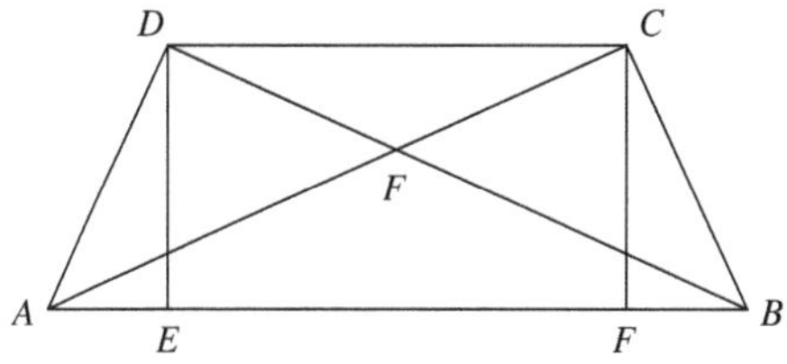

Again, in Δ^{le} ABC,

$$CF = \frac{AC \cdot BC}{2R_2}$$

where R_2 is the circum radius of the ΔABC.

Also,

$$AE = \sqrt{AD^2 - DE^2}$$

and

$$FB = \sqrt{BC^2 - CF^2}$$

(२) विषमभुजान्तस्त्रिभुजे प्रकल्प्य कर्णों भुवौ तदाबाधे ।
पृथगूर्ध्वाधरखण्डे कर्णयुतौ कर्णयोरधरे ॥
त्रिभुजे भुजौ तु भूमिस्तल्लम्बो लम्बकधरं खण्डम् ।
ऊर्ध्वमवलम्बकखंडं लम्बयोगार्धमधरोनम् ॥

(Br. Sp. Si. XII. 30–31)

i.e., "In the triangles inside a scalene quadrilateral, treat the diagonals as the bases and calculate the 'abadhas' which happen to be the upper and lower parts of the diagonals separately; in the triangle below the point of intersection of the diagonals, the lower parts of the diagonals are sides and the base of the quadrilateral is the base of the triangle."

Brahmagupta's most significant contribution to the geometry of cyclic quadrilateral is the following two theorems.

Theorem 1: Theorem on the area of a cyclic quadrilateral

स्थूलफलं त्रिचतुर्भुजबाहुप्रतिबाहुयोगदलघातः ।
भुजयोगार्धचतुष्टयभुजोनघातात् पदं सूक्ष्मम् ॥

(Br. Sp. Si. XII. 21)

The area A of a cyclic quadrilateral is given by

$$A = \sqrt{(s-a)(s-b)(s-c)(s-d)}$$

where a, b, c, and d are the four sides of the cyclic quadrilateral and where $2s = a + b + c + d$. The area B of a triangle is given by

$$B = \sqrt{s(s-a)(s-b)(s-c)}$$

where a, b, and c are three sides of the triangle and where $2s = a + b + c$. As mentioned before, the formula for the area of the triangle is called Heron's formula. The formula for the area of a cyclic quadrilateral was rediscovered by W. Snell in AD 1619, nearly thousand years later. The proof of this theorem has been given by Jyesthedeva.

Theorem 2: Theorem on the diagonals of a cyclic quadrilateral

Brahmagupta stated this theorem as follows:

"Divide mutually the sum of the products of the adjacent sides (of the cyclic quadrilateral) and multiply the resulting quotients by the sum of the product of the opposite sides (of the cyclic quadrilateral). The square roots of these products will be the diagonals of the cyclic quadrilateral."

Let a, b, c, d be the sides of the cyclic quadrilateral $ABCD$ as in the figure. Then AC and BD are two diagonals of cyclic quadrilateral $PQRS$.

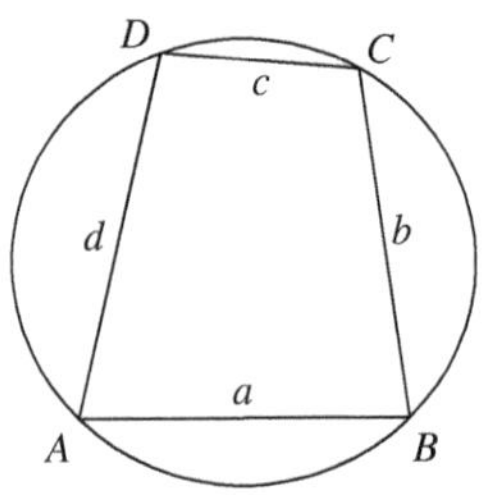

Then,

$$BD = \left\{\frac{(ab+cd)(ac+bd)}{ad+bc}\right\}^{\frac{1}{2}}$$

and

$$AC = \left\{\frac{(ad+bc)(ac+bd)}{ab+cd}\right\}^{\frac{1}{2}}$$

It is possible that Brahmagupta obtained the above results from the special cyclic quadrilaterals he has constructed on page 189. From the cyclic quadrilateral $ABCD$,

$$AB \cdot BC + AD \cdot DC = qcra + rbpc = cr(aq + bp) = cr \cdot BD$$

$$AB \cdot AD + CB \cdot CD = qcrb + rapc = cr(bq + ap) = cr \cdot AC$$

So,

$$\frac{AB \cdot BC + AD \cdot DC}{AB \cdot AD + CB \cdot CD} = \frac{BD}{AC} \tag{1}$$

Also, $AB \cdot CD + AD \cdot CB = qc \cdot pc + rb \cdot ra = pqc^2 + r^2ab$

$$\begin{aligned}
&= pq(a^2 + b^2) + (p^2 + q^2)ab = pqa^2 + pqb^2 + p^2ab + q^2ab \\
&= (pqa^2 + p^2ab) + (pqb^2 + q^2ab) \\
&= ap[aq + bp] + bq[bp + aq] \\
&= (aq + bp)(ap + bq) = BD \cdot AC
\end{aligned}$$

So,

$$AB \cdot CD + AD \cdot CB = AC \cdot BD \tag{2}$$

We observe that (2) verifies Ptolemy's theorem.
Multiplying (1) and (2), we get

$$BD^2 = \frac{AB \cdot BC + AD \cdot DC}{AB \cdot AD + CB \cdot CD}(AB \cdot CD + AD \cdot CB)$$

Dividing (2) by (1), we get

$$\frac{AB \cdot CD + AD \cdot CB}{AB \cdot BC + AD \cdot DC}(AB \cdot AD + CB \cdot CD) = AC^2$$

If $AB = a$, $BC = b$, $CD = c$, and $DA = d$, we have

$$BD^2 = \left(\frac{ab + cd}{ad + bc}\right)(ac + bd)$$

and

$$AC^2 = \left(\frac{ac + bd}{ab + cd}\right)(ad + bc)$$

So,

$$BD = \left\{\left(\frac{ab+cd}{ad+bc}\right)(ac+bd)\right\}^{\frac{1}{2}}$$

and

$$AC = \left\{\left(\frac{ac+bd}{ab+cd}\right)(ad+bc)\right\}^{\frac{1}{2}}$$

A rigorous proof of Brahmagupta's above theorem has been given by Jyesthedeva.

Shadow Problems

In chapter "sankucchādijnānadhāya" (chapter treating problems involving gnomon, shadow, etc.), Brahmagupta has included a section on how to determine the height and distance of the objects by observing their reflection in water.

Lemma 1:

युतदृष्टि गृहोच्च्यहृता ह्यन्तरभूमिदृ गौच्च्यसंगुणिता ।
फलभूर्न्यस्ते तोये प्रतिरूपाग्रं गृहस्य नरात् ॥

(Br. Sp. Si. XIX. 17)

i.e., "when the distance between the man and the house is divided by the sum of the heights of the house and the man's eyes and then when this quotient is multiplied by the height of his eyes, the tip of the image of the house will be seen, when the reflecting water is at a distance equal to the above product."

Let AB be the house (the object), let PQ be the height of the man's eyes. Let R be the reflecting point. Then, the man will be able to see the tip of the image, when

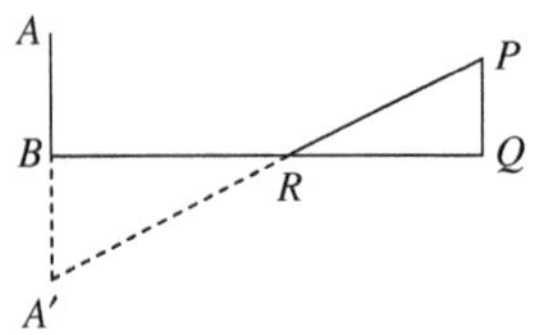

$$RQ = \frac{BQ \cdot PQ}{AB + PQ}$$

Proof: Now Δ^{les} $A'BR$ and PQR are similar.

So,

$$\frac{A'B}{PQ} = \frac{BR}{RQ} = K \quad \text{(say)} \qquad \text{(i)}$$

Adding 1 to both sides, we get

$$1 + \frac{A'B}{PQ} = \frac{BR}{RQ} + 1$$

$$\frac{PQ + A'B}{PQ} = \frac{BR + RQ}{RQ} = \frac{BQ}{RQ}$$

So,

$$RQ = \frac{BQ \cdot PQ}{A'B + PQ} = \frac{BQ \cdot PQ}{AB + PQ}$$

Also from (i),

$$AB = A'B = \frac{PQ \cdot BR}{RQ}$$

i.e., the height of house = $(PQ \cdot BR)/RQ$.

Lemma 2 (Brahmagupta): By observing from two different distances, the height and distance of the object can be found.

प्रथमद्वितीयनृजलान्तरेणोद्धृता जलापसृतिः ।
दृष्ट्यौच्चगुणोच्छायस्तोयान्तृजलान्तरगुणा भूः॥

(Br. Sp. Si. XIX. 19)

i.e., "The distance between the first and the second positions of water is divided by the difference between the distances of the man from the water. When the resulting quotient is multiplied by the height of the man's eyes, it gives the height of the house. On the other hand, when the above quotient is multiplied by the distance between the man and the water, it gives the distance between the water and house."

In the figure, let AB be the house. Let PQ and RS be the two positions of the man. Let C and D be two points of reflections.

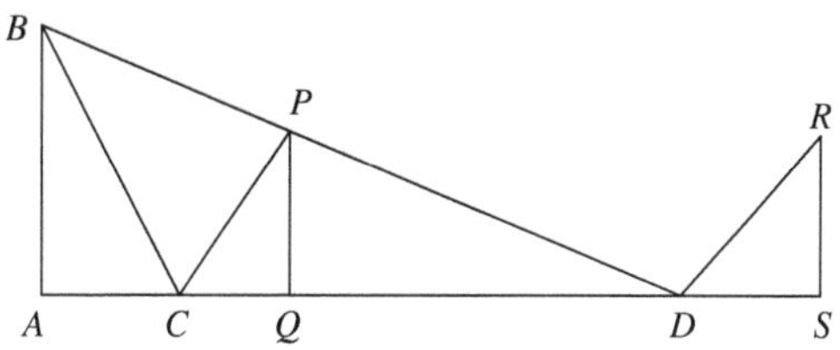

Then,

i. $AB = \dfrac{CD \cdot PQ}{DS - CQ}$

ii. $AC = \dfrac{CD \cdot CQ}{DS - CQ}$

Proof: From Lemma 1,

$$AB = \frac{PQ \cdot AC}{CQ} \Rightarrow \frac{AB}{PQ} = \frac{AC}{CQ}$$

Again from Lemma 1,

$$AB = \frac{RS \cdot AD}{DS} \Rightarrow \frac{AB}{RS} = \frac{AD}{DS}$$

Since $PQ = RS$, we have

$$\frac{AB}{PQ} = \frac{AC}{CQ} = \frac{AB}{RS} = \frac{AD}{DS} = K \text{ say } (*)$$

So, $AC = K \cdot CQ$ and $AD = K \cdot DS$

$$AD - AC = K(DS - CQ)$$

$$\frac{AD - AC}{DS - CQ} = K$$

So,

$$K = \frac{AB}{PQ} = \frac{AD - AC}{DS - CQ} = \frac{CD}{DS - CQ}$$

i.e.,

$$AB = \frac{PQ \cdot CD}{DS - CQ}$$

which proves (i).

Also from (*),

$$\frac{AC}{CQ} = K = \frac{CD}{DS - CQ}$$

Thus,

$$AC = \frac{CQ \cdot CD}{DS - CQ}$$

which proves (ii).

Brahmagupta also dealt with a problem combining shadow and reflection to obtain the height at which the light from a given source at a known height reflected from the water surface between the given source and the wall.

Lemma 2:

ज्ञातैश्छायापूरुषैर्विज्ञाते तोयकुडथयोर्विवरे ।
कुडद्ये त्र्प्रर्कतेजसो यो वेत्यारूढिं स तन्त्त्रज्ञः ॥

(Br. Sp. Si. XIX. 8)

i.e., "How can the ascent of the sun's rays on a wall from the given ratio of the shadow to the object and the distance between the water and the wall be calculated?"

The solution is the following:

छायापुरूषछिन्नं जलकुड् यान्त्तरमवाप्तमारूढिः।

(Br. Sp. Si. XIX. 20)

i.e., "The distance between the water and the wall divided by the ratio of the shadow to the object gives the height of the ascent."

Let *SM* be the incident ray striking the reflective surface at *M*. Let *MA* be the reflected ray striking the wall *BL* at *A*.

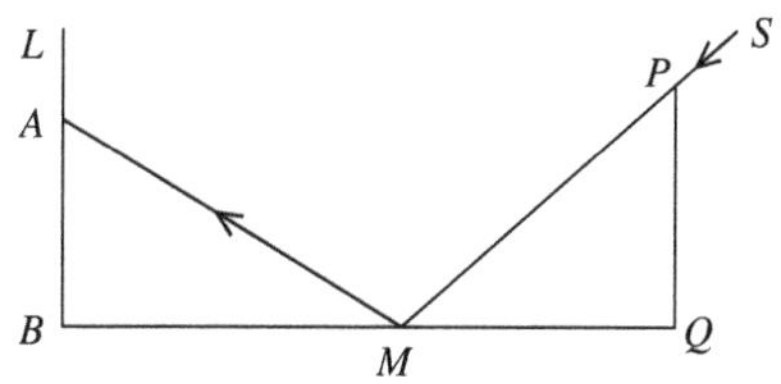

Let *PQ* be the gnomon in the path of incident say *SM*.

Here, *QM* is the shadow. Then,

$$AB = \frac{PQ \cdot BM}{MQ}$$

Proof: Now, Δ^{les} *ABM* and *PQM* are similar.

So, we have

$$\frac{AB}{PQ} = \frac{BM}{MQ}$$

So,

$$AB = \frac{PQ \cdot BM}{MQ}$$

Interpolation Formula

Brahmagupta initiated a new branch of Mathematics, Interpolation theory. When he was 67 years old, Brahmagupta, as mentioned earlier, wrote another Expository astronomical book called "Khandakhadyaka" in AD 665. In Chapter IX of this book, he introduces a new method of obtaining from a given table of sines consisting of tabulated values of six angles at equal intervals of 15°. Brahmagupta takes the radius R of the circle as 150 instead of 3438 used by Suryasiddhanta and Aryabhata I. Brahmagupta's work on Interpolation theory has been thoroughly discussed by P.C. Sengupta [122]. The English translation of Brahmagupta's "Khandakadyaka" with Puthrudakasvami's commentary has been brought out by P.C. Sengupta in 1941 and Brahmagupta's "Khandakadyaka" with Bhattotpala's commentary has been brought out in two volumes by Bina Chatterjee in 1970. The rule Brahmagupta employed to form this table is equivalent to the formula

$$f(a+xh)=f(a)+x\frac{\Delta f(a)+\Delta f(a-h)}{2}+\frac{x^2}{2!}[\Delta f(a-h)-\Delta f(a)]$$

i.e.,

$$f(a+xh)=f(a)+x\frac{\Delta f(a)+\Delta f(a-h)}{2}+\frac{x^2}{2!}\Delta^2 f(a-h)$$

up to second-order differences. So, Brahmagupta was the first mathematician in the history of world mathematics to introduce second-order difference interpolation nearly 10 centuries before the rediscovery of this formula which is presently known as the Newton–Sterling formula.

Brahmagupta has given the following table of the first and the second differences for R sine values ($R = 150$) at intervals of 15° of the arc.

Arc (°)	*R* sine	First Difference	Second Difference
0	0		
		39	
15	39		−3
		36	
30	75		−5
		31	
45	106		−7
		24	
60	130		−9
		15	
75	145		−10
		5	
90	150		

We will now give a few examples of finding R sine values ($R = 150$) of intermediate angles, with the help of Brahmagupta's table of R sines.

Example 1

Find the value of $R \sin 67°30'$ ($R = 150$).

Let $f(\theta) = R \sin \theta$, where $R = 150$.

Let $a = 60°$, then $f(60°) = R \sin 60° = 130$.

Let $h = 15°$ and

$$x = \frac{7(1/2)}{15} = \frac{1}{2}$$

Then,

$$xh = \frac{1}{2}(15°) = 7\frac{1°}{2} = 7°30'$$

$$a + xh = 60° + 7°30' = 67°30'$$

Now,

$$f(a + xh) = f(a) + x\frac{\Delta f(a) + \Delta f(a-h)}{2} + \frac{x^2}{2!}\Delta^2 f(a-h)$$

i.e.,

$$f(67°30') = f(60°) + \frac{1}{2}\frac{\Delta f(60°) + \Delta f(60° - 15°)}{2} + \frac{x^2}{2!}\Delta^2 f(60° - 15°)$$

$$R \sin 67°30' = R \sin 60° + \frac{1}{4}[\Delta(R \sin 60°) + \Delta(R \sin 45°)] + \frac{1^2}{8}\Delta^2(R \sin 45°)$$

$$R \sin 67°30' = 130 + \frac{1}{4}[15 + 24] + \frac{1}{8}(-9)$$

$$= 130 + \frac{39}{4} - \frac{9}{8} = 130 + \frac{78 - 9}{8} = 130 + \frac{69}{8}$$

$$= 138\frac{5}{8} = 138 \cdot 625$$

So, $138 \cdot 625$ is the value of $150 \sin 67°30'$.

(Continued)

Example 1 (cont'd)

Thus,

$$\sin 67°30' = \frac{138 \cdot 625}{150} = 0.924166$$

Actual value of $\sin 67°30' = 0.9239$ from modern sine tables.
This shows that Brahmagupta was reasonably accurate.

Example 2

Find the value of $R \sin 33°45'$ ($R = 150$).
Let $f(\theta) = R \sin \theta$

$$f(33°45°) = R \sin 33°45° = R \sin 33\frac{3°}{4}$$

Let

$$a = 30°, \quad h = 15°, \quad x = \frac{3\frac{3}{4}}{15} = \frac{1}{4}$$

So,

$$a + xh = 30° + \frac{1}{4} \cdot 15° = 30° + 3\frac{3°}{4} = 33\frac{3°}{4}$$

Now,

$$f(a + xh) = f(a) + x\frac{\Delta f(a) + \Delta f(a-h)}{2} + \frac{x^2}{2!}\Delta^2 f(a-h)$$

$$f\left(33\frac{3°}{4}\right) = f(30°) + \frac{1}{4}\frac{\Delta f(30°) + \Delta f(15°)}{2} + \frac{(1/4)^2}{2}\Delta^2 f(15°)$$

$R \sin 33°45'$

$$= R \sin 30° + \frac{1}{8}[\Delta(R \sin 30°) + \Delta(R \sin 15°)] + \frac{1}{32}\left[\Delta^2(R \sin 15°)\right]$$

$$= 75 + \frac{1}{8}[31 + 36] + \frac{1}{32}[-5]$$

$$= 75 + \frac{67}{8} - \frac{5}{32} = 75 + \frac{268 - 5}{32}$$

$$= 75 + \frac{263}{32} = 75 + 8\frac{7}{32} = 83\frac{7}{32}$$

$$= 83.21875$$

This value of 83.21875 = 150 sin 33°45′
So,

$$\sin 33°45' = \frac{83.21785}{150} = 0.55478$$

The actual value of sin 33°45′ from modern tables is 0.5532. This again shows that Brahmagupta's table of *R* sines is reasonably accurate.

Example 3

Find the value of *R* sin 48°20′.
Let $f(\theta) = R \sin \theta$

$$f(48°20') = R \sin 48°20'$$

Let

$$a = 45°, \quad h = 15°, \quad x = \frac{3\frac{1}{3}}{15} = \frac{\frac{10}{3}}{15} = \frac{2}{9}$$

$$a + xh = 45° + \frac{2}{9} \cdot 15° = 45° + \frac{30°}{9} = 45° + 3\frac{1°}{3}$$

$$= 48\frac{1°}{3} = 48°20'$$

Now,

$$f(a + xh) = f(a) + x\frac{\Delta f(a) + \Delta f(a - h)}{2} + \frac{x^2}{2!}\Delta^2 f(a - h)$$

(*Continued*)

Example 3 (cont'd)

$$f(48°20') = f(45°) + \frac{2}{9}\frac{\Delta f(45°) + \Delta f(30°)}{2} + \frac{\left(\frac{2}{9}\right)^2}{2}\Delta^2 f(30°)$$

$$R\sin 48°20' = R\sin 45° + \frac{1}{9}\left[[\Delta(R\sin 45°) + \Delta(R\sin 30°)] + \frac{2}{81}\Delta^2(R\sin 30°)\right]$$

$$= 106 + \frac{1}{9}[24 + 31] + \frac{2}{81}[-7]$$

$$= 106 + \frac{55}{9} - \frac{14}{81} = 106 + \frac{495 - 14}{81} = 106 + \frac{481}{81}$$

$$= 106 + 5\frac{76}{81} = 111\frac{76}{81} = 111.93827$$

This value of $111.93827 = 150 \sin 48°20'$.
So,

$$\sin 48°20° = \frac{111.93827}{150} = 0.74625$$

The actual value of $\sin 48°20'$ from modern sine table is 0.7470, which shows that Brahmagupta's table of *R* sine is reasonably accurate.

Bhaskara II was greatly influenced by the work of Brahmagupta. In his admiration of Brahmagupta's talent and ingenuity, Bhaskara II showered the title, "ganaka cakra cūdamani" ("the gem in the circle of mathematicians"). Not only Brahmagupta richly deserves the title showered on him by Bhaskara II but he also deserves much bigger title, "the gem in the circle of mathematicians in the world history of mathematics." In estimating the mathematical work of Brahmagupta, one has to remember that he belonged to seventh century, 14 centuries before our present time.

8 Sridhara and Prthudakaswami Caturveda

Date of Compositions of Sridhara

A mathematician of some reputation after Brahmagupta is Sridhara. Sridhara was not an astronomer, unlike most of the ancient Indian mathematicians. Sridhara deserves enormous credit of dispensing with astronomy and providing mathematics an independent separate status. In AD 750, Sridhara authored a book on Arithmetic called "Pātiganita Sara." This book contains 300 verses. He also wrote another mathematical work called "Trisatika." Sridhara also wrote a treatise on Algebra. This book on Algebra is not available now, perhaps it has been lost. Fortunately, relevant part of it has been preserved in quotations by later mathematicians Bhaskara II in "Bijaganita, page 61," Jnanaraja (AD 1503) in his "Bijaganita," and "Suryadasa" (AD 1541) in his commentary on Bhaskara II's "Bijaganita."

We will now give some of Sridhara's contributions to mathematics.

Solution to Quadratic Equation

Sridhara's method "Caturāhatavargasamai rūpaih paksadvayam gunayet, sanskrti verse Avyaktavargarūpairyuktan paksau tato mūlam"

This is the Sridhara's rule as stated by Jnanaraja and Suryadasa. According to Krsna (AD 1580) and Rāmakrsna (AD 1648), the second line of this verse should be the following:

> "*purvāvyaktasya krteh samarūpāni ksipet tayoreva*"

i.e., "Multiply both sides of the equation by four times the coefficient of the square of the unknown and add to both sides the square of the coefficient of the unknown. Then extract the square root."

Suppose we are given $ax^2 + bx = c$.

Multiplying by $4a$, we get $4a^2x^2 + 4abx = 4ac$.

Adding b^2 to both sides, we get $4a^2x^2 + 4abx + b^2 = 4ac + b^2$

i.e., we have $(2ax + b)^2 = 4ac + b^2$.

Mathematical Achievements of Pre-modern Indian Mathematicians. DOI: http://dx.doi.org/10.1016/B978-0-12-397913-1.00008-9

So,

$$2ax + b = \sqrt{4ac + b^2}$$

$$x = \frac{\sqrt{4ac + b^2}}{2a} - \frac{b}{2a} \quad (^*)$$

Application of the Quadratic Formula in Arithmetic Progression

An application of this has been given in Sridhara's "Triśatikā," in finding the number of terms of an arithmetic progression (AP): (Triś, R. 41)

If S is the sum of the first n terms of the AP $a + (a + b) + (a + 2b) + \cdots$, then

$$S = \frac{n}{2}\{2a + (n-1)b\}$$

$$2S = bn^2 + n(2a - b)$$

Then, from (*), we get

$$n = \frac{\sqrt{4b(2s) + (2a-b)^2} - (2a - b)}{2b}$$

i.e.,

$$n = \frac{\sqrt{8bs + (2a-b)^2} - (2a - b)}{2b}$$

We find the following examples of Sridhara in Trisatika.

Example 1

"The third part of a necklace of pearls broken in an amorous struggle fell to the ground; its fifth part was on the couch; the sixth part was saved by the wench; and the tenth part was taken by her lover. Six pearls remained strung. How many pearls were there in the necklace?"

(Trisatika, p. 14)

Let x be the number of pearls in the necklace.

Then,

$$\frac{x}{3}+\frac{x}{5}+\frac{x}{6}+\frac{x}{10}+6=x$$

Multiplying by 30, we get

$$10x+6x+5x+3x+180=30x$$

$$24x+180=30x$$

i.e.,

$$30x-24x=180$$

$$6x=180 \Rightarrow x=30$$

Thus, there are 30 pearls in the necklace.

Example 2

"One-twelfth part of a pillar, as multiplied by 1/30 part thereof, was found to be under water, 1/20 of the remainder, as multiplied by 3/16 thereof, was found to be buried in the mire below, and 20 hasta of the pillar were found above the water. If the length of the pillar is integer, can you give me the length of the pillar?"

(Trisatika, p. 13 and G.S.S. p. 49)

Let x basta be the length of the pillar.
Then, we have

$$\left(x-\frac{x}{12}\cdot\frac{x}{30}\right)-\frac{1}{20}\left(x-\frac{x}{12}\cdot\frac{x}{30}\right)\left(x-\frac{x}{12}\cdot\frac{x}{30}\right)\frac{3}{16}=20$$

i.e.,

$$\left(x-\frac{x^2}{360}\right)-\frac{3}{320}\left(x-\frac{x^2}{360}\right)^2=20$$

Let

$$y=x-\frac{x^2}{360}$$

(*Continued*)

Example 2 (cont'd)

Then, we have

$$y - \frac{3}{320}y^2 = 20$$

Multiplying by 320, we get

$$320y - 3y^2 = 6400$$

i.e.,

$$3y^2 - 320y + 6400 = 0$$

$$y = \frac{320 \pm \sqrt{(320)^2 - 4 \cdot 3 \cdot 6400}}{6}$$

$$y = \frac{320 \pm 8\sqrt{1600 - 1200}}{6} = \frac{320 \pm 8 \cdot 20}{6}$$

$$y = \frac{320 \pm 160}{6} = \frac{160}{6} \text{ or } \frac{480}{6}$$

i.e.,

$$y = \frac{80}{3} \text{ or } 80$$

(a) Let $y = 80$
Then,

$$x - \frac{x^2}{360} = 80$$

i.e.,

$$360x - x^2 = 360 \cdot 80$$

$$x^2 - 360x + 28,800 = 0$$

$$x = \frac{360 \pm \sqrt{(360)^2 - 4 \cdot 28,800}}{2}$$

$$= \frac{360 \pm 10\sqrt{1296 - 1152}}{2} = \frac{360 \pm 10\sqrt{144}}{2}$$

$$x = \frac{360 \pm 120}{2} = 240 \text{ or } 120$$

(b) Let $y = \frac{80}{3}$

Then,

$$x - \frac{x^2}{360} = \frac{80}{3}$$

$$360x - x^2 = 120 \cdot 80$$

i.e.,

$$x^2 - 360x + 9600 = 0$$

$$x = \frac{360 \pm \sqrt{(360)^2 - 4(9600)}}{2} = \frac{360 \pm 10\sqrt{1296 - 384}}{2}$$

$$x = \frac{360 \pm 10\sqrt{912}}{2} = 180 \pm 5\sqrt{912}$$

$$x = 180 \pm 10\sqrt{228}$$

Since x is an integer, we can discard these values of x.
Hence, $x = 240$ or 120.

Sridhara gave the following example on AP.

Example 3

The first term of an AP is 20. The sum of the first seven terms is 245. What is the common difference in this AP?

(*Continued*)

Example 3 (cont'd)

Let d be the common difference of the AP.
If a is the term and S_n is the sum of the first n terms of the AP, then

$$S_n = \frac{n}{2}\{2a + (n-1)d\}$$

$$S_7 = 245 = \frac{7}{2}\{2 \cdot 20 + (7-1)d\}$$

$$70 = 40 + 6d$$

$$6d = 30$$

$$d = 5$$

Thus, common difference d of the AP is 5.

Sridhara's formula for the volume of sphere is the following:

गोलव्यासघनार्धं स्वाष्टादशभागसंयुतं गणितम् ।

(T. S. 56)

i.e., "The sum of half the cube of the diameter of sphere and 18th part of itself is the volume of the sphere."

That is, volume of sphere

$$= \frac{d^3}{2} + \frac{1}{2}\frac{d^3}{18} = \frac{19}{36}d^3$$

where d is the diameter of the sphere.

This can be obtained from the correct formula for the volume of a sphere as follows:

$$\pi = \sqrt{10} = \sqrt{9+1} = \sqrt{3^2+1} = 3 + \frac{1}{2 \cdot 3} \quad \text{(by Heron's formula)}$$

So,

$$\pi = \frac{19}{6}$$

Then, volume of sphere

$$= \frac{\text{surface area of the sphere} \times \text{diameter}}{6}$$

So, volume of the sphere

$$= \frac{\pi d^2 \cdot d}{6} = \frac{\sqrt{10} \cdot d^3}{6} = \frac{19}{6}\frac{d^3}{6} = \frac{19d^3}{36}$$

Sridhara also gave a formula for the volume V of the frustum of a cone:

$$V = \frac{\pi}{24} \cdot h\{d_1^2 + d_2^2 + (d_1 + d_2)^2\}$$

where h is the height and d_1 and d_2 are the diameter of the base and the top. In his view, the conical frustum is a well.

मुखतलतद्योगानां वर्गैक्यकृतेः पदं दशगुणायाः।
वेधगुणं चतुरन्वितविंशति–भवतं फलं कूपे ॥

i.e., "Add to the sum of the squares of diameters of the top and the bottom, the square of the sum of these two diameters. Multiply the square of this sum by 10. Square root of this product multiplied by depth and divided by 24 gives the volume of a circular well."

If h is the depth, d_1 and d_2 are the diameters of the top and the bottom, then the volume v is given by

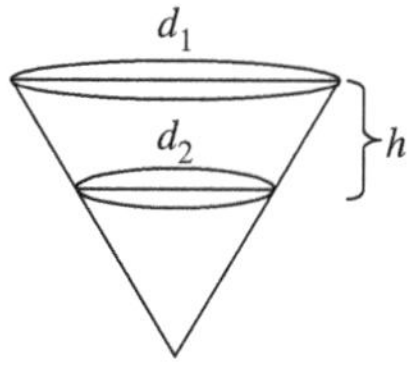

$$v = \frac{h}{24}\sqrt{10\{(d_1^2 + d_2^2) + (d_1 + d_2)^2\}^2}$$

Proof: Let ABC be the cone and $ABDE$ be the frustum of the cone. Let CF be the altitude of the cone.

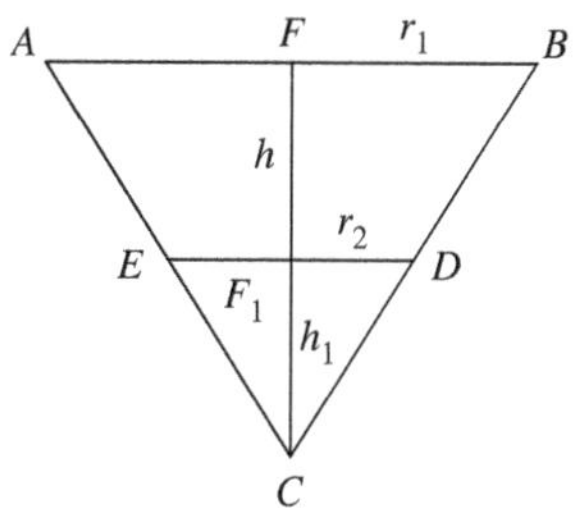

Let r_1 and r_2 be the radii of the top and bottom of the conical frustum. Let h be the height of the conical frustum.

Then,

$$\frac{r_1}{r_2} = \frac{h_1 + h}{h_1}$$

from similar triangles BFC and DF_1C.

Hence,

$$r_1 h_1 = r_2 h_1 + r_2 h$$

$$h_1(r_1 - r_2) = r_2 h \Rightarrow h_1 = \frac{r_2 h}{r_1 - r_2}$$

v = volume of the frustum of the cone $ABDE$

= volume of cone ABC − volume of cone BEC

$$v = \frac{1}{3}\sqrt{10}r_1^2(h + h_1) - \frac{1}{3}\sqrt{10}r_2^2 h_1$$

$$= \frac{1}{3}\sqrt{10}\left[r_1^2\left(h + \frac{r_2 h}{r_1 - r_2}\right) - r_2^2\frac{r_2 h}{r_1 - r_2}\right]$$

$$= \frac{1}{3}\frac{\sqrt{10}h}{r_1 - r_2}\left[r_1^2(r_1 - r_2) + r_1^2 r_2 - r_2^3\right]$$

$$= \frac{1}{3}\sqrt{10}h\frac{\left[r_1^3 - r_1^2 r_2 + r_1^2 r_2 - r_2^3\right]}{r_1 - r_2}$$

$$= \frac{1}{3}\sqrt{10}h\frac{\left[r_1^3 - r_2^3\right]}{r_1 - r_2} = \frac{1}{3}\sqrt{10}h\left[r_1^2 + r_2^2 + r_1 r_2\right]$$

$$= \frac{1}{3}\sqrt{10}h\left[\left(\frac{d_1}{2}\right)^2 + \left(\frac{d_2}{2}\right)^2 + \left(\frac{d_1}{2}\right)\left(\frac{d_2}{2}\right)\right]$$

$$= \frac{1}{24}\sqrt{10}h\left[2d_1^2 + 2d_2^2 + 2d_1 d_2\right]$$

$$= \frac{1}{24}\sqrt{10}h\left[(d_1^2 + d_2^2) + (d_1 + d_2)^2\right]$$

Since $\pi = \sqrt{10}$, this becomes

$$v = \frac{\pi h}{24}\left\{(d_1^2 + d_2^2) + (d_1 + d_2)^2\right\}$$

Prthudakaswami Caturveda

In about AD 860, Prthudakaswami Caturveda wrote a masterly commentary on Brahmagupta's "Brahmasphuta Siddhānta." This commentary contains numerous examples to illustrate the mathematical assertions made in the "Brahmasphuta Siddhanta." According to Kaye [71], Prthudakaswami Caturveda is said to have proved "Ptolemy's theorem."

Example 1

A horse was purchased by nine dealers in partnership, whose contributions were one up to nine and was sold by them for five less than five hundred. What is each man's share of sales proceed?

Let the contributions of the nine partners be

$$x,\ 2x,\ 3x,\ 4x,\ 5x,\ 6x,\ 7x,\ 8x,\ 9x$$

Then,

$$x + 2x + 3x + 4x + 5x + 6x + 7x + 8x + 9x = 500 - 5 = 495$$

$$45x = 495$$

$$x = \frac{495}{45} = 11$$

Then, the share of sales proceed of the nine partners are

11, 22, 33, 44, 55, 66, 77, 88, 99

Example 2

Four colleges, containing equal number of students, were invited to take part in a feast. A fifth, a half, a third, and a quarter of the total number of students in each college came from the respective colleges to the feast and added to one,

(*Continued*)

Example 2 (cont'd)

two, three, and four, their number was 87 or with those deducted their number was 67. Find the number of students who came to the feast from each college.

Let x be the number of students in each of the four colleges.

Then,

$$\left(\frac{x}{5}+1\right)+\left(\frac{x}{2}+2\right)+\left(\frac{x}{3}+3\right)\left(\frac{x}{4}+4\right)=87$$

or

$$\left(\frac{x}{5}-1\right)+\left(\frac{x}{2}-2\right)+\left(\frac{x}{3}-3\right)+\left(\frac{x}{4}-4\right)=67$$

which means

$$\frac{x}{5}+\frac{x}{2}+\frac{x}{3}+\frac{x}{4}=(87-10) \quad \text{or} \quad (67+10)=77$$

Multiplying by 60, we get

$$12x+30x+20x+15x=77\cdot 60$$

i.e.,

$$77x=77\cdot 60$$

$$x=60$$

The number of students who came from each college is $x/5$, $x/2$, $x/3$, and $x/4$, respectively, i.e., 12, 30, 20, and 15, respectively.

So, 12, 30, 20, and 15 students came to the feast from each college, respectively.

As an illustration of his rule to solve quadratic equation, Brahmagupta has stated two examples (Examples 1 and 2 stated below), where we have to solve practically the same equation.

Example 1

"The square root of the residue of the revolution of the sun less 2 is diminished by 1, multiplied by 10 and added 2. When will this be equal to the residue of the revolution of the sun less 1?"

(B.S.S. xviii. 49)

Prthudakaswami solves this problem the following way.

Let the residue of the revolution of the sun be $x^2 + 2$.

Then, we have

$$10\left(\sqrt{(x^2+2)-2}\ -1\right)+2=(x^2+2)-1$$

i.e.,

$$10(x-1)+2=x^2+1$$

$$x^2-10x=-9$$

Brahmagupta's solution for the quadratic $ax^2+bx=c$ is

$$x=\pm\sqrt{ac-\left(\frac{b}{2}\right)^2}-\frac{b}{2}$$

Here,

$$x=\pm\sqrt{25-9}+5=\pm4+5=9\ \text{or}\ 1$$

Brahmagupta took the first value 9.

Example 2

"When will the square of one-fourth the residue of the exceeding months less three be equal to the residue of the exceeding months?"

Prthudakaswami took $4x$ as the residue of the exceeding months.

Then, we have

$$\left\{\frac{1}{4}(4x)-3\right\}^2=4x$$

i.e.,

$$(x-3)^2=4x$$

i.e.,

$$x^2-10x=-9$$

As before in Example 1, we get $x=9$ or 1.

In this case, Brahmagupta took the second value 1.

The above two examples clearly illustrate that Brahmagupta considers sometimes the positive square root and sometimes the negative square root. These two examples unmistakably demonstrate that Brahmagupta knew that a quadratic equation has two roots.

Solution of a Remainder Problem Involving Simultaneous Indeterminate Equations

Example

Find a number which leaves remainders 5, 4, 3, and 2 when divided by 6, 5, 4, and 3, respectively.

Let N be the number.

Then, we have $N = 6x + 5 = 5y + 4 = 4z + 3 = 3w + 2$.

$$6x + 5 = 5y + 4$$

So,

$$6x + 1 = 5y$$

We now solve this indeterminate equation

$$5)6(1 = q_1$$
$$\underline{5}$$
$$1$$

The number of quotients is one and is odd.

We now form the following table:

1	1	$Y = 1 \equiv 1 \pmod 5$
1	1	$X = 1 \equiv 1 \pmod 6$
0	0	

So, $(x, y) = (5 - 1, 6 - 1)$ is the least solution of $6x + 1 = 5y$, i.e., $(x, y) = (4, 5)$ is the least solution of $6x + 1 = 5y$.

General solution:

$$x = 5t + 4$$

$$y = 6t + 5$$

where t is any arbitrary integer.

$$N = 6x + 5 = 6(5t + 4) + 5 = 30t + 24 + 5$$

$$N = 30t + 29$$

But, $N = 4z + 3$.
Hence, $4z + 3 = 30t + 29$, i.e., $2z = 15t + 13$.
We now solve this indeterminate equation

$$2)15(7 = q_1$$
$$\underline{14}$$
$$1$$

The number of quotients is one and so is odd.
We now form the following table:

7	91	$Y = 91 \equiv 1 (\text{mod } 15)$
13	13	$X = 13 \equiv 1 (\text{mod } 2)$
0	0	

So, $(t, z) = (2 - 1, 15 - 1)$ is the least solution of $2z = 15t + 13$.
So, $(t, z) = (1, 14)$ is the least solution of $2z = 15t + 13$.
General solution:

$$t = 2m + 1$$

$$z = 15m + 14$$

where m is any arbitrary integer.

$$N = 4z + 3 = 4(15m + 14) + 3$$

$$N = 60m + 59$$

where m is any arbitrary integer.
Least value of $N = 59$.
The last condition is identically satisfied, since 59 divided by 3 leaves a remainder 2.

Solutions of $Nx^2 \pm c = y^2$.
Brahmagupta has stated the following result: If $(p\ q)$ be a rational solution of $Nx^2 \pm c = y^2$ and if $(\alpha\ \beta)$ be a solution of $Nx^2 + 1 = y^2$, then $x = p\beta \pm q\alpha$, $y = q\beta \pm Np\alpha$ is a solution of $Nx^2 \pm c = y^2$.

(B.B.S. XVII. 66)

Proof: $(p\ q)$ is a solution of $Nx^2 \pm c = y^2 \Rightarrow (p\ q\ \ \pm c)$.
$(\alpha\ \beta)$ is a solution of $Nx^2 + 1 = y^2 \Rightarrow (\alpha\ \beta\ 1)$.

$$(p\ q\ \ \pm c)\circ(\alpha\ \beta\ 1) = (p\beta \pm q\alpha\ q\beta \pm p\alpha N\ \ \pm c)$$

which means $(p\beta \pm q\alpha\ q\beta \pm p\alpha N)$ is a solution of $Nx^2 \pm c = y^2$.

Brahmagupta has also stated the following result:

If $(\alpha\ \beta)$ is a root of $Mn^2x^2 \pm c = y^2$ and if $(\alpha'\ \beta')$ is a root of $Mu^2 \pm c = y^2$, then $\alpha = \alpha'/n$ and $\beta = \beta'$.

Proof: $(\alpha\ \beta)$ is a root of $Mn^2x^2 \pm c = y^2$.

So,

$$Mn^2\alpha^2 \pm c = \beta^2 \tag{1}$$

i.e., $M(n\alpha)^2 \pm c = \beta^2$, which means that $(n\alpha\ \beta)$ is a solution of $Mu^2 \pm c = y^2$. But $(\alpha^1\ \beta^1)$ is a solution of $Mu^2 \pm c = y^2$.

Here,

$$n\alpha = \alpha^1 \quad \text{and} \quad \beta = \beta^1$$

i.e.,

$$\alpha = \frac{\alpha^1}{n} \quad \text{and} \quad \beta = \beta^1$$

Rational Isosceles Triangle with a Given Altitude

If a is the altitude, Brahmagupta has stated that the sides and the base of a rational isosceles triangle are, respectively

$$\frac{1}{2}\left(\frac{a^2}{m} + m\right), \quad \frac{1}{2}\left(\frac{a^2}{m} + m\right)$$

and

$$\frac{a^2 - m^2}{m}$$

where m is any rational number.

Proof: Let ABC be an isosceles triangle with $AB = AC$. Let AD be the altitude from A on the base BC.

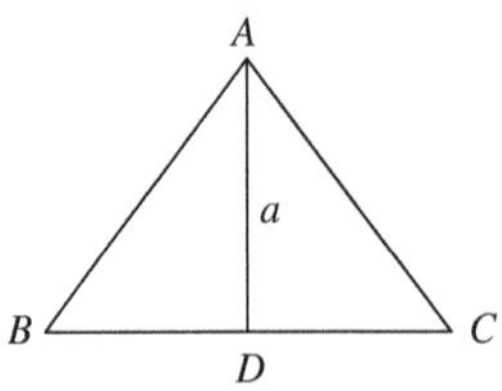

Let $AD = a$.

Let $AB = x$ and $BD = y$.
In the right triangle ADB,

$$AD^2 = AB^2 - BD^2$$

i.e.,

$$a^2 = x^2 - y^2 = (x+y)(x-y)$$

Let

$$x - y = m \quad (1)$$

Then,

$$x + y = \frac{a^2}{m} \quad (2)$$

Adding (1) and (2), we get

$$2x = m + \frac{a^2}{m} \Rightarrow x = \frac{1}{2}\left(m + \frac{a^2}{m}\right)$$

Subtracting (1) from (2), we get

$$2y = \frac{a^2}{m} - m \Rightarrow y = \frac{1}{2}\left(\frac{a^2}{m} - m\right)$$

Thus,

$$AB = AC = \frac{1}{2}\left(m + \frac{a^2}{m}\right)$$

$$BC = BD + DC = 2BD = 2y = \left(\frac{a^2}{m} - m\right) = \frac{a^2 - m^2}{m}$$

An Example of Constructing an Integral Isosceles Triangle

Prthudakaswami took $a = 8$ and $m = 4$, then the equal sides of the isosceles triangle are of length

$$= \frac{1}{2}\left(\frac{a^2}{m} + m\right) = \frac{1}{2}\left(\frac{64}{4} + 4\right) = 10$$

$$\text{Base} = \frac{a^2 - m^2}{m} = \frac{64 - 16}{4} = \frac{48}{4} = 12$$

Prthudakaswami obtained integral isosceles triangle with sides (10, 10, 12).

Rational Scalene Triangle with a Given Altitude

If m is the altitude, Brahmagupta has stated that the sides and the base of a rational scalene triangle are respectively

$$\frac{1}{2}\left(\frac{m^2}{p} + p\right), \quad \frac{1}{2}\left(\frac{m^2}{q} + q\right)$$

and

$$\frac{1}{2}\left(\frac{m^2}{p} - p\right) + \frac{1}{2}\left(\frac{m^2}{q} - q\right)$$

where m, p, and q are any rational numbers.

Proof: Let ABC be a rational scalene triangle with altitude $AD = m$.

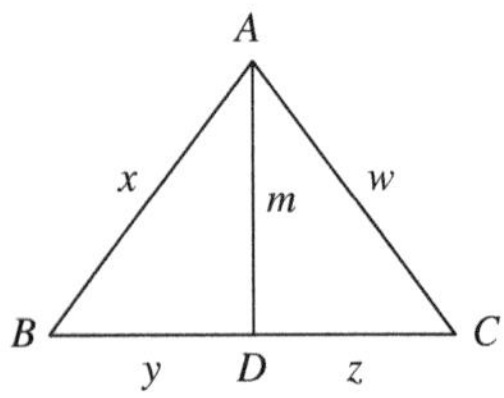

Let $AB = x$ and $BD = y$.

$$m^2 = AD^2 = AB^2 - BD^2$$

$$= x^2 - y^2 = (x + y)(x - y)$$

Let

$$x - y = p \tag{1}$$

where p is any rational number.

Then,

$$x + y = \frac{m^2}{p} \tag{2}$$

Adding (1) and (2), we get

$$2x = p + \frac{m^2}{p}$$

and hence

$$x = \frac{1}{2}\left(p + \frac{m^2}{p}\right)$$

Subtracting (1) from (2), we get

$$2y = \frac{m^2}{p} - p$$

So,

$$y = \frac{1}{2}\left(\frac{m^2}{p} - p\right) = BD$$

Similarly, let $AC = w$ and $DC = z$.
In the right triangle ADC,

$$AD^2 = AC^2 - DC^2$$

i.e., $m^2 = w^2 - z^2 = (w + z)(w - z)$.
Let

$$w - z = q \tag{3}$$

where q is any rational number.
Then,

$$w + z = \frac{m^2}{q} \tag{4}$$

Adding (3) and (4), we get

$$2w = q + \frac{m^2}{q}$$

i.e.,

$$w = \frac{1}{2}\left(q + \frac{m^2}{q}\right)$$

Subtracting (3) from (4), we get

$$2z = \frac{m^2}{q} - q$$

i.e.,

$$z = \frac{1}{2}\left(\frac{m^2}{q} - q\right) = DC$$

Hence,

$$AB = x = \frac{1}{2}\left(p + \frac{m^2}{p}\right)$$

$$AC = w = \frac{1}{2}\left(q + \frac{m^2}{q}\right)$$

$$BC = BD + DC = \frac{1}{2}\left(\frac{m^2}{p} - p\right) + \frac{1}{2}\left(\frac{m^2}{q} - q\right)$$

An Example of Constructing an Integral Scalene Triangle

Prthudakaswami took $m = 12$, $p = 6$, and $q = 8$.

Then, the sides of the scalene triangle are

$$\frac{1}{2}\left(\frac{m^2}{p} + p\right) = \frac{1}{2}\left(\frac{144}{6} + 6\right) = \frac{1}{2}(24 + 6) = 15$$

$$\frac{1}{2}\left(\frac{m^2}{q} + q\right) = \frac{1}{2}\left(\frac{144}{8} + 8\right) = \frac{1}{2}(18 + 8) = 13$$

$$\text{Base} = \frac{1}{2}\left(\frac{m^2}{p} - p\right) + \frac{1}{2}\left(\frac{m^2}{q} - q\right)$$

$$= \frac{1}{2}\left(\frac{144}{6} - 6\right) + \frac{1}{2}\left(\frac{144}{8} - 8\right) = \frac{1}{2}(24 - 6) + \frac{1}{2}(18 - 8)$$

$$= \frac{1}{2}(18) + \frac{1}{2}(10) = 9 + 5 = 14$$

Altitude $= m = 12$

$$\text{Area} = \frac{1}{2}\text{base} \times \text{height} = \frac{1}{2}(14) \cdot 12 = 84 \text{ sq. units}$$

Construction of a rational Isosceles Trapezium

On page 191, Brahmagupta has given an isosceles trapezium, whose sides, diagonals, altitudes, segments, and areas are all rational numbers.

Construct the rectangle $ABNM$ such that $AM = BN = 2mm$, where m and n are any rational numbers.

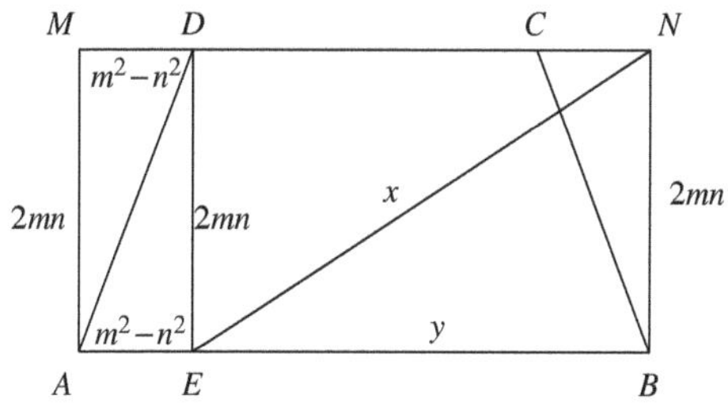

Take points C and D on NM such that

$$CN = MD = m^2 - n^2$$

Draw DE perpendicular from D on AB.
Then, $DE = 2mn$ and $AE = m^2 - n^2$.
Join EN. Then $ABCD$ is an isosceles trapezium such that

$$BE = \frac{1}{2}\left[\frac{4m^2n^2}{p} - p\right]$$

$$\text{Base } AB = \frac{1}{2}\left[\frac{4m^2n^2}{p} - p\right] + (m^2 - n^2)$$

$$\text{Face } CD = \frac{1}{2}\left[\frac{4m^2n^2}{p} - p\right] - (m^2 - n^2)$$

$$AD = BC = (m^2 + n^2)$$

Proof: In the right triangles *AMD* and *BCN*,

$$AD^2 = BC^2 = (m^2 - n^2)^2 + (2mn)^2$$

$$AD^2 = BC^2 = m^4 + 2m^2n^2 + n^4 = (m^2 + n^2)^2$$

So, $AB = BC = m^2 + n^2$.
Let $EN = x$ and $EB = y$.
Then, in the right triangle *EDN*,

$$DE^2 = EN^2 - DN^2 = EN^2 - EB^2 = x^2 - y^2$$

Hence,

$$x^2 - y^2 = DE^2 = (2mn)^2 = 4m^2n^2$$

i.e., $(x + y)(x - y) = 4m^2n^2$.
Let

$$x - y = p \tag{1}$$

where p is any rational number.
Then,

$$x + y = \frac{4m^2n^2}{p} \tag{2}$$

Subtracting (1) from (2), we get

$$2y = \frac{4m^2n^2}{p} - p$$

$$y = \frac{1}{2}\left[\frac{4m^2n^2}{p} - p\right]$$

Hence,

$$y = EB = \frac{1}{2}\left[\frac{4m^2n^2}{p} - p\right]$$

Now, $AB = AE + EB$.
So,

$$AB = (m^2 - n^2) + \frac{1}{2}\left[\frac{4m^2n^2}{p} - p\right]$$

Also, $DC = DN - CN = EB - CN$.

$$DC = \frac{1}{2}\left[\frac{4m^2n^2}{p} - p\right] - (m^2 - n^2)$$

Segments of the base are

$$AE = m^2 - n^2 \quad \text{and} \quad EB = \frac{1}{2}\left[\frac{4m^2n^2}{p} - p\right]$$

Prthudakaswami constructed the following isosceles trapezium:
He took $m = 3$, $n = 2$, and $p = 6$.
Then,

$$EB = \frac{1}{2}\left[\frac{4m^2n^2}{p} - p\right] = \frac{1}{2}\left[\frac{4 \cdot 9 \cdot 4}{6} - 6\right]$$

$$EB = \frac{1}{2}[24 - 6] = \frac{1}{2}(18) = 9$$

$$AE = m^2 - n^2 = 3^2 - 2^2 = 9 - 4 = 5$$

Thus,

Base $AB = EB + AE = 9 + 5 = 14$

Face $CD = EB - CN = EB - AE = 9 - 5 = 4$

Also,

Altitude $DE = 2mm = 2(3)(2) = 12$

Area of the trapezium

$$= \frac{1}{2}[AB + CD]DE$$

$$= \frac{1}{2}[14 + 4]12 = \frac{1}{2}(18)(12) = 108 \text{ sq. units}$$

9 Mahavira

The most celebrated mathematician of the ninth century AD is Mahaviracharya, well known as Mahavira (AD 817–875). Mahavira was a Jain by religion. He was in the Royal Court of Rastrakuta King, Amoghavarsha Nṛpatuṅga who ruled from AD 815 to AD 874. His capital was called Mānyakheta near Gulbarga in Karnataka. He wrote his work called "Ganita Sara Sangraha" (G.S.S.) in AD 850. This work was found about a century ago. It was edited and translated to English by Prof. M. Rangacharya of Presidency College of Madras University. This English edition was published in 1912 by the Government of Madras. This book has been very popular and widely used in the south India. This work was translated into Telugu in the eleventh century by Paviluri Mallana.

'Ganita Sara Sangraha' is a text on pure mathematics. This text contains nine chapters of about 1100 slokas dealing with arithmetic, algebra, geometry, and mensuration. The subject matter is presented like a present-day textbook without a chapter on decimals. Decimals are not of Indian origin and the ancient Indians did not know them. In 1963, this work was translated into Hindi by Prof. L.C. Jain and published by Jaina Sanskrit Samraksha Singha of Sholapur.

Arithmetic

Garland Numbers

The following products resemble a garland, i.e., they give the same numbers whether read from left to right or from right to left.

$$139 \times 109 = 15151$$

$$27994681 \times 441 = 12345654321$$

$$14287143 \times 7 = 100010001$$

$$333333666667 \times 33 = 11000011000011$$

$$12345679 \times 9 = 111111111$$

$$152207 \times 73 = 11111111$$

Mathematical Achievements of Pre-modern Indian Mathematicians. DOI: http://dx.doi.org/10.1016/B978-0-12-397913-1.00009-0

$11011011 \times 91 = 1002002001$

$142857143 \times 7 = 1000000001$

Mahavira called these numbers garland numbers.

Algebra

To find the cube of a natural number, Mahavira has provided several formulae based on arithmetic progressions as well as formulae based on some elementary algebra:

i. $a + 3a + 5a + \cdots$ up to a terms $= a^3$.
ii. $a^2 + (a-1)[1 + 3 + 5 + 7 + \cdots]$ up to a terms $= a^3$.
iii. $a(a+b)(a-b) + b^2(a-b) + b^3 = a^3$.
iv. $3[1 \cdot 2 + 2 \cdot 3 + 3 \cdot 4 + \cdots + (a-1)a] + a = a^3$.

Another important and interesting continuation of Mahavira is to express a given fraction as the sum of several fractions which have unity as numerators. Mahavira has called such expressions "rupakamsa rasi" (unit fractions):

(i) $1 = \frac{1}{2} + \frac{1}{3} + \frac{1}{3^2} + \frac{1}{3^3} + \cdots + \frac{1}{3^{n-2}} + \frac{1}{2 \cdot 3^{n-2}}$ (G.S.S. Sl. 75)

e.g., $n = 6$

$$1 = \frac{1}{2} + \frac{1}{3} + \frac{1}{3^2} + \cdots + \frac{1}{3^4} + \frac{1}{2 \cdot 3^4}$$

(ii) To express the sum of $(2n-1)$ (i.e., odd number)
Unit fractions:

$$1 = \frac{1}{2 \cdot 3 \cdot \frac{1}{2}} + \frac{1}{3 \cdot 4 \cdot \frac{1}{2}} + \frac{1}{4 \cdot 5 \cdot \frac{1}{2}} + \frac{1}{5 \cdot 6 \cdot \frac{1}{2}} + \cdots + \frac{1}{(2n-1)(2n)\frac{1}{2}} + \frac{1}{2n \cdot \frac{1}{2}}$$

(G.S.S. Sl. 77)

e.g., $n = 3$

$$1 = \frac{1}{3} + \frac{1}{6} + \frac{1}{10} + \frac{1}{15} + \frac{1}{3}$$

(iii) To express a given unit fraction as the sum of r unit fractions.

$$\frac{1}{n} = \frac{1}{n(n+1)} + \frac{1}{(n+1)(n+2)} + \cdots + \frac{1}{(n+r-2)(n+r-1)} + \frac{1}{n+r-1}$$

(G.S.S. Sl. 78)

e.g., $n = 5$, $r = 4$

$$\frac{1}{5} = \frac{1}{30} + \frac{1}{42} + \frac{1}{56} + \frac{1}{8}$$

(iv) To express any fraction as the sum of unit fractions:

(G.S.S. Sl. 80)

Let p/q be a given fraction such that $p < q$.
Find i so that $(q + i)/p$ is an integer $= r$.
Then,

$$\frac{p}{q} = \frac{rp}{rq} = \frac{q+i}{rq} = \frac{1}{r} + \frac{i}{rq}$$

Similarly, Find j so that $(rq + j)/i$ is an integer $= s$.
Then,

$$\frac{i}{rq} = \frac{is}{rqs} = \frac{rq+j}{rqs} = \frac{1}{s} + \frac{j}{rqs} \text{ and so on}$$

Since, $j < i$, the process ends after a finite number of steps.

(v) To express a unit fraction as the sum of unit fractions:
First method:

$$\frac{1}{ab} = \frac{1}{a(a+b)} + \frac{1}{b(a+b)} \qquad b \geq 1$$

Second method:

$$\frac{1}{a} = \frac{1}{ab} + \frac{1}{(ab/(b-1))}$$

where b is chosen so that a is divisible by $b - 1$.

Permutations and Combinations

We have seen in Chapter 3, Jains called permutations and combinations as "prastarana" and "vikalpa." Mahavira was the world's first mathematician to give the general formula

$$_nC_r = \binom{n}{r} = \frac{n(n-1)(n-2)\cdots(n-r+1)}{1 \cdot 2 \cdot 3 \cdots r}$$

Geometry

Inscribed Circle in a Triangle

परिधेः पादेन भजेदनायतक्षे त्रसूक्ष्मगणितं तत् ।
क्षे त्राभ्यन्तरवृत्ते विष्कम्भोऽयं विनिर्दिष्टः ॥

(G.S.S. VII. 223½)

i.e., The area of any rectilinear figure other than a rectangle divided by one-fourth of the perimeter gives the diameter of the inscribed circle. This means, the area of a triangle divided by 1/2 its perimeter gives the radius of the inscribed circle.

Let *I* be the center of the circle inscribed in the triangle *ABC*. Join *AI*, *CI*, and *BI*. Let *r* be the radius of the inscribed circle.

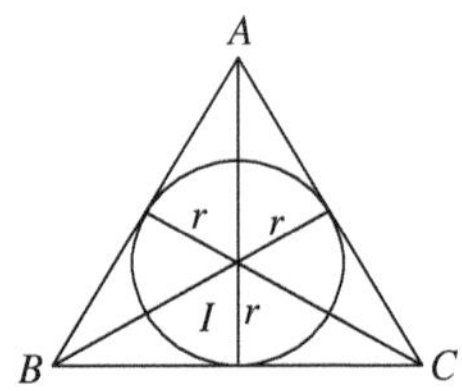

Now, area of Δ^{le}

$$AIB = \frac{1}{2}AB \cdot r$$

Area of Δ^{le}

$$AIC = \frac{1}{2}AC \cdot r$$

Area of Δ^{le}

$$BIC = \frac{1}{2}BC \cdot r$$

So, area of Δ^{le} *AIB* + area of Δ^{le} *AIC* + area of Δ^{le} *BIC*

$$= \frac{1}{2}r(AB + AC + BC)$$

i.e., area of $ABC = (1/2)r$ (perimeter of the Δ^{le} *ABC*).
Here,

$$r = \frac{\text{Area of } \Delta^{le}\, ABC}{(1/2)\text{perimeter of the } \Delta^{le}\, ABC}$$

Area of an Annulus

Mahavira gave the area A of an annulus as $A = (d \pm t)\pi t$, where d is the inner or outer diameter and t the width of the annuals.

Proof:

(i) Let d be the diameter of the inner circle.

Since t is the width of the annulus, the diameter D of the outer circle is given by $D = d + 2t$.

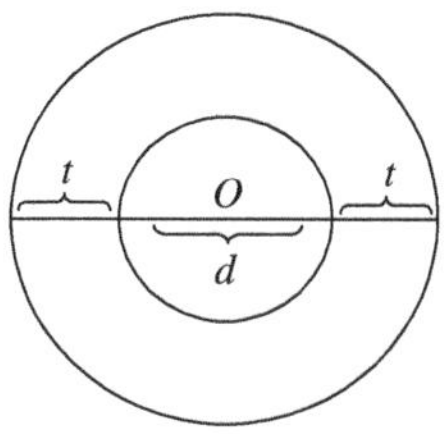

Now, A = area of the annulus

= Area of the outer circle−area of the inner circle

$$= \frac{\pi}{4}D^2 - \frac{\pi}{4}d^2 = \frac{\pi}{4}(D^2 - d^2)$$

$$= \frac{\pi}{4}(D + d)(D - d)$$

$$= \frac{\pi}{4}[(d + 2t + d)(d + 2t - d)]$$

$$= \frac{\pi}{4}(2d + 2t)2t = \pi(d + t)t = \pi t(d + t)$$

(ii) Let d be the diameter of the outer circle and D be the diameter of the inner circle.

Then, $D = d - 2t$.

Area of the annulus = area of the outer circle−area of the inner circle.

$$A = \frac{\pi}{4}d^2 - \frac{\pi}{4}D^2 = \frac{\pi}{4}(d^2 - D^2) = \frac{\pi}{4}(d + D)(d - D)$$

$$A = \frac{\pi}{4}(d + d - 2t)(d - (d - 2t))$$

$$= \frac{\pi}{4}(2d - 2t)2t = \pi(d - t)t = \pi t(d - t)$$

Mahavira and later by Narayana Pandita have given correct formula to find the area of the space enclosed by three or four equal, mutually touching circles.

(i) The area A for the space between three mutual touching equal circles is the difference of the area of an equilateral triangle of side equal to the diameter and half the area of the circle.

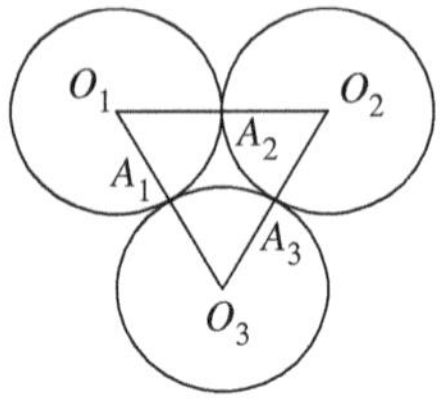

Proof: Let O_1, O_2, and O_3 be the centers of the circle. Let the circles touch at A_1, A_2, and A_3. Let r be the radius of each circle.

First, we note that the triangle $O_1O_2O_3$ is an equilateral triangle with each side = 2, radius of the circle = $2r$.

A = shaded area

$$= \text{Area of the equilateral } \Delta^{\text{le}} O_1O_2O_3$$
$$- \text{ Area of section } O_1A_1A_2$$
$$- \text{ Area of section } O_2A_2A_3$$
$$- \text{ Area of the sector } O_3A_1A_3$$

$$= \text{Area of the equilateral } \Delta^{\text{le}} O_1O_2O_3 - \left[\frac{1}{2}r^2\frac{\pi}{3} + \frac{1}{2}r^2\frac{\pi}{3} + \frac{1}{2}r^2\frac{\pi}{3}\right]$$

$$A = \text{Area of the equilateral } \Delta^{\text{le}} O_1O_2O_3 - \frac{1}{2}r^2\pi$$

$$A = \text{Area of the equilateral } \Delta^{\text{le}} O_1O_2O_3 - \frac{1}{2}\text{area of the circle}$$

(ii) The area A of the space enclosed by four equal mutually touching circles of diameter d is given by

$$A = d^2 - \frac{\pi d^2}{4}$$

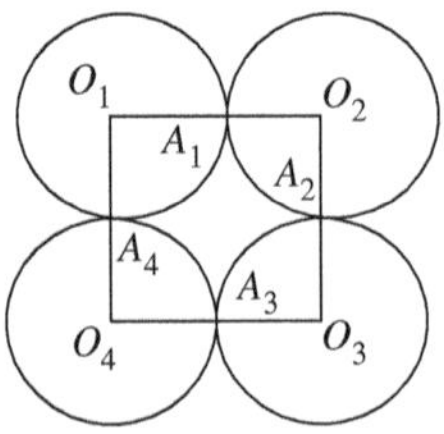

Proof: Let $O_1O_2O_3$, and O_4 be the centers of each of the four circles and let the circles touch each mutually at A_1, A_2, A_3, and A_4.

Let A = shaded area.

Then,

$$
\begin{aligned}
A = {} & \text{Area of square } O_1O_2O_3O_4 \\
& - [\text{Area of the sector } OA_1A_4 \\
& + \text{Area of the sector } O_2A_1A_2 \\
& + \text{Area of the sector } A_2O_3A_3 \\
& + \text{Area of the sector } A_3O_4A_4]
\end{aligned}
$$

Since each of the inner angles of the four sectors are $= 90^\circ = \pi/2$, the area of each sector

$$= \frac{1}{2}r^2 \cdot \frac{\pi}{2} = \frac{1}{2}\left(\frac{d}{2}\right)^2 \frac{\pi}{2} = \frac{d^2}{16}\pi$$

The side of the square $O_1O_2O_3O_4$ is d, the diameter of each circle.
Thus,

$$A = d^2 - 4 \cdot \frac{d^2\pi}{16} = d^2 - \frac{d^2\pi}{4}$$

Mahavira has also attempted to give a formula for the area of space enclosed by any number of equal circles:

रज्जवधैकृतित्त्र्यंशो बाहुविभक्तो निरेकवाहुगु णः।
सर्वेषामश्रवतां फलं हि बिंबान्तरे चतुर्थाशः ॥

(G.S.S. VII. 39)

i.e. "When the quotient of one-third the square of the semiperimeter divided by the number of sides is multiplied by the number of sides minus one gives the area of all regular figures with corners (or rectilinear sides). One-fourth of the above area is the area enclosed between the circles."

This means, if a polygon has n sides each equal to a, its area

$$= \frac{(na/2)^2}{3}\frac{n-1}{n} = \frac{S^2}{3}\frac{n-1}{n}$$

where S is the semiperimeter.

The above formula is true only if triangles, squares, and circles are considered as polygon with infinite number of sides and if approximate values are taken.

Rational Right Triangles

Mahavira has given a thorough examination of rational figures under the heading "Janyavyavahara." First, he provides general solution for rational right triangle:

वगंविशेषः कोटिः संवर्गो द्विगुणितो भवेद् बाहुः।
वर्गस्रमासः कर्णश्चायतचतुरश्रजन्यस्य ॥

(G.S.S. VII.90½)

i.e., "The difference of the squares of two elements ('bijas') is the upright ('koti'), twice their product is the base ('Bhuja') and the sum of the squares is the diagonal ('karṇa') of the generated rectangle."

A rational right triangle can be produced by the choice of two arbitrary positive integers m and n as the elements ("bijas"). Then, we observe that the sides of right triangle are $m^2 - n^2$, $2mn$, $m^2 + n^2$. We can observe that Mahavira like his predecessors Brahmagupta and the vedic scholars of "Sulvasutra" mentioned of the rectangle instead of right triangle.

Given a side containing a right angle, he gave the following rule:

> *"The 'sankramana' between an optional divisor of the square of the given upright ('Koti') or the base ('Bhuja') and the respective quotient is the diagonal and the base or the upright."*

He gave two rules for the selection of elements ("bijas").

First rule:

The "sankramana" between a rational devisor of the upright and the quotient are the elements.

Second rule:

The "sankramana" between any rational divisor of half the given side and the quotient are the elements.

Let a be the given side containing the right triangle. Let p be an optional number.

Then, according to the first rule, the elements are

$$m = \frac{1}{2}\left[\frac{a}{P} + P\right] \text{ and } n = \frac{1}{2}\left[\frac{a}{P} - P\right]$$

Then, the sides of the right triangle are as follows:

$$2mn = 2 \cdot \frac{1}{2}\left[\frac{a}{P} + P\right]\frac{1}{2}\left[\frac{a}{P} - P\right] = \frac{1}{2}\left[\frac{a^2}{P^2} - P^2\right]$$

$$m^2 - n^2 = \left\{\frac{1}{2}\left(\frac{a}{P} + P\right)\right\}^2 - \left\{\frac{1}{2}\left(\frac{a}{P} - P\right)\right\}^2$$

$$= \frac{1}{4}\left[\left(\frac{a}{P} + P\right)^2 - \left(\frac{a}{P} - P\right)^2\right] = \frac{1}{4}\left[4\frac{a}{P} \cdot P\right] = a$$

$$m^2 + n^2 = \left\{\frac{1}{2}\left(\frac{a}{P} + P\right)\right\}^2 + \left\{\frac{1}{2}\left(\frac{a}{P} - P\right)\right\}^2$$

$$= \frac{1}{4}\left[\left(\frac{a}{P} + P\right)^2 + \left(\frac{a}{P} - P\right)^2\right] = \frac{1}{4} \cdot 2\left[\frac{a^2}{P^2} + P^2\right]$$

$$= \frac{1}{2}\left(\frac{a^2}{P^2} + P^2\right)$$

So, the sides of the right triangle are

$$a,\ \frac{1}{2}\left[\frac{a^2}{p^2} - p^2\right] \text{ and } \frac{1}{2}\left[\frac{a^2}{P^2} + P^2\right]$$

According to the second rule, if q is an optional number, the elements are q and $a/2q$.

If $m = q$ and $n = a/2q$, the sides of the right triangle are

$$2mn = 2q\left(\frac{a}{2q}\right) = a$$

$$m^2 - n^2 = q^2 - \left(\frac{a}{2q}\right)^2 = q^2 - \frac{a^2}{4q^2}$$

$$m^2 + n^2 = q^2 + \left(\frac{a}{2q}\right)^2 = q^2 + \frac{a^2}{4q^2}$$

For the solution of a right triangle given the hypotenuse, Mahavira has suggested three rules. They are the following:

First rule:

"The square root of half the sum and difference of the diagonal (hypotenuse) and the square of an optional number give the elements."

Given the hypotenuse c, let p be an optional number. Then, the elements are

$$m = \sqrt{\frac{1}{2}(c + p^2)} \text{ and } n = \sqrt{\frac{1}{2}(c - p^2)}$$

Hence, the sides of the right triangle are

$$2mn = 2\sqrt{\frac{1}{2}(c + p^2)}\sqrt{\frac{1}{2}(c - p^2)}$$

$$= \sqrt{(c + p^2)(c - p^2)} = \sqrt{c^2 - p^4}$$

$$m^2 - n^2 = \frac{1}{2}(c + p^2) - \frac{1}{2}(c - p^2) = \frac{1}{2}(2p^2) = p^2$$

$$m^2 + n^2 = \frac{1}{2}(c + p^2) + \frac{1}{2}(c - p^2) = \frac{1}{2}(2c) = c$$

So, the sides of the right triangle are p^2, $\sqrt{c^2 - p^4}$, and c.

Second rule:

"Square root of the difference of the squares of the diagonal (hypotenuse) and an optional number and the optional number will be the upright and the base."

If c is the given hypotenuse and p an optional number, then the sides of the right triangle are p, $\sqrt{c^2 - p^2}$, and c.

It is noteworthy to remark that the above two sets of solution of a right triangle are not totally correct, unless p is so chosen to make $\sqrt{c^2 - p^4}$ and $\sqrt{c^2 - p^2}$ rational.

Third rule:

"Form different right triangles from the elements. Divide the three sides of these right triangles by their diagonal (hypotenuse). Multiplying by the given hypotenuse, we will get the three sides of the right triangle."

Having obtained the general solution of rational right triangle with sides $m^2 - n^2$, $2mn$, $m^2 + n^2$ where m and n are rational numbers and if x, y, c are the sides of the required right triangle, then

$$\frac{x}{m^2 - n^2} = \frac{y}{2mn} = \frac{c}{m^2 + n^2}$$

So,

$$x = \frac{(m^2 - n^2)c}{m^2 + n^2} \quad \text{and} \quad y = \frac{2mnc}{m^2 - n^2}$$

Hence, the three sides of the required right triangle are

$$\frac{c(m^2 - n^2)}{m^2 + n^2}, \ \frac{2mnc}{m^2 - n^2} \ \text{and } c$$

To demonstrate his method, Mahavira gave four right triangles (39,52), (25,60), (33,56), and (16,63) having the same diagonal (hypotenuse) 65.

(G.S.S. 123–124½)

Proof: (i) Let $m = 2$, $n = 1$, $c = 65$

Then,

$$\frac{m^2 - n^2}{m^2 + n^2}c = \frac{4 - 1}{4 + 1}(65) = \frac{3}{5}(65) = 39$$

$$\frac{2mn}{m^2 + n^2}c = \frac{2(2)(1)}{4 + 1} \cdot 65 = \frac{4}{5}(65) = 52$$

(ii) Let $m = 3$, $n = 2$, $c = 65$
Then,

$$\frac{m^2 - n^2}{m^2 + n^2}c = \frac{9 - 4}{9 + 4} \cdot 65 = \frac{5}{13}(65) = 25$$

$$\frac{2mn}{m^2 + n^2}c = \frac{2(3)(2)}{9 + 4} \cdot 65 = \frac{12}{13}(65) = 60$$

(iii) Let $m = 7$, $n = 4$, $c = 65$
Then,

$$\frac{m^2 - n^2}{m^2 + n^2}c = \frac{49 - 16}{49 + 16} \cdot 65 = \frac{33}{65}(65) = 33$$

$$\frac{2mn}{m^2 + n^2}c = \frac{2(7)(4)}{49 + 16} \cdot 65 = 56$$

(iv) Let $m = 9$, $n = 7$, $c = 65$
Then,

$$\frac{m^2 - n^2}{m^2 + n^2}c = \frac{81 - 49}{81 + 49} \cdot 65 = \frac{32}{130}(65) = 16$$

$$\frac{2mn}{m^2 + n^2}c = \frac{2(9)(7)}{81 + 49} \cdot 65 = \frac{126}{130}(65) = 63$$

It is noteworthy to point out that this method was rediscovered in Europe by Leonardo Fibonacci of Pisa (1202) and Vita more than three centuries later.

Areas and Sides

Mahavira described a method to obtain rational triangles (or squares) in which the area will be equal to any multiple of a side, a diagonal or perimeter, or any linear combination of two or more of them.

This means, we have to solve

$$x^2 + y^2 = z^2$$

$$ax + by + cz = dxy \quad (*) \qquad \text{(G.S.S. VII, 112 ½)}$$

where a, b, c, and d are rational numbers with $d \neq 0$.

His method is as follows:

Start with any rational solution of

$$p^2 + q^2 = r^2 \quad (**)$$

We have to calculate the value of

$$ap + bq + cr = Q \text{ (say)}$$

Then,

$$\frac{x}{p} = \frac{y}{q} = \frac{z}{r} = K = \frac{Q}{dpq}$$

$$\Rightarrow x = \frac{Q}{dq}, \quad y = \frac{Q}{dp}, \quad z = \frac{rQ}{dpq}$$

We will now give two examples given by Mahavira to illustrate his above method.

Example 1

"In a rectangle, the area is numerically equal to the perimeter. In another rectangle, the area is numerically equal to the diagonal. What are the sides in each case!" *(G.S.S. VII. 115½)*

Solution

Let x and y be the sides of the rectangle and z be the diagonal of the rectangle.

Then, we have to solve two separate simultaneous equations, namely

$$\begin{aligned} x^2 + y^2 &= z^2 \\ 2(x+y) &= xy \end{aligned} \tag{1}$$

and

$$\begin{aligned} x^2 + y^2 &= z^2 \\ xy &= z \end{aligned} \tag{2}$$

Let $p = s^2 - t^2$, $q = 2st$, and $r = s^2 + t^2$ be the solution of $p^2 + q^2 = r^2$.

Comparing the second equation of (1) with $ax + by + cz = dxy$, we get $a = b = 2$, $c = 0$, $d = 1$.

Then, $Q = ap + bq + cr = 2(s^2 - t^2) + 2(2st) + 0$.

$$Q = 2s^2 + 4st - 2t^2$$

$$x = \frac{Q}{dq} = \frac{2s^2 + 4st - 2t^2}{2st}, \quad y = \frac{Q}{dp} = \frac{2s^2 + 4sd - 2t^2}{s^2 - t^2}$$

$$z = \frac{rQ}{dpq} = \frac{(s^2 + t^2)(2s^2 + 4st - 2t^2)}{2(s^2 - t^2)st}$$

Comparing the second equation of (2), with $ax + by + cz = dxy$, we get, $a = 0, b = 0, c = 1, d = 1$.

Then,

$$Q = ap + bq + cr = 0 + 0 + 1 \cdot (s^2 + t^2) = s^2 + t^2$$

Now,

$$X = \frac{Q}{dq} = \frac{s^2 + t^2}{2st}, \quad Y = \frac{Q}{dp} = \frac{s^2 + t^2}{s^2 - t^2}$$

and

$$Z = \frac{rQ}{dpq} = \frac{(s^2 + t^2)(s^2 + t^2)}{2st(s^2 - t^2)} = \frac{(s^2+t^2)^2}{2st(s^2 - t^2)}$$

Example 2

"Find a rectangle in which twice the diagonal thrice the base and four times the upright and twice the perimeter are together equal to area."
(G.S.S. vii. 119½)

Solution

Let x and y be the base and upright of the rectangle and z be the diagonal at the rectangle.

Then, we have to solve

$$x^2 + y^2 = z^2$$

$$2z + 3x + 4y + 2(2x + 2y) = xy$$

i.e., $7x + 8y + 2z = xy$.

Comparing this equation with $ax + by + cz = dxy$, we get

$$a = 7, b = 8, c = 2, d = 1$$

Now, $Q = ap + bq + cr$.

If $p = S^2 - t^2$, $q = 2st$, and $r = s^2 + t^2$, then $Q = 7(s^2 - t^2) + 8(2st) + 2(s^2 + t^2)$.

$$= 9s^2 + 16st - 5t^2$$

(*Continued*)

Example 2 (cont'd)

Now,

$$x = \frac{Q}{dq} = \frac{9s^2 + 16st - 5t^2}{2st}$$

$$y = \frac{Q}{dp} = \frac{9s^2 + 16st - 5t^2}{s^2 - t^2}$$

$$Z = \frac{rQ}{dpq} = \frac{(s^2 + t^2)(9s^2 + 16st - 5t^2)}{2st(s^2 - t^2)}$$

Mahavira has also obtained rational right triangles, when the sum of sides multiplied by arbitrary rational numbers is given.

Let x, y be sides of the right triangle and let z be the hypotenuse. Then, we have to solve

(i) $x^2 + y^2 = z^2$

(ii) $ax + by + cz = l$, where a, b, c, and l are all rational numbers. His method is as follows:

Let (p,q,r) be a solution of (i).

Let $ap + bq + cr = Q$.

Then,

$$\frac{x}{p} = \frac{y}{q} = \frac{z}{r} = \frac{l}{Q}$$

So,

$$x = \frac{pl}{Q}, \quad y = \frac{ql}{Q}, \quad z = \frac{rl}{Q}$$

Example 1

"If the perimeter of a rectangle is unity, tell me quickly, what are its base and upright."
(G.S.S. vii. 118½)

Solution

Let x and y be sides of the rectangle. Let z be the diagonal of the rectangle.

We have to solve

$$x^2 + y^2 = z^2 \quad (1)$$

$$2(x + y) = 1 \quad (2)$$

Let (p,q,r) be a solution of (1), where

$$p = s^2 - t^2, \quad q = 2st, \quad r = s^2 + t^2$$

Let $Q = ap + bq + cr$.
Comparing (2) with $ax + by + cz = l$,

$$a = 2, \quad b = 2, \quad c = 0, \quad l = 1$$

Then, $Q = 2p + 2q + Or = 2(s^2 - t^2) + 2(2st) = 2s^2 + 4st - 2t^2$.
Now,

$$x = \frac{Pl}{Q} = \frac{(s^2 - t^2) \cdot 1}{2s^2 + 4st - 2t^2} = \frac{s^2 - t^2}{2s^2 + 4st - 2t^2} = \frac{s^2 - t^2}{2(s^2 + 2st - t^2)}$$

$$y = \frac{ql}{Q} = \frac{2st \cdot 1}{2s^2 + 4st - 2t^2} = \frac{st}{s^2 + 2st - t^2}$$

$$z = \frac{rl}{Q} = \frac{s^2 + t^2}{2s^2 + 4st - 2t^2} = \frac{s^2 + t^2}{2(s^2 + 2st - t^2)}$$

Example 2

"Find a rectangle in which twice the diagonal, thrice the base and four times the upright, and the perimeter together is equal to unity."
(G.S.S. 119½)

Solution
Let x be the base, y be the upright, and z be the diagonal of a rectangle. Then, we have to solve

$$x^2 + y^2 = z^2 \quad (1)$$

$$2z + 3x + 4y + 2(x + y) = 1 \quad (2)$$

(*Continued*)

Example 2 (cont'd)

Now, (2) becomes

$$2z + 5x + 6y = 1 \tag{3}$$

Let (p,q,r) be a solution of (1), where

$$p = s^2 - t^2,\quad q = 2st,\quad \text{and } r = s^2 + t^2$$

Comparing (3) with $ax + by + cz = l$, we get $a = 5$, $b = 6$, $c = 2$, and $l = 1$. Hence,

$$Q = ap + bq + cr$$

$$= 5(s^2 - t^2) + 6(2st) + 2(s^2 + t^2)$$

$$Q = 7s^2 + 12st - 3t^2$$

Then,

$$x = \frac{pl}{Q} = \frac{(s^2 - t^2)}{7s^2 + 12st - 3t^2},\quad y = \frac{ql}{Q} = \frac{(2st)}{7s^2 + 12st - 3t^2}$$

and

$$z = \frac{rl}{Q} = \frac{(s^2 + t^2)}{7s^2 + 12st - 3t^2}$$

Pairs of Rectangles

Mahavira examined the following general problem.

To find (x,y) and (u,v) representing the base and the upright respectively of two rectangles satisfying the relation.

$$a(x + y) = b(u + v)*$$

$cxy = duv$, where a, b, c, and d are known integers.

Let

$$s = \frac{a}{b} \text{ and } t = \frac{d}{c}$$

Case 1: Let $s \geq 1$ and $t > 1$

Let $y = ms^2t^2$.

$$x = 3[y - mt + 1] = 3[ms^2t^2 - mt + 1]$$

Substituting in (*), we get

$$u + v = \frac{2a}{2b}(x + y) = s(x + y)$$

$$u + v = s[ms^2t^2 + 3(ms^2t^2 - mt + 1)]$$

$$= s[4ms^2t^2 - 3mt + 3]$$

$$uv = \frac{c}{d}xy = \frac{3}{t}\left[ms^2t^2\right]\left[ms^2t^2 - mt + 1\right]$$

Now, $(u - v)^2 = (u + v)^2 - 4uv$

$$= s^2[4ms^2t^2 - 3mt + 3]^2 - 12ms^2t[ms^2t^2 - mt + 1]$$

$$= s^2[(4ms^2t^2 - 3mt + 3)^2 - 12mt(ms^2t^2 - mt + 1)]$$

$$= s^2[16m^2s^4t^4 + 9m^2t^2 + 9 - 24m^2s^2t^3 + 24ms^2t^2 - 18mt$$

$$- 12m^2s^2t^3 + 12m^2t^2 - 12mt]$$

$$= s^2[16m^2s^4t^4 + 21m^2t^2 - 30mt - 36m^2s^2t^3 + 9 + 24ms^2t^2]$$

$$= s^2\left[16m^2s^4t^4 + \frac{81}{4}m^2t^2 + 9 - 36m^2s^2t^3 - 27mt + 24ms^2t^2\right.$$

$$\left. + \frac{3}{4}m^2t^2 - 3mt\right]$$

$$= s^2\left[\left(4ms^2t^2 - \frac{9}{2}mt + 3\right)^2 + \frac{3}{4}mt(mt - 4)\right]$$

Choose the arbitrary number m such that

$$mt - 4 = 0 \Rightarrow m = \frac{4}{t}$$

Then,

$$(u-v)^2 = s^2\left(4ms^2t^2 - \frac{9}{2}mt+3\right)^2$$

where

$$m = \frac{4}{t}$$

Thus,

$$u - v = s\left(4ms^2t^2 - \frac{9}{2}mt + 3\right)$$

where

$$m = \frac{4}{t}$$

$$= s\left[4 \cdot \frac{4}{t}s^2t^2 - \frac{9}{2}\frac{4}{t} \cdot t + 3\right]$$

$$u - v = s[16s^2t - 18 + 3] = s[16s^2t - 15] \quad \text{(i)}$$

We have $u + v = s[4ms^2t^2 - 3mt + 3]$

$$= s\left[4 \cdot \frac{4}{t}s^2t^2 - 3 \cdot \frac{4}{t}t + 3\right]$$

$$= s[16s^2t - 12 + 3] = s[16s^2t - 9] \quad \text{(ii)}$$

Adding (i) and (ii), we get

$$2u = s[16s^2t - 15 + 16s^2t - 9]$$

$$2u = s[32s^2t - 24] \Rightarrow u = s[16s^2t - 12]$$

Subtracting (i) from (ii), we get

$$2v = s[16s^2t - 9 - (16s^2t - 15)]$$

$$2v = s(6) = 6s$$

So, $v = 3s$.

Substituting in the values of x and y, we have

$$y = ms^2t^2 = \frac{4}{t}s^2t^2 = 4s^2t = 4\frac{a^2}{b^2}\cdot\frac{d}{c} = \frac{4a^2d}{b^2c}$$

$$x = 3[y - mt + 1] = 3\left[4s^2t - \frac{4}{t}t + 1\right]$$

$$= 3\left[4s^2t - 3\right] = 3\left[4\frac{a^2}{b^2}\frac{d}{c} - 3\right]$$

$$= 3\left[\frac{4a^2d}{b^2c} - 3\right]$$

Case 2: $t = 1$

Mahavira has suggested that in this case let $y = ms^2t^2 = ms^2$ and $x = y - 1 = ms^2 - 1$.

Bibhushana Dutta [29] has shown that the above restriction is unnecessary.

Dutta takes, for $s \geq 1$ and $t \geq 1$

$$y = ms^2t^2 \text{ and } x = y - 1$$

where m is an arbitrary number.

Substituting in (*) we get

$$u + v = \frac{2a}{2b}(x + y) = s(x + y) = s\left[ms^2t^2 - 1 + ms^2t^2\right]$$

i.e., $u + v = s(2ms^2t^2 - 1)$

$$uv = \frac{c}{d}xy = \frac{1}{t}xy = \frac{1}{t}\left[ms^2t^2 - 1\right]\left[ms^2t^2\right]$$

$$= ms^2t[ms^2t^2 - 1]$$

Now, $(u - v)^2 = (u + v)^2 - 4uv$

$$= s^2(2ms^2t^2 - 1)^2 - 4ms^2t[ms^2t^2 - 1]$$

i.e., $(u - v)^2 = s^2[(2ms^2t^2 - 1)^2 - 4mt(ms^2t^2 - 1)]$

$$= s^2[4m^2s^4t^4 - 4ms^2t^2 + 1 - 4m^2s^2t^3 + 4mt]$$

$$(u-v)^2 = s^2[(4m^2s^4t^4 - 4ms^2t^2 + 1 - 4m^2s^2t^3 + m^2t^2 + 2mt] - [m^2t^2 - 2mt]$$

i.e., $(u-v)^2 = s^2[(2ms^2t^2 - mt - 1)^2 + mt(2 - mt)]$
Choose the arbitrary number m such that

$$2 - mt = 0 \Rightarrow 2 = mt \Rightarrow m = \frac{2}{t}$$

Then,

$$(u-v)^2 = s^2(2ms^2t^2 - mt - 1)^2$$

where

$$m = \frac{2}{t}$$

So,

$$u - v = s(2ms^2t - mt - 1) \quad \text{(iii)}$$

We have

$$u + v = s(2ms^2t^2 - 1) \quad \text{(iv)}$$

Adding (iii) and (iv), we get

$$2u = s[2ms^2t^2 - mt - 1 + 2ms^2t^2 - 1]$$

$$2u = s[4ms^2t^2 - mt - 2]$$

So,

$$u = s\left[2ms^2t^2 - \frac{mt}{2} - 1\right]$$

Subtracting (iii) from (iv), we get

$$2v = s[2ms^2t^2 - 1 - (2ms^2t - mt - 1)]$$

$$2v = s[mt]$$

$$v = \frac{mts}{2}$$

Hence, substituting

$$m = \frac{2}{t}, \quad u = s\left[2 \cdot \frac{2}{t} s^2 t^2 - \frac{1}{2}\frac{2}{t} t - 1\right]$$

$$u = s[4s^2 t - 2] = 2s[2s^2 t - 1]$$

$$= \frac{2a}{b}\left[\frac{2a^2 t}{b} - 1\right] = \frac{2a}{b}\left[\frac{2a^2}{b^2}\frac{d}{c} - 1\right]$$

$$v = \frac{2}{t}\frac{ts}{2} = s = \frac{a}{b}$$

Now, substituting in x and y, we get

$$y = ms^2t^2 = \frac{2}{t} s^2 t^2 = 2s^2 t = 2\frac{a^2}{b^2}\frac{d}{c} = \frac{2a^2 d}{b^2 c}$$

and

$$x = y - 1 = \frac{2a^2 d}{b^2 c} - 1$$

Isosceles Triangles with Integral Sides

Mahavira gave the following rule for isosceles triangles with integral sides:

Let m and n be two integers such that $m \neq n$. If the equal sides are of length $m^2 + n^2$, then the base of this isosceles triangle is either (i) $2(m^2 - n^2)$ or (ii) $4mn$.

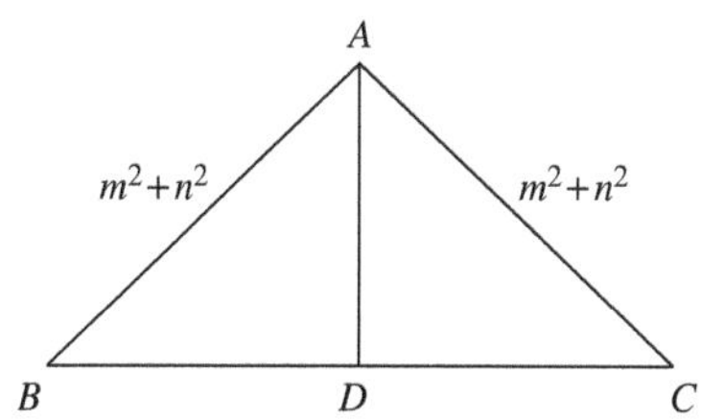

Proof: Let ABC be an isosceles triangle with equal sides $AC = AB = m^2 + n^2$. Draw AD perpendicular to the base BC. Let $AD = p$.

If $BD = x$, then $BD = DC = x \Rightarrow BC = 2x$.

In the right triangle ABD,

$$BD^2 = AB^2 - AD^2 = (m^2 + n^2)^2 - p^2$$

Then,

$$x = BD = \sqrt{(m^2+n^2)^2 - p^2}$$

Case 1: Let $p = 2mn$

Then,

$$x = \sqrt{(m^2 + n^2) - 4m^2n^2} = \sqrt{m^4 + n^4 - 2m^2n^2}$$

$$x = \sqrt{(m^2-n^2)^2} = (m^2 - n^2)$$

Base $BC = 2x = 2(m^2 - n^2)$.

Case 2: Let $p = m^2 - n^2$

Then,

$$x = \sqrt{(m^2 + n^2) - (m^2-n^2)^2} = \sqrt{4m^2n^2}$$

$$x = 2mn$$

Base $BC = 2x = 4mn$.

Mahavira gave the following rule for obtaining two rational isosceles triangles, whose perimeters as well as their areas related in given proportions.

Let s_1 and s_2 be the perimeters of the two triangles and let A_1 and A_2 be their respective areas.

Let

$$\frac{s_1}{s_2} = \frac{m}{n} \text{ and } \frac{A_1}{A_2} = \frac{p}{q}$$

where m, n, p, q are all known integers.

The sides of an isosceles triangle are

$$m_*^2 + n_*^2,\ m_*^2 + n_*^2 \text{ and } 4m_*n_*$$

where m_* and n_* are two arbitrary integers.

For the first isosceles triangle, let $m_* = 6ak_1$ and $n_* = (2a - 1)k_1$.

For the second isosceles triangle, let $m_* = (4a + 1)k_2$ and $n_* = (4a - 2)k_2$.

Then, the sides of the first isosceles triangle are

$$m_*^2 + n_*^2 = (6ak_1)^2 + \{(2a-1)k_1\}^2 = k_1^2\{36a^2 + 4a^2 - 4a + 1\}$$

$$= k_1^2(40a^2 - 4a + 1)$$

$$4m_*n_* = 4(6ak_1) \cdot (2a - 1)k_1 = 24ak_1^2(2a - 1)$$

Perimeter of the first isosceles triangle $= s_1$.

$$s_1 = 2(m_*^2 + n_*^2) + 4m_*n_*$$

$$= 2k_1^2(40a^2 - 4a + 1) + 24ak_1^2(2a - 1)$$

$$= 2k_1^2[40a^2 - 4a + 1 + 12a(2a - 1)]$$

$$s_1 = 2k_1^2[64a^2 - 16a + 1] = 2k_1^2(8a-1)^2$$

Area of the first isosceles triangle $= A_1$.

$$A_1 = \frac{1}{2}\text{base} \times \text{height(altitude)}$$

$$= \frac{1}{2}(4m_*n_*)(m_*^2 - n_*^2)$$

$$= 2(6ak_1)(2a - 1)k_1[(6ak_1)^2 - \{k_1(2a-1)\}^2]$$

$$= 12ak_1^4(2a - 1)[36a^2 - (4a^2 - 4a + 1)]$$

$$= 12ak_1^4(2a - 1)(32a^2 + 4a - 1)$$

$$A_1 = 12ak_1^4(2a - 1)(8a - 1)(4a + 1)$$

Now, the sides of the second isosceles triangle are

$$m_*^2 + n_*^2 = \{(4a+1)k_2\}^2 + \{(4a-2)k_2\}^2$$

$$= k_2^2[16a^2 + 8a + 1 + 16a^2 - 16a + 4]$$

$$= k_2^2[32a^2 - 8a + 5]$$

$$4m_*n_* = 4(4a + 1)k_2(4a - 2)k_2$$

$$= 4(4a + 1)(4a - 2)k_2^2$$

The perimeter of the second isosceles triangle $= s_2$.

$$s_2 = 2(m_*^2 + n_*^2) + 4m_*n_*$$

$$= 2k_2^2(32a^2 - 8a + 5) + 4k_2^2(16a^2 - 4a - 2)$$

$$= 2k_2^2[32a^2 - 8a + 5 + 2(16a^2 - 4a - 2)]$$

$$s_2 = 2k_2^2(64a^2 - 16a + 1) = 2k_2^2(8a-1)^2$$

The area of second isosceles triangle $= A_2$

$$= \frac{1}{2}\text{base} \times \text{altitude}$$

$$= \frac{1}{2}(4m_*n_*)(m_*^2 - n_*^2)$$

$$= 2m_*n_*(m_*^2 - n_*^2)$$

$$= 2(4a + 1)k_2(4a - 2)k_2[\{(4a+1)k_2\}^2 - \{(4a-2)k_2\}^2]$$

$$= 2k_2^2(4a + 1)(4a - 2)k_2^2[16a^2 + 8a + 1 - (16a^2 - 16a + 4)]$$

$$= 2k_2^4(4a + 1)(4a - 2)[24a - 3]$$

$$= 6k_2^4(4a + 1)(4a - 2)(8a - 1)$$

Now,

$$\frac{s_1}{s_2} = \frac{m}{n}$$

$$\frac{2k_1^2(8a-1)^2}{2k_2^2(8a-1)^2} = \frac{m}{n} \Rightarrow \frac{k_1^2}{k_2^2} = \frac{m}{n}$$

We can choose $k_1^2 = m$ and $k_2^2 = n$.
Now,

$$\frac{A_1}{A_2} = \frac{p}{q}$$

i.e.,

$$\frac{12ak_1^4(2a - 1)(8a - 1)(4a + 1)}{6k_2^4(4a + 1)(4a - 2)(8a - 1)} = \frac{p}{q}$$

i.e.,

$$a\frac{k_1^4}{k_2^4}=\frac{p}{q} \quad a\frac{m^2}{n^2}=\frac{p}{q}$$

So,

$$a=\frac{n^2P}{m^2q}$$

Hence, the equal sides of the first isosceles triangle are each equal to $(6ak_1)^2+\{(2a-1)k_1\}^2$

$$=k_1^2[(6a)^2+(2a-1)^2]$$

$$=m\left[\left(\frac{6n^2p}{m^2q}\right)^2+\left\{2\frac{n^2p}{m^2q}-1\right\}^2\right]$$

The base of the first isosceles triangle

$$=24ak_1^2(2a-1)$$

$$=24m\left(\frac{n^2p}{m^2q}\right)\left[\frac{2n^2p}{m^2q}-1\right]$$

Altitude of the first isosceles triangle

$$=m_*^2-n_*^2=(6ak_1)^2-\{(2a-1)k_1\}^2$$

$$=k_1^2\{(6a)^2-(2a-1)^2\}$$

$$=m\left\{\left(6\frac{n^2p}{m^2q}\right)^2-\left(\frac{2n^2P}{m^2q}-1\right)^2\right\}$$

Equal sides of the second isosceles triangle are equal to

$$\{(4a+1)k_2\}^2+\{(4a-2)k_2\}^2$$

$$=k_2^2[(4a+1)^2+(4a-2)^2]$$

$$=n\left[\left(4\frac{n^2p}{m^2q}+1\right)^2+\left(\frac{4n^2p}{m^2q}-2\right)^2\right]$$

The base of the second isosceles triangle is equal to $4k_2^2(4a+1)(4a-2)$

$$= 4n\left[\left(4\frac{n^2 p}{m^2 q}+1\right)\left(4\frac{n^2 p}{m^2 q}-2\right)\right]$$

Altitude of the second isosceles triangle

$$= m_*^2 - n_*^2 = \{(4a+1)k_2\}^2 - \{(4a-2)k_2\}^2$$

$$= k_2^2[(4a+1)^2 - (4a-2)^2]$$

$$= n\left[\left(4\frac{n^2 p}{m^2 q}+1\right)^2 - \left(\frac{4n^2 p}{m^2 q}-2\right)^2\right]$$

It is noteworthy to observe that if we multiply the above values by m^4q^2, we get pairs of isosceles triangle of integral sides.

An Important Special Case

Let $m = n = p = q = 1$.

First isosceles triangle:

$$\text{Equal side} = m\left\{\left(\frac{6n^2 p}{m^2 q}\right)^2 + \left(\frac{2n^2 p}{m^2 q}-1\right)^2\right\}$$

$$= \left\{(6)^2 + \left(\frac{2(1)}{1}-1\right)^2\right\} = 36 + 1 = 37$$

$$\text{Base} = \frac{24n^2 pm}{m^2 q}\left[\frac{2n^2 p}{m^2 q}-1\right]$$

$$= 24[2-1] = 24$$

Perimeter $= 2(37) + 24 = 74 + 24 = 98.$

$$\text{Altitude} = m\left\{\left(\frac{6n^2 p}{m^2 q}\right)^2 - \left(\frac{2n^2 p}{m^2 q}-1\right)^2\right\}$$

$$= \{36 - 1\} = 35$$

$$\text{Area} = \frac{1}{2}\text{base} \times \text{altitude} = \frac{1}{2}(24)(35) = 420 \text{ square units}$$

Second isosceles triangle:

$$\text{Equal side} = n\left\{\left(\frac{4n^2 p}{m^2 q} + 1\right)^2 + \left(\frac{4n^2 p}{m^2 q} - 2\right)^2\right\}$$

$$= \{(4+1)^2 + (4-2)^2\} = 25 + 4 = 29$$

$$\text{Base} = 4n\left[\left(\frac{4n^2 p}{m^2 q} + 1\right)\left(\frac{4n^2 p}{m^2 q} - 2\right)\right]$$

$$= 4[(4 + 1)(4 - 2)] = 40$$

$$\text{Altitude} = n\left[\left(\frac{4n^2 p}{m^2 q} + 1\right)^2 - \left(\frac{4n^2 p}{m^2 q} - 2\right)^2\right]$$

$$= [(4+1)^2 - (4-2)^2] = 25 - 4 = 21$$

$$\text{Area} = \frac{1}{2}\text{base} \times \text{altitude} = \frac{1}{2}(40)(21) = 420 \text{ square units}$$

Perimeter = 2(29) + 40 = 58 + 40 = 98.

This particular case was treated by Frans Van Schooten Jr. (AD 1657) and J.H. Ran (AD 1697).

Quadrilaterals

Mahavira divided quadrilaterals into five different types:

- **i.** "Visama" (unequal sides)—mostly the cyclic quadrilateral
- **ii.** "Sama" (all sides equal)—the rhombus and the square
- **iii.** "Dvidsama" (pairs of opposite sides equal)—the parallelogram and the rectangle
- **iv.** "Dvisama" (two sides equal)—the isosceles trapezium
- **v.** "Trisama" (three sides equal)—the trapezium with three equal sides.

Mahavira stated the following for the construction of an isosceles trapezium: (G.S.S. VII. 99½)

"The sum of the perpendicular of the first generated rectangle and the perpendicular of the second rectangle which is generated from the rational factors of the half

the base of the first will be the face; their difference will be the face. The smaller of the diagonals (of the generated rectangles) will be the flank side. The base of either rectangle will be the altitude."

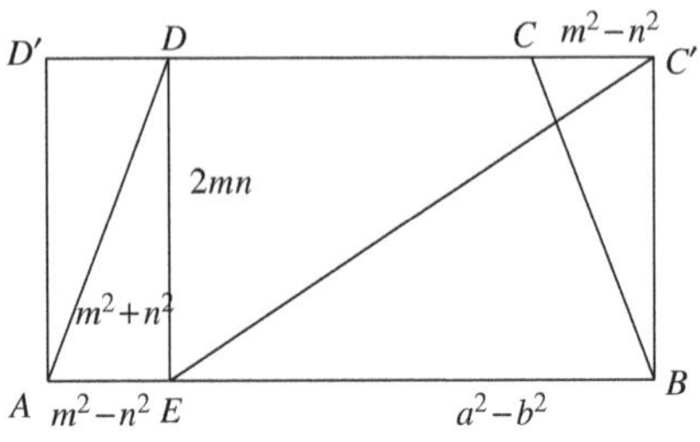

Figure 1

The first rectangle generated from m and n is the rectangle $DD'AE$, where perpendicular $AE = m^2 - n^2$, base $DE = 2mn$, and $AD = m^2 + n^2$.

Now, half of the base $DE = mn$.

If a and b are two rational factors of mn, then $ab = mn$.

The second rectangle formed from these two factors a and b is the rectangle $EBC'D$, where altitude $EB = a^2 - b^2$, base $BC^1 = 2ab = 2mn$, and $BD = EC' = a^2 + b^2$, if $m^2 + n^2 < a^2 + b^2$. By judiciously juxtaposing these two rectangles, we will obtain our trapezium.

Now, $ABCD$ is the required trapezium, where C is a point on DC'such that $AE = C'C = m^2 - n^2$.

Base AB of the trapezium $ABCD = AE + EB = m^2 - n^2 + (a^2 - b^2)$.

Face CD of the trapezium $ABCD = C'D - CC' = BE - CC'$.

Face $CD = a^2 - b^2 - (m^2 - n^2)$.

If $m^2 - n^2 < a^2 - b^2$, then altitude $DE = 2mn = 2ab$.

Area of the trapezium $ABCD$ = area of Δ^{le} ABC + area of Δ^{le} ADC

$$= \frac{1}{2}AB \cdot ED + \frac{1}{2}CD \cdot BC'$$

$$= \frac{1}{2}\left[(m^2 - n^2) + (a^2 - b^2)\right]2mn$$

$$+ \frac{1}{2}\left[(a^2 - b^2) - (m^2 - n^2)\right]2mn, \quad \text{if } m^2 - n^2 < a^2$$

Area of the trapezium $ABCD = (a^2 - b^2)2mn = 2ab(a^2 - b^2)$.

The rationale for the necessity of the conditions $m^2 + n^2 < a^2 + b^2$ and $m^2 - n^2 < a^2 - b^2$ will be obvious Figure 1 and 2. When the above conditions are reversed so that $m^2 - n^2 > a^2 - b^2$ and $m^2 + n^2 > a^2 + b^2$, the sides of the trapezium $ABCD$ from figure 2 will be as follows:

$$AB = AE + EB$$

$$= (m^2 - n^2) + (a^2 - b^2) \text{ if } m^2 - n^2 > a^2 - b^2$$

$$CD = CD' - D'D = AE - D'D$$

$$= m^2 - n^2 - (a^2 - b^2), \text{ if } m^2 - n^2 > a^2 - b^2$$

$$C'D = CD + CC' = CD + D'D = CD' = AE = m^2 - n^2$$

$$CE = 2ab = 2mn = AD'$$

$$AC = m^2 + n^2 \quad BD = m^2 + n^2 \quad BC = a^2 + b^2 = AD$$

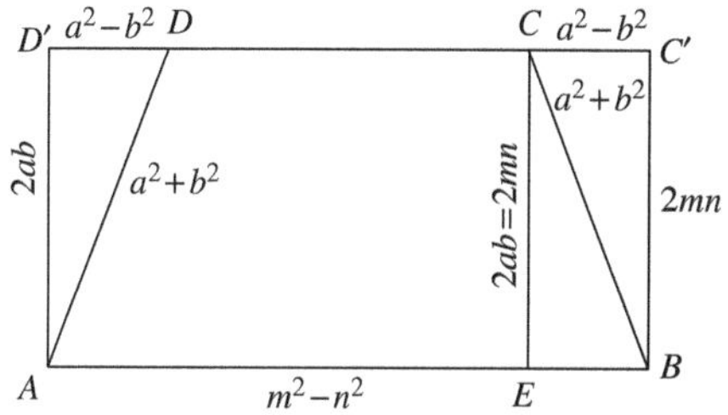

Figure 2

Area of the trapezium $ABDC$ = area of ABC + area of ADC

$$= \frac{1}{2}AB \cdot CE + \frac{1}{2}CD \cdot AD'$$

$$= \frac{1}{2}\left[(m^2 - n^2) + (a^2 - b^2)\right]2ab + \frac{1}{2}\left[(m^2 - n^2) - (a^2 - b^2)\right]2ab$$

$$= (m^2 - n^2)2ab = (m^2 - n^2)2mn$$

Mahavira provided the following method for determining the face, base, and equal sides of an isosceles trapezium having an area and altitude equal to those of another isosceles trapezium with known dimension. His method is as follows: "Perform the 'visama-sankramana' between the square of the perpendicular of the given isosceles trapezium. The bigger result will give the equal sides of the required isosceles trapezium. Add the smaller result to the sum of the base and face of the given isosceles trapezium. One half of that will give the base of the required isosceles trapezium. Subtract the smaller result from the sum of the base and face of the given isosceles trapezium. One half of that will give the face of the required isosceles trapezium."

(G.S.S. VII. 173½–174½)

Let $A'B'CD'$ be the given isosceles trapezium with base $A'B' = a'$, face $C'D' = b'$, equal sides $A'D' = B'C' = c'$, and height $D'E' = h'$.

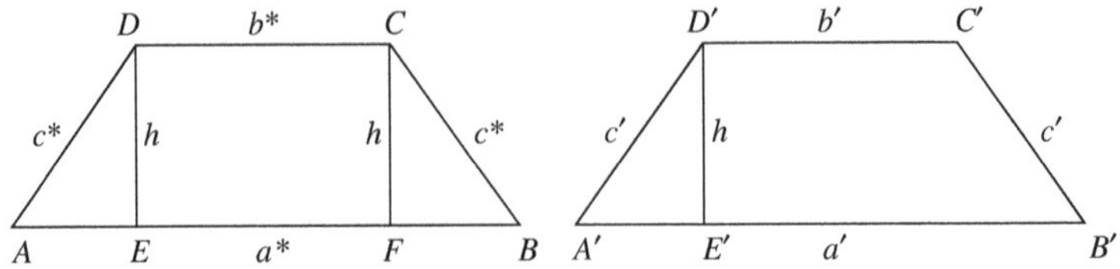

Let $ABCD$ be the required isosceles trapezium with the base $AB = a^*$, face $CD = b^*$, equal sides $AC = BD = c^*$, and height $DE = h$. Now, both trapeziums are equal in area and height of the two trapeziums is also equal.

Then, $h = h'$.

Area of the trapezium

$$A'BC'D' = \frac{1}{2}(a' + b')h$$

Area of the trapezium

$$ABCD = \frac{1}{2}(a^* + b^*)h$$

Since these two are equal, we get

$$a^* + b^* = a' + b' \tag{1}$$

In the right triangles ADE and BFC,

$$AE = FB$$

Now, $AB = AE + EF + FB = EF + 2AE = CD + 2AE$.

So, $a^* = b^* + 2AE$.

$$2AE = a^* - b^* \Rightarrow AE = \frac{a^* - b^*}{2} = FB$$

In the right Δ^{le} AED,

$$DE^2 = AD^2 - AE^2$$

$$h^2 = c^{*2} - \left(\frac{a^* - b^*}{2}\right)^2 = \left[c^* - \frac{a^* - b^*}{2}\right]\left[c^* + \frac{a^* - b^*}{2}\right]$$

Let

$$c^* - \left(\frac{a^* - b^*}{2}\right) = d \tag{2}$$

where d is an optional number.

Then,

$$c^* + \frac{a^* - b^*}{2} = \frac{h^2}{d} \tag{3}$$

Adding (2) and (3), we get

$$2c^* = d + \frac{h^2}{d} \Rightarrow c^* = \frac{d + (h^2/d)}{2} \tag{4}$$

Subtracting (2) from (3), we get

$$a^* - b^* = \frac{h^2}{d} - d \tag{5}$$

But from (1), $a^* + b^* = a' + b'$.
Adding (1) and (5), we get

$$2a^* = \left(\frac{h^2}{d} - d\right) + (a' + b')$$

$$\Rightarrow a^* = \frac{1}{2}\left[\left(\frac{h^2}{d} - d\right) + a' + b'\right] \tag{6}$$

Subtracting (5) from (1), we get

$$2b^* = a' + b' - \left(\frac{h^2}{d} - d\right)$$

i.e.,

$$b^* = \frac{1}{2}\left[a' + b' - \left(\frac{h^2}{d} - d\right)\right] \tag{7}$$

We have stated earlier that if m, n, a, b are rational numbers such that $m^2 \pm n^2 < a^2 \pm b^2$,

Face $C'D' = b' = (a^2 - b^2) - (m^2 - n^2)$

Base $A'B' = a' = (a^2 - b^2) + (m^2 - n^2)$

Equal side $c' = m^2 + n^2$

Altitude $= 2mn = 2ab$

Substituting these values of a',b' in (6) and (7), we get

$$a^* = \frac{1}{2}\left[\frac{h^2}{d} - d\right] + \frac{1}{2}(a' + b') = \frac{1}{2}\left(\frac{h^2}{d} - d\right) + a^2 - b^2$$

$$b^* = \frac{1}{2}(a' + b') - \frac{1}{2}\left(\frac{h^2}{d} - d\right) = a^2 - b^2 - \frac{1}{2}\left(\frac{h^2}{d} - d\right)$$

$$a^* = (a^2 - b^2) + \frac{1}{2}\left[\frac{4a^2b^2}{d} - d\right], \quad b^* = (a^2 - b^2) - \frac{1}{2}\left[\frac{4a^2b^2}{d} - d\right]$$

and

$$c^* = \frac{1}{2}\left(\frac{h^2}{d} + d\right) \quad \text{from (4)}$$

$$= \frac{1}{2}\left[\frac{4a^2b^2}{d} + d\right]$$

Also, we have stated earlier that if $m^2 \pm n^2 > a^2 \pm b^2$,

Base $A'B' = a' = (m^2 - n^2) + (a^2 - b^2)$

Face $C'D' = b' = (m^2 - n^2) - (a^2 - b^2)$

Equal side $= c' = m^2 + n^2$ and altitude $= 2mn = 2ab$

So, from (6) and (7), we get

$$a^* = \frac{1}{2}\left(\frac{h^2}{d} - d\right) + \frac{1}{2}(a' + b')$$

$$= \frac{1}{2}\left(\frac{h^2}{d} - d\right) + (m^2 - n^2) = (m^2 - n^2) + \frac{1}{2}\left[\frac{4m^2n^2}{d^2} - d\right]$$

$$b^* = \frac{1}{2}(a' + b') - \frac{1}{2}\left(\frac{h^2}{d} - d\right)$$

$$= (m^2 - n^2) - \frac{1}{2}\left(\frac{h^2}{2} - d\right) = (m^2 - n^2) - \frac{1}{2}\left[\frac{4m^2n^2}{d} - d\right]$$

$$c^* = \frac{1}{2}\left(\frac{h^2}{d} + d\right) = \frac{1}{2}\left[\frac{4m^2n^2}{d} + d\right]$$

Mahavira agreed with Brahmagupta's rule to find a trapezium with three equal sides. He pointed out a method of obtaining it. He stated the following:

भुजपदहतबीजान्तरहृतजन्यधनाप्तभागहाराभ्याम् ।
तद्भुजकोटिभ्यां च द्विसम इव त्रिसमचतुरश्रे ॥

(G.S.S. VII. 101½)

i.e., "To find a trapezium with three equal sides, proceed as in the case of the isosceles trapezium with the rectangles generated from the side and upright as well as from the quotient of the area of the rectangle divided by the product of the difference of its elements and square root of its side and its divisor."

The generated rectangle with m and n is given by $m^2 - n^2$, $2mn$, $m^2 + n^2$.

$$\frac{\text{Area}}{(m-n)\sqrt{\text{upright}}} = \frac{2mn(m^2-n^2)}{(m-n)\sqrt{2mn}} = \sqrt{2mn}(m+n)$$

Let

$$m_1 = (m+n)\sqrt{2mn}, \quad n_1 = (m-n)\sqrt{2mn}$$

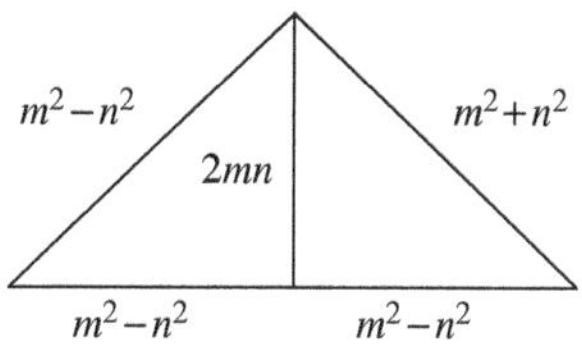

The rectangle generated from m_1 and n_1 is

$$m_1^2 - n_1^2 = 2mn[(m+n)^2 - (m-n)^2]$$

$$= 2mn[4mn] = 8m^2n^2$$

$$\text{Diagonal} = m_1^2 + n_1^2 = 2mn[(m+n)^2 + (m-n)^2]$$

$$= 2mn[2m^2 + 2n^2] = 4mn(m^2 + n^2)$$

$$2m_1n_1 = 2 \cdot 2mn(m^2 - n^2) = 4mn(m^2 - n^2)$$

So, the rectangle generated by m_1, n_1 is

$$8m^2n^2, \;\; 4mn(m^2 - n^2), \;\; \text{and} \;\; 4mn(m^2 + n^2) \quad (*)$$

Let

$$m_2 = 2mn, \quad n_2 = m^2 - n^2$$

The rectangle generated by m_2 and n_2 is

$$m_2^2 + n_2^2 = (2mn)^2 + (m^2 - n^2)^2 = (m^2 + n^2)^2$$

$$2m_2n_2 = 4mn(m^2 - n^2)$$

$$m_2^2 - n_2^2 = 4m^2n^2 - (m^2 - n^2)^2 = 6m^2n^2 - m^4 - n^4 \quad \text{if } (m^2 - n^2) < 2mn$$

Hence, the rectangle generated by m_2 and n_2 is

$$6m^2n^2 - m^4 - n^4, \quad 4mn(m^2 - n^2), \quad \text{and } (m^2 + n^2)^2 \quad (**)$$

Juxtaposition of rectangle (*) and (**) will yield Brahmagupta's trapezium $ABCD$ with three equal sides $CD = DA = BC$.

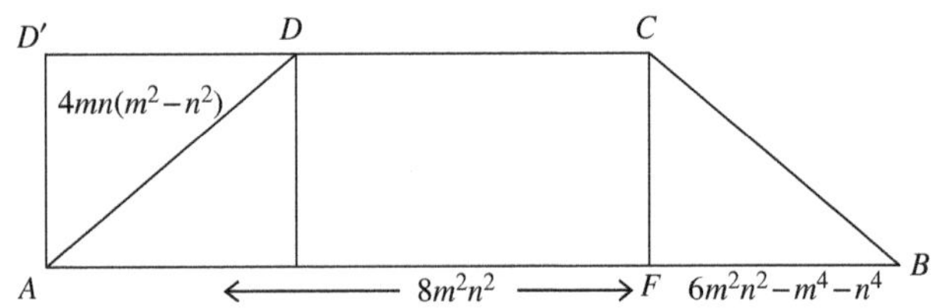

Construct the rectangle $AFCD'$ such that

$$AF = 8m^2n^2 \quad \text{and} \quad AD' = 4mn(m^2 - n^2)$$

Continue AF to B such that $BF = 6m^2n^2 - m^4 - n^4$.
Take a point D on CD' such that $DD' = FB$.
Join AD and BC. Then, $ABCD$ is our required trapezium.
We will now show $AD = BC = CD$.

$$BC^2 = BF^2 + FC^2 = D'D^2 + AD'^2 = AD^2$$

So, $BC = AD$.
Now, $BC^2 = BF^2 + CF^2 = BF^2 + AD^2$

$$= (6m^2n^2 - m^4 - n^4)^2 + \{4mn(m^2 - n^2)\}^2$$

$$= \{4m^2n^2 - (m^2 - n^2)^2\}^2 + 16m^2n^2(m^2 - n^2)^2$$

$$= 16m^4n^4 - 8m^2n^2(m^2 - n^2)^2$$

$$+ (m^2 - n^2)^4 + 16m^2n^2(m^2 - n^2)^2$$

$$= 16m^4n^4 + 8m^2n^2(m^2 - n^2)^2 + (m^2 - n^2)^4$$

$$= \{4m^2n^2 + (m^2 - n^2)^2\}^2 = (m^2 + n^2)^4$$

$$BC = (m^2 + n^2)^2$$

$$\text{Base } AB = AF + FB$$
$$= 8m^2n^2 + 6m^2n^2 - m^4 - n^4$$
$$= 14m^2n^2 - m^4 - n^4$$

$$\text{Face } CD = CD' - D'D = AF - FB$$
$$= 8m^2n^2 - [6m^2n^2 - m^4 - n^4] \quad \text{if } m^2 - n^2 < 2$$
$$= m^4 + n^4 + 2m^2n^2 = (m^2 + n^2)^2$$

$AD = BC = \text{face } CD = 2m^2n^2 + m^4 + n^4 = (m^2 + n^2)^2$
Altitude $CF = 4mn(m^2 - n^2)$.
Area of trapezium $ABCD = CF(AB + CD)/2 = 4mn(m^2 - n^2)8m^2n^2$

$$= 32m^3n^3(m^2 - n^2)$$

Rational Scalene Triangles

For the construction of rational scalene triangles, Mahavira proposed the following rule:

> *"Half the base of the generated rectangle is divided by an optional number. With this divisor and the quotient as 'bijas' (elements), another rectangle can be generated. The sum of the uprights of these two rectangles will be the base of the required scalene triangle; the two diagonals will be its sides and the base of either rectangle will be its altitude."*

Let m, n be two rational numbers. With these two as "bijas" (elements), let the rational rectangle $EBFA$ formed from them be $m^2 - n^2$, $2mn$, $m^2 + n^2$.

Base of this rectangle $= 2mn$.

Let a, b be any two rational factors of half the base mn.

Then, $ab = mn$.

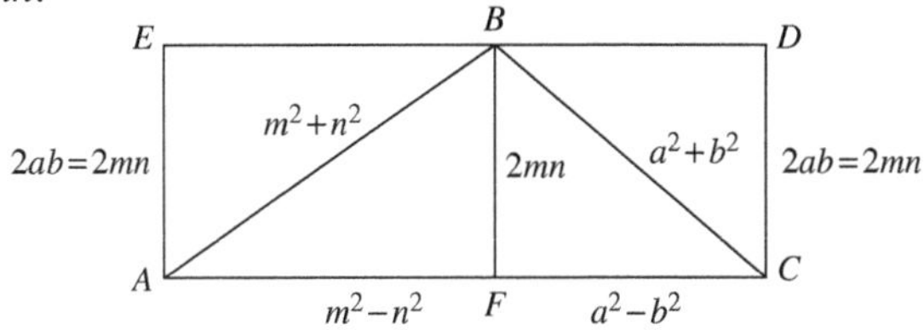

Figure 1

with a and b as "bijas" (element), let the rectangle generated be $BDCF$, $a^2 - b^2$, $2ab$, $a^2 + b^2$.

Juxtapose these two rectangles so that they do not overlap (see figure 1). Then, the sides of the required scalene triangle are (a^2+b^2), (m^2+n^2), and $(a^2-b^2)+(m^2-n^2)$.

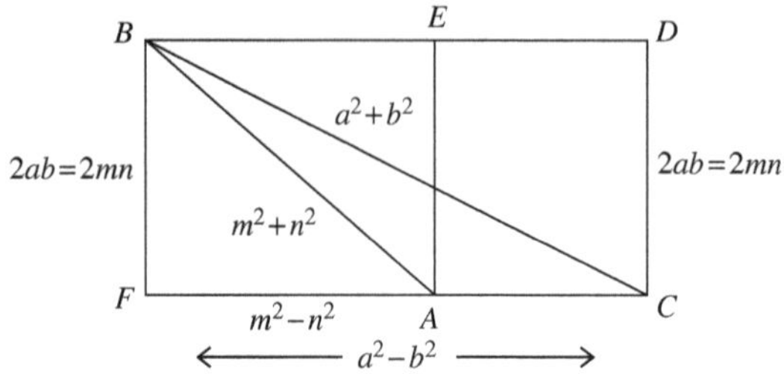

Figure 2

We note that the two rectangles can be juxtaposed to overlap. In this case, the sides of the required scalene triangle ABC are

$$a^2+b^2,\ m^2+n^2,\ (a^2-b^2)-(m^2-n^2)\quad \text{if } m^2-n^2<a^2-b^2\quad \text{(see figure 2).}$$

Rational Cyclic Quadrilateral

Mahavira gave the following formula for the sides of a cyclic quadrilateral.

Let the rectangle generated from m and n be smaller than that from a and b. Then, the sides of a cyclic quadrilateral are $(m^2-n^2)(a^2+b^2)(m^2+n^2)$, $(a^2-b^2)(m^2+n^2)$, $2mn(a^2+b^2)(m^2+n^2)$, $2ab(m^2+n^2)^2$ and two diagonals are

(G.S.S. VII. 103½)

$$\{2ab(m^2-n^2)+2mn(a^2-b^2)\}(m^2+n^2)$$

and

$$\{(m^2-n^2)(a^2-b^2)+4mnab\}(m^2+n^2)$$

The two altitudes are

$$\{2ab(m^2-n^2)+2mn(a^2-b^2)\}2mn$$

$$\{(m^2-n^2)(a^2-b^2)+4mnab\}(m^2-n^2)$$

Area A of this cyclic quadrilateral is given by

$$A=\frac{1}{2}\left\{2ab(m^2-n^2)+2mn(a^2-b^2)\right\}$$

$$\{(m^2-n^2)(a^2-b^2)+4mnab\}(m^2+n^2)^2$$

Mahavira gave a very clear rule to determine the circum diameter of a cyclic quadrilateral as well as of a triangle:

श्रुतिरवलम्बभक्ता पार्श्वभुजघ्ना चतुर्भुजे त्रिभुजे।
भुजघातो लम्बहृतो भवेद्बहिर्वृत्तविष्कम्भः ॥

(G.S.S. VII. 213)

In the case of a cyclic quadrilateral, the diameter of the circumscribing circle is the diagonal divided by the altitude and multiplied by the flank. In the case of a triangle, the circum diameter is the product of the two sides divided by the altitude.

Proof:

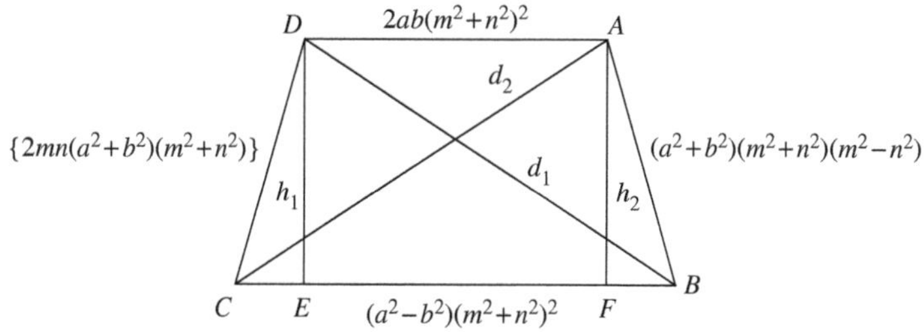

Let $ABCD$ be a quadrilateral with sides

$$a^* = AB = (a^2 + b^2)(m^2 + n^2)(m^2 - n^2)$$

$$BC = (a^2 - b^2)(m^2 + n^2)^2 = b^*$$

$$c^* = CD = 2mn(a^2 + b^2)(m^2 + n^2) \text{ and}$$

$$DA = 2ab(m^2 + n^2)^2 = d^*$$

$$BD = d_1 = \{2ab(m^2 - n^2) + 2mn(a^2 - b^2)\}\{m^2 + n^2\} \text{ and}$$

$$AC = d_2 = \{(m^2 - n^2)(a^2 - b^2) + 4mnab\}\{m^2 + n^2\}$$

Let the altitudes

$$DE = h_1 = \{2ab(m^2 - n^2) + 2mn(a^2 - b^2)\}2mn \text{ and}$$

$$AF = h_2 = \{(m^2 - n^2)(a^2 - b^2) + 4mnab\}\{m^2 - n^2\}$$

Now,

$$d_1 = \{(2ab(m^2 - n^2) + 2mn(a^2 - b^2)\}(m^2 + n^2)$$

$$h_1 = \{2ab(m^2 - n^2) + 2mn(a^2 - b^2)\}2mn$$

Let R_1 and R_2 be the diameters of the circumscribing circles of the triangles BCD and ABC, respectively.

Now,

$$R_1 = \frac{BD \cdot DC}{DE}$$

i.e.,

$$D_1 = \frac{2mn(a^2+b^2)(m^2+n^2)}{h_1} d_1$$

$$= 2mn(a^2+b^2)(m^2+n^2)\frac{(m^2+n^2)}{2mn}$$

$$= (a^2+b^2)(m^2+n^2)^2$$

Again,

$$d_2 = \{(m^2-n^2)(a^2-b^2)+4mnab\}\{m^2+n^2\}$$

$$h_2 = \{(m^2-n^2)(a^2-b^2)+4mnab\}\{m^2-n^2\}$$

Again,

$$R_2 = \frac{AC \cdot AB}{AF}$$

i.e.,

$$D_2 = \frac{d_2}{h_2}(a^2+b^2)(m^2+n^2)(m^2-n^2)$$

$$= \frac{m^2+n^2}{m^2-n^2}(a^2+b^2)(m^2+n^2)(m^2-n^2)$$

$$= (a^2+b^2)(m^2+n^2)^2$$

Thus, $R_1 = R_2$.
Here, $ABCD$ is a cyclic quadrilateral.

Mahavira provided a method to form rational triangles and quadrilaterals inscribed in a circle of a given diameter. His rule is as follows:

> *"Divide the given diameter of the circle by the computed diameter of the circle circumscribing only generated figure of the same kind. The product of this quotient and the sides of the generate figure will give the sides of the required figure."*

This means, we should first generate a rational inscribed triangle or a rational cyclic quadrilateral. Next, we have to compute the diameter of the circumscribing circle. Then, we have to divide the given diameter by newly computed diameter. The product of the quotient of this division and the dimensions of the sides of the generated figure will give the dimensions of the required kind of figure.

We have shown in the earlier page that the sides of a rational scalene triangle are proportional to m^2+n^2, a^2+b^2, $a^2-b^2 \pm (m^2-n^2)$ and its altitude is proportional to $2mn = 2ab$, m, n being optional rational numbers. Since the diameter of the circumscribing circle = product of two sides/altitude, the diameter D^* of the circumscribing circle about the above scalene triangle is proportional to

$$\frac{(m^2+n^2)(a^2+b^2)}{2mn}$$

Then, the sides of rational triangle inscribed in circle of given diameter D will be the product of D/D^* and the dimensions of the generated scalene triangle. This means that the sides of a rational triangle inscribed in a circle of given diameter D will be

$$\frac{2mn}{a^2+b^2}D,\quad \frac{2mn}{m^2+n^2}D,\quad \frac{2mnD[a^2-b^2 \pm (m^2-n^2)]}{(m^2+n^2)(a^2+b^2)}$$

and its altitude

$$= D \cdot \frac{2mn}{((m^2+n^2)(a^2+b^2))/2mn}$$

$$= D \frac{(2mn)^2}{(m^2+n^2)(a^2+b^2)}$$

The dimensions of the sides of a rational cyclic quadrilateral as given by Mahavira are on page 266. The diameter of the circumcircle D^* is (a^2+b^2) $(m^2+n^2)^2$.

Then, according to Mahavira's above rule, the sides of a rational cyclic quadrilateral inscribed in a circle of diameter D are found as follows:

$$D \cdot \frac{a^*}{D^*} = \frac{D \cdot AB}{D^*} = D \cdot \frac{(a^2+b^2)(m^2+n^2)(m^2-n^2)}{(a^2+b^2)(m^2+n)^2}$$

$$= D \cdot \frac{m^2-n^2}{m^2+n^2}$$

$$D \cdot \frac{b^*}{D^*} = D \cdot \frac{BC}{D^*} = D \cdot \frac{(a^2-b^2)(m^2+n^2)^2}{(a^2+b^2)(m^2+n^2)^2}$$

$$= D \cdot \frac{(a^2 - b^2)}{(a^2 + b^2)}$$

$$D \cdot \frac{c^*}{D^*} = D \cdot \frac{CD}{D^*} = D \cdot \frac{2mn(a^2 + b^2)(m^2 + n^2)}{(a^2 + b^2)(m^2 + n^2)^2} = D \cdot \frac{2mn}{(m^2 + n^2)}$$

$$D \cdot \frac{d^*}{D^*} = D \cdot \frac{DA}{D^*} = D \cdot \frac{2ab(m^2 + n^2)^2}{(a^2 + b^2)(m^2 + n^2)^2} = D \cdot \frac{2ab}{(a^2 + b^2)}$$

$$D\frac{2mn}{m^2 + n^2}, \; D\frac{m^2 - n^2}{m^2 + n^2}, \; D\frac{2ab}{a^2 + b^2}, \; D\frac{a^2 - b^2}{a^2 + b^2}$$

Its diagonals are

$$\left\{2ab(m^2 - n^2) + 2mn(a^2 - b^2)\right\} \frac{D}{(a^2 + b^2)(m^2 + n^2)}$$

and

$$\left\{(m^2 - n^2)(a^2 - b^2) + 4abmn\right\} \frac{D}{(a^2 + b^2)(m^2 + n^2)}$$

Its area

$$= \frac{D^2}{2(a^2 + b^2)(m^2 + n^2)} \left\{2ab(m^2 - n^2) + 2mn(a^2 - b^2)\right\}$$

$$\{(m^2 - n^2)(a^2 - b^2) + 4abmn\}$$

Rational Quadrilateral with a Given Area

To obtain all rational cyclic rectangles having a given area A, he stated the following:

> *"In the case of a square, a side of the square is the square root of the area* A. *In the case of a rectangle, on optional number* m *and the quotient of* A/m *will respectively be the upright and the base of the rectangle."*
>
> (1. G.S.S. VII. 146 2. G.S.S. VII. 148)

i. If A be the area of a square, then $l = \sqrt{A}$ gives the four equal sides perpendicular to each other.

ii. If A be the area of a rectangle and if m be an optional number, the $l = A/m$ will be the base and m will be the upright.

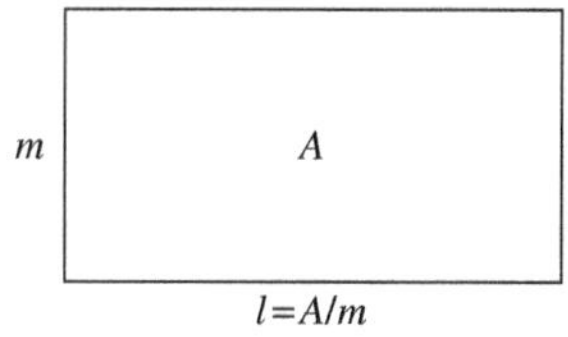

With regard to forming a rational cyclic isosceles trapezium, Mahavira stated the following rule:

> *"Multiply the area* A *by the square of an optional number. From this product, subtract the area of the generated rectangle. Divide this difference by the base of the generated rectangle and divide this quotient by the optional number. This gives the face of the isosceles trapezium; add twice the upright of the generated rectangle to the above quotient and divide this sum by the optional number. This gives the base. The base of the generated rectangle divided by the optional number gives the altitude. Divide by the diagonal of the generated rectangle by the optional number. That will give the two equal flank sides."*
>
> (G.S.S. VII. 145)

Let $m^2 - n^2$, $2mn$, $m^2 + n^2$ be the upright, base, and diagonal of the generated rectangle.

Now, the area of the generated rectangle $= (m^2 - n^2)2mn$.

Let p be an optional number such that $p^2A > 2mn(m^2 - n^2)$.

Thus, face of the required isosceles trapezium

$$= \frac{1}{p}\left\{\frac{p^2A - 2mn(m^2 - n^2)}{2mn}\right\}$$

Base of the required isosceles trapezium

$$= \frac{1}{p}\left\{\frac{p^2A - 2mn(m^2 - n^2)}{2mn} + 2(m^2 - n^2)\right\}$$

$$= \frac{1}{p}\left\{\frac{p^2A - 2mn(m^2 - n^2) + 2(2mn)(m^2 - n^2)}{2mn}\right\}$$

$$= \frac{1}{p}\left\{\frac{p^2A + 2mn(m^2 - n^2)}{2mn}\right\}$$

Altitude of the required isosceles trapezium $= 2mn/p$.

Equal flank sides of the required isosceles trapezium $= (m^2 + n^2)/p$.

Cyclic Quadrilateral of Given Area

For the construction of a cyclic quadrilateral of given area, Mahavira proposed the following rule:

> *"Break up the square of the area into four arbitrary rational factors. Subtract each factor separately from half the sum of these factors. The remainders are the four unequal sides of the cyclic quadrilateral."*

Let A be the area of the cyclic quadrilateral.
Let $A^2 = pqrs$, where p, q, r, and s are four arbitrary rational numbers.
Let

$$a = \frac{1}{2}(p+q+r+s) - p$$

$$b = \frac{1}{2}(p+q+r+s) - q$$

$$c = \frac{1}{2}(p+q+r+s) - r$$

$$d = \frac{1}{2}(p+q+r+s) - s$$

Let $2s^* = a+b+c+d = 2(p+q+r+s) - (p+q+r+s)$

$$= p+q+r+s$$

$$s^* - a = \frac{p+q+r+s}{2} - a = p$$

$$s^* - b = \frac{p+q+r+s}{2} - b = q$$

$$s^* - c = \frac{p+q+r+s}{2} - c = r$$

$$s^* - d = \frac{p+q+r+s}{2} - d = s$$

Also $A^2 = pqrs = (s^*-a)(s^*-b)(s^*-c)(s^*-d)$ or

$$A = \sqrt{(s^*-a)(s^*-b)(s^*-c)(s^*-d)}$$

Here, a, b, c, and d are the four unequal sides of the cyclic quadrilateral.

Cyclic Trapezium of Given Area with Three Equal Sides

For the construction of an inscribed trapezium with three equal sides of given area, Mahavira gave the following rule:

> *"Divide the square of the given area by the cube of an arbitrary rational number. Add to this the optional number. Half of this sum gives the three equal sides of a rational trapezium. On the other hand, subtract the value of the equal sides from twice the optional number. That will give the base. We get the altitude by dividing the area by the optional number."*
>
> (G.S.S. VI. 60)

Let A be the area of the rational trapezium and let p be a rational number. Then, the three equal sides

$$= \frac{1}{2}\left[\frac{A^2}{p^3} + p\right]$$

Base

$$= 2p - \frac{1}{2}\left[\frac{A^2}{p^3} + p\right]$$

Altitude

$$= \frac{A}{p}$$

Proof: Let $ABCD$ be a cyclic trapezium with $AB = a$, $BC = b$, $CD = c$ and $DA = d$.
Here, $BC = CD = AD \Rightarrow b = c = d$.
The area A of a cyclic quadrilateral with sides a, b, c, d is given by

$A^2 = (s-a)(s-b)(s-c)(s-d)$, where
$2s = a + b + c + d$.

Since in this case, $b = c = d$, (1) becomes

$$A^2 = (s-a)(s-b)^3$$

where

$$2s = a + 3b$$

Let $p = s - b$.

There $A^2 = (s - a)p^3$.
Hence,

$$\frac{A^2}{p^3} = (s - a)$$

Now,

$$\frac{1}{2}\left[\frac{A^2}{p^3} + p\right] = \frac{1}{2}[s - a + s - b]$$

$$= \frac{1}{2}[2s - a - b] = \frac{1}{2}[(a + 3b) - a - b] = \frac{2b}{2} = b$$

Also,

$$2p - \frac{1}{2}\left[\frac{A^2}{p^3} + p\right] = 2p - b = 2(s - b) - b$$

$$= 2s - 2b - b = 2s - 3b = a$$

A = area of $ABCD = 1/2(AB + CD)h$, if h is the length of perpendicular drawn from D on AB.

$$A = \frac{1}{2}(a + b)h \Rightarrow h = \frac{2A}{a + b} = \frac{2A}{2p - b + b} = \frac{2A}{2p} = \frac{A}{p}$$

Mahavira provided a rule for calculating the base and the altitude of the sections with proportions areas, where a given isosceles trapezium is divided by lines drawn parallel to the parallel sides.

खण्डयुति–भक्ततलमुखकृत्यन्तरगुणित–खण्डमुखवर्गयुतम्।
मूलमधस्तलमुखयुतदलहृतलब्धं च लम्बकः क्रमशः॥

(G.S.S. VII. 175½)

i.e., "The product of the part and the difference of the squares of the base and the face are divided by the sum of the parts. Square root of the sum of this resulting quotient and the square of the face is the base of the section. The area of the section divided by the sum of the bottom side of the section and the base is the altitude of the section."

Let $ABCD$ be an isosceles trapezium with $BC = AD$. Let this trapezium $ABCD$ be divided into parts with area in the ratio 1/m.

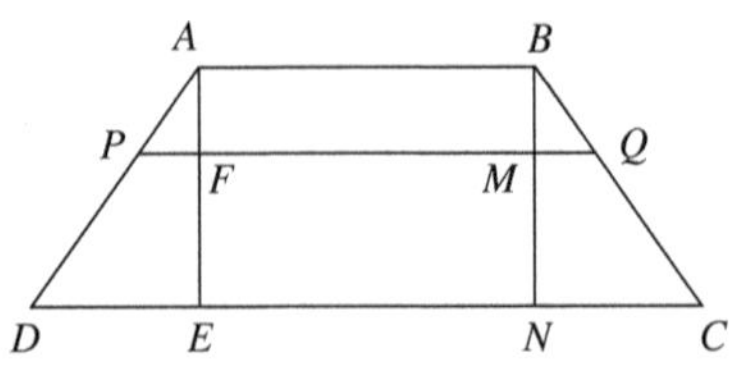

Area of $PABQ$/area of $PQCD = l/m$ by the line PQ drawn parallel to CD. Draw AE perpendicular to DC from A, cutting PQ at F. Let $AB = s$, $DC = t$, and $PQ = r$.

Draw BN perpendicular to CD, cutting PQ at M.
Now, $AD = BC$, $AE = BN$, and $AF = BM$.
From right triangles ADE and BNC,

$$DE^2 = AD^2 - AE^2 = BC^2 - BN^2 = NC^2$$

So, $DE = NC$.
Now, triangles APF and ADE are similar.
So,

$$\frac{AF}{AE} = \frac{PF}{DE} \tag{1}$$

Similarly, triangles BMQ and BNC are similar.
So,

$$\frac{MQ}{NC} = \frac{BM}{BN} = \frac{AF}{AE}$$

So,

$$\frac{AF}{AE} = \frac{PF}{DE} = \frac{MQ}{NC} = \frac{BM}{BN} = k \text{ (say)}$$

So, $MQ = kNC = kDE$ and $PF = kDE$.
Now, $r = PQ = PF + FM + MQ = kDE + s + KDE = s + 2kDE$.

$$t = DC = DE + EN + NC = DE + s + DE = s + 2DE$$

Hence,

$$\frac{r-s}{t-s} = \frac{2kDE}{2DE} = k = \frac{AF}{AE}$$

Now,

$$\frac{r^2 - s^2}{t^2 - s^2} = \frac{(r+s)}{(t+s)} \cdot \frac{(r-s)}{(t-s)} = \frac{(r+s)}{t+s}\frac{AF}{AE}$$

$$= \frac{\text{Area of } PABQ}{\text{Area of } ABCD} = \frac{l}{l+m}$$

So,

$$r^2 - s^2 = \frac{l}{l+m}(t^2 - s^2)$$

$$r^2 = \frac{l}{l+m}(t^2 - s^2) + s^2$$

$$r = \sqrt{\frac{l}{l+m}(t^2 - s^2) + s^2}$$

Altitude

$$AF = \frac{\text{Area of } APBQ}{(1/2)(PQ + AB)} = \frac{\text{Area of } APBQ}{(1/2)(\text{face} + \text{bottom side of } APBQ)}$$

Shadow Problem

Mahavira has dealt with an interesting problem of inclination ("Avanti") of an inclined object knowing the length of shadow cast by it.

छायावर्गाच्छोध्या नरभाकृति गुणितशङ्कृतः
सैकनरच्छायाकृतिगुणिता छायाकृतेः शोध्या ॥
तन्मूलं छायायां शोध्य नरभानवर्गरूपेण ।
भागं हृत्वा लब्धं स्तम्भस्यावनतिरेव स्यात् ॥

(G.S.S. IX. 32–33)

i.e., "The square of the length of the pillar is multiplied by the square of the ratio of the shadow to the object. This product is subtracted from the square of the length of the shadow. The difference obtained is multiplied by the above square of the ratio increased by one. The result obtained is again subtracted from the square of the shadow of the pillar. The square root of this new difference is subtracted from the length of shadow and is divided once again by the square of the shadow object ratio increased by one. This will determine the inclination of the pillar."

Here, "Avanti" (inclination) is taken to be projection of the slanting object on the horizontal. Let PQ be the slanting pillar. Let QR be the shadow cast on the horizontal by the pillar PQ.

Draw PS perpendicular drawn from P on QR.

So, RS is the "Avanti" of the pillar.

To prove

$$QS = \frac{QR - \sqrt{QR^2 - (QR^2 - s^2PQ^2)(s^2 + 1)}}{s^2 + 1}$$

where s is the ratio of the shadow to the object.

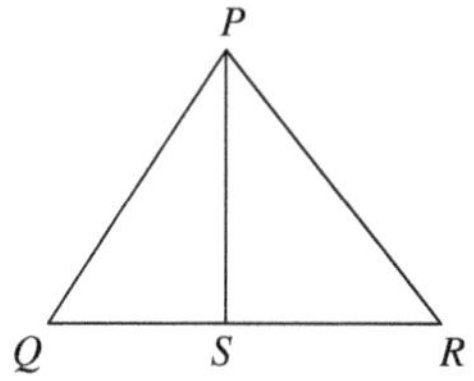

Now, SR is the shadow of the vertical PS.
So,

$$\frac{SR}{PS} = s \Rightarrow SR = s \cdot PS = s\sqrt{PQ^2 - QS^2} \qquad \text{(i)}$$

Also, $SR = QR - QS$.
So, $SR^2 = (QR - QS)^2 = s^2 \cdot (PQ^2 - QS^2)$ from (i).

Thus, we have

$$QR^2 - 2QR \cdot QS + QS^2 = s^2PQ^2 - s^2QS^2$$

i.e., $QS^2(s^2 + 1) - 2QR \cdot QS = s^2PQ^2 - QR^2$.

Dividing by $s^2 + 1$, we get

$$QS^2 - \frac{2QRQS}{s^2 + 1} = \frac{s^2PQ^2 - QR^2}{s^2 + 1}$$

Adding $QR^2/(s^2 + 1)$ to both sides, we get

$$\left(QS - \frac{QR}{s^2+1}\right)^2 = \frac{s^2PQ^2 - QR^2}{(s^2+1)^2} + \frac{QR^2}{(s^2+1)^2}$$

i.e.,

$$\left(QS - \frac{QR}{s^2+1}\right)^2 = \frac{(s^2 + 1)[s^2PQ^2 - QR^2] + QR^2}{(s^2+1)^2}$$

Taking square root, we have

$$QS - \frac{QR}{s^2 + 1} = \pm \frac{\sqrt{(s^2 + 1)[s^2PQ^2 - QR^2] + QR^2}}{s^2 + 1}$$

Hence,

$$QS = \frac{QR \pm \sqrt{(s^2 + 1)[s^2PQ^2 - QR^2] + QR^2}}{s^2 + 1}$$

i.e.,

$$QS = \frac{QR \pm \sqrt{QR^2 - (s^2+1)[QR^2 - s^2PQ^2]}}{s^2+1}$$

Since $QS < QR$, we discard the positive square root and take the negative square root.

Thus,

$$QS = \frac{QR - \sqrt{QR^2 - (s^2+1)[QR^2 - s^2PQ^2]}}{s^2+1}$$

Quadratic Indeterminate Equation

Mahavira in his Ganita Sara Sangraha has given two problems which involve equations of this form.

Example 1 (G.S.S. vi, page 284)

Find the increase or decrease of two numbers a and b so that the product of the resulting numbers will be equal to a given number d.

i.e., $(a \pm x)(b \pm y) = d$.

Case 1: Consider, $(a + x)(b + y) = d$

Take

$$x = \frac{d - ab}{d + b}$$

Then,

$$\left[a + \frac{d - ab}{d + b}\right][b + y] = d$$

$$\left(\frac{ad + ab + d - ab}{d + b}\right)(b + y) = d$$

$$\frac{d(a+1)}{d+b}[b + y] = d$$

So,

$$b+y=\frac{(d+b)}{a+1}\Rightarrow y=\frac{d+b}{a+1}-b$$

$$=\frac{d+b-ab-b}{a+1}=\frac{d-ab}{a+1}$$

So,

$$x=\frac{d-ab}{d+b},\quad y=\frac{d-ab}{a+1}$$

is a solution of $(a+x)(b+y)=d$.

Case 2: Consider $(a-x)(b-y)=d$

Let

$$x=\frac{-d+ab}{b+1}$$

Then, we have

$$\left(a-\frac{(-d+ab)}{b+1}\right)(b-y)=d$$

$$\frac{(ab+a+d-ab)}{b+1}(b-y)=d$$

$$\left(\frac{a+d}{b+1}\right)(b-y)=d$$

$$b-y=\frac{d(b+1)}{a+d}$$

$$y=b-\frac{d(b+1)}{a+d}=\frac{ab+bd-bd-d}{a+d}$$

$$=\frac{ab-d}{a+d}$$

So,

$$x=\frac{ab-d}{b+1},\quad y=\frac{ab-d}{a+d}$$

is a solution of $(a-x)(b-y)=d$.

Example 2 (G.S.S. VI. page 35)

Let x be the capital invested and let y be the period of time in months, then the interest will be mxy, where m is the rate of interest per month.

Then, the problem is to solve $mxy + x + y = p$.

Let $y = Q$.

Then, we have $mxQ + x + Q = p$.

$$x(mQ + 1) = p - Q$$

$$x = \frac{p - Q}{mQ + 1}$$

So,

$$y = Q \text{ and } x = \frac{p - Q}{mQ + 1}$$

is the solution of $mxy + x + y = p$ simultaneous linear equations.

Mahavira gave the following two examples.

Example 1 (Mahavira)

"The price of 9 citrans and 7 fragrant woodapples taken together is 107. The price of 7 citrons and 9 fragrant woodapples is 101. Oh mathematician, tell me the price of a citron and a fragrant woodapple."

Solution

Let x be the price of each citron and y be the price of each woodapple.

Then, we have (i) $9x + 7y = 107$ and (ii) $7x + 9y = 101$.

Adding (i) and (ii), we get $16x + 16y = 208 \Rightarrow x + y = 13$.

Subtracting (ii) from (i), we get $2x - 2y = 6 \Rightarrow x - y = 3$.

So, $2x = 16 \Rightarrow x = 8$ and $2y = 10 \Rightarrow y = 5$.

Or, in general, if $ax + by = m$, $bx + ay = n$, then the solution is

$$x = \frac{am - bn}{a^2 - b^2} \text{ and } y = \frac{an - bm}{a^2 - b^2}$$

Example 2

"A wizard having mystic powers, observing an ongoing cockfight, spoke individually to both owners. To one he said, 'if your bird wins, you give me your stake of money and if your bird does not win, I shall give you two-thirds of that.' With the other, he promised the same way to give three-fourths. His gain will be 12 coins from both of them. Tell me, O, mathematician, the stake of money of each of the cock owners."

Solution

Let x and y coins be the stake of two cockowners, respectively.

Then, we have

$$x - \frac{3}{4}y = 12 \text{ and } y - \frac{2}{3}x = 12$$

Or, in general,

$$x - \frac{c}{d}y = p \text{ and } y - \frac{a}{b}x = q$$

i.e.,

$$xd - cy = pd \quad (1)$$

$$by - ax = bq \quad (2)$$

Multiplying (1) by a, we get $axd - acy = apd$.
Multiplying (2) by d, we get $-axd + bdy = bqd$.
Adding, we get $y(bd - ac) = d(ap + bq)$.

$$y = \frac{d(ap + bq)}{bd - ac}$$

$$xd = pd + cy = pd + \frac{cd(ap + bq)}{bd - ac}$$

$$x = p + \frac{c(ap + bq)}{bd - ac} = \frac{pbd - pac + pac + cbq}{bd - ac}$$

$$x = \frac{(pd + cq)b}{bd - ac}$$

Mahavira has given a following example on system of linear equations.

Example

"Four merchants were asked individually by the custom officer about the total value of their commodities. The first Merchant leaving out the worth of his own commodity, stated it to be 22, the second to be 23, the third to be 24 and fourth 27, each of them deducting the worth of their own commodity. Oh friend, tell me the worth of the commodities of the four merchants individually."

Solution

Let x_1, x_2, x_3, and x_4 be the worth of the commodities of the four merchants, respectively.

Then, we have

$$x_2 + x_3 + x_4 = 22 \quad \text{(i)}$$

$$x_1 + x_3 + x_4 = 23 \quad \text{(ii)}$$

$$x_1 + x_2 + x_4 = 24 \quad \text{(iii)}$$

$$x_1 + x_2 + x_3 = 27 \quad \text{(iv)}$$

Adding (i), (ii), and (iii), we get

$$3(x_1 + x_2 + x_3 + x_4) = 22 + 23 + 24 + 27 = 96$$

$$\Rightarrow x_1 + x_2 + x_3 + x_4 = 32$$

$$\Rightarrow x_1 = 10, \;\; x_2 = 9, \;\; x_3 = 8, \;\; x_4 = 5 \text{ from(i), (ii), (iii), and (iv)}$$

A More Generalized System of Linear Equations

Consider the following system:

$$b_1 \sum_{r=1}^{n} x_r - c_1 x_1 = a_1 \quad \text{(i)}$$

$$b_2 \sum_{r=1}^{n} x_r - c_2 x_2 = a_2 \quad \text{(ii)}$$

$$b_r \sum_{r=1}^{n} x_r - c_r x_r = a_r$$

$$\vdots$$

$$b_n \sum_{r=1}^{n} x_r - c_n x_n = a_n \qquad (n)$$

Then,

$$\sum_{r=1}^{n} x_r = \frac{\sum_{r=1}^{n}(a_r/c_r)}{\sum_{n=1}^{n}(b_r/c_r) - 1}$$

and

$$x_r = \frac{b_r}{c_r} \frac{\sum_{r=1}^{n}(a_r/c_r)}{\sum_{r=1}^{n}(b_r/c_r) - 1} - \frac{a_r}{c_r}$$

Divide (i) by c_1, (ii) by c_2,... (n) by c_n and then adding, we get

$$\sum_{r=1}^{n} x_r \left[\frac{b_1}{c_1} + \frac{b_2}{c_2} + \cdots + \frac{b_n}{c_n}\right] - \sum_{r=1}^{n} x_r = \sum_{r=1}^{n} \frac{a_r}{c_r}$$

$$\sum_{r=1}^{n} x_r \left[\sum_{r=1}^{n} \frac{b_r}{c_r} - 1\right] = \sum_{r=1}^{n} \frac{a_r}{c_r}$$

$$\sum_{r=1}^{n} x_r = \frac{\sum_{r=1}^{n}(a_r/c_r)}{\{\sum_{r=1}^{n}(b_r/c_r) - 1\}}$$

From (i),

$$c_1 x_1 = b_1 \sum_{r=1}^{n} x_r - a_1$$

$$x_1 = \frac{b_1}{c_1} \sum_{r=1}^{n} x_r - \frac{a_1}{c_1}$$

$$x_1 = \frac{b_1}{c_1} \frac{\sum_{r=1}^{n}(a_r/c_r)}{[\sum_{r=1}^{n}(b_r/c_r)] - 1} - \frac{a_1}{c_1}$$

similarly,

$$x_2 = \frac{b_2}{c_2}\left[\frac{\sum_{r=1}^{n}(a_r/c_r)}{\left(\sum_{r=1}^{n}(b_r/c_r)\right)-1}\right] - \frac{a_2}{c_2}$$

$$x_r = \frac{b_r}{c_r}\left[\frac{\sum_{r=1}^{n}(a_r/c_r)}{\left(\sum_{r=1}^{n}(b_r/c_r)\right)-1}\right] - \frac{a_r}{c_r} \text{ and so on}$$

Mahavira has given the following example.

Example

"Three merchants borrowed money mutually from one another. The first one borrowing 4 from the second and 5 from the third, became twice rich as the other two. The second one borrowing 4 from the first and 6 from the third, became thrice as rich as the other two. The third merchant borrowing 5 from the first and 6 from the second, became five times as rich as the other two. Oh mathematician, tell me the amount that each of the merchants had initially."

Solution

Suppose the three merchants had x, y, z amount of money initially. Then, we have

i. $x + 4 + 5 = 2[(y - 4) + z - 5] = 2y + 2z - 18$
ii. $y + 4 + 6 = 3[(x - 4) + (z - 6)] = 3x + 3z - 30$
iii. $z + 5 + 6 = 5[(x - 5) + (y - 6)] = 5x + 5y - 55$

From (i), we get $2y + 2z - x = 27 \Rightarrow 2(x + y + z) - 3x = 27$.
From (ii), we get $3x + 3z - y = 40 \Rightarrow 3(x + y + z) - 4y = 40$.
From (ii), we get $5x + 5y - z = 66 \Rightarrow 5(x + y + z) - 6z = 66$.
Hence, $a_1 = 27, a_2 = 40, a_3 = 66; b_1 = 2, b_2 = 3, b_3 = 5; c_1 = 3, c_2 = 4, c_3 = 6$.

$$x = \frac{(b_1/c_1)\sum_{r-1}^{3}(a_r/c_r)}{\left(\sum_{r-1}^{3}(b_r/c_r)\right)-1} = \frac{(b_1/c_1)((a_1/c_1)+(a_2/c_2)+(a_3/c_3))}{((b_1/c_1)+(b_2/c_2)+(b_3/c_3))-1} - \frac{a_1}{c_1}$$

$$y = \frac{(b_2/c_2)\sum_{r-1}^{3}(a_r/c_r)}{\left(\sum_{r-1}^{3}(b_r/c_r)\right)-1} = \frac{(b_2/c_2)((a_1/c_1)+(a_2/c_2)+(a_3/c_3))}{((b_1/c_1)+(b_2/c_2)+(b_3/c_3))-1} - \frac{a_2}{c_2}$$

$$z = \frac{(b_3/c_3)\sum_{r-1}^{3}(a_r/c_r)}{\left(\sum_{r-1}^{3}(b_r/c_r)\right)-1} = \frac{(b_1/c_1)((a_1/c_1)+(a_2/c_2)+(a_3/c_3))}{((b_1/c_1)+(b_2/c_2)+(b_3/c_3))-1} - \frac{a_3}{c_3}$$

From the result on page 285, we have

$$x = \frac{(2/3)((27/3) + (40/4) + (66/6))}{((2/3) + (3/4) + (5/6)) - 1} - \frac{27}{3}$$

$$x = \frac{(2/3)(9 + 10 + 11)}{(8 + 9 + 10 - 12)/12} - 9 = \frac{20}{(15/12)} - 9$$

$$x = 20\left(\frac{4}{5}\right) - 9 = 16 - 9 = 7$$

$$y = \frac{3}{4}\frac{((27/3) + (40/4) + (66/6))}{((2/3) + (3/4)(5/6)) - 1} - \frac{40}{4}$$

$$y = \frac{3}{4}\frac{30}{(5/4)} - 10 = \frac{3}{4}(24) - 10 = 8$$

$$z = \frac{5}{6}\frac{30}{(5/4)} - 11 = \frac{5}{6}(24) - 11 = 9$$

General problem (Mahavira)

"Suppose there are n merchants. Suppose $a_{r_1}, a_{r_2}, a_{r_3}, \ldots, a_{r,\,r-1}, a_{r,r+1}, \ldots, a_r, n$ be the amount borrowed by the rth merchant from the others. Let X_r be the amount he had initially. (Here, $r = 1, 2, \ldots, n$). By doing so, he becomes b_r times richer than the others. Find out how much money each of n merchants had initially."

Solution

We have

$$X_r + \sum_{\substack{m=1 \\ m \neq r}}^{n} a_{r,\,m} = b_r \left[\sum_{\substack{m=1 \\ m \neq r}}^{n} X_m - \sum_{\substack{n=1 \\ m \neq r}}^{n} a_{r,m} \right] (r = 1, 2, \ldots, n)$$

(*Continued*)

General problem (Mahavira) (cont'd)

So,

$$X_r + (b_r + 1) \sum_{\substack{m=1 \\ m \neq r}}^{n} a_{r,m} = b_r \left[\sum_{m=1}^{n} X_m - X_r \right] (r = 1, 2, \ldots, n)$$

Let

$$K_r = (b_r + 1) \sum_{\substack{m=1 \\ m \neq r}}^{n} a_{r,m}$$

Then, we have

$$X_r + K_r = b_r \sum_{m=1}^{n} X_m - b_r X_r \ (r = 1, 2, \ldots, n)$$

$$X_r(1 + b_r) + K_r = b_r \sum_{m=1}^{n} X_n$$

$$X_r + \frac{K_r}{1 + b_r} = \frac{b_r}{1 + b_r} \sum_{m=1}^{n} X_m \ (r = 1, 2, \ldots, n)$$

Adding we get,

$$\sum_{r=1}^{n} X_r + \left[\frac{K_1}{1 + b_1} + \frac{K_2}{1 + b_2} + \cdots + \frac{K_n}{1 + b_n} \right]$$

$$= \sum_{r=1}^{n} X_r \left(\frac{b_1}{r + b_1} + \frac{b_2}{1 + b_2} + \cdots + \frac{b_n}{1 + b_n} \right)$$

$$\sum_{r=1}^{n} X_r + \sum_{r=1}^{n} \frac{K_r}{1 + b_r} = \sum_{r=1}^{n} X_r \left[\sum_{r=1}^{n} \frac{b_r}{1 + b_r} \right]$$

$$\sum_{r=1}^{n} \frac{K_r}{1 + b_r} = \sum_{n=1}^{n} X_r \left[\sum_{r=1}^{n} \frac{b_r}{1 + b_r} - 1 \right]$$

So,

$$\sum_{r=1}^{n} X_r = \frac{\sum_{r=1}^{n}(K_r/(1+b_r))}{\left(\sum_{r=1}^{n} b_r/(1+b_r)\right)-1}$$

Thus,

$$X_r = \frac{b_r}{1+b_r}\sum_{r=1}^{n} X_r - \frac{K_r}{1+b_r}$$

$$= \frac{b_r}{1+b_r}\frac{\sum_{r=1}^{n}(K_r/(1+b_r))}{\left(\sum_{r=1}^{n} b_r/(1+b_r)\right)-1} - \frac{K_r}{1+b_r}\ (r=1,\ 2,\ldots,n)$$

Certain interest problems dealt by Mahavira lead to simple simultaneous equations involving unknowns.

Examples (Mahavira, G.S.S. VI. 37, 39, 42)

Certain capital amount $C_1, C_2, \ldots, C_n$ are lent out at the same interest rate r for different periods of time $t_1, t_2, \ldots, t_n$ if $i_1, i_2, \ldots, i_n$ are the interest accrued by these capitals, respectively, where $i_k = rt_kC_k/100$, $k = 1, 2, \ldots,$ then, if $i_1 + i_2 + \cdots + i_n = I$ and if C_r and t_r are known for $r = 1, 2, \ldots, n$, we have

i. $i_K = \dfrac{IC_kt_k}{C_1t_1 + C_2t_2 + \cdots + C_nt_n}$, $k = 1, 2, \ldots, n$

ii. If $C_1 + C_2 + \cdots + C_n = C$, if i_r and t_r are known for $r = 1, 2, \ldots, n$, then

$$C_K = \frac{C_i i_k/t_K}{(i_1/t_1) + (i_2/t_2) + (i_n/t_n)}$$

iii. If $t_1 + t_2 + \cdots + t_n = T$ if i_K and C_K are known for $k = 1, 2, \ldots, n$, then

$$t_k = \frac{T_{i_k/C_k}}{(i_1/C_1) + (i_2/C_2) + \cdots + (i_n/C_n)} \text{ for } k = 1,\ 2, \ldots n$$

Proof of (i)

$$i_k = \frac{rt_kC_k}{100} \text{ for } k = 1, 2, \ldots, n$$

(Continued)

Examples (Mahavira, G.S.S. VI. 37, 39, 42) (cont'd)

Then, $I = i_1 + i_2 + \cdots + i_n$

$$= \frac{r}{100}[t_1C_1 + t_2C_2 + \cdots + t_nC_n]$$

So,

$$\frac{I}{t_1C_1 + t_2C_2 + \cdots + t_nC_n} = \frac{r}{100} \Rightarrow i_K = \frac{It_KC_K}{t_1C_1 + t_2C_2 + \cdots + t_nC_n}$$

Proof of (ii)

$$i_K = \frac{rt_KC_K}{100} \text{ for } k = 1, 2, \ldots, n$$

Then,

$$C_K = \frac{100i_K}{rt_K} \text{ for } k = 1, 2, \ldots, n$$

Then, $C = C_1 + C_2 + C_3 + \cdots + C_n.$

$$= \frac{100}{r}\left[\frac{i_1}{t_1} + \frac{i_2}{t_2} + \cdots + \frac{i_n}{t_n}\right]$$

So,

$$\frac{100}{r} = \frac{C}{(i_1/t_1) + (i_2/t_2) + \cdots + (i_n/t_n)}$$

Now,

$$C_K = \frac{100}{r}\frac{i_K}{t_K} = \frac{i_K}{t_K}\left[\frac{C}{(i_1/t_1) + (i_2/t_2) + \cdots + (i_n/t_n)}\right]$$

Proof of (iii)

$$i_k = \frac{r}{100}[t_kC_k] \text{ for } k = 1, 2, \ldots, n$$

So,

$$t_k = \frac{100i_k}{rC_k}$$

Then, $T = t_1 + t_2 + t_3 + \cdots + t_n$

$$= \frac{100}{r}\left[\frac{i_1}{C_1} + \frac{i_2}{C_2} + \cdots + \frac{i_n}{C_n}\right]$$

So,

$$\frac{T}{(i_1/C_1) + (i_2/C_2) + \cdots + (i_n/C_n)} = \frac{100}{r} = \frac{t_k C_K}{i_K} \text{ for } k = 1, 2, \ldots, n$$

Then,

$$t_K = \frac{i_K}{C_K} \frac{T}{(i_1/C_1) + (i_2/C_2) + \cdots + (i_n/C_n)}$$

Mahavira gave the following problems which require the solution of a quadratic.

Example 1

"One-fourth of a herd of camels was seen in a forest; twice the square root of the heard had gone to the slopes; three times five camels were on the bank of a river. How many camels were there in the heard?"
(G.S.S. iv. 34)

Solution
Let x be the number of camels in the herd, so we have

$$\frac{1}{4}x + 2\sqrt{x} + 15 = x \Rightarrow \frac{3}{4}x - 2\sqrt{x} - 15 = 0$$

Let $y = \sqrt{x}$.
Then, we have

$$\frac{3}{4}y^2 - 2y - 15 = 0$$

$$3y^2 - 8y - 60 = 0$$

$$y = \frac{8 \pm \sqrt{64 + 4 \cdot 3 \cdot 60}}{6}$$

$$y = \frac{8 \pm \sqrt{64 + 720}}{6} = \frac{8 \pm \sqrt{784}}{6} = \frac{8 \pm 28}{6}$$

(Continued)

Example 1 (cont'd)

So,

$$y = \frac{36}{6} \text{ or } \frac{-20}{6}$$

$y = 6$, discarding the negative value, since $y > 0$.
So, $\sqrt{x} = 6 \Rightarrow x = 36$.
So, the heard had 36 camels.

Example 2

"One-third of a herd of elephants and three times the square root of the remaining part of the heard were seen on the mountain slopes and in a lake a tusker (a male elephant) was seen along with three female elephants. How many elephants were there in the head."
(G.S.S. iv. 41)

Solution

Let x be number of elephants in the herd.

Then, we have

$$\frac{x}{3} + 3\sqrt{\frac{2x}{3}} + 4 = x$$

i.e.,

$$\frac{2x}{3} - 3\sqrt{\frac{2x}{3}} - 4 = 0$$

Let

$$y = \sqrt{\frac{2x}{3}}$$

Then, we have $y^2 - 3y - 4 = 0$.

$$y = \frac{3 \pm \sqrt{9 + 16}}{2} = \frac{3 \pm 5}{2}$$

$$\Rightarrow y = 4 \text{ or } y = -1.$$

Since $y > 0$, $y = -1$ can be discarded.So,

$$y = 4 \Rightarrow \sqrt{\frac{2x}{3}} = 4$$

Squaring, we get

$$\frac{2x}{3} = 16$$

$$2x = 48 \Rightarrow x = 24$$

So, there were 24 elephants in the herd.

Example 3

"One-sixteenth of a group of peacocks multiplied by itself was sitting on the mango tree. One-ninth of the remainder multiplied by itself together with 14 were sitting on the 'tamala tree.' How many peacocks were there in the group?"

Solution

Let x be the number of peacocks.

We have

$$\left(\frac{x}{16}\right)^2 + \left(\frac{1}{9}\frac{15x}{16}\right)^2 + 14 = x$$

i.e.,

$$\frac{x^2}{256} + \frac{1}{81}\frac{225x^2}{256} + 14 = x$$

Multiply by (81)(256), we get

$$81x^2 + 225x^2 + 290,304 = 20,736x$$

$$306x^2 - 20,736x + 290,304 = 0$$

i.e.,

$$17x^2 - 1152x + 16,128 = 0$$

(*Continued*)

Example 3 (cont'd)

$$x = \frac{1152 \pm \sqrt{(1152)^2 - 4(17)(16,128)}}{34} = \frac{1152 \pm \sqrt{230,400}}{34}$$

$$x = \frac{1152 \pm 480}{34} = \frac{1632}{34} \text{ or } \frac{672}{34}$$

$$= 48 \text{ or } \frac{336}{17}$$

$x = 48$, since x is an integer. So, there were 48 peacocks in the group.

Mahavira: Certain problems lead to the equations

$$\left(\frac{a}{b}x \pm d\right)^2 + c = x$$

Then, the solution x is given as

$$x = \left\{\left(\frac{b}{2a} + d\right) \pm \sqrt{\left(\frac{b}{2a} + d\right)^2 - d^2 - c}\right\} \div \frac{a}{b}$$

Proof:

$$\left(\frac{a}{b}x \pm d\right)^2 + c = x \text{ means } \frac{a^2}{b^2}x^2 \pm \frac{2axd}{b} + d^2 + c = x$$

i.e.,

$$\frac{a^2}{b^2}x^2 \pm \left(\frac{2ad}{b} - 1\right)x + d^2 + c = 0$$

Dividing by $(2ad/b) - 1$, we get

$$\frac{a^2}{b^2}\frac{x^2}{((2ad/b) - 1)} \pm x + \frac{d^2 + c}{((2ad/b) - 1)} = 0$$

This is of the form $(A/B)x^2 \pm x + K = 0$, where

$$A = a^2, \quad B = b^2\left(\frac{2ad}{b} - 1\right) = 2adb - b^2$$

and

$$K = \frac{d^2 + c}{((2ad/b) - 1)} = \frac{(d^2 + c)b}{2ad - b}$$

Then,

$$x = \frac{\pm (B/A) \pm \sqrt{((B/A) - 4K)(B/A)}}{2}$$

i.e.,

$$x = \frac{\pm(b(2ad-b))/a^2 \pm \sqrt{\{(b(2ad-b))/a^2 - 4((d^2+c)b)/(2ad-b)\}(b(2ad-b))/a^2}}{2}$$

$$= (b/a)\frac{[\pm(2ad-b)/a \pm \sqrt{\{(b(2ad-b))/a^2 - (4(d^2+c)b)/(2ad-b)\}(2ad-b)/b}]}{2}$$

$$= \frac{b}{a}\left[\pm\left(\frac{b}{2a} - d\right) \pm \sqrt{\frac{(2ad-b)^2}{4a^2} - (d^2+c)}\right]$$

$$= \frac{b}{a}\left[\pm\left(\frac{b}{2a} - d\right) \pm \sqrt{\left(\frac{b}{2a} - d\right)^2 - (d^2+c)}\right]$$

Mahavira has considered certain equations of higher degree in dealing with problems involving geometric series. They are of the following types:

i. $ax^n = b$.

ii. $a\frac{x^n - 1}{x - 1} = q$.

where a is the first term of a geometric progression (GP), b is the $(n + 1)$th term ("gunadhana"), q its sum, and x is the unknown common factor.

To solve (i), Mahavira pointed out that the unknown number x which on multiplication by itself as many times as the number of terms is equal to the $(n + 1)$ term b ("Gundhana") divided by the first term is the common ratio, i.e., $x = \sqrt[n]{b/a}$. Mahavira however did not indicate how to find such a root.

(Continued)

Example 3 (cont'd)

Mahavira gave the solution of the equation of type (ii) in this way: "common ratio is that unknown by which the sum divided by first term a is divisible again and again, diminishing unity every time."

We will demonstrate below his method with two examples.

Example 1

"The first term of a geometric progression is 5, number of times is 4, and the sum is 200. Find the common ratio."

$$5\frac{x^4-1}{x-1}=200 \Rightarrow \frac{x^4-1}{x-1}=40$$

$$(x^3+x^2+x+1)=40$$

If we have the divisor 3,

$$\frac{40-1}{3}=\frac{39}{3}=13 \quad \frac{13-1}{3}=\frac{12}{3}=4$$

$$\frac{4-1}{3}=\frac{3}{3}=1 \quad \frac{1-1}{3}=\frac{0}{3}=1$$

We observe that the number 40 is exhausted on four successive divisions by 3. So, the common ratio $x=3$.

Example 2

"The first term of a geometric progression is 3, the number of times is 6, and the sum is 11,718. Find the common ratio."

Here,

$$3\frac{x^6-1}{x-1}=11,718 \quad \frac{x^6-1}{x-1}=3906$$

$$x^5+x^4+x^3+x^2+x+1=3906$$

If we have the divisor 5,

$$\frac{3906-1}{5}=\frac{3905}{5}=781$$

$$\frac{781-1}{5}=\frac{780}{5}=156$$

$$\frac{156-1}{5}=\frac{155}{5}=31$$

$$\frac{31-1}{5}=\frac{30}{5}=6$$

$$\frac{6-1}{5}=\frac{5}{5}=1$$

$$\frac{1-1}{5}=\frac{0}{5}=1$$

We observe that the number 3906 is exhausted on six successive divisions by 5.

Thus, the common ratio $x=5$.

Mahavira stated that the solution of a quadratic equation of the form $x-A\sqrt{Bx}=R$ is

$$x=\left(\frac{A+\sqrt{A^2+(4R/B)}}{2}\right)^2\cdot B$$

Proof: Let $y=\sqrt{x}$.

Then, $x-A\sqrt{Bx}=R$ becomes

$$y^2-A\sqrt{B}y-R=0$$

So,

$$y=\frac{A\sqrt{B}\pm\sqrt{\left(A\sqrt{B}\right)^2+4R}}{2}$$

(*Continued*)

Example 2 (cont'd)

So,

$$y = \frac{A\sqrt{B} + \sqrt{(A\sqrt{B})^2 + 4R}}{2}, \quad \text{since } y > 0$$

$$y = \frac{A\sqrt{B} + \sqrt{B(A^2 + (4R/B))}}{2}$$

$$= \frac{\left(A + \sqrt{A^2 + (4R/B)}\right)\sqrt{B}}{2}$$

So,

$$x = y^2 = \left(\frac{A + \sqrt{A^2 + (4R/B)}}{2}\right)^2 \cdot B$$

Mahavira has examined some of equation of having the general form

$$a\sqrt{b \cdot x} + a_2\sqrt{b_2\left(x - a_1\sqrt{bx}\right)} + a_3\sqrt{b_3\left\{(x - a_1\sqrt{bx}) - a_2\sqrt{(x-a)\sqrt{bx}}\right\}}$$

$$+ \cdots + R = x$$

or

$$(x - a_1\sqrt{bx}) - a_2\sqrt{b_2\left(x - a_1\sqrt{bx}\right)}$$

$$-a_3\sqrt{b_3\left\{x - a_1\sqrt{bx} - a_2\sqrt{b_2(x - a_1\sqrt{bx})}\right\}}$$

$$- \cdots - R = x$$

Mahavira pointed out that if there are r terms on the left side, on rationalization, we will get an equation of second degree in x with proper substitution; this can be reduced to the quadratic equation of the form $x - A\sqrt{Bx} = R$, which we examined earlier.

Mahavira illustrated the above assertion with two following examples.

Example 1

"Of a head of elephants, nine times the square root of two-thirds plus six times the square root of three-fifths of the remainder entered a deep forest; the remaining 24 elephants were swimming in a near lake. How many elephants were there in the herd."

Solution

Let x be the number of elephants in the herd.

Then, we have

$$9\sqrt{\frac{2x}{3}}+6\sqrt{\frac{3}{5}\left[x-9\sqrt{\frac{2x}{3}}\right]}+24=x$$

Let

$$y=x-9\sqrt{\frac{2x}{3}}$$

Then, we have

$$y-6\sqrt{\frac{3}{5}y}=24$$

So, here $A=6$, $R=24$, and $B=3/5$.

Then, by the formula on page 295

$$y=\left(\frac{A+\sqrt{A^2+(4R/B)}}{2}\right)^2\cdot B$$

$$y=\left\{\frac{[6+\sqrt{36+(96/(3/5))}]^2}{2}\right\}\cdot\frac{3}{5}$$

$$=\left(\frac{(6+\sqrt{36+160})}{2}\right)^2\cdot\frac{3}{5}$$

$$=\left(\frac{(6+\sqrt{196})}{2}\right)^2\frac{3}{5}=\left(\frac{(6+14)}{2}\right)^2\frac{3}{5}=100\cdot\frac{3}{5}=60$$

(Continued)

Example 1 (cont'd)

Now,

$$y = x - 9\sqrt{\frac{2x}{3}}$$

So, we get

$$x - 9\sqrt{\frac{2x}{3}} = 60$$

Here,

$$A = 9, \quad B = \frac{2}{3}, \quad R = 60$$

So,

$$x = \left(\frac{A + \sqrt{A^2 + (4R/B)}}{2}\right)^2 \cdot B$$

$$x = \left\{\frac{9 + \sqrt{81 + (240/(2/3))}}{2}\right\}^2 \cdot \frac{2}{3}$$

i.e.,

$$x = \left(\frac{9 + \sqrt{81 + 360}}{2}\right)^2 \frac{2}{3}$$

$$x = \left(\frac{9 + \sqrt{441}}{2}\right)^2 \frac{2}{3} = \left(\frac{9 + 21}{2}\right)^2 \cdot \frac{2}{3}$$

i.e.,

$$x = (15)^2 \cdot \frac{2}{3} = (225) \cdot \frac{2}{3} = 150$$

So, the number of elephants in the heard is 150.

Example 2

"Four times the square root of half of a group of boars went into forest, where tigers hunt; twice the square root of tenth part of the remainder multiplied by 4 went into a cave, nine times the square root of half of the remainder went to the bank of a deserted river, boars numbering seven times eight were seen in a neighboring forest. Tell me the number of boars in that group."

Solution

Let x be number of boards in the group.

Then, we have

$$4\sqrt{\frac{x}{2}}+8\sqrt{\frac{1}{10}\left(x-4\sqrt{\frac{x}{2}}\right)}$$
$$+9\sqrt{\frac{1}{2}\left[x-\left\{4\sqrt{\frac{x}{2}}+8\sqrt{\frac{1}{10}\left(x-4\sqrt{\frac{x}{2}}\right)}\right\}\right]}+56=x \quad (*)$$

Let

$$y=x-4\sqrt{\frac{x}{2}}$$

Then, (*) becomes

$$-56+y-8\sqrt{\frac{1}{10}y}-9\sqrt{\left\{y-8\sqrt{\frac{1}{10}y}\right\}\frac{1}{2}}=0 \quad (**)$$

Let

$$z=y-8\sqrt{\frac{y}{10}}$$

Then, (**) becomes

$$z-9\sqrt{\frac{1}{2}z}=56 \quad (***)$$

Here,

$$A=9,\ B=\frac{1}{2},\ R=56$$

(*Continued*)

Example 2 (cont'd)

Then,

$$z=\left(\frac{A+\sqrt{A^2+(4R/B)}}{2}\right)^2 B$$

So,

$$z=\left(\frac{9+\sqrt{81+(224/(1/2))}}{2}\right)^2\cdot\frac{1}{2}=\left(\frac{9+\sqrt{81+448}}{2}\right)^2\cdot\frac{1}{2}$$

Hence,

$$z=\left(\frac{9+\sqrt{529}}{2}\right)^2\frac{1}{2}=\left(\frac{9+23}{2}\right)^2\cdot\frac{1}{2}=(256)\frac{1}{2}=128$$

Now,

$$z=y-8\sqrt{\frac{y}{10}}$$

So, we have

$$y-8\sqrt{\frac{y}{10}}=128$$

Here,

$$A=8,\ \ B=\frac{1}{10},\ \ R=128$$

Hence,

$$y=\left(\frac{A+\sqrt{A^2+(4R/B)}}{2}\right)^2\cdot B$$

$$=\left(\frac{8+\sqrt{64+(512/(1/10))}}{2}\right)^2\cdot\frac{1}{10}=\left(\frac{8+\sqrt{5184}}{2}\right)^2\cdot\frac{1}{10}$$

So,

$$y = \left(\frac{8+72}{2}\right)^2 \cdot \frac{1}{10} = 1600 \cdot \frac{1}{10} = 160$$

$$y = x - 4\sqrt{\frac{x}{2}}$$

$$x - 4\sqrt{\frac{x}{2}} = 160$$

$$A = 4, \quad R = 160, \quad B = \frac{1}{2}$$

i.e.,

$$x = \left\{\frac{4+\sqrt{(16+640/(1/2))}}{2}\right\}^2 \cdot \frac{1}{2}$$

i.e.,

$$x = \left\{\frac{4+\sqrt{(16+1280)}}{2}\right\}^2 \frac{1}{2} = \left(\frac{4+\sqrt{1296}}{2}\right)^2 \cdot \frac{1}{2}$$

i.e.,

$$x = \left(\frac{4+36}{2}\right)^2 \cdot \frac{1}{2} = (20)^2 \cdot \frac{1}{2}$$

i.e.,

$$x = 400 \cdot \frac{1}{2} = 200$$

So, the number of boars in that group = 200.

Simultaneous Quadrate Equations

(a) Mahavira gave the solution of $x + y = a$ (i) $xy = b$ (ii) as

$$x = \frac{1}{2}\left(a + \sqrt{a^2 - 4b}\right), \quad y = \frac{1}{2}\left(a - \sqrt{a^2 - 4b}\right)$$

Proof:

$$(x-y)^2 = (x+y)^2 - 4xy = a^2 - 4b$$

So,

$$x - y = \sqrt{a^2 - 4b} \qquad \text{(iii)}$$

But, $x + y = a$ from (i)
Adding we get

$$2x = a + \sqrt{a^2 - 4b} \Rightarrow x = \frac{1}{2}\left(a + \sqrt{a^2 - 4b}\right)$$

Subtracting (iii) from (i), we get $2y = a - \sqrt{a^2 - 4b}$.i.e.,

$$y = \frac{1}{2}\left(a - \sqrt{a^2 - 4b}\right)$$

(b) Mahavira gave the solution of $x^2 + y^2 = c$; $xy = b$ as

$$x = \frac{1}{2}\left[\sqrt{c + 2b} + \sqrt{c - 2b}\right]$$

$$y = \frac{1}{2}\left[\sqrt{c + 2b} - \sqrt{c - 2b}\right]$$

Proof:

$$x = \frac{1}{2}[(x + y) + (x - y)]$$

$$= \frac{1}{2}\left[\sqrt{(x+y)^2} + \sqrt{(x-y)^2}\right]$$

$$= \frac{1}{2}\left[\sqrt{x^2 + y^2 + 2xy} + \sqrt{x^2 + y^2 - 2xy}\right]$$

$$= \frac{1}{2}\left[\sqrt{c + 2b} + \sqrt{c - 2b}\right]$$

$$y = \frac{1}{2}[(x + y) - (x - y)] = \frac{1}{2}\left[\sqrt{(x+y)^2} - \sqrt{(x-y)^2}\right]$$

$$= \frac{1}{2}\left[\sqrt{x^2 + y^2 + 2xy} - \sqrt{x^2 + y^2 - 2xy}\right]$$

$$= \frac{1}{2}\left[\sqrt{c + 2b} - \sqrt{c - 2b}\right]$$

Mahavira: Solution of $(1 - (a/b))x - c\sqrt{x} = d\ (*)$ is

$$x = \left\{\frac{(c/2)}{1-(a/b)} + \sqrt{\left\{\frac{(c/2)}{1-(a/b)}\right\}^2 + \frac{d}{1-(a/b)}}\right\}^2$$

Proof: Dividing by

$\left(1 - \frac{a}{b}\right)$ in $(*)$, we get

$$x - \frac{c}{(1-(a/b))}\sqrt{x} = \frac{d}{1-(a/b)}$$

Here,

$$A = \frac{c}{1-(a/b)}, \quad B = 1, \quad R = \frac{d}{1-(a/b)}$$

Now,

$$x = \left\{A + \sqrt{A^2 + \frac{4R}{B}}/2\right\}^2 B$$

i.e.,

$$x = \left\{\left(\frac{c}{1-(a/b)} + \sqrt{\left(\frac{c}{1-(a/b)}\right)^2 + 4\frac{d}{1-(a/b)}}\right)/2\right\}^2 \cdot 1$$

i.e.,

$$x = \left\{\frac{(c/2)}{1-(a/b)} + \sqrt{\left(\frac{(c/2)}{1-(a/b)}\right)^2 + \frac{d}{1-(a/b)}}\right\}^2$$

Simultaneous Quadratic Equations

Mahavira has also tackled certain other problems which involve simultaneous quadratic equations of the form

$$u + x = a \quad \text{(i)}$$

$$u + y = b \quad \text{(ii)}$$

and

$$urw = \alpha x \quad \text{(iii)}$$

$$usw = \alpha y \quad \text{(iv)}$$

He pointed out that the solution of the above equations is

$$x = \frac{a-b}{r-s} \cdot r, \quad y = \frac{a-b}{r-s} \cdot s, \quad \text{and } w = \left(\frac{a-b}{rb-sa}\right)\alpha$$

Proof:

$$\frac{urw}{usw} = \frac{r}{s} = \frac{\alpha x}{\alpha y} = \frac{x}{y} = \frac{a-u}{b-u} \quad \text{from (i) and (ii)}$$

$$\frac{r}{s} = \frac{a-u}{b-u} \text{ gives } rb - ru = as - su$$

i.e., $su - ru = as - rb$.

$$u(s-r) = as - rb$$

$$u = \frac{as-rb}{s-r} = \frac{rb-as}{r-s}$$

But

$$x = a - u = a - \frac{(rb-as)}{r-s} = \frac{ar - as - rb + as}{r-s}$$

i.e.,

$$x = \frac{ar-rb}{r-s} = r\left(\frac{a-b}{r-s}\right)$$

Now,

$$y = b - u = b - \frac{(rb-as)}{r-s} = \frac{br - bs - rb + as}{r-s} = \frac{s(a-b)}{r-s}$$

Also,

$$w = \frac{\alpha x}{ur} = \frac{\alpha(r/(r-s))(a-b)}{r((rb-as)/(r-s))} = \frac{\alpha(a-b)}{rb-as}$$

In the above equations (i) and (iv), x and y represent interest accrued on the principal u for period r and s respectively and w is the interest rate per α. Mahavira observed this result in the following way:

> *"When the difference of the mixed sums* a *and* b *multiplied by each other's periods* r *and* s *is divided by the difference of the period, the quotient obtained is known as the principal* u.*"*

i.e.,

$$u = \frac{br-as}{r-s}$$

Mahavira also examined the problems of the form

$$u + x = p \text{ and } uxw = dm$$

$$u + y = q \tag{1}$$

$$uyw = dn \tag{2}$$

He stated that the solutions of the above equations are

$$x = \left(\frac{p-q}{m-n}\right)m, \quad y = \left(\frac{p-q}{m-n}\right)n, \quad \text{and } w = \frac{d(m-n)^2}{(p-q)(mq-np)}$$

Proof: From (2),

$$\frac{uxw}{uyw} = \frac{x}{y} = \frac{dm}{dn} = \frac{m}{n} \tag{3}$$

From (1), $x = p - u$ and $y = q - u$.
So,

$$\frac{x}{y} = \frac{p-u}{q-u} = \frac{m}{n} \quad \text{from (3)}$$

i.e., $pn - un = mq - mu$.

So, $mu - un = mq - pn$.
i.e., $u(m - n) = mq - pn$.
So,

$$u = \frac{mq - pn}{m - n}$$

From (1),

$$x = p - u = p - \frac{mq - pn}{m - n} = \frac{pm - pn - mq + pn}{m - n}$$

So,

$$x = \frac{pm - mq}{m - n} = \frac{m(p - q)}{m - n}$$

Again from (1),

$$y = q - u = q - \frac{mq - pn}{m - n} = \frac{mq - qn - mq + pn}{m - n}$$

So,

$$y = \frac{pn - qn}{m - n} = n\left(\frac{p - q}{m - n}\right)$$

Now, from (2),

$$w = \frac{dm}{ux} = \frac{dm}{((mq - pn)/(m - n))((p - q)/(m - n))m}$$

So,

$$w = \frac{d(m-n)^2}{(mq - pn)(p - q)}$$

Here, x and y represent the period for which the principal u is lent out at the interest rate w per d, and m and n are respective interest.

Mahavira pointed out "when the difference of the mixed sums p and q are multiplied by each other's interests namely n and m and then divided by the difference of interests, the quotient so obtained is the principal."

Indeterminate Equation of First Order $by = ax + c$

Mahavira slightly modified the methods of his predecessors Aryabhata I and Brahmagupta.

In any of these two methods, form the table as before leaving the first quotient q. Then, the top figure in the very last column of the table proposed by Aryabhata I will give the value of x. We now illustrate his modification by a few examples.

Example 1

Solve in integers the equation $137x + 10 = 60y$

As before,

$$60)137(2 = q$$
$$\underline{120}$$
$$r_1 = 17)60(3 = q_1$$
$$\underline{51}$$
$$r_2 = 9)17(1 = q_2$$
$$\underline{9}$$
$$r_3 = 8)9(1 = q_3$$
$$\underline{8}$$
$$r_4 = 1$$

The quotients leaving q is q_1, q_2, and q_3 and the number is 3, an odd number. Last residue $r_4 = 1$ and the penultimate residue is $r_3 = 8$.

Find a positive inter m_1 to find the quotient q^* such that

$$m_1 r_4 - 10 = r_3 q^*$$

$$m_1 \cdot 1 - 10 = 8q^*$$

So, $m_1 = 18, \quad q^* = 1$

Form the following table as before

	3	3	3	130
	1	1	37	37
	1	19	19	19
$m_1 =$	18	18	18	
$q^* =$	1	1		

$X = 130 \quad 130 \equiv 10 (\text{mod } 60)$

(Continued)

Example 1 (cont'd)

So, $x = \alpha = 10$.
Substituting $\alpha = 10$ in $137x + 10 = 60y$, we get

$$137(10) + 10 = 60y$$

$$1380 = 60y \Rightarrow y = \frac{1380}{60} = 23$$

which is exactly the solutions we have obtained by the earlier two methods. General solution is $x = 60m + 10$; $y = 137m + 23$.

Example 2

Solve in integers $63x + 7 = 23y$ (1)

$$\begin{array}{l} 23)63(2 = q \\ \quad \underline{46} \\ r_1 = 17)23(1 = q_1 \\ \qquad \underline{17} \\ \qquad r_2 = 6)17(2 = q_2 \\ \qquad\quad \underline{12} \\ \qquad\quad r_3 = 5)6(1 = q_3 \\ \qquad\qquad \underline{5} \\ \qquad\qquad r_4 = 1 \end{array}$$

The quotients leaving q is q_1, q_2, and q_3 and so the number of quotients is 3, which is an odd number. Here, the last residue is 1 and the penultimate residue is 5. We form the table as usual.

First, we have to find a positive integer m_1 to find the positive integer quotient q^* such that

$$m_1 r_4 - 7 = r_3 q^*$$

i.e., $m_1 \cdot 1 - 7 = 5q^*$

Choosing $m_1 = 12$, we get $q^* = 1$.

Now, we form the following table:

	1	1	1	51
	2	2	38	38
	1	13	13	13
$m_1 =$	12	12	12	
$q^* =$	1	1		

$X = 51$, now $X = 51 \equiv 5 (\text{mod } 23)$. Hence, $x = \alpha = 5$.
Substituting $x = 5$ in the equation $63x + 7 = 23y$, we get

$$63(5) + 7 = 23y$$

So, $23y = 315 + 7 = 322$.

$$y = \frac{322}{23} = 14$$

$x = 5$, $y = 14$ is the least solution of $63x + 7 = 23y$.
General solution is $x = 23m + 5$ $y = 63m + 14$.

Example 3

Solve in integers $83x + 15 = 17y$

$$17)83(4 = q$$
$$\underline{68}$$
$$r_1 = 15)17(1 = q_1$$
$$\underline{15}$$
$$r_2 = 2)15(7 = q_2$$
$$\underline{14}$$
$$r_3 = 1$$

The quotients leaving q are q_1 and q_2 and the number of quotients is 2 and so even. Here, the last residue is 1 and the penultimate residue is 2.

First, we have to find a positive integer m such that

(Continued)

Example 3 (cont'd)

$$m_1 r_3 + 15 = r_2 q^* \Rightarrow m_1 \cdot 1 + 15 = 2q^*$$

Selecting $m_1 = 1$, we get $q^* = 8$.
We now form the following table:

	1	1	16
	7	15	15
$m_1 =$	8	8	
$q^* =$	1	1	

$$X = 16 \equiv 16(\text{mod}17)$$

Substituting $x = \alpha = 16$ in the equation $83x + 15 = 17y$, we get $(83)(16) + 15 = 17y$, i.e., $17y = 1328 + 15 = 1343$.

$$y = \frac{1343}{17} = 79$$

$x = 16$ and $y = 79$ is the least solution of $83x + 15 = 17y$.
General solution is $x = 17m + 16$; $y = 83m + 79$.
Mahavira obtained general solution of a special case of the equation $ax + by = c$.

Lemma: If $ax + by = c(a + b)$, then the general solution of this equation is $x = c \pm (m/a)$ and $y = c \pm (m/b)$, where m is an arbitrary integer.

Proof: We have $ax + by = c(a + b) = ca + cb$.i.e., $a(x - c) = b(c - y) = \pm m$, where m is an arbitrary integer.

$$a(x - c) = \pm m \Rightarrow x - c = \pm \frac{m}{a} \Rightarrow x = c \pm \frac{m}{a}$$

$$b(c - y) = \pm m \Rightarrow c - y = \pm \frac{m}{b} \Rightarrow y = c \pm \frac{m}{b}$$

where m is an arbitrary integer.

Mahavira illustrated the lemma with the following example.

Example

"The colors of two pieces of gold of weight 16 and 10 are unknown, but the color of the mixture is 4. Can you tell me the color of each piece of gold?"

Solution

Let x and y be the colors of two pieces of gold.

Then, $16x + 10y = 4(16 + 10)$.

So, $16x + 10y = 104$ or $8x + 5y = 52 = 13c$, where $c = 4$.

Then, $8x - 8c = 5c - 5y$.

$8(x - c) = 5(c - y) = \pm m$, where m is an arbitrary integer.

$$8(x - c) = \pm m \Rightarrow x = c \pm \frac{m}{8} = 4 \pm \frac{m}{8}$$

$$5(c - y) = \pm m \Rightarrow y = c \pm \frac{m}{5} = 4 \pm \frac{m}{5} \text{ is the general solution}$$

A generalized case of conjunct pulvarizer ("Samslista Kuttaka") is a system of indeterminate equations of first degree, where the devisors as well as the multipliers are different like the following system:

$$b_1y_1 = a_1x \pm c_1$$

$$b_2y_2 = a_2x \pm c_2 \qquad (*)$$

$$b_ny_n = a_nx \pm c_n$$

Mahavira and later Sripati (AD 1039) have given a method of finding the general solution of such a system. Their method is as follows.

Consider the first two equations

$$b_1y_1 = a_xx \pm c_1 \qquad (1)$$

$$b_2y_2 = a_2x \pm c_2 \qquad (2)$$

Let α_1 be the least value of x in (1) by the method of pulvarizer. Let $b_1m_1 + \alpha_1$ be the general value of x satisfying (1), when m_1 is an arbitrary integer. Similarly, find the general solution $b_2m_2 + \alpha_2$ for the value of x in (2).

If (1) and (2) have the same value of x, we must have

$$b_1m_1 + \alpha_1 = b_2m_2 + \alpha_2$$

(*Continued*)

Example (cont'd)

Then,

$$b_2 m_2 = b_1 m_1 + (\alpha_1 - \alpha_2) \quad (3)$$

Solve (3) using once again the method of pulverization. Let m^* be the least value found for m_1. Substituting $m_1 = m^*$ in $b_1 m_1 + \alpha_1$ of x, we get the value of x as $b_1 m_1^* + \alpha_1 = b_2 m_2 + \alpha_2$ which satisfy both equations. Equate the general value of x derived from this with the value of x derived from the third equation and continue this process.

To demonstrate his method, Mahavira has proposed several problems. Here, we will solve one of his problems using his method.

Example

"Five heaps of fruits along with two more fruits are divided equally between 9 travelers; six heaps of fruits along with four more fruits were equally divided among eight. Four heaps along with one more fruit were divided among seven. Tell me number of fruits in each help."

Solution

Let x be the number of fruits in each heap. Let y_1, y_2, y_3 be the total number of fruits received by each traveler in the three distributions.

Then, we have

$$9y_1 = 5x + 2 \quad (1)$$

$$8y_2 = 6x + 4 \quad (2)$$

$$7y_3 = 4x + 1 \quad (3)$$

We first solve (1), i.e., $9y_1 = 5x + 2$ by the pulverization method.

$$9)5(0 = q$$
$$\underline{0}$$
$$r_1 = 5)9(1 = q_1$$
$$\underline{5}$$
$$r_2 = 4)5(1 = q_2$$
$$\underline{4}$$
$$r_3 = 1$$

The quotients leaving q are q_1 and q_2. So, the number of quotients is 2 and even. Here, the last residue r_3 is 1 and the penultimate residue r_2 is 4.

First we have to find a positive integer m_1^* such that

$$m_1^* r_3 + 2 = r_2 q^*$$

$$m_1^* \cdot 1 + 2 = 4q^*$$

Selecting $m_1^* = 2$, we get $q^* = 1$.
We now form the following table:

	1	1	5
	1	3	3
$m_1^* =$	2	2	2
$q^* =$	1	1	

$$X = 5 \equiv 5(\text{mod} 9)$$

Least value $x = \alpha = 5$.
So, general value of $x = 9m_1 + 5$, where m_1 is an arbitrary integer.
Next we solve (2), namely $8y_2 = 6x + 4$.
Dividing by 2, we get $4y_2 = 3x + 2$.

$$4)3(0 = q$$
$$\underline{0}$$
$$r_1 = 3)4(1 = q_1$$
$$\underline{3}$$
$$r_2 = 1$$

The quotient leaving q is q_1. So, the number of quotient is 1, which is an odd number. Here, the last residue r_2 is 1 and the penultimate residue r_1 is 3.
We have to find a positive integer m_2^* such that

$$m_2^* r_2 - 2 = r_1 q^*$$

$$m_2^* \cdot 1 - 2 = 3q^*$$

Choosing $m_2^* = 5$, we get $q^* = 1$. We now form the following table:

	1	6
$m_2^* =$	5	5
$q^* =$	1	1

$X = 6 \equiv 2(\text{mod } 4)$; least value of $x = \alpha \equiv 2$.
So, the general value of $x = 4m_2 + \alpha = 4m_2 + 2$.
If (1) and (2) have the same value for x, then

(*Continued*)

Example (cont'd)

$$9m_1 + 5 = 4m_2 + 2 \quad \text{i.e.,} \quad 9m_1 = 4m_2 - 3$$

We now solve this equation $9m_1 = 4m_2 - 3$.

$$9)4(0 = q$$
$$\underline{0}$$
$$r_1 = 4)9(2 = q_1$$
$$\underline{8}$$
$$r_2 = 1$$

Quotient leaving q is q_1 and so the number of quotients is 1, which is an odd number. Here, the last residue $r_2 = 1$ and the penultimate residue $r_1 = 4$. Now we have to find a positive integer m_3^* such that

$$m_3^* r_2 + 3 = r_1 q^*$$

$$m_3^* \cdot 1 + 3 = 4q^*$$

Choosing

$$m_3^* = 1, \quad q^* = 1$$

We now form the following table:

2	3	$x = 3$
1	1	$x = 3 \equiv 3 \pmod 9$
1	1	

So, $\alpha =$ the least value of $m_2 = 3$.

So, the general value of m_2 is $m_2 = 9m_3 + 3$, where m_3 is an arbitrary integer.

Now, we solve $7y_3 = 4x + 1$.

$$7)4(0 = q$$
$$\underline{0}$$
$$r_1 = 4)7(1 = q_1$$
$$\underline{4}$$
$$r_2 = 3)4(1 = q_2$$
$$\underline{3}$$
$$r_3 = 1$$

The number of quotients leaving q is 2 and so even. The last residue r_3 is 1 and the penultimate residue r_2 is 3. We have to find on integer m_4^* such that $r_3 m_4^* + 1 = r_2 q^*$, i.e., $m_4^* + 1 = 3q^*$.

So, $m_4^* = 2,\ q^* = 1$.

We now form the following table:

1	1	5	
1	3	3	$x = 5 \equiv 5 (\text{mod } 7)$
2	2	2	
1	1		

So, the least value of x is 5.

General value of $x = 7m_4 + 5$, where m_4 is an arbitrary positive integer.

But, general value of $x = 4m_2 + 2$.

General value of $x = 4m_2 + 2 = 4(9m_3 + 3) + 2 = 36m_3 + 14$.

Equating the two general values of x, we get

$$5 + 7m_4 = 36m_3 + 14$$

i.e., $7m_4 = 36m_3 + 9$.

$$36m_3 = 7m_4 - 9$$

$$\begin{array}{c} 36)7(0 = q \\ \underline{0} \\ r_1 = 7)36(5 = q_1 \\ \underline{35} \\ r_2 = 1 \end{array}$$

Leaving q, the number of quotients is 1, which is odd.

Here, the last residue $r_2 = 1$ and the penultimate residue $r_1 = 7$. Now, we have to find on integer m_5^* such that

$$r_2 m_5^* + 9 = r_1 q^* \quad \Rightarrow m_5^* + 9 = 7q^*$$

If $m_5^* = 5,\ q^* = 2$, we now form the following table:

	5	27	$x = 27 \equiv 27 (\text{mod } 36)$
$m_5^* =$	5	5	
$q^* =$	2	2	

So, the least value of $m_4 = 27$.

(*Continued*)

Example (cont'd)

Substituting $m_4 = 27$ in $36m_3 = 7m_4 - 9$.
We get $36m_3 = 7(27) - 9 = 18 \Rightarrow m_3 = 5$.
So, general value of $m_3 = 7m + 5$, where m is an arbitrary integer.
But general value of $x = 36m_3 + 14 = 36[7m + 5] + 14 = 194 + 252m$.
So, the least value of $x = 194$.

10 Aryabhata II and Sripati

Aryabhata II (Tenth Century)

Lemma 1: If $(\alpha\ \beta)$ is a solution of $ax \pm 1 = by$, then $(c\alpha\ c\beta)$ is a solution of $ax \pm c = by$.

Proof: $(\alpha\ \beta)$ is a solution of $ax \pm 1 = by$.
So, $a\alpha \pm 1 = b\beta$.

Multiplying by c, we get

$$ac\alpha \pm c = b\beta c$$

i.e., $(c\alpha\ c\beta)$ is a solution of $ax \pm c = by$.

Lemma 2: If $(\alpha\ \beta)$ is a solution of $by' = a'x \pm 1$, where $y' = y/d$, then $(c'\alpha\ c\beta)$ is a solution of $ax \pm c = by$, where $a = da'$, $c = dc'$.

Proof: If $(\alpha\ \beta)$ is a solution of $by' = a'x \pm 1$, where $y' = y/d$

$$b\beta = a'\alpha \pm 1$$

Multiplying by dc', we get

$$bdc'\beta = da'c'\alpha \pm dc'$$

$$bc\beta = ac'\alpha \pm c$$

So, $(c'\alpha\ c\beta)$ is a solution of $by = ax \pm c$.

Lemma 3: Let $b = db'$, $c = dc'$. Then, if $(\alpha\ \beta)$ is a solution of $b'y = ax' \pm 1$, where $x' = x/d$, then $(c\alpha\ c'\beta)$ is a solution of $by = ax \pm c$.

Proof: $(\alpha\ \beta)$ is a solution of $b'y = ax' \pm 1$.
So, $b'\beta = a\alpha \pm 1$.

Multiplying by dc'

$$dc'b'\beta = dc'a\alpha \pm dc'$$

$$b(c'\beta) = a(c\alpha) \pm c$$

So, $(c\alpha\ c'\beta)$ is a solution of $by = ax \pm c$.

Mathematical Achievements of Pre-modern Indian Mathematicians. DOI: http://dx.doi.org/10.1016/B978-0-12-397913-1.00010-7

Lemma 4: Let $a = a'd$, $c = c'd$, $b = b'd'$, $c' = c''d'$. Then, if $(\alpha\ \beta)$ is a solution of $b'y' = a'x' \pm 1$, where

$$x' = \frac{x}{d}, \quad y' = \frac{y}{d}$$

then

$$\left(\frac{c\alpha d'}{a\alpha'\alpha - bd\beta}\ \frac{cd\beta}{ad'\alpha - bd\beta}\right)$$

is a solution of $by = ax \pm c$.

Proof: $(\alpha\ \beta)$ is a solution of $b'y' = a'x' \pm 1$.

So, $b'\beta = a'\alpha \pm 1$.

Multiplying by dd', we get

$$dd'b'\beta = dd'a'\alpha \pm dd'$$

i.e.,

$$bd\beta = a(d'\alpha) \pm dd' \tag{1}$$

Dividing by dd', we get

$$\frac{bd\beta}{dd'} = \frac{a\alpha d'}{dd'} \pm 1$$

Multiplying by c, we get

$$\frac{cbd\beta}{dd'} = \frac{ac\alpha d'}{dd'} \pm c \tag{2}$$

Now, from (1), we have

$$ad'\alpha - bd\beta = \pm dd' \tag{3}$$

Substituting (3) in (2), we get

$$\frac{cbd\beta}{ad'\alpha - bd\beta} = \frac{ac\alpha d'}{ad'\alpha - bd\beta} \pm c$$

$$b\left(\frac{cd\beta}{ad'\alpha - bd\beta}\right) = a\left(\frac{c\alpha d'}{ad'\alpha - bd\beta}\right) \pm c$$

So,

$$\left(\frac{c\alpha d'}{ad'\alpha - bd\beta}\ \frac{cd\beta}{ad'\alpha - bd\beta}\right)$$

is a solution of $by = ax \pm c$.

Here, we observe that

$$c = c''dd'$$

$$c = c''(ad'\alpha - bd\beta) \quad \text{from (3)}$$

Thus, the values of x and y are integers.

समचतुरस्त्रे ऽर्धसमे वाभीबाश्रवणवर्गोनात् ।
सर्वभुजवर्गयोगमूलं कर्णो रितीयः स्यात् ॥

(Ma. Si. XIV. 81–82)

i.e., "In a parallelogram or a rhombus, the diagonal will be equal to the square root of the difference between the sum of squares of all the sides and the square of the other diagonal."

Proof: Let $ABCD$ be a parallelogram. Let AM and BN be altitudes from A and B, respectively, on DC. Join the diagonals AC and BD.

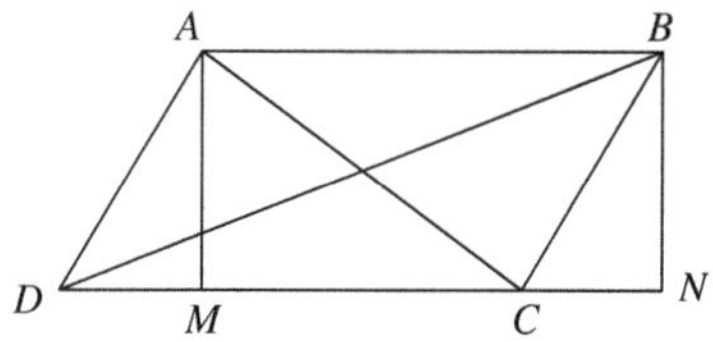

In the right Δ^{le} DBN,

$$BD^2 = DN^2 + BN^2$$

$$= (DC + CN)^2 + BN^2 \quad (1)$$

In the right Δ^{le} BCN,

$$BN^2 = BC^2 - CN^2 \quad (2)$$

Substituting (2) in (1), we get

$$BD^2 = (DC + CN)^2 + BC^2 - CN^2 \quad (3)$$

Again, in the right Δ^{le} ACM,

$$AC^2 = AM^2 + MC^2$$

$$AC^2 = AM^2 + (DC - DM)^2 \quad (4)$$

Also, in the right Δ^{le} ADM,

$$AM^2 = AD^2 - DM^2 \tag{5}$$

Substituting (5) in (4), we get

$$AC^2 = AD^2 + (DC-DM)^2 - DM^2$$

Since $CN = DM$, we have

$$AC^2 = AD^2 + (DC-CN)^2 - CN^2 \tag{6}$$

Adding (3) and (6), we get

$$AC^2 + BD^2 = AD^2 + BC^2 + (DC+CN)^2 + (DC-CN)^2 - 2CN^2$$

$$= AD^2 + BC^2 + DC^2 + CN^2 + DC^2 + CN^2 - 2CN^2$$

$$= AD^2 + BC^2 + 2DC^2$$

$$AC^2 + BD^2 = 2BC^2 + 2AB^2, \quad \text{since } AD = BC \text{ and } DC = AB$$

So, $AC^2 = 2BC^2 + 2AB^2 - BD^2$, if BD is given or $BD^2 = 2BC^2 + 2AB^2 - AC^2$, if AC is given.

Area of a Rhombus

Āryabhata II has given the following formula for the area of a rhombus:

श्रुतिघातः समचतुरस्रे त्र्यधितः फलं स्यात् ।

(Ma. Si. XIV. 82)

i.e., "The area of a rhombus or a square is equal to half the product of the two diagonals."

Proof: Let $ABCD$ be a parallelogram. Join the diagonal AC and BD. Let O be the point of intersection of the diagonals AC and BD.

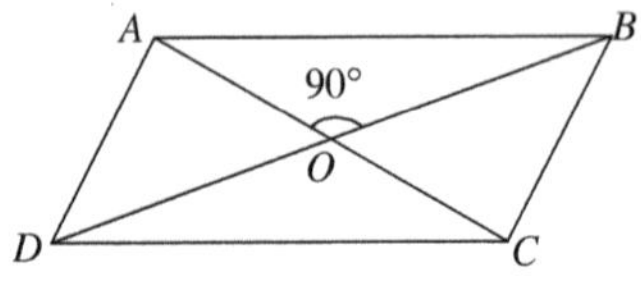

Now,

$$\text{Area of } \Delta^{le}\ ABD = \frac{1}{2} BD \cdot OA$$

$$\text{Area of } \Delta^{le}\ BCD = \frac{1}{2} BD \cdot OC$$

$$\text{Area of rhombus } ABCD = \frac{1}{2} BD(OA + OC) = \frac{1}{2} BD \cdot AC$$

From the above proof, it becomes evident that Aryabhata I was aware that the diagonals of a rhombus bisect each other at right angles.

Area of a Triangle, Square, Rectangle, and Isosceles Trapezium

Aryabhata II gives correct formula for the area of a triangle, a square, a rectangle, and an isosceles trapezium.

त्रिभुजचतुर्भुजफलानयनाय
वदनक्षितियोगदलं लम्बहतं जायते गणितम् ।
त्रिभुजे समचतुरस्रेऽर्धसमे वा कर्णभेदेऽपि॥
शृंगाटके न नियमाद्विषमचतुर्बाहुके च न प्रायः ।
याम्योत्तरलम्बैक्यार्द्ध क्वा (कु + त्र्या) स्यैक्यार्धताडितं निकटम् ॥

(Ma. Si. XIV. 78–79)

i.e., "The area of a triangle, a square, a rectangle, and an isosceles trapezium is equal to one-half of the product of the altitude and the sum of the base and face."

Proof: Let $ABCD$ be an isosceles trapezium, such that $AD = BC$.

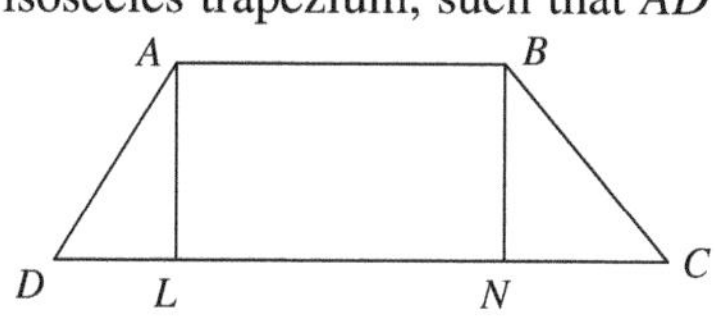

Draw AL and BN $\perp^{rs}$ from A and B on DC.

Then,

$$\text{Area of } ABCD = \frac{1}{2}(AB + DC)AL$$

Since $AL = BN$,

$$DL^2 = AD^2 - AL^2 = BC^2 - BN^2 = CN^2$$

Hence, $DL = CN$.

Area of trapezium $ABCD$

$$= \text{Area of the rectangle } ABNL + \text{Area of } \Delta ADL + \text{Area of } \Delta BNC$$

$$= AB \cdot AL + \frac{1}{2} DL \cdot AL + \frac{1}{2} CN \cdot BN$$

$$= AB \cdot AL + \frac{1}{2} DL \cdot AL + \frac{1}{2} CN \cdot AL$$

$$= AL\left(AB + \frac{1}{2} DL + \frac{1}{2} CN\right)$$

$$= \frac{1}{2} AL(2AB + DL + CN) = \frac{1}{2} AL(AB + LN + DL + CN)$$

$$= \frac{1}{2} AL(AB + DC)$$

In the case of a triangle, the face is zero and the formula for the area reduces to half the product of the base and the altitude.

Area of a Quadrilateral

चतुरस्रेऽत्रत्यत्रिभुजद्वयफलयुतिर्गणितम् ।
तत्रत्यकयोः कर्णो भूः स्यादितरे भुजाश्च चत्वारः॥

(Ma. Si. XIV. 68)

i.e., "In the case of a quadrilateral, its area is equal to the sum of the areas of two triangles, with the diagonal as the base."

Proof: Let $ABCD$ be a quadrilateral and let BD be a diagonal.

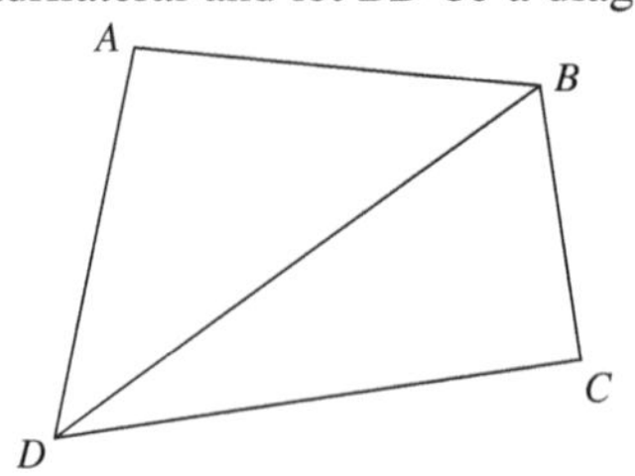

Then,

$$\text{Area of the quadrilateral } ABCD = \text{Area of the triangle } ABD + \text{Area of the triangle } BCD$$

Diagonals of a Quadrilateral

The diagonals of a quadrilateral are computed in terms of the adjacent sides of the quadrilateral and the altitude at the end of that diagonal.

Let $ABCD$ be a quadrilateral with AC and BD as diagonals. Draw AM and CN perpendicular to DC and AB (produced), respectively.

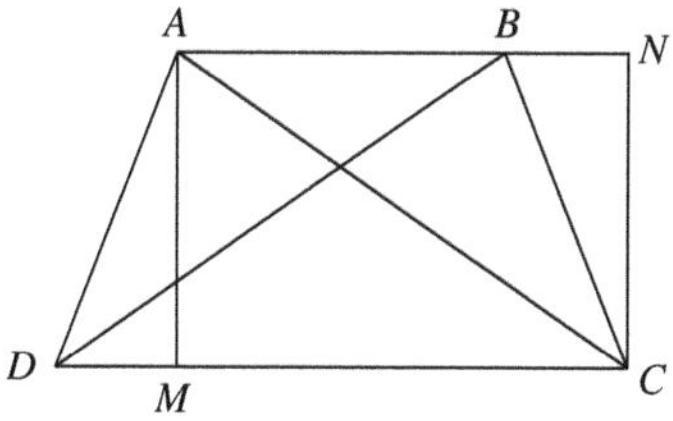

Then,

$$AC = \sqrt{AM^2 + (DC - \sqrt{AD^2 - AM^2})^2}$$

or

$$AC = \sqrt{NC^2 + (AB + \sqrt{BC^2 - NC^2})^2}$$

Proof: In the right Δ^{le} ADM,

$$DM^2 = AD^2 - AM^2$$

i.e.,

$$DM = \sqrt{AD^2 - AM^2}$$

Again, in the right Δ^{le} ACM,

$$AC^2 = AM^2 + MC^2 = AM^2 + (DC - DM)^2$$

So,

$$AC = \sqrt{AM^2 + (DC - DM)^2}$$

i.e.,

$$AC = \sqrt{AM^2 + (DC - \sqrt{AD^2 - AM^2})^2}$$

Similarly, in the right Δ^{le} CBN,

$$BN^2 = BC^2 - CN^2$$

i.e.,

$$BN = \sqrt{BC^2 - CN^2}$$

In the right Δ^{le} ANC,

$$AC^2 = AN^2 + NC^2$$

But

$$AN = AB + BN$$
$$AN = AB + \sqrt{BC^2 - CN^2}$$

So,

$$AC^2 = (AB + \sqrt{BC^2 - CN^2})^2 + NC^2$$

Thus,

$$AC = \sqrt{NC^2 + \left(AB + \sqrt{BC^2 - CN^2}\right)^2}$$

Calculation of the Second Diagonal of a Quadrilateral, Given One Diagonal

Aryabhata II's method is as follows:

Let the given diagonal be the base of two triangles formed by it in the given quadrilateral. First, calculate the altitudes from the end of the second diagonal on the given diagonal. Then, the second diagonal can be calculated in terms of the above altitudes and the "abadhas" formed by them on the given diagonal.

इष्टश्रवणं वसुधां परिकल्प्य त्रिभुजयोरुभयोः।
लम्बाबाधाः साध्याः स्थाप्य कर्णाग्रमूलयोबांधाः।
क्षे त्त्रद्वयबाधान्तरवर्गाल्लम्बैक्यवर्गयुतात् ।
मूलं द्वितीयकर्णश्चतुश्राणां च सर्वेषाम् ॥

(Ma. Si. XIV. 86–87)

i.e., "The second diagonal is equal to the square root of the sum of squares of the sum of two altitudes and the square of the difference between the two 'abadhas.'"

Let $ABCD$ be the given quadrilateral. Let AC be the given diagonal. Let BM and ND be the altitudes on AC from B and D, respectively. Let DP be the perpendicular drawn from D on BM.

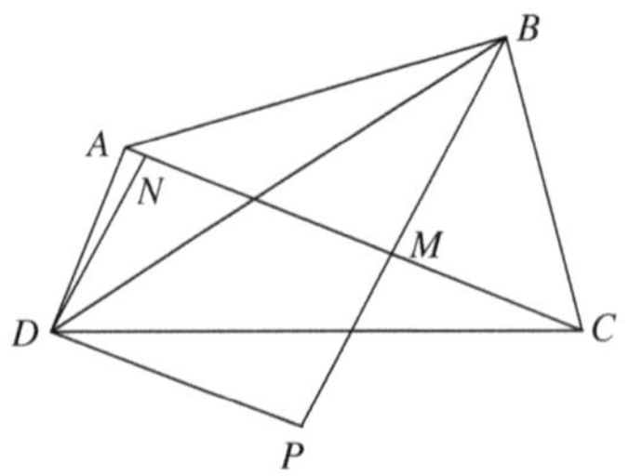

Then, $BD^2 = (DN + BM)^2 + (AM - AN)^2$.

Proof: Now, $DP = MN$ and $DN = PM$.

In the right Δ^{le} DBP,

$$\begin{aligned} BD^2 &= DP^2 + BP^2 = (PM + BM)^2 + DP^2 \\ &= (DN + BM)^2 + MN^2 \end{aligned}$$

i.e.,

$$BD^2 = (DN + BM)^2 + (AM - AN)^2$$

Arc of a Circle

Aryabhata II restated the formula for the length of arc $= a = \sqrt{6h^2 + c^2}$, where h is the height and c is the length of the chord.

From this formula,

$$h = \sqrt{\frac{a^2 - c^2}{6}} \quad \text{and} \quad c = \sqrt{a^2 - 6h^2}$$

Arybhata II's formula for the approximate area of the segment is as follows:

ज्याबाणैक्यदलेष्वोर्घातात् स्वघ्नात् स्वनवमभागयुतात् ।
यन्मूलं तत् स्थूलं क्षे त्रफलं कार्मुके भवति ॥

(Ma. Si. XIV. 89)

i.e., "Half the height is multiplied by the sum of the chord and the height. Square of this product is added to 1/9 of this product itself. Square root of the resulting quantity is the area of the segment."

That is, area of the segment

$$= \sqrt{\left\{\frac{h(c+h)}{2}\right\}^2 + \frac{1}{9}\frac{h(c+h)}{2}}$$

where h is the height and c is the chord.

Volume of a Sphere

Aryabhata II gave the following formula for the volume of the sphere:

कन्दुकपिण्डस्य घनो दलितः स्वाष्टादशांशसंयुक्तः ।
धनहस्ता......

(Ma. Si. XIV. 108)

The volume of the sphere is equal to the sum of half the cube of the diameter and 1/18 of its part.

If d is the diameter of the cube,

$$\text{Volume of the sphere} = \frac{1}{2}d^3 + \frac{1}{18}\left(\frac{1}{2}d^3\right)$$
$$= \frac{1}{2}d^3 + \frac{1}{36}d^3 = \frac{19}{36}d^3 = \frac{19}{36}d^3$$
$$= \frac{19}{36}(2r)^3, \quad \text{where } r \text{ is the radius of the sphere}$$
$$= \frac{38}{9}r^3, \quad \text{where } r \text{ is the radius of the sphere}$$

Value of π

Aryabhata II insisted the accurate ("Sūksma") value of π.

This volume of the sphere

$$= \frac{38}{9}r^3 = \frac{22}{7} \cdot \frac{133}{99}r^3$$

Volume of the sphere

$$= \frac{133}{99}\pi r^3$$

according to Aryabhata II.

The correct value of the volume of the sphere

$$= \frac{4}{3}\pi r^3 = \frac{132}{99}\pi r^3$$

Sripati (AD 1025)

Sripati is an astronomer and flourished in the eleventh century. He composed "Ganita tilaka," "Siddhanta Siromani," and several other astronomical treatises and manuals. In his work "Ganita tilaka," he completely dispensed off with geometry and algebra.

It was well known to the earlier ancient Indian astronomers that four corrections have to be applied to mean positions to determine the true positions of the planets. One of these corrections is called "udayantara samskara." This is the correction that has to be applied to obtain the planetary positions at the true midnight or true sunrise caused by the obliquity of the ecliptic with the celestial equator. Sripati is the first ancient Indian astronomer to give this correction and named it "Yatāsāva."

We will now proceed with his mathematical contributions.

Solution of a Quadratic Equation

Sripati provided two methods of solving quadratic equation. His first method is identically the same as that of his predecessor, Sridhara. The second method is as follows:

"Multiply the given equation by the coefficient of the square of the unknown and add the square of half of the coefficient of the unknown. Extract the square root. Then, decrease it by half of the coefficient of the unknown and divide by the coefficient of the square of the unknown. This quotient obtained is the value of the unknown."

The given equation is $ax^2 + bx = c$ (Si. Śe. xiv. 17–19).

The coefficient of the square of unknown is a and the coefficient of the unknown is b.

So, the procedure is as follows:

Multiplying by a, we get $a^2x^2 + abx = ac$.

Adding $(b/2)^2$, we have

$$a^2x^2 + abx + \left(\frac{b}{2}\right)^2 = ac + \frac{b^2}{4}$$

i.e.,

$$\left(ax + \frac{b}{2}\right)^2 = ac + \frac{b^2}{4}$$

Taking the square root, we obtain

$$ax + \frac{b}{2} = \sqrt{ac + \frac{b^2}{4}}$$

i.e.,

$$ax = \sqrt{ac + \frac{b^2}{4}} - \frac{b}{2}$$

Thus,

$$x = \frac{\sqrt{ac + (b^2/4)} - (b/2)}{a}$$

Construction of Rational Quadrilateral

For the construction of rational quadrilaterals, Sripati stated the following:

"Take two right triangles. The 'bhuja' (base) and 'koti' (upright) of the right triangles should be reciprocally multiplied by the hypotenuses. Of the products obtained, the greatest is the base; the smallest is the face and the remaining two will be two flank sides of a quadrilateral with unequal sides."

For illustration, Sripati took two right triangles (3 4 5) and (5 12 13) to get the quadrilateral (25, 39, 60, 52), using his procedure:

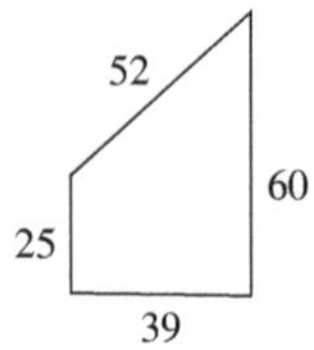

Rational Solution of "Vargaprakrti" Equation

Sripati provided an elegant method to obtain a rational solution of the "Vargaprakrti" equation,

$$Nx^2 + 1 = y^2$$

where N is a nonsquare integer.

Sripati's Theorem

If m is a suitably chosen rational number, then

$$\left(\frac{2m}{m^2 - N} \quad \frac{m^2 + N}{m^2 - N}\right)$$

is a rational solution of the "Vargaprakrti" equation $Nx^2 + 1 = y^2$, where N is a nonsquare integer.

(Si. Sc. XIV. 33)

Proof: Let m^2 be a suitably chosen rational number.

We have the identity,

$$N \cdot 1^2 + (m^2 - N) = m^2 \qquad \text{(i)}$$

This means $(\alpha \quad \beta) = (1 \quad m)$ is a solution of the equation $Nx^2 + K = y^2$, where $K = m^2 - N$.

Then, by Brahmagupta's corollary on page 165,

$$(2\alpha\beta \quad \beta^2 + N\alpha^2) = (2m \quad m^2 + N)$$

is a solution of the equation,

$$Nx^2 + K^2 = y^2$$

where

$$K = m^2 - N$$

i.e.,

$$N(2m)^2 + (m^2 - N)^2 = (m^2 + N)^2 \quad \text{(ii)}$$

Dividing (ii) by $(m^2 - N)^2$, we get

$$N\left(\frac{2m}{m^2 - N}\right)^2 + 1 = \left(\frac{m^2 + N}{m^2 - N}\right)^2$$

which means that

$$\left(\frac{2m}{m^2 - N}\ \frac{m^2 + N}{m^2 - N}\right)$$

is a rational solution of $Nx^2 + 1 = y^2$, where m is a suitably chosen rational number.

Bakshali Equation

Sripati provided a solution of the Bakshali equation

$$axy = bx + cy + d \quad (*)$$

Theorem

If m is an arbitrarily chosen number, then the solution of the equation (*) $axy = bx + cy + d$ is either

$$x = \frac{1}{a}(m + c), \quad y = \frac{1}{a}\left(\frac{bc + ad}{m}\right) + \frac{b}{a} \quad \text{(i)}$$

or

$$y = \frac{1}{a}(m + b), \quad x = \frac{1}{a}\left(\frac{bc + ad}{m}\right) + \frac{c}{a} \quad \text{(ii)}$$

Proof: (i) Let m be an arbitrarily chosen number.

Let

$$x = \frac{1}{a}(m + c)$$

(*) can be written as $x(ay - b) = cy + d$. So,

$$x = \frac{cy + d}{ay - b} = \frac{1}{a}(m + c)$$

Cross multiplying, we get

$$acy + ad = amy - bm + cay - bc$$

$$ad = amy - bm - bc$$

i.e.,

$$amy = ad + bc + bm$$

$$y = \frac{ad + bc + bm}{am} = \frac{1}{a}\left(\frac{ad + bc}{m}\right) + \frac{b}{a}$$

Proof: (ii) Let

$$y = \frac{1}{a}(m + b)$$

(*) Can be written as $y(ax - c) = bx + d$.

$$y = \frac{bx + d}{ax - c} = \frac{m + b}{a}$$

Cross multiplying, we get

$$abx + ad = amx + abx - cm - cb$$

$$ad = amx - cm - cb$$

So,

$$amx = ad + bc + cm$$

$$x = \frac{ad + bc + cm}{am} = \frac{1}{a}\left(\frac{ad + bc}{m}\right) + \frac{c}{a}$$

Sripati provided the formula for the area of a trapezium as well as a method for the computation of the circumradius of an isosceles trapezium.

चतुर्भुजेष्टश्रुतिपार्श्वबाहुवधस्य चार्द्धं खलु लम्बभक्तम् ।
त्र्यतुल्यबाहोः प्रतिबाहुभाग (बाहु?) वर्गैक्यमूलस्य दलं हि यद्वा ॥

(Si. Se. p. 86)

i.e., "In an isosceles trapezium, if half the product of the given diagonal and the side adjacent to is divided by its altitude, the resulting quotient is the circumradius."

11 Bhaskara II

The most celebrated mathematician and an excellent astronomer of ancient India is Bhaskaracharya or commonly known as Bhaskara II. As we have seen, he is the second ancient mathematician with this name. He wrote his famous work "Siddhanta Siromani" in AD 1150. He mentioned in his celebrated work that he was born in the year AD 1114, he belonged to a place called Bijjada Bida near Sahyadri mountains and his father was a saintly and scholarly Brahmin of Sandilya gotra by name Mahadeva. This means that Bhaskara II wrote his famous work when he was 36 years old. The place Bijjada Bida is the modern Bijapur in Karnataka state.

According to Epigraphia Indica, Vol. I, page 340, a stone inscription was found at a place called Patan about 10 miles southwest of Chalisgoan in Maharashtra, on which it is inscribed that Changadeva, grandson of Bhaskara II, was an astronomer at the court of King Singahana of the Yadava dynasty, who ruled at Devagiri from AD 1210 to 1237 and that Changadeva built a monastery in Patan for illuminating the works of Bhaskara II and his descendents. It also said that this monastery received endowments from King Saideva of Nikhumba dynasty for its maintenance and King Jaitrapala invited Lakshmidhara, a son of Bhaskara II from Patan. This is all we know about the life history of Bhaskara II and his descendents.

The "Siddhanta Siromani" consists of four parts with names "Leelavati," "Bijaganitha" (algebra), "Goladhaya" (the chapter on celestial globe), and "Grahaganitha" (mathematics of planets). The first two parts essentially are texts on mathematics and the last two parts relate to astronomy. His work exhibits a good amount of poetic skill. He wrote "Vasana Bhashya," a commentary in prose, which explains mathematics occurring in his poems, with a number of illustrative examples. In AD 1183, when he was 69 years old, he also wrote a shorter astronomical text called "Karanakutuhala."

Bhaskara II acknowledged at the end of "Bijaganitha" that he has gathered and simplified the algebraical material from the works of his predecessors, Brahmagupta, Sridhara, and Padmanabha. Unfortunately, the algebraical works of Sridhara and Padmanabha are unavailable now. As a result, it has not been possible to estimate to what extent Bhaskara II is indebted to them. Although Bhaskara II did not mentioned the name of Mahavira, a fellow predecessor Karnataka mathematician, there are vivid instances, which exhibit that he got inspiration from the work of Mahavira.

"Leelavati" is mainly arithmetic. However, it contains some mensuration, geometry dealing with problems on right triangles, and a chapter on "Kuttaka." The chapter on "Kuttaka" is repeated in the second chapter of "Bijaganitha" with

Mathematical Achievements of Pre-modern Indian Mathematicians. DOI: http://dx.doi.org/10.1016/B978-0-12-397913-1.00011-9

noteworthy differences. The rationale for this repetition appears to be that the intention of Bhaskara II is that a reader of "Leelavati" may be satisfied with the mechanical application of the method, whereas a reader of "Bijaganitha" may likely to understand the theory behind the method.

"Leelavati has been divided into 13 chapters by later scribers. It commences with a prayer to Lord Vidya Ganapathi and consists of number system, the eight operations of arithmetic, fractions, zero, rule of three, interest, progression, geometry, shadow problems, permutations, and "Kuttaka."

An interesting conception of infinity that any number other than zero divided by zero gives infinity, occurs first time in the history of mathematics in Bhaskara II's work. He dealt with zero and infinity in the first chapter of "Bijaganitha." He made the statement that $\infty \pm k = \infty$.

In the "Goladhyaya," Bhaskara II has dealt extensively with theoretical astronomy, improving the formulae and the methods used by his predecessors and explaining the reasons for the formulae he employed as well as expounding clearly the eccentric and epicyclic theories for the motion of planets. In "Yantradhyaya," Bhaskara II has introduced a large number of astronomical instruments.

In the "Grahaganitha," Bhaskara II has made an exhaustive study of problems like the lunar and solar eclipses, risings, settings, and conjunction of planets and determination of the mean and true positions of planets.

The text "Karanakutuhala" contains 139 slokas. In it, Bhaskara II has taken the sunrise of "Phalguna" of saka year 1104 (AD 1182), Thursday as epoch for purpose of computations, which is now popularly known as "Grahagama Kutuhala."

Bhaskara II is the first ancient Indian astronomer to introduce the equation of motion, which is at present called "Evection."

Bhaskara II called the subject of permutations and combinations as "Anka Pasha." The Jains had called this as "Vikalpa." Bhaskara II wrote down the general value of ${}_np_r$ and ${}_nc_r$ which had been given earlier by his predecessor, Mahavira. The theorem that the number of permutation of r things of which k_1 are of one kind, k_2 of another kind, etc, is $(r!/k_1!k_2!\ldots)$ is given first by Bhaskara II in the history of world mathematics.

He has given the following problem, which is not elementary.

Example (Bhaskara II)

A number has 5 digits and the sum of the numbers forming the digits is 13. If zero is not a digit, find the total numbers of number possible.

Solution

91111, 82111, 72211, 73111, 64111, 63211, 62221, 55111, 54211, 53311, 53221, 52222, 44311, 44221, 43321 are possible numbers.

Permutation of 91111 gives the numbers 91111, 19111, 11911, 11191, 11119, and so gives five numbers.

Similarly 52222 and 13333 give rise to 5 numbers each. 55111 and 22333 give rise to 10 numbers each.

82111, 73111, 64111, 43222, and 61222 give rise to 20 numbers each.

72211, 53311, 44221, and 44311 give rise to 30 numbers each.

63211, 54211, 53221, and 43321 give rise to 60 numbers each.

Total number of numbers $= 3 \times 5 + 2 \times 10 + 5 \cdot 20 + 4 \cdot 30 + 4 \cdot 60$

$$= 15 + 20 + 100 + 120 + 240 = 495$$

Bhaskara II also gave a formula for the following general problem: Let n be the number of digits (zero is not a digit). Let s be the sum of the n digits. Then, if $9 + n > s$, the total number of possible digits is $s^{-1}C_{n-1}$. Bhaskara II, however, did not give the proof. The following is the proof:

Consider the product of n equal factors.

$$(x + x^2 + x^3 + \cdots + x^n +)(x + x^2 + x^3 + \cdots + x^n +)(x + x^2 + x^3 + \cdots + x^n + \cdots)$$
$$\cdots n \text{ factors}$$

In this product, a term in x^s corresponds to different ways of choosing a set of indices, one from each factor such that their sum is s. Since $s < n + 9$, no power of higher than 9 can be considered from any factor. Then, the required number is equal to the coefficient of x^s in the expansion $(x + x^2 + x^3 + \cdots)^n = x^n(1 + x + x^2 + \cdots)^n$ or equivalently the coefficient of x^{s-n} in the expansion of $(1 + x + x^2 + \cdots)^n = \{(1 - x)^{-1}\}^n = (1 - x)^{-n}$.

$$\text{So, the required coefficient} = \frac{n(n+1)(n+2)\cdots(n+s-n-1)}{(s-n)!}$$
$$= \frac{n(n+1)(n+2)\cdots(s-1)}{(s-n)!}$$
$$= \frac{(s-1)!}{(n-1)!(s-n)!} = {}_{s-1}C_{n-1}$$

The following two examples are found in Bhaskara II's "Bijaganitha."

Example 1

Find four different numbers whose sum is equal to the sum of their squares.

(Continued)

Example 1 (cont'd)

Solution

We observe that $1 + 2 + 3 + 4 = 10$; $1^2 + 2^2 + 3^2 + 4^2 = 30$.

Hence, altering the numbers in the ratio

$$\frac{10}{30} = \frac{1}{3}$$

we get the required numbers.

Thus,

$$\frac{1}{3} + \frac{2}{3} + \frac{3}{3} + \frac{4}{3} = 3\frac{1}{3} = \left(\frac{1}{3}\right)^2 + \left(\frac{2}{3}\right)^2 + \left(\frac{3}{3}\right)^2 + \left(\frac{4}{3}\right)^2$$

So, the required four numbers are

$$\frac{1}{3}, \frac{2}{3}, 1, \frac{4}{3}$$

Example 2

Find four different numbers whose sum of squares is equal to the sum of their cubes.

We observe that

$$1^2 + 2^2 + 3^2 + 4^2 = 30$$

and

$$1^3 + 2^3 + 3^3 + 4^3 = 1 + 8 + 27 + 64 = 100$$

Hence, altering the numbers in the ratio

$$\frac{30}{100} = \frac{3}{10}$$

we get the required numbers.

$$\left(\frac{3}{10}\right)^2 + \left(\frac{6}{10}\right)^2 + \left(\frac{9}{10}\right)^2 + \left(\frac{12}{10}\right)^2 = \frac{27}{10} = \left(\frac{3}{10}\right)^3 + \left(\frac{6}{10}\right)^3 + \left(\frac{9}{10}\right)^3 + \left(\frac{12}{10}\right)^3$$

So, the required four different numbers are

$$\frac{3}{10}, \frac{6}{10}, \frac{9}{10}, \frac{12}{10} \text{ or } \frac{3}{10}, \frac{3}{5}, \frac{9}{10}, \text{and } \frac{6}{5}$$

Bhaskara II's "Ista-Karma" rule (rule of supposition) is as follows:

"Whenever a required number is multiplied or divided by whatever fraction of the number is found to have been increased or diminished, assume an optional number; on it perform the same operations in accordance with the statement of the problem; multiply the given number in the statement of the problem by the assumed number and then divide this product by the number which resulted from the above operation. The quotient will be the required number."

This may be illustrated with the following example:

"What is that number which multiplied by five, diminished by its third part and then divided by ten, will become together with its one-third, half and one-fourth parts, equal to 68?"

To illustrate thus, let us take $x = 5$.

Then,

$$\frac{5 \times 5 - (5 \cdot 5/3)}{10} + \frac{5}{3} + \frac{5}{2} + \frac{5}{4} = \frac{75 - 25}{30} + \frac{5}{3} + \frac{5}{2} + \frac{5}{4}$$

$$= \frac{5}{3} + \frac{5}{3} + \frac{5}{2} + \frac{5}{4} = \frac{20 + 20 + 30 + 15}{12} = \frac{85}{12}$$

$$(68 \times 5) \div \frac{85}{12} = \frac{340 \times 12}{85} = 4 \cdot 12 = 48$$

Bhaskara II gave the following example.

Example 3

"One person possesses 300 coins and 6 horses. Another possesses 10 horses (each of similar value) and in addition, he has a debt of 100 coins. But they are equally rich. Can you tell me the price of a horse?"
(BBI, 444)

Solution

Let the cost of a horse be x coins.

Worth of one person $= 300 + 6x$.

Worth of another person $= 10x - 100$.

Since they are equally rich, we have

(*Continued*)

Example 3 (cont'd)

$$300 + 6x = 10x - 100$$

$$300 + 100 = 10x - 6x$$

$$400 = 4x$$

i.e.,

$$x = \frac{400}{4} = 100$$

The price of each horse is 100 coins.

Bhaskara II gave the following example on simultaneous equations.

Example 4

"One says, give me a hundred, friend, I will become twice as rich as you. The other replies, if you give me ten, I will become six times rich as you. Tell me what is the amount that these two people had."

Solution

Let x and y be the amount of the two persons, respectively.

Then,

$$x + 100 = 2(y - 100) \quad (1)$$

$$y + 10 = 6(x - 10) \quad (2)$$

Bhaskara II gave two methods to solve this example:

First method:

Assume $x = 2z - 100$ and $y = z + 100$.

Then, we observe that this satisfies (1).

Substituting these values in (2), we get

$$z + 100 + 10 = 6(2z - 100 - 10)$$

$$z + 110 = 12z - 660$$

$$11z = 770 \Rightarrow z = 70$$

Hence, $x = 140 - 100 = 40$ and $y = 70 + 100 = 170$.

Second method:
From (1), we have $x = 2y - 200 - 100 = 2y - 300$.
From (2),

$$x = \frac{y+10}{6} + 10 = \frac{y+70}{6}$$

So,

$$2y - 300 = \frac{y+70}{6}$$

or

$$12y - 1800 = y + 70 \Rightarrow 11y = 1870 \Rightarrow y = 170$$

Then,

$$x = \frac{y+70}{6} = \frac{170+70}{6} = 40$$

Bhaskara II has given the same rule as Brahmagupta for simultaneous equations involving several variables.

Example 1

"Eight rubies, ten emeralds, and a hundred pearls which are in your earring were purchased by me for you separately at an equal amount. The price rates of the three types of gems together were three less than half a hundred. Tell me dear lady, if you are a skilled mathematician, the price of each of these three gems."

Solution
Let x, y, z be the price of a ruby, an emerald, and a pearl, respectively.
Let w be the amount to buy these rubies, emeralds, and pearls.
Then,

$$w = 8x = 10y = 100z \tag{1}$$

$$x + y + z = 50 - 3 = 47 \tag{2}$$

(Continued)

Example 1 (cont'd)

From (1),

$$x = \frac{w}{8}, \quad y = \frac{w}{10}, \quad z = \frac{w}{100}$$

Substituting these in (2), we get

$$\frac{w}{8} + \frac{w}{10} + \frac{w}{100} = 47$$

Multiplying by 200, we get

$$25w + 20w + 2w = 9400$$

$$47w = 9400 \Rightarrow w = \frac{9400}{47} = 200$$

Then,

$$x = \frac{w}{8} = \frac{200}{8} = 25, \quad y = \frac{w}{10} = \frac{200}{10} = 20, \quad \text{and } z = \frac{w}{100} = \frac{200}{100} = 2$$

Example 2 (Bijaganitha, page 52)

"Tell three numbers which become equal when added with their half, one-fifth and one-ninth parts and each of which diminished by those parts of the other two, leaves 60 as the remainder."

Solution

Let x, y, and z be the three numbers.

Then,

$$x + \frac{x}{2} = y + \frac{y}{5} = z + \frac{z}{9} \tag{1}$$

Also,

$$\left.\begin{aligned} x - \frac{y}{5} - \frac{z}{9} &= 60 \\ y - \frac{x}{2} - \frac{z}{9} &= 60 \\ z - \frac{x}{2} - \frac{x}{5} &= 60 \end{aligned}\right\} \tag{2}$$

From (1),

$$\frac{3x}{2} = \frac{6y}{5} = \frac{10z}{9} = w \text{ (say)}$$

Then,

$$x = \frac{2w}{3}, \quad y = \frac{5w}{6}, \quad \text{and} \quad z = \frac{9w}{10}$$

From the first of the equations in (2),

$$\frac{2w}{3} - \frac{w}{6} - \frac{w}{10} = 60$$

Multiplying by 30, we get

$$20w - 5w - 3w = 1800 = 12w = 1800$$
$$= w = 150$$

Then,

$$x = \frac{2}{3}w = \frac{2}{3}(150) = 100, \quad y = \frac{5}{6}w = \frac{5}{6}(150) = 125$$

and

$$z = \frac{9}{10}w = \frac{9}{10}(150) = 135$$

Example 3 (Leelavati, page 22)

"Three portions x, y, and z of the sum of money i were lent out at different rate of interest r_1, r_2, and r_3, for three periods of time t_1, t_2, and t_3, respectively. The interest that each of them earned was the same. What were the three portions of money".

Let x, y, and z be the three portions. Then,

$$x + y + z = i \tag{i}$$

(*Continued*)

Example 3 (Leelavati, page 22) (cont'd)

Interest earned by each of these three portions

$$\frac{xr_1t_1}{100} = \frac{yr_2t_2}{100} = \frac{zr_3t_3}{100} = w \text{ (say)} \quad \text{(ii)}$$

Then,

$$x = \frac{100w}{r_1t_1}, \quad y = \frac{100w}{r_2t_2}, \quad z = \frac{100w}{r_3t_3}$$

Substituting in (1), we get

$$\frac{100w}{r_1t_1} + \frac{100w}{r_2t_2} + \frac{100w}{r_3t_3} = i$$

Then,

$$w = \frac{i}{(100/r_1t_1) + (100/r_2t_2) + (100/r_3t_3)}$$

Then,

$$x = \frac{100w}{r_1t_1} = \frac{i}{r_1t_1[(1/r_1t_1) + (1/r_2t_2) + (1/r_3t_3)]}$$

$$y = \frac{100w}{r_2t_2} = \frac{i}{r_2t_2[(1/r_1t_1) + (1/r_2t_2) + (1/r_3t_3)]}$$

$$z = \frac{100w}{r_3t_3} = \frac{i}{r_3t_3[(1/r_1t_1) + (1/r_2t_2) + (1/r_3t_3)]}$$

Quadratic Equations

Hindus had known early that the quadratic equation has two roots. In this connection, Bhaskara II mentioned the following rule given by an ancient mathematician Padmanabha, whose treatise on algebra has now been lost. He said, "If the square root of the known side of the quadratic be less than negative absolute term occurring in the square root of the unknown side, then taking it negative as well as positive, we get two values of the unknown." Bhaskara II with the following examples demonstrated that the quadratic equation has two roots.

Example 1 (Bijaganitha, page 65)

वनांतराले प्लवगाष्टभागः संवर्गितोवल्गाति जातरागः
फूत्कार नादप्रतिनादरुष्टा दृष्टागिरौ द्वादशते कियंतः

"The eighth part of a group of monkeys, squared was skipping in the forest. Another 12 were on the hill, screaming. How many monkeys were there in that group."

Solution

Let x be the number of monkeys in the group.

Then,

$$\left(\frac{x}{8}\right)^2 + 12 = x$$

Multiplying by 64, we get $x^2 - 64x = -768$.

Adding $(32)^2$ on both sides, we get

$$x^2 - 64x + 32^2 = -768 + 32^2 = -768 + 1024$$

$$(x-32)^2 = 256$$

So, $x - 32 = \pm 16$.

Here, $|16| < |32|$.

Hence both negative and positive signs have to be taken.

So, $x = 32 + 16$ and $x = 32 - 16$.

Thus, $x = 48$ or 16.

Example 2 (Bijaganitha, page 65ff)

यूथात् चांशकस्त्रयूनो वर्गितोगंह्वरंगतः
दृष्टः शाखामृगः शाखामारूडो वदतेकति

"The fifth part of monkeys leaving out three squared has entered a cave; one has climbed a tree and sitting there. Tell me, how many monkeys are there in that group."

Solution

Let x be the number of monkeys in the group.

Then, we have

$$\left(\frac{1}{5}x - 3\right)^2 + 1 = x$$

(Continued)

Example 2 (Bijaganitha, page 65ff) (cont'd)

$$\frac{1}{25}x^2 - \frac{6x}{5} + 9 + 1 = x$$

Multiplying by 25, we get

$$x^2 - 30x + 250 = 25x \text{ or } x^2 - 55x = -250$$

Multiplying by 4, we get

$$4x^2 - 220x = -1000 \Rightarrow 4x^2 - 220x + 55^2 = -1000 + 55^2$$

i.e.,

$$(2x-55)^2 = -1000 + 55^2 = -1000 + 3025 = 2025$$

$$(2x - 55) = \pm 45$$

We observe that $|45| < |55|$.
So, we consider both positive and negative signs.
So, $2x = 45 + 55$ and $2x = 55 - 45$.
$x = 50$ and $x = 5$.
Since $(x/5) - 3 < 0$, we discard the value $x = 5$.
So, $x = 50$.

Example 3 (Bijaganitha, page 664)

"The shadow of gnomon of 12 fingers being decreased by a third part of the hypotenuse is equal to 14 fingers. Oh mathematician, tell me quickly, how much is the shadow."

Solution
Let x be the shadow and y be the hypotenuse.
Then,

$$x - \frac{y}{3} = 14 \Rightarrow 3x - y = 42$$

or
$y = 3x - 42.$

Now, $y^2 = 12^2 + x^2 = 144 + x^2$.

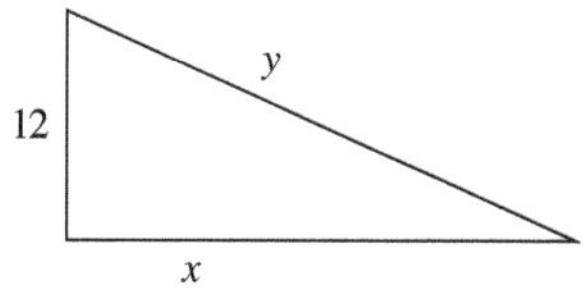

So, $(3x - 42)^2 = 144 + x^2$.

$$9x^2 - 252x + 1764 = 144 + x^2$$

i.e.,
$8x^2 - 252x = 144 - 1764 = -1620$.
Multiplying by 2, we get

$$16x^2 - 504x = -3240$$

$$(4x-63)^2 - 63^2 = -3240$$

$$(4x-63)^2 = -3240 + 3969 = 729 = (27)^2$$

$$4x - 63 = \pm 27$$

Here, $|27| < |63|$.
So, considering both positive and negative signs, we get

$$4x - 63 = 27 \text{ and } 4x - 63 = -27$$

So,

$$4x = 90$$

and

$$4x = 36 \Rightarrow x = \frac{45}{2} \text{or } 9$$

Since $x > 14$, we have

$$x = \frac{45}{2}$$

Bhaskara II has also given the following problems.

Bhasakara II (Leelavati)

Example 1

"A dazzling pearl necklace of a young lady was pulled and torn in a love quarrel ('Mithuna Kalaha') and the pearls were scattered all over. One-third of the number of pearls were found on the bed. One-sixth was picked by the lovely lady, one-tenth were collected by the lover, and six were still hanging in the chain of the necklace. How many pearls were there in the necklace?"

Solution

Let x be the number of pearls in the necklace.

Then, we have

$$\frac{x}{3} + \frac{x}{6} + \frac{x}{10} + 6 = x$$

Multiplying by 30, we get

$$10x + 5x + 3x + 180 = 30x$$
$$30x - 18x = 180$$
$$12x = 180 \Rightarrow x = \frac{180}{12} = 15$$

Example 2

अलिकुलदलमूलं मालतीं यातमष्टौ
निखिलनवमभागाश्चालिनी भृंगमेकम् ।
निशिपरिमललुब्धं पद्ममध्ये निरूद्धम्।
प्रतिरणति रणन्तं ब्रूहि कान्तेऽलिसंख्याम् ॥

"The square root of half a swarm of bees flow to flower bushes to suck honey from the flowers. Eight-nines of entire swam flew behind them. One lady bee could not escape from a lotus flower which closed at night. From inside, she was hamming in response to a humming call from male bee. Oh lady, tell me how many bees were there in the entire swarm."

Solution

Let x be the number of bees in the entire swarm.

Then, we have

$$\sqrt{\frac{x}{2}} + \frac{8}{9}x + 2 = x$$

i.e.,

$$\sqrt{\frac{x}{2}}+2=\frac{1}{9}x \Rightarrow 9\sqrt{\frac{x}{2}}=x-18$$

Squaring both sides, we get

$$81\left(\frac{x}{2}\right)=(x-18)^2=x^2-36x+324$$

Multiplying by 2, we get

$$81x=2x^2-72x+648$$

i.e.,

$$2x^2-153x+648=0$$

$$(2x-9)(x-72)=0 \Rightarrow x=\frac{9}{2} \text{ or } x=72$$

We can discard $x=(9/2)$, since x has to be an integer. Thus $x=72$.

Example 3

यातंहंसकुलस्य मूलदशकं मेघागमे मानसं ।
प्रोड्डियस्थलपद्मिनीवनमगादष्टांशकोऽम्भस्तटात् ॥
बाले बालमृणालशालिनि जलेकेलीक्रियालालसम् ।
दृषृंहंसयुगत्रयञ्च सकलां यूथस्य सङ्ख्यां वद॥

"Oh girl, on the advent of rainy season, ten times the square root of a number of swans who were swimming in a lake flew away to Manasa Sarovar and one-eighth flew to a forest called Sthala Padmini. Three remained in the lake enjoying themselves swimming from one side to the other. Tell me, how many swans were there in the lake."

Solution

Let x be the number of swans swimming in the lake.

Then, we have

$$10\sqrt{x}+\frac{x}{8}+6=x$$

(Continued)

Example 3 (cont'd)

i.e.,

$$10\sqrt{x} = \frac{7x}{8} - 6$$

Multiplying by 8, we get

$$80\sqrt{x} = 7x - 48$$

Squaring both sides, we get

$$6400x = (7x-48)^2 = 49x^2 - 672x + 2304$$

i.e.,

$$49x^2 - 7072x + 2304 = 0$$

$$(49x - 16)(x - 144) = 0$$

$$= x = 144 \text{ and } x = \frac{16}{49}$$

Since x has to be an integer, we can discard the value 16/49. Hence $x = 144$.

Example 4

बाले मरालकुलमूलदलानि सप्त
तीरे विलासभरमन्थरगाण्यपश्यम्।
कुर्वच्चकेलिकलहं कलहंसयुग्मम्
शेषंजले वद मरालकुलप्रमाणम् ॥

"Oh girl, seven times half of the square root of a number of swans swimming a lake flew to the bank of the lake. The remaining two continued swimming in the lake. Tell me how many swans were there swimming in the lake."

Solution

Let x be the number of swans swimming in the lake.

Then, we have

$$\frac{7}{2}\sqrt{x} + 2 = x$$

Multiplying by 2, we get

$$7\sqrt{x} = 2x - 4$$

Squaring both sides, we get

$$49x = (2x-4)^2 = 4x^2 - 16x + 16$$

or

$$4x^2 - 65x + 16 = 0$$

$$(x - 16)(4x - 1) = 0$$

So,

$$x = 16 \text{ and } x = \frac{1}{4}$$

Since x is an integer, we can discard the value $x = 1/4$.
Hence $x = 16$.

Example 5

यदिसम भुवि वेगुद्विन्त्रिपारिप्रमागौ
रागाक पवनवेगादेकदेशे सभग्नः
भुविनृपभित हस्तेष्वंगलग्नं तदग्रं
कथयकतिषु भूलादेषभग्नः करेषु

"Oh mathematician, a bamboo, 32 cubit long, is standing on a level ground. A blast of wind breaks it and it touches the ground 16 cubits from its foot. Can you tell me at what height the bamboo broke."

Solution

In the figure, let O be the point where the bamboo broke. Let AB be the bamboo before breaking.

Let the end of broken bamboo touch the ground at C.
Then, $AC = 16$.
Let $AO = x$.

$$OC = OB = AB - OA = 32 - x$$

(*Continued*)

Example 5 (cont'd)

In the right Δ^{le} AOC,

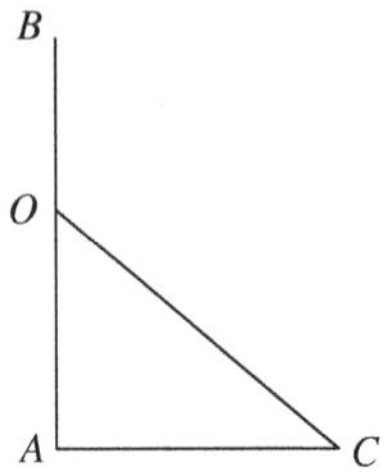

We have $OC^2 = OA^2 + AC^2$

$$(32-x)^2 = x^2 + (16)^2$$

$$x^2 - 64x + 1024 = x^2 + 256$$

$$\Rightarrow 64x = 1024 - 256 = 768$$

$$\Rightarrow x = \frac{768}{64} = 12 \text{ cubits}$$

Example 6

त्रस्तिस्तंभतले बिलं तदुपरि क्रिद्या शिखंद्दी स्थितः
स्तंभेहस्तनवोच्छिते त्रिगुरितस्तंभप्रमारांतरे
दृष्टहिं विलमात्रजन् तमपतत् तिर्यक्सतरयोपरि
क्षिप्रं ब्रूहि तयोर्विलात् कतिमितैत्साम्येनगत्योयुंतिः॥

"A peacock is perched on the top of a pillar 9 cubic feet high. There is a snake hole at the bottom of the pillar. When the peacock sees a snake moving toward its hole at a distance 27 units from the hole, the peacock pounces on it. If the speed of the snake and the peacock are the same, at what distance from the snake hole, the peacock pounces on the snake?"

Solution

Let BC be the pillar. Then, $BC = 9$. Let D be the position of the snake, when peacock sees it. Then, $BD = 27$. Let A be the position of the snake when the peacock pounces on it. Let $AB = x$.

The peacock is perched at C, when it sees the snake at D. Since their speeds are the same and since A is the position where the peacock pounces on the snake, the distance traversed by the peacock, namely CA = the distance traversed by the snake, namely DA.

$$CA = DA = DB - AB = 27 - x$$

In the right triangle ABC,

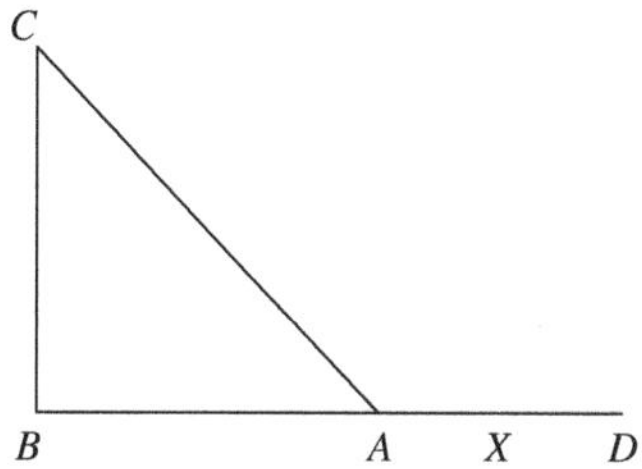

$$CA^2 = AB^2 + BC^2$$

$$(27 - x)^2 = x^2 + 9^2$$

$$729 - 54x + x^2 = x^2 + 81$$

$$54x = 729 - 81 = 648$$

$$x = \frac{648}{54} = 12$$

So, the peacock pounces on the snake when the snake is at a distance of 12 cubits from its hole at the bottom of the pillar.

Example 7

वृक्षात् हस्तशतोच्छ्रयाच्छतयुगेवापी: कपि: कोऽप्यगात्
उत्तीर्याथपरोद्रुसं श्रुतिपथेनो ड्डीयकिंचिद्द्रुमात्
जातेवं समतातयोर्यदिगता वुड्डायमानं कियत्
विद्वन् चेत् सुपरिश्रमोस्ति गरितेन्दिप्रं तदाचच्क्षामे

"A well is situated at a distance of 200 cubits from the foot of a tree of height 100 cubits. Two monkeys were sitting at the top of the tree. One monkey climbs down the tree and reaches the well. The other monkey jumps some distance high above the tree and lands on the well. Oh learned man, if you know mathematics well, tell me how high the second monkey jumped high above the tree if the two monkeys travel the same distance to reach the well."

(*Continued*)

Example 7 (cont'd)

Solution

Let BC be the tree and A be the well. So, $AB = 200$ and $BC = 100$. The two monkeys were sitting at C, the top of the tree.

Distance traversed by the first monkey $= CB + BA = 100 + 200 = 300$.

Let CD be the jump of the second monkey high above the tree.

Then, distance traversed by the second monkey $= CD + DA$.

Let the jump CD be x.

So,

$$CD + DA = CB + BA = 300$$

$$DA = 300 - CD = 300 - x$$

In the right triangle ABD,

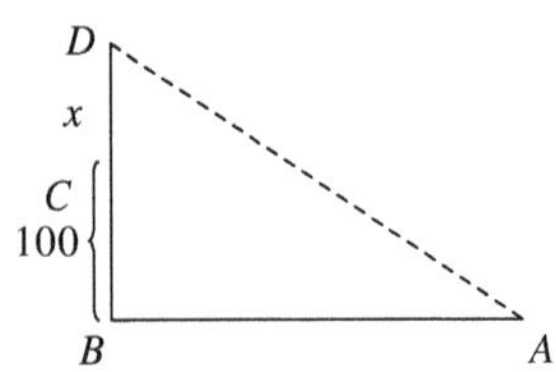

$$DA^2 = BD^2 + AB^2$$

$$(300-x)^2 = (100+x)^2 + (200)^2$$

$$90,000 - 600x + x^2 = 10,000 + 200x + x^2 + 40,000$$

$$90,000 - 50,000 = 600x + 200x$$

$$800x = 40,000$$

$$x = \frac{4000}{800} = 50 \text{ cubits}$$

Example 8

This example refers to the battle between Partha (Arjuna) and Karna in the great epic "Mahabharatha." The description given here about this battle is just an imaginary one and is not found in the epic.

पार्थ यार्णवधायमार्गणगणं कृद्धोरणेःरान्दधो।
तरुयार्थत निवार्य तच्छरगणं मूलैश्चतुर्भिर्हयान् ॥
शरव्यं षड्भिरथेषुभिस्त्रिभिरपिछत्रंछवजंकार्मुकम्।
थिच्छेदास्यशिरः शरेणकति ते यानर्जुनः सन्दधे ॥

"In the Mahabharatha battle, Partha in anger shot a stream of arrows to kill Karna. With half the number of arrows, he neutralized Karna's arrows; killed Karna's horses with four times the square root of number of arrows (in the above stream), striked Salya (Karna's Sarathi) with six arrows; demolished Karna's umbrella, flag, and bow with three arrows and removed Karna's head with one arrow. What is the number of arrows Partha shot in that stream?"

Solution

Let x be the number of arrows in the stream of arrows shot by Partha.

Then, we have

$$\frac{x}{2} + 4\sqrt{x} + 6 + 3 + 1 = x$$

$$4\sqrt{x} = \frac{x}{2} - 10 \text{ or } 8\sqrt{x} = x - 20$$

Squaring both sides, we get

$$64x = (x-20)^2 = x^2 - 40x + 400$$

i.e.,

$$x^2 - 104x + 400 = 0$$

$$(x - 100)(x - 4) = 0$$

So,

$$x = 100 \text{ or } x = 4$$

Since

$$x = 4 < 10, \text{it is inadmissible}$$

Hence,

$$x = 100$$

Example 9

चक्र कौंचाकुलितसलिले कपि दृष्टं तटाके
तोयादूर्ध्य कमलकलिंकाग्रं वितस्तिप्रमागां
मंदंमंदं चलिंतमनिलेनाहृतं हस्तयुग्मे
तस्मिन्मग्नं गगककथय दिप्रमंभः प्रमागं

"In a tank surrounded by birds, a lotus flower of height one palm $\left(\frac{1}{2}\text{ cubit}\right)$ has blossomed. Suddenly, due to wild wind, it sinks in the water 2 cubits away. Quickly tell me mathematician, what is the depth of water in the tank?"

Solution

Let PQ be the lotus plant. Let PS be the positions of the lotus after the wild wind. Let RS be the water level and then $RS = 2$ cubits. Then, PR is the depth of water and RQ is the height of the flower = (1/2) cubit.

Let PR be x.

Then,

$$PS = PQ = x + \frac{1}{2}$$

In the right triangle RSP, $PS^2 = RP^2 + RS^2$.

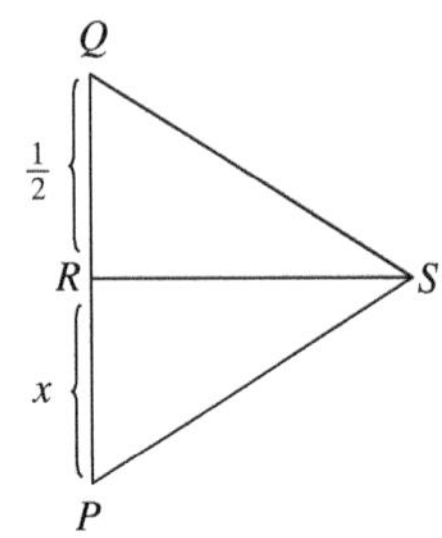

So,

$$\left(x + \frac{1}{2}\right)^2 = x^2 + 2^2$$

$$x^2 + x + \frac{1}{4} = x^2 + 4$$

$$x = 4 - \frac{1}{4} = 3\frac{3}{4} \text{ cubits}$$

Example 10

"Two bamboos of heights 15 and 10 ft are standing on a level ground. Tight strings are tied from the tip of either bamboo to the foot of the other. What is the height of the point of intersection of the two strings above the ground."

Solution

Let *PQ* and *RS* be two bamboos of height 15 and 10 ft, respectively. Let *PS* and RQ be the two strings tied from the tip of one bamboo to the foot of the other. Let *T* be the point of intersection of the two strings.

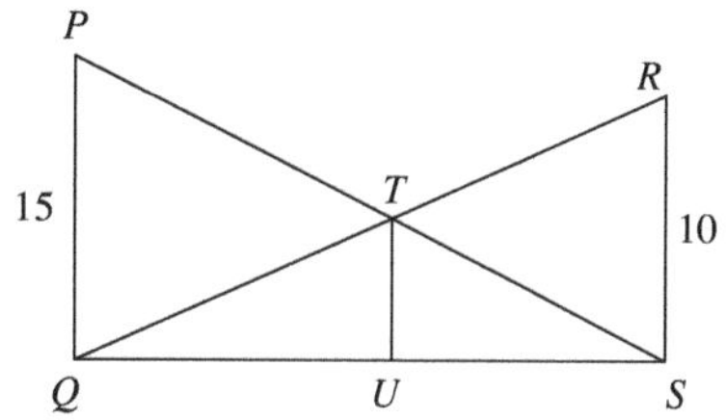

Let *TU* be the perpendicular from *T* on *QS* (the level ground).

Now, Δ^{le} *STU* and *SPQ* are similar.
So,

$$\frac{TU}{PQ} = \frac{US}{QS} \Rightarrow \frac{TU}{15} = \frac{US}{QS} \tag{1}$$

Again, Δ^{les} *QTU* and *QRS* are similar.
So,

$$\frac{TU}{RS} = \frac{QU}{QS} \Rightarrow \frac{TU}{10} = \frac{QU}{QS} \tag{2}$$

Add (1) and (2), we get

$$TU\left[\frac{1}{15} + \frac{1}{10}\right] = \frac{1}{QS}[US + QU] = \frac{QS}{QS} = 1$$

Hence,

$$TU\left[\frac{2+3}{30}\right] = 1 \Rightarrow \frac{TU}{6} = 1 \Rightarrow TU = 6$$

(Continued)

Example 10 (cont'd)

Equations of Higher Degrees

The ancient Indians were not very successful is solving cubic and biquadratic equations. Bhaskara II used the method of "madyamaharana" ("elimination of the middle"). The method consists of the reduction of the original equation to simple and quadratic equations by employing useful transformation and introduction of auxiliary quantities. Bhaskara II has provided the following two examples, one a cubic equation and another a biquadratic equation, where such reduction is possible.

Example 1 (Bhaskara II, Bijaganitha, page 64f)

"Can you give me the number, Oh learned man, which when multiplied by 12 and increased by the cube of the number is equal to the sum of 35 and six times the square of the number?"

Solution

Let x be the number.

Then, we have $12x + x^3 = 35 + 6x^2$ or $x^3 - 6x^2 + 12x = 35$.

Subtracting 8 from both sides, we get

$$x^3 - 6x^2 + 12x - 8 = 35 - 8 = 27$$

i.e.,

$$(x-2)^3 = 27 = 3^3$$

Hence,

$$x - 2 = 3$$

or

$$x = 2 + 3 = 5$$

Example 2 (Bhaskara II, Bijaganitha, page 64f)

"Can you tell me the numbers which when multiplied by 200 and added to the square of the number and then multiplied by 2 and the resulting number is subtracted from the fourth power of the number is equal to one myriad (ten thousand) minus unity?"

Solution

Let x be the required number.

Then, we have

$$x^4 - 2(200x + x^2) = 10,000 - 1$$

$$x^4 - 2x^2 - 400x = 9999$$

Adding $4x^2 + 400x + 1$ to both sides, we get

$$x^4 - 2x^2 - 400x + 4x^2 + 400x + 1 = 999 + 4x^2 + 400x + 1$$

i.e.,

$$x^4 + 2x^2 + 1 = 4x^2 + 400x + 10,000$$

$$(x^2+1)^2 = (2x+100)^2$$

Taking the square root,

$$x^2 + 1 = \pm(2x + 100)$$

Case 1: If $x^2 + 1 = 2x + 100$, we have

$$x^2 - 2x + 1 = 100$$

$$(x-1)^2 = 10^2 \Rightarrow x - 1 = \pm 10 \Rightarrow x = 11 \text{ or } -9$$

Case 2: If $x^2 + 1 = -(2x + 100)$, then we have

$$x^2 + 2x + 1 = -100$$

$(x + 1)^2 = -100$, which is not possible.

Thus, $x = 11$ or -9.

Bhaskara II's Modification of Solving $ax + c = by$

Bhaskara II's method is as follows.

Bhaskara II's First Method

Work out the mutual division to the finish until the last remainder is one. Take the first column of the sequence of quotients along with additive c and 0 attached at the end. The rest of the process is as usual. Hence, his method involves in taking $x_n = 0$ and $y_n = c$ in equation (2n) of I on page 130, which becomes

$$X_{n-1} + c_{q_{n-1}} + 0$$

This method gives the solution of $ax + c = by$ if the number of quotients is even. Also, this method solves the equation $ax - c = by$ if the number of quotients is odd.

The following examples will illustrate his method.

Example 1

Solve in positive integers the equation $100x + 90 = 63y$.

Solution

$$
\begin{array}{l}
63)100(1 = q_1 \\
\underline{63} \\
\quad 37)63(1 = q_2 \\
\quad \underline{37} \\
\qquad 26)37(1 = q_3 \\
\qquad \underline{26} \\
\qquad\quad 11)26(2 = q_4 \\
\qquad\quad \underline{22} \\
\qquad\qquad 4)11(2 = q_5 \\
\qquad\qquad \underline{8} \\
\qquad\qquad\quad 3)4(1 = q_6 \\
\qquad\qquad\quad \underline{3} \\
\qquad\qquad\quad 1
\end{array}
$$

Here the number of quotients are $q_1 = 1$, $q_2 = 1$, $q_3 = 1$, $q_4 = 2$, $q_5 = 2$, $q_6 = 1$, and so even.

We now form the following table:

1	1	1	1	1	1	2430
1	1	1	1	1	1530	1530
1	1	1	1	900	900	900
2	2	2	630	630	630	
2	2	270	270	270		
1	90	90	90			
90	90	90				
0	0	0				

So,

$$Y = 2430 \text{ and } X = 1530$$

Hence,

$$X = 1530 \equiv 18(\text{mod } 63)$$

$$Y = 2430 \equiv 30(\text{mod } 100)$$

$x = 18$, $y = 30$ is the least solution.

General solution is $x = 63m + 18$; $y = 100m + 30$, where m is an arbitrary integer.

Example 2

Find the integral solution of the equation $60x - 3 = 13y$.

Carrying the mutual division, we get

Solution

$13)60(4 = q_1$
$\underline{52}$
$8)13(1 = q_2$
$\underline{8}$
$5)8(1 = q_3$
$\underline{5}$
$3)5(1 = q_4$
$\underline{3}$
$2)3(1 = q_5$
$\underline{2}$
1

Here the quotients are $q_1 = 4$, $q_2 = 1$, $q_3 = 1$, $q_4 = 1$, $q_5 = 1$ and the number of quotients is odd.

Now, we now form the following table:

4	4	4	4	4	69
1	1	1	1	15	15
1	1	1	9	9	9
1	1	6	6	6	
1	3	3	3		
3	3	3			
0	0				

(*Continued*)

Example 2 (cont'd)

So,

$$Y = 69 \text{ and } X = 15$$

$$X = 15 \equiv 2 (\text{mod } 13)$$

$$Y = 69 \equiv 9 (\text{mod } 60)$$

$x = 2$ and $y = 9$ is the least solution of $6x - 3 = 13y$

General solution is $x = 13m + 2$; $y = 60m + 9$, where m is an arbitrary integer.

Example 3

Solve in positive integers, the equation $37x + 18 = 20y$.

Solution

$$\begin{array}{l} 20)37(1 = q_1 \\ \quad \underline{20} \\ \quad 17)20(1 = q_2 \\ \qquad \underline{17} \\ \qquad 3)17(5 = q_3 \\ \qquad\quad \underline{15} \\ \qquad\quad 2)3(1 = q_4 \\ \qquad\qquad \underline{2} \\ \qquad\qquad 1 \end{array}$$

So, the quotients are $q_1 = 1$, $q_2 = 1$, $q_3 = 5$, and $q_4 = 1$. The number of quotients is even.

So we now form the following table:

1	1	1	1	234
1	1	1	126	126
5	5	108	108	108
1	18	18	18	
18	18	18		
0	0	0		

So,

$$Y = 234; \quad X = 126$$

$$X = 126 \equiv 6 (\text{mod } 20)$$

$$Y = 234 \equiv 12 (\text{mod } 37)$$

So $x = 6$ and $y = 12$ is the least solution.
So, the general solution is

$$x = 20m + 6 \quad \text{and} \quad y = 37m + 12$$

where m is an arbitrary integer.

Example 4

Solve in positive integers the equation $193x - 7 = 41y$.

Solution

41)193(4 = q_1
<u>164</u>
29)41(1 = q_2
<u>29</u>
12)29(2 = q_3
<u>24</u>
5)12(2 = q_4
<u>10</u>
2)5(2 = q_5
<u>4</u>
1

So, the quotients are $q_1 = 4$, $q_2 = 1$, $q_3 = 2$, $q_4 = 2$, and $q_5 = 2$. The number of quotients is odd.

We now form the following table:

4	4	4	4	4	560
1	1	1	1	119	119
2	2	2	84	84	84
2	2	35	35	35	
2	14	14	14		
0	7	7			
0	7				

(Continued)

Example 4 (cont'd)

So,

$$Y = 560; \quad X = 119$$

$$Y = 560 \equiv 174 (\text{mod } 193)$$

$$X = 119 \equiv 37 (\text{mod } 41)$$

$$x = 37, \quad y = 174 \text{ is the least solution}$$

General solution is $x = 41m + 37$; $y = 193m + 174$, where m is an arbitrary integer.

Lemma (Bhaskara II): If (α,β) is a solution of $ax - c = by$, $(a,b,c > 0)$, then $x = (b - \alpha) + bm$ and $y = (a - \beta) + am$ is the general solution of $ax + c = by$.

Proof of this lemma was given by Krsna (AD 1580). His proof has been given on page 689 in this book.

To illustrate this lemma, we give the following example.

Solve in positive integers the equation $193x - 7 = 41y$. We showed that $(\alpha\ \beta) = (37\ 174)$ is a solution of the equation $193x - 7 = 41y$. Then, from the above lemma, $x = b - \alpha = 41 - 37 = 4$ and $y = a - \beta = 193 - 174 = 19$ is the solution of $193x + 7 = 41y$.

General solution is $x = 41m + 4$; $y = 193m + 19$.

Example (Bhaskara II)

Solve in positive integers the equation $221x + 65 = 195y$ (*) and from this solution, solve the equation $221x - 65 = 195y$.

Solution

Our given equation is $221x + 65y = 195y$ (*). Since 221, 65, and 195 are divisible by 13, we will divide (*) by 13. We get

$$17x + 5 = 15y$$

Doing the mutual division as before, we have

$$
\begin{array}{l}
15)17(1 = q_1 \\
\underline{15} \\
\quad 2)15(7 = q_2 \\
\quad \underline{14} \\
\quad\ \ 1
\end{array}
$$

So, the quotients are $q_1 = 1$ and $q_2 = 1$. So, the number of quotients is even.

We now form the following table:

1	1	40
7	35	35
5	5	5
0	0	

So,

$$Y = 40; \quad X = 35$$

$$Y = 40 \equiv 6 (\text{mod } 17)$$

$$X = 35 \equiv 5 (\text{mod } 15)$$

$$x = 5, \quad y = 6 \text{ is the least solution}$$

General solution of $221x + 65 = 195y$ is

$$x = 15m + 5 \text{ and } y = 17m + 6$$

From the previous Bhaskara II's lemma on page 360, we have

$$x = b - \alpha = 15 - 5 = 10$$

$$y = a - \beta = 17 - 6 = 11$$

So, $x = 10$, $y = 11$ is the least solution of $221x - 65 = 195y$.
The general solution of $221x - 65 = 195y$ is given by

$$x = 15m + 10 \text{ and } y = 17m + 11$$

Lemma: If (α,β) is an integral solution of $ax + c = by$ and if k is an integer, then $(\alpha,k\beta)$ is an integral solution of $kax + kc = by$.

Proof: Since (α,β) is a solution of $ax + c = by$, we have $a\alpha + c = b\beta$. Multiplying by k, we get $ka\alpha + kc = kb\beta$, which mean that $(\alpha,k\beta)$ is an integral solution of $kax + kc = by$.

Using this lemma, Bhaskara II gave his second method of finding integral solution of $kax + kc = by$.

Bhaskara II's Second Method

Example 1

Solve the equation $100x + 90 = 63y$.

Now, 10 is the common factor of dividend 100 and additive 90. Dividing these numbers by 10, we get a new dividend 10 and new additive 10.

Now, we solve $10x + 9 = 63y$ as before.

Solution

$$
\begin{array}{l}
63)10(0 = q_1 \\
\quad\ \underline{0} \\
\quad 10)63(6 = q_2 \\
\qquad\ \ \underline{60} \\
\qquad\quad 3)10(3 = q_3 \\
\qquad\qquad\ \underline{9} \\
\qquad\qquad\ 1
\end{array}
$$

Now the quotients are $q_1 = 0$, $q_2 = 1$, and $q_3 = 1$, and so the number of quotients is odd.

We now form the table as usual:

0	0	0	27
6	6	171	171
3	27	27	27
9	9	9	
0	0		

$$Y = 27;\ \ X = 171;\ \ X = 171 \equiv 45(\text{mod } 63);\ \ Y = 27 \equiv 7(\text{mod } 10)$$

Since the number of quotients is odd, we find $\alpha = 63 - 45 = 48$ and $y = 10 - 7 = 3$.

We substitute this value α of x in the given equation, then, $100x + 90 = 100(18) + 90 = 63y$.

$$1890 = 63y \Rightarrow y = 30 = \beta$$

So, $(\alpha\ \beta) = (18\ 30)$ is a solution of $100x + 90 = 63y$. Alternatively, if we multiply the discarded value 3 of y by the common factor 10, we get 30, which is the value of y in the given equation.

So, general solution of $100x + 90 = 63y$ is $x = 18 + 63m$ and $y = 30 + 100m$, where m is an arbitrary integer.

Example 2

Solve in integers the equation $111x + 54 = 20y$ (*).

Solution

In this case, the dividend 111 and the additive 54 have a common factor 3. Dividing these numbers by 3, we get a new dividend 37 and a new additive 18.

So we solve in integers the equation $37x + 18 = 20y$. We solved this equation on page 359, and found $y = 12$ and $x = 6$ as the least solution.

Since in this case, the number of quotients is even, add these numbers to the corresponding divided 37 and the devisor 20.

We get $x = 6 + 20 = 26$ and $y = 12 + 37 = 49$.

Substitute the value $x = 26$ in the given equation to get the value of y.

We then have

$$(111)(26) + 54 = 20y$$

$$20y = 2886 + 54 = 2940$$

$$y = 147$$

So, $(\alpha\ \beta) = (26\ 147)$ is the least solution of (*). Alternatively, multiply the value 49 for y by the common factor 3, we get 147.

So, in either case $(\alpha\ \beta) = (26\ 147)$ is the least solution of $111x + 54 = 20y$ and the general solution is

$$x = 26 + 20m; \quad y = 147 + 111m$$

where m is an arbitrary integer.

Example 3

Solve in positive integers the equation $55x + 15 = 1$.

Solution

In this case, 5 is a common factor of the dividend 55 and additive 15. Dividing these numbers by 5, we get a new dividend 11 and a new additive 3.

(*Continued*)

Example 3 (cont'd)

We now solve $11x + 3 = 19y$ (*)

$19)11(0 = q_1$
$\underline{0}$
$11)19(1 = q_2$
$\underline{11}$
$8)11(1 = q_3$
$\underline{8}$
$3)8(2 = q_4$
$\underline{6}$
$2)3(1 = q_5$
$\underline{2}$
1

So the quotient the $q_1 = 0$, $q_2 = 1$, $q_3 = 1$, $q_4 = 2$ and $q_5 = 1$

Hence, the number of quotients are odd.

We now form the following table:

0	0	0	0	0	12
1	1	1	1	21	21
1	1	1	12	12	12
2	2	9	9	9	
1	3	3	3		
3	3	3			
0	0				

Hence,

$$Y = 12 \text{ and } X = 21$$

$$Y = 12 \equiv 1 (\text{mod } 11)$$

$$X = 21 \equiv 2 (\text{mod } 19)$$

So $(x\ y) = (2\ 1)$ is the least solution of $11x - 3 = 19y$.

Since the number of quotients is odd, we subtract 1 from 11 and 2 from 19 to get $x = 17$ and $y = 10$. Hence $(x\ y) = (17\ 10)$ is the least solution of $11x + 3 = 19y$.

Substituting $x = 17$ in (*), we get $55(17) + 15 = 19y$.

$$935 + 15 = 19y = 19y = 950 = y = 50$$

Or alternatively, multiplying the value 10 for y by the common factor 5, we get $y = 50$.

So $\alpha = 17$, $\beta = 50$ is the least solution of $55x + 15 = 19y$.

So, the general solution is

$$x = 17 + 19m, \quad y = 50 + 55m$$

Lemma: If (α,β) is an integral solution of $ax + c = by$ and if k is an integer, then $(k\alpha,\beta)$ is an integral solution of $ax + kc = kby$.

Proof: Since (α,β) is a solution of $ax + c = by$, we have $a\alpha + c = b\beta$. Multiplying by k, we get $ka\alpha + kc = kb\beta$, which means that $(k\alpha,\beta)$ is an integral solution of $ax + kc = kby$.

Using this lemma, Bhaskara II gave his third method of finding integral solutions of $ax + kc = kby$.

Bhaskara II's Third Method

Example 1

Solve in integers $100x + 90 = 63y$. Here 9 is the greatest common factor between the additive 90 and divisor 63. Dividing these numbers by 9, we get a new devisor 7 and a new additive 10.

So, we solve the equation $100x + 10 = 7y$.

Solution

$7)100(14 = q_1$
$\underline{7}$
30
$\underline{28}$
$2)7(3 = q_2$
$\underline{6}$
1

So, the quotients are $q_1 = 14$ and $q_2 = 3$, and so the number of quotients is even. So, we now form the following table:

14	14	430
3	30	30
10	10	10
0	0	

(Continued)

Example 1 (cont'd)

So,

$$Y = 430; \quad X = 30$$

Hence,

$$Y = 430 \equiv 30(\text{mod } 100)$$

$$X = 30 \equiv 2(\text{mod } 7)$$

Here,

$$y = 30, \quad x = 2$$

Substituting $y = 30$ in the given equation (*), we get

$$63(30) = 100x + 90$$

$$1890 - 90 = 100x = 100x = 1800 = x = 18$$

So, $x = 18$, $y = 30$ is the least solution of (*) Alternatively, multiply the value $x = 2$ by the common factor 9, we get $x = 18$.

General solution is $x = 18 + 63m$; $y = 30 + 60m$, where m is an arbitrary integer.

Example 2

Find the integral solution of $77x + 25 = 45y$ (*).

Solution

Now, the greatest common factor of additive 25 and the divisor 45 is 5. Dividing these numbers by 5, we get a new devisor 9 and a new additive 5.

So, we solve $77x + 5 = 9y$.

$$9)77(8 = q_1$$
$$\underline{72}$$
$$5)9(1 = q_2$$
$$\underline{5}$$
$$4)5(1 = q_3$$
$$\underline{4}$$
$$1$$

So, the quotients are $q_1 = 8$, $q_2 = 1$, and $q_3 = 1$, and the number of quotients is odd. So, we now form the following table:

8	8	8	85
1	1	10	10
1	5	5	5
5	5	5	
0	0		

So, $Y = 85$; $X = 10$.

So, $Y = 85 \equiv 8 \pmod{77}$.

$$X = 10 \equiv 1 \pmod 9$$

Hence, $(x,y) = (1,8)$ are the least solution of $77x - 5 = 9y$.

Since the number of quotients is odd, $x = 9 - 1 = 8$, $y = 77 - 8 = 69$ are solutions of $77x + 5 = 9y$.

Substituting $y = 69$ in the given equation (*), we get

$$45(69) = 77x + 25 = 77x = (45)(69) - 25$$

i.e., $77x = 3105 - 25 = 3080$.

So,

$$x = \frac{3080}{77} = 40$$

So, $x = 40$, $y = 69$ is the least solution of (*), Alternatively, multiply the value 8 for x by the common factors 5, we get $x = 40$.

So, the general solution of (*) is $x = 40 + 45m$; $y = 69 + 77m$, where m is an arbitrary integer.

Example 3

Find the integral solutions of $43x + 16 = 20y$ (*).

Solution

Here, the greatest common factor of the additive 16 and the divisor 20 is 4. Dividing these numbers by 4, we get a new devisor 5 and a new additive 4.

(Continued)

Example 3 (cont'd)

We now solve the equation 43x + 4 = 5y.

$$
\begin{array}{l}
5)43(8 = q_1 \\
\underline{40} \\
\quad 3)5(1 = q_2 \\
\quad \underline{3} \\
\quad\quad 2)3(1 = q_3 \\
\quad\quad \underline{2} \\
\quad\quad 1
\end{array}
$$

So, the quotients are $q_1 = 8$, $q_2 = 1$, and $q_3 = 1$, and the number of quotients is odd.

We now form the following table:

8	8	8	68
1	1	8	8
1	4	4	4
4	4	4	
0	0		

So,

$$Y = 68; \quad X = 8$$

$$Y = 68 \equiv 25 (\text{mod } 43)$$

$$X = 8 \equiv 3 (\text{mod } 5)$$

So, $y = 25$ and $x = 3$ is a solution of $43x - 4 = 5y$.
Since the number of quotients is odd,

$$\beta = 43 - 25 = 18 \text{ and } \alpha = 5 - 3 = 2$$

Substituting $y = \beta = 18$ in the given equation (*), we get

$$43x + 16 = 20(18) = 360$$

$$43x = 360 - 16 = 344$$

$$x = \frac{344}{43} = 8$$

Alternatively, multiply the value 2 for x by the greatest common factor 4, we get $x = 8$.

So, $x = 8$, $y = 18$ is the least solution of (*).

General solution is $x = 8 + 20m$; $y = 18 + 43m$, where m is an arbitrary positive integer.

Lemma: If (α,β) is an integral solution of $ax + c = by$ and if j and k are integers, then $(j\alpha,k\beta)$ is an integral solution of $kax + jkc = jby$.

Proof: If (α,β) is a solution of $ax + c = by$, then we have $a\alpha + c = b\beta$.

Multiplying by jk, we get $jka\alpha + jkc = jkb\beta$, which means that $(j\alpha,k\beta)$ is an integral solution $akx + jkc = jby$.

Using this lemma, Bhaskara II gave his fourth method to find integral solutions of $akx + jkc = jby$.

Bhaskara II's Fourth Method

Example 1

Solve in integers $100x + 90 = 63y$ (*).

Solution

Divide the divisor 63 and the additive 90 by their greatest common factor 9. Then, we get 7 and 10. Again divide the dividend 100 and the new additive 10 by their greater common factor 10. We get 10 and 1.

Now, we solve the equation 10x + 1 = 7y.

$$
\begin{array}{l}
7)10(1 = q_1 \\
\quad \underline{7} \\
\quad 3)7(2 = q_2 \\
\qquad \underline{6} \\
\qquad 1
\end{array}
$$

The quotients are $q_1 = 1$ and $q_2 = 2$, and the number of quotients is even.

We now form the following table:

1	1	3
2	2	2
1	1	1
0	0	

(*Continued*)

Example 1 (cont'd)

So,

$$Y = 3,\ X = 2;\ \ Y \equiv 3 (\text{mod } 10);\ \ X \equiv 2 (\text{mod } 7)$$

Multiplying the x value 2 by the greatest common factor 9 of the original divisor 63 and the original additive 90, we get $x = 18$.

Similarly, multiplying the y value 3 by the greatest common factor 10 of the original dividend 100 and the original additive 90, we get $y = 30$.

Alternatively, substitute $x = 18$ or $y = 30$ in (*) and get the values of y and x, respectively.

We get $y = 30$ and $x = 18$.

Hence, $x = 18$, $y = 30$ is the least solution of (*) and the general solution is $x = 18 + 63m$; $y = 30 + 100m$, which is exactly what we had before on page 357.

Example 2

Solve in integer the equation $64x + 60 = 21y$ (*).

The greatest common factor of the divisor 21 and the additive 60 is 3. Dividing these numbers by 3, we get a new divisor 7 and new additive 20. The greatest common factor of the original dividend 64 and the new additive 20 is 4. Dividing these numbers by 4, we get new dividend 16 and a new additive 5.

So, we now solve the equation $16x + 5 = 7y$.

$$
\begin{array}{l}
7)16(2 = q_1 \\
\ \ \underline{14} \\
\ \ \ \ 2)7(3 = q_2 \\
\ \ \ \ \ \ \ \underline{6} \\
\ \ \ \ \ \ \ 1
\end{array}
$$

So, the quotients are $q_1 = 2$ and $q_2 = 3$, and so the number of quotients is even.

We now form the following table:

2	2	35
3	15	15
5	5	5
0	0	

So, $Y = 35$; $X = 15$.

Also, $Y = 35 \equiv 3 \pmod{16}$; $X = 15 \equiv 1 \pmod{7}$.

So, $x = 1$, $y = 3$ is the least solution of $16x + 5 = 7y$.

Multiplying the x value 1 by the greater common factor 3 of the original additive 60 and to the original divisor 21, we get $x = 3$. Again multiplying the y value 3 by the greatest common factor 4 of the original dividend 64 and the reduced additive 20, we get $y = 12$.

$x = 3$ and $y = 12$ is the least solution of (*).

General solution is $x = 3 + 21m$; $y = 12 + 64m$, where m is an arbitrary integer.

Example 3

Solve in integers the equation $153x + 45 = 70y$ (*). The greatest common factor of the divisor 70 and the additive 45 is 5. Dividing these two numbers by 5, we get a new divisor 14 and a new additive 9. The greatest common factor of the dividend 153 and the new additive 9 is 3. Dividing these two numbers by 3, we get a new dividend 51 and a new additive 3.

We now solve the equation $51x + 3 = 14y$.

$14)51(3 = q_1$
$\underline{42}$
$9)14(1 = q_2$
$\underline{9}$
$5)9(1 = q_3$
$\underline{5}$
$4)5(1 = q_4$
$\underline{4}$
1

So, the quotients are $q_1 = 3$, $q_2 = 1$, $q_3 = 1$, and $q_4 = 1$, and the number of quotients is even.

We now form the following table:

3	3	3	3	33
1	1	1	9	9
1	1	6	6	6
1	3	3	3	
3	3	3		
0	0			

(*Continued*)

Example 3 (cont'd)

So, $Y = 33$; $X = 9$.

$$Y = 33 \equiv 33(\text{mod}51)$$

$$X = 9 \equiv 9(\text{mod}14)$$

Now, multiplying the x value 9 by the greatest common factor 5 of the original additive 45 and the original divisor 70, we get $x = 45$.

Again multiplying the y value 33 by the greatest common factor 3 of the original dividend 153 and the new additive 9, we get $y = 99$.

$x = 45$ and $y = 99$ is the least solution of the equation (*).

General solution is $x = 45 + 70m$; $y = 99 + 153m$, where m is an arbitrary integer.

Solution of $by + ax = \pm c$:

The solution of the equation of the form $by + ax = \pm c$ was in general treated by ancient Indian mathematicians as solving $by = -ax \pm c$ or $by = ax \pm c$, where $a < 0$.

Lemma 1 (Bhaskara II): If $(\alpha\ \beta)$ is a solution of $by = ax - c$, then $(\alpha - b,\ a - \beta)$ is a solution of $by = -ax + c$.

Proof: If $(\alpha\ \beta)$ is a solution of $by = ax - c$, then we have

$$b\beta = a\alpha - c \Rightarrow -b\beta = c - a\alpha$$

Adding ab to both sides, we get $c - a\alpha + ab = -b\beta + ab$.

So, $c - a(\alpha - b) = b(a - \beta)$, which means $(\alpha - b\ a - \beta)$ is a solution of $by = -ax + c$.

Lemma 2: If $(\alpha\ \beta)$ is a solution of $by = ax + c$, then $(\alpha\ -\beta)$ is a solution of $-by = ax + c$ or $by = -ax - c$.

Proof: $(\alpha\ \beta)$ is a solution of $by = ax + c$.

So, $b\beta = a\alpha + c$.

Then, $-b\beta = -a\alpha - c$.

This means $(\alpha\ -\beta)$ is a solution of $by = -ax - c$.

Lemma 3: If (α β) is a solution of $ax + c = by$, then $x = \alpha - b$ and $y = a - \beta$ is a solution if $ax + by + c = 0$.

Proof: If (α β) is a solution of $ax + c = by$, we have $a\alpha + c = b\beta$.
Subtracting ab from both sides, we get

$$a\alpha + c - ab = b\beta - ab$$

$$a(\alpha - b) + c = b(\beta - a) \text{ or } a(\alpha - b) + c + b(a - \beta) = 0$$

which shows that $x = \alpha - b$, $y = a - \beta$ is a solution of $ax + c + by = 0$.

Lemma 4: If (α β) is a solution of $by = ax + c$, then ($-\alpha$ β) is a solution of $by = -ax + c$.

Proof: (α β) is a solution of $by = ax + c$ means $b\beta = a\alpha + c = -b\beta = -a\alpha - c$, which means ($-\alpha$ β) is a solution of $b\beta = a\alpha + c \Rightarrow -b\beta = -a\alpha - c$.

Bhaskara II has given the following two examples.

Example 1 (Bhaskara II, Bijaganitha, page 2980)

Find the integral solutions of $13y = -60x \pm 3$.
First, we solve the equation $13y = 60x - 3$.

$13)60(4 = q_1$
$\underline{52}$
$8)13(1 = q_2$
$\underline{8}$
$5)8(1 = q_3$
$\underline{5}$
$3)5(1 = q_4$
$\underline{3}$
$2)3(1 = q_5$
$\underline{2}$
1

(*Continued*)

Example 1 (Bhaskara II, Bijaganitha, page 2980) (cont'd)

So, the number of quotients is 5 and hence odd. We now form the following table:

4	4	4	4	4	69
1	1	1	1	15	15
1	1	1	9	9	9
1	1	6	6	6	
1	3	3	3		
3	3				
0	0				

So, $Y = 69 \equiv 9(\text{mod } 60)$.

$$X = 15 \equiv 2(\text{mod } 13)$$

So, $(\alpha\ \beta) = (2\ 9)$ is a solution of $13y = +60x - 3$.
Then,

$$x = b - \alpha = 13 - 2 = 11$$

$$y = a - \beta = 60 - 9 = 51$$

Then, $(11\ 51)$ is a solution of $13y = 60x + 3$.
Now, from Bhaskara II's previous lemma,

$$x = \alpha - b = 2 - 13 = -11$$

$$y = a - \beta = 60 - 9 = 51$$

So, $(x\ y) = (-\ 11\ 51)$ is a solution of $13y = -60x + 3$.

General solution of $13y = -60x + 3$ is $x = 13m - 11$ and $y = 60m + 51$, where m is an arbitrary integer.

Also, if $(11\ 51)$ is a solution of $13y = 60x + 3$, then $(11\ \ -51)$ is a solution of $13y = -60x - 3$.

General solution of $13y = -60 - 3$ is

$$x = 11 + 13m;\ \ y = -51 + 60m$$

where m is an arbitrary integer.

Example 2 (Bhaskara II, Bijaganitha page 30)

Solve in integer the equation $-11y = 18x \pm 10$.

We will first solve the equation $11y = 18x + 10$.

$$\begin{array}{l} 11)18(1 \\ \underline{11} \\ \quad 7)11(1 \\ \quad \underline{7} \\ \qquad 4)7(1 \\ \qquad \underline{4} \\ \qquad\quad 3)4(1 \\ \qquad\quad \underline{3} \\ \qquad\quad 1 \end{array}$$

Here the number of quotients is 4 and so even.

We now form the following table:

1	1	1	1	50
1	1	1	30	30
1	1	20	20	20
1	10	10	10	10
10	10	10		
0	0			

$$Y = 50 \equiv 14(\text{mod } 18); \quad X = 30 \equiv 8(\text{mod } 11)$$

$(\alpha\ \beta) = (8\ 14)$ is a solution of $11y = 18x + 10$.

Then, $(8\ \ -14)$ is a solution of $-11y = 18x + 10$.

General solution of $-11y = 18x + 10$ is $x = 11m + 8$; $y = 18m - 14$, where m is an arbitrary integer.

Now, if $(\alpha\ \beta) = (8\ 14)$ is a solution of $11y = 18x + 10$, then

$$x = b - \alpha = 11 - 8 = 3$$

$y = a - \beta = 18 - 14 = 4$ is a solution of $11y = 18x - 10$.

Thus, if $(3\ 4)$ is a solution of $11y = 18x - 10$, then $(3\ \ -4)$ is a solution of $-11y = 18x - 10$.

General solution of $-11y = 18x - 10$ is $x = 3 + 11m$; $y = -4 + 18m$, where m is an arbitrary integer.

Bhaskara gave the following example to demonstrate how to solve a linear equation involving more than two variables.

Example (Bhaskara II, Bijaganitha, page 77)

"One person has flawless five rubies, eight sapphires, and seven pearls, and 90 coins. Another person has seven rubies, nine sapphires and six pearls of the same kind, and 62 coins. If both are equally wealthy, tell me, Oh algebraist, the price of each gem."

Let x, y, z be the price of a ruby, a sapphire, and a pearl, respectively.

Then, we have

$$5x + 8y + 7z + 90 = 7x + 9y + 6z + 62$$

$$7x - 5x = 8y + 7z + 90 - 9y - 6z - 62$$

$$2x = z - y + 28 \quad (*)$$

Method 1: Suppose $z = 1$.

Then, $2x = 1 - y + 28 = 29 - y$.

We observe that if $y = 1$, $2x = 29 - 1 = 28 \Rightarrow x = 14$.

(y x) = (1 14) is a solution of $2x = 29 - y$.

General solution is $x = 14 - m$ and $y = 1 + 2m$ and $z = 1$, where m is an arbitrary integer.

Method 2: Let $x = 3z$ and $y = 2z$, then (*) becomes

$$2(3z) = z - 2z + 28$$

$$6z = -z + 28 = 7z = 28 = z = 4$$

Then

$$x = 3z = 12 \text{ and } y = 2z = 8$$

Method 3: Let $y = 5$, $z = 3$, then (*) becomes

$$2x = -5 + 3 + 28 = 26$$

$$x = 13$$

$$x = 13, \quad y = 5, \quad \text{and} \quad z = 3$$

In the following examples, Bhaskara II has demonstrated a method of solving simultaneous indeterminate equations of first order.

Example 1 (Bhaskara II, Bijaganitha, page 79)

"Four merchants have 5, 3, 6, and 8 horses, respectively; camels 2, 7, 4, and 1, respectively; mules 8, 2, 1, and 3, respectively; 7, 1, 2, and 1 oxen, respectively. If all the four merchants are equally wealthy, tell me the price of each animal."

Let x, y, z, w be the price of a horse, a camel, a mule, and an oxen, respectively. Let P be the wealth of each of them

$$\text{Wealth of the first merchant} = 5x + 2y + 8z + 7w = P \quad \text{(i)}$$

$$\text{Wealth of the second merchant} = 3x + 7y + 2z + w = P \quad \text{(ii)}$$

$$\text{Wealth of the third merchant} = 6x + 4y + z + 2w = P \quad \text{(iii)}$$

$$\text{Wealth of the fourth merchant} = 8x + y + 3z + w = P \quad \text{(iv)}$$

Subtracting (ii) from (i), we get

$$2x - 5y + 6z + 6w = 0$$

$$\Rightarrow x = \frac{1}{2}(5y - 6z - 6w) \quad \text{(v)}$$

Subtracting (ii) from (iii), we get

$$3x - 3y - z + w = 0 \Rightarrow x = \frac{1}{3}(3y + z - w) \quad \text{(vi)}$$

Subtract (iii) from (iv), we get $2x - 3y + 2z - w = 0$

$$\Rightarrow x = \frac{1}{2}(3y - 2z + w) \quad \text{(vii)}$$

From (v) and (vi), we get

$$\frac{1}{2}(5y - 6z - 6w) = \frac{1}{3}(3y + z - w)$$

So,

$15y - 18z - 18w = 6y + 2z - 2w.$

Hence,

$$9y = 20z + 16w \Rightarrow y = \frac{1}{9}(20z + 16w) \quad \text{(viii)}$$

(Continued)

Example 1 (Bhaskara II, Bijaganitha, page 79) (cont'd)

From (vi) and (vii), we have

$$\frac{1}{3}(3y + z - w) = \frac{1}{2}(3y - 2z + w)$$

So,

$$6y + 2z - 2w = 9y - 6z + 3w$$

So,

$$3y = 8z - 5w \Rightarrow y = \frac{1}{3}(8z - 5w) \qquad \text{(ix)}$$

So, from (viii) and (ix), we have

$$y = \frac{1}{9}(20z + 16w) = \frac{1}{3}(8z - 5w)$$

i.e.,

$$20z + 16w = 24z - 15w$$

$$4z = 31w \Rightarrow z = \frac{31w}{4}$$

Let $w = 4m$, where m is an arbitrary integer.
Then,

$$z = \frac{31}{4}w = \frac{31}{4}(4m) = 31m$$

From (viii),

$$y = \frac{1}{9}(20z + 16w)$$

$$= \frac{1}{9}[20 \cdot (31m) + 16(4m)] = \frac{1}{9}(620m + 64m)$$

$$= \frac{1}{9}(684m) = 76m$$

From (vi),

$$x = \frac{1}{3}(3y + z - w) = \frac{1}{3}(228m + 31m - 4m)$$

$$= \frac{1}{3}(255m) = 85m$$

So, the general solutions is $x = 85m$, $y = 76m$, $z = 31m$, and $w = 4m$, where m is an arbitrary integer.

Example 2 (Bhaskara II, Bijaganitha, page 91)

"Give me quickly, Oh mathematician the number which when multiplied by 23 and separately divided by 60 and 80 leaves remainders, whose sum is 100."

Let x be the required number.

When $23x$ is divided by 60 and 80, respectively, let the corresponding quotients be q_1 and q_2, and the corresponding remainders be r_1 and r_2.

Then,

$$23x = q_1(60) + r_1 \Rightarrow x = \frac{1}{23}(60q_1 + r_1) \quad (1)$$

Also,

$$23x = q_2(80) + r_2 \Rightarrow x = \frac{1}{23}(80q_2 + r_2) \quad (2)$$

Also,

$$r_1 + r_2 = 100 \quad (3)$$

Adding (1) and (2), we get

$$2x = \frac{60q_1 + 80q_2 + r_1 + r_2}{23} = \frac{60q_1 + 80q_2 + 100}{23}$$

$$x = \frac{30q_1 + 40q_2 + 50}{23} \quad (*)$$

(Continued)

Example 2 (Bhaskara II, Bijaganitha, page 91) (cont'd)

Bhaskara II pointed out that in order to solve (*), the remainders r_1 and r_2 be such that $r_1 < 60$ and $r_2 < 90$.

In this example, suppose $r_1 = 40$ and $r_2 = 60$.

Then, we have

$$60q_1 + 40 = 80q_2 + 60$$

So, dividing by 20, we get $3q_1 + 2 = 4q_2 + 3$ or $3q_1 = 4q_2 + 1$.

Now, we solve this indeterminate equation.

$$\begin{array}{c} 3)4(1 \\ \underline{3} \\ 1 \end{array}$$

So, the number of quotients is 1 and hence odd.

We now form the following table:

1	1	So, $Y = 1 \equiv 1 \pmod 4$
1	1	$X = 1 \equiv 1 \pmod 3$
0	0	

So, $(\alpha\ \beta) = (1\ 1)$ is a solution of $3q_1 = 4q_2 - 1$.

Then, $q_2 = b - \alpha = 3 - 1 = 2$.

$$q_1 = a - \beta = 4 - 1 = 3$$

So, $(q_2\ q_1) = (2\ 3)$ is a solution of $3q_1 = 4q_2 + 1$.

So, the general solution of $3q_1 = 4q_2 + 1$ is $q_2 = 2 + 3m$ and $q_1 = 3 + 4m$, where m is an arbitrary integer.

Now,

$$x = \frac{30q_1 + 40q_2 + 50}{23}$$

$$= \frac{30(3 + 4m) + 40(2 + 3m) + 50}{23}$$

i.e.,

$$= \frac{120m + 120m + 90 + 80 + 50}{23}$$

$$x = \frac{240m + 220}{23}$$

where m is an arbitrary integer.

We now have to solve the equation $23x = 240m + 220$.

$$\begin{array}{l} 23)240(10 \\ \underline{23} \\ 10)23(2 \\ \underline{20} \\ 3)10(3 \\ \underline{9} \\ 1 \end{array}$$

So, the number of quotient is 3 and so odd.

We now form the following table:

10	10	10	16060
2	2	1540	1540
3	660	660	660
220	220	220	
0	0		

$$Y = 16,060; \quad X = 1540$$

$$Y = 16,060 \equiv 220(\text{mod } 240); \; X = 1540 \equiv 22(\text{mod } 23)$$

So, $m = 22$ and $x = 220$ is a solution of $23x = 240m - 220$.
Hence, $m = 23 - 22 = 1$ and $x = 240 - 220 = 20$.
So, $(m\ x) = (1\ 20)$ is a solution of $23x = 240m + 220$.
Thus, $x = 240n + 20$; $m = 23n + 1$ is the general solution, where n is an arbitrary integer.
Similarly for the other values of r_1 and r_2, we get different solutions.

Example 3 (Bhaskara II, Bijaganitha, page 857)

"What is that number, which leaves remainders 5, 4, 3, 2 when divided by 6, 5, 4, 3, respectively?"

Let N be the required number.

Let x, y, z, and w be the respective quotients, when the number N is divided by 6, 5, 4, and 3 and leaves remainders 5, 4, 3, and 2.

Then, $N = 6x + 5 = 5y + 4 = 4z + 3 = 3w + 2$.

(*Continued*)

Example 3 (Bhaskara II, Bijaganitha, page 857) (cont'd)

Now, $4z+3=3w+2 \Rightarrow 4z=3w-1$.
We now solve the equation

$$\begin{array}{l} 4)3(0 \\ \ \ \underline{0} \\ \ \ 3)4(1 \\ \ \ \ \ \underline{3} \\ \ \ \ \ 1 \end{array}$$

So, the number of quotients is 2 and is even.

We now form the following table:

0	0	1
1	1	1
1	1	1
0	0	

So, $Y=1\equiv 1(\text{mod } 3)$; $X=1\equiv 1(\text{mod } 4)$.
Hence, $w=1$, $z=1$ is the least solution of $4z=3w+1$.
So, $w=\alpha=4-1=3$ and $\beta=3-1=2$ is the least solution of $4z=3w-1$.
Since $(w\ z)=(3\ 2)$ is a solution of $4z=3w-1$.
Now, the general solution of $4z=3w-1$ is

$$w=4m+3$$

$$z=3m+2$$

where m is an arbitrary integer.
Now, the general solution of $N=4z+3=4(3m+2)+3=12m+11$.
But, $N=12m+11=5y+4$.
So, $5y=12m+7$.
We now solve this indeterminate equation.

$$\begin{array}{l} 5)12(2 \\ \ \ \underline{10} \\ \ \ \ 2)5(2 \\ \ \ \ \ \ \underline{4} \\ \ \ \ \ \ 1 \end{array}$$

The number of quotients is 2 and so even.

We now form the following table:

2	2	35
2	14	14
7	7	7
0	0	

So, $Y = 35 \equiv 11(\text{mod } 12)$.

$$X = 14 \equiv 4(\text{mod } 5)$$

$(m\ y) = (4\ 11)$ is a solution of $5y = 12m + 7$. Now, the general solution of $5y = 12m + 7$ is

$$m = 5t + 4$$

$$y = 12t + 11$$

where t is an arbitrary integer.

But, $N = 5y + 4 = 5(12t + 11) + 4 = 60t + 55 + 4 = 60t + 59$ is the general solution of N, where t is an arbitrary integer.

Least value of $N = 59$. This value of $N = 59$ satisfying the other condition, namely it leaves a remainder 5, when divided by 6.

Thus, the general value of $N = 60t + 59$, when t is an arbitrary integer.

Bhaskara II solved the same problem, starting from x and y instead of z and w. We will show his work below:

$N = 6x + 5 = 5y + 4 \Rightarrow 5y = 6x + 1$. We now solve the equation.

$$\begin{array}{c} 5)6(1 \\ \underline{5} \\ 1 \end{array}$$

The number of quotients is 1 and so odd. We now form the following table:

1	1
1	1
0	0

So, $y = 1 \equiv 1(\text{mod } 6)$; $x = 1 \equiv 1(\text{mod } 5)$.

So, $(x,y) = (1,1)$ is solution of $5y = 6x - 1$.

Hence, $x = \alpha = 5 - 1$ and $y = \beta = 6 - 1 = 5$ is a solution of $5y = 6x + 1$.

So, $(x\ y) = (\alpha\ \beta) = (4\ 5)$ is a solution of $5y = 6x + 1$.

Thus, the general solution of $5y = 6x + 1$ is

$$x = 5m + 4$$

$$y = 6m + 5$$

where m is an arbitrary integer.

(*Continued*)

Example 3 (Bhaskara II, Bijaganitha, page 857) (cont'd)

Now, $N = 5y + 4 = 5(6m + 5) + 4 = 30m + 29$ is the general solution of N.
But, $N = 30m + 29 = 4z + 3$.
So, $4z = 30m + 26$, i.e. $2z = 15m + 13$
We now solve this indeterminate equation.

$$\begin{array}{l} 2)15(7 \\ \underline{14} \\ \;\;1 \end{array}$$

The number of quotients is 1 and so odd.

We now form the following table:

7	91
13	13
0	0

So, $y = 91 \equiv 1(\text{mod } 15)$; $x = 13 \equiv 1(\text{mod } 2)$.
Then, $(m\ z) = (1\ 1)$ is a solution of $2z = 15m - 13$.
Hence, $m = \alpha = 2 - 1 = 1$ and $z = 15 - 1 = 14$ is a solution of

$$2z = 15m + 13$$

General solution of $2z = 15m + 13$ is $z = 15t + 14$; $m = 2t + 1$, where t is an arbitrary integer.

Now,

$$N = 4z + 3 = 4(15t + 14) + 3$$

$N = 60t + 59$, where t is an arbitrary integer and the least value of $N = 59$, which satisfies the last condition namely 59 divided by 3 leaves remainder 2.

So, $N = 60t + 59$ is the general solution, where t is an arbitrary integer.

Bhaskara II agreed with the method of Aryabhata for solving system of indeterminate equations of first degree. He gave the following example to demonstrate that.

Example (Bhaskara II, Bijaganitha, page 35, Leelavati, page 82)

Find the integral solutions of the system of equations

$$63y_1 = 5x - 7 \quad (1)$$

$$63y_2 = 10x - 14 \quad (2)$$

Adding (1) and (2), we get

$$63(y_1 + y_2) = 15x - 21$$

Let $y_1 + y_2 = y$. Then, we have

$$63y = 15x - 21 \Rightarrow 21y = 5x - 7$$

We will now solve this indeterminate equation.

```
21)5(0
   0
   5)21(4
     20
      1
```

The number of quotients is 2 and so even.

We now form the following table:

0	0	7
4	28	28
7	7	7
0	0	

So, $Y = 7 \equiv 2 (\text{mod } 5)$; $X = 28 \equiv 7 (\text{mod } 21)$.
Hence, $x = 7$, $y = 2$ is a solution of $21y = 5x + 7$.
So, $\alpha = 21 - 7 = 14$ and $\beta = 5 - 2 = 3$ is a solution of $21y = 5x - 7$.
So, the general solution of $21y = 5x - 7$ is

$$x = 21m + 14; \quad y = 5m + 3$$

Least value of $x = 14$ and least value of $y = y_1 + y_2 = 3$. If we let $x = 14$ in (1), we get

$$63y_1 = 5(14) - 7 = 63 \Rightarrow y_1 = 1$$

(Continued)

Example (Bhaskara II, Bijaganitha, page 35, Leelavati, page 82) (cont'd)

If we let $x = 14$ in (2), we get

$$63y_2 = 10(14) - 14 = 140 - 14 = 126$$

So, $y_2 = 2$.
Also, it satisfies $y_1 + y_2 = 1 + 2 = 3 = y$.
We will now solve (1), namely $63y_1 = 5x - 7$.

```
63)5(0
   0
   5)63(12
     60
      3)5(1
        3
        2)3(1
          2
          1
```

The number of quotients is 4 and so even.

We now form the following table:

0	0	0	0	14
12	12	12	175	175
1	1	14	14	14
1	7	7	7	
7	7	7		
0	0			

So, $Y = 14 \equiv 4 \pmod 5$; $X = 175 \equiv 49 \pmod{63}$.
$(x\ y_1) = (49\ 4)$ is a solution of $63y_1 = 5x + 7$.
So, $x = \alpha = 63 - 49 = 14$ and $y = \beta = 5 - 4 = 1$ is a solution of $63y_1 = 5x - 7$.

General solution of $63y_1 = 5x - 7$ is $x = 14 + 63m$; $y_1 = 1 + 5m$, where m is any arbitrary integer.

Next we solve (2), namely $63y_2 = 10x - 14$.

```
63)10(0
    0
   10)63(6
      60
       3)10(3
          9
          1
```

The number of quotients is 3 and so odd. We now form the table:

0	0	0	42
6	6	266	266
3	42	42	42
14	14	14	
0	0		

So, $Y = 42 = 2(\text{mod } 10)$; $X = 266 \equiv 14(\text{mod } 63)$.
$(x\ y_2) = (14\ 2)$ is a solution of $63y_1 = 10x - 14$.
General solution of $63y_1 = 10x - 14$ is

$$x = 63m_1 + 14; \quad y = 10m_1 + 2$$

where m_1 is an arbitrary integer.

Example (Bhaskara II, Leelavati)

"A number when multiplied by 7 and divided by 62 leaves a remainder 3. The same number when multiplied by 6 and divided by 101 leaves a remainder 5. The same number when multiplied by 8 and divided by 17 leaves a remainder 9. Can you tell me the number?"

Let x be the number. Let y_1, y_2, y_3 be the corresponding quotients.
Then,

$$\frac{7x}{62} = y_1 + \frac{3}{62}; \quad \frac{6x}{101} = y_2 + \frac{5}{101}; \frac{8x}{17} = y_3 + \frac{9}{17}$$

So, we have

$$62y_1 + 3 = 7x \tag{i}$$

$$101y_2 = 6x - 5 \tag{ii}$$

$$17y_3 = 8x - 9 \tag{iii}$$

We now solve (i), namely $62y_1 = 7x - 3$.

(*Continued*)

Example (Bhaskara II, Leelavati) (cont'd)

62)7(0
0
7)62(8
56
6)7(1
6
1

The number of quotients is 3 and so odd.

We now form the following table:

0	0	0	3
8	8	27	27
1	3	3	3
3	3	3	
0	0		

So, $Y = 3 \equiv 3(\text{mod } 7)$; $X = 27 \equiv 27(\text{mod } 62)$.
Hence, $x = 27$ and $y_1 = 3$ is a solution of $62y_1 = 7x - 3$.
So, the general solution of the equation $62y_1 = 7x - 3$ is $x = 62m + 27$ and $y_1 = 7m + 3$, where m is an arbitrary integer.
From equation (2), we have $101y_2 = 6x - 5$.
Substituting $x = 62m + 27$, we get

$$101y_2 = 6(62m + 27) - 5$$

i.e., $101y_2 = 372m + 162 - 5$.

$$101y_2 = 372m + 157$$

But, $157 \equiv 56(\text{mod } 101)$.
So, we solve $101y_2 = 372m + 56$.

101)372(3
303
69)101(1
69
32)69(2
64
5)32(6
30
2)5(2
4
1

The number of quotient is 5 and so odd.

We now form the following table:

3	3	3	3	3	8456
1	1	1	1	2296	2296
2	2	2	1568	1568	1568
6	6	728	728	728	
2	112	112	112		
56	56	56			
0	0				

So, $Y = 8456 \equiv 272(\text{mod } 372)$.

$$X = 2296 \equiv 74(\text{mod } 101)$$

So, $(m\ y_2) = (\alpha\ \beta) = (74\ 272)$ is a solution of $101y_2 = 372m - 56$.

So, $y_2 = 372 - 272 = 100$ and $m = 101 - 74 = 27$ is a solution of $101y_2 = 372m + 56$.

So, the general solution of $101y_2 = 372m + 56$ is $y_2 = 372n + 100$ and $m = 27 + 101n$, where n is an arbitrary integer.

But, the general solution of $x = 62m + 27$.

So, $x = 62(27 + 101n) + 27$.

i.e., $x = 6262n + 1674 + 27$.

i.e., $x = 6262n + 1701$.

Equation (3) is $17y_3 = 8x - 9$.

Substituting the value of x, we get

$$17y_3 = 8(6262n + 1701) - 9$$

i.e., $17y_3 = 50{,}096n + 13{,}599$.

$$13,599 \equiv 16(\text{mod } 17)$$

So, we solve $17y_3 = 50{,}096n + 16$.

(*Continued*)

Example (Bhaskara II, Leelavati) (cont'd)

17)50096(2946
34
160
153
79
68
116
102
14)17(1
14
3)14(4
12
2)3(1
2
1

The number of quotients is 4 and so is even. We now form the following table:

2946	2946	2946	2946	292896
1	1	1	96	96
4	4	80	80	80
1	16	16	16	
16	16	16		
0	0			

So, $Y = 282{,}896$; $X = 96$.

$$X = 96 \equiv 11(\text{mod}17)$$

$$y = 28,2896 \equiv 32,416(\text{mod } 50,096)$$

So, $n = 11$ and $y_3 = 32{,}416$ is a solution of the equation $17y_3 = 50{,}096n + 16$.

So, the general solution of $17y_3 = 50{,}096n + 16$ is $n = 17p + 11$ and $y_3 = 50{,}096p + 32{,}416$, where p is an arbitrary integer.

But,

$$x = 6262n + 1701$$

$$= 6262(17p + 11) + 1701$$

$$= 106,454p + 68,882 + 1701$$

$$= 106,454p + 70,583$$

where p is a arbitrary integer.

Least value of x is 70,583.

From (1),

$$62y_1 = 7x - 3 = 7(106,454p + 70,583) - 3$$

$$= 7(106,454p) + 494,078$$

So,

$$y_1 = 12,019p + 7969$$

From (2),

$$101y_2 = 6(70,583) - 5$$

$$101y_2 = 423,493$$

$$y_2 = \frac{423,493}{101} = 4193$$

Least value of $y_2 = 4193$.

From (3),

$$17y_3 = 8x - 9$$

$$17y_3 = 8(70,583) - 9$$

$$17y_3 = 564,655$$

$$y_3 = \frac{564,655}{17} = 33,215$$

Least value of $y_3 = 33{,}215$ and least value of $y_1 = 7969$.

Example (Bhaskara II)

Solve in integers the system of equations

$$62y_1 = 7x + 3 \tag{1}$$

$$101y_2 = 6x + 5 \tag{2}$$

$$17y_3 = 8x + 9 \tag{3}$$

In the previous example, we found that $x = 27$ and $y_1 = 3$ is the least solution of $62y_1 = 7x - 3$. So, $x = 62 - 27 = 35$ and $y_1 = 7 - 3 = 4$ is a solution of (1), namely $62y_1 = 7x + 3$.

General solution of $62y_1 = 7x + 3$ is $x = 62m_1 + 35$ and $y_1 = 7m_1 + 4$, where m_1 is an arbitrary integer.

From (2), we have $101y_2 = 6x + 5$.

Substituting the value of x in this equation, we get

$$101y_2 = 6(62m_1 + 35) + 5$$

i.e.,

$$101y_2 = 372m_1 + 215 = 101(y_2 - 2) = 372m_1 + 13$$

So, we now solve $101y_2^* = 372m_1 + 13$, where $y_2^* = y_2 - 2$

$$\begin{array}{l} 101)372(3 \\ \quad \underline{303} \\ \quad\; 69)101(1 \\ \quad\quad\;\; \underline{69} \\ \quad\quad\;\; 32)69(2 \\ \quad\quad\quad\;\; \underline{64} \\ \quad\quad\quad\;\; 5)32(6 \\ \quad\quad\quad\quad\; \underline{30} \\ \quad\quad\quad\quad\; 2)5(2 \\ \quad\quad\quad\quad\quad \underline{4} \\ \quad\quad\quad\quad\quad 1 \end{array}$$

So, the number of quotients is 5 and so odd.
We now form the following table:

3	3	3	3	3	1963
1	1	1	1	533	533
2	2	2	364	364	364
6	6	169	169	169	
2	26	26	26		
13	13	13			
0	0				

So, $Y = 1963 \equiv 103 (\text{mod } 372)$; $X = 533 \equiv 28 (\text{mod } 101)$.
So, $m_1 = 28$ and $y_2^* = 103$ is a solution of $101y_2^* = 372m_1 - 13$.
So, $m_1 = 101 - 28 = 73$ and $y_2^* = 372 - 103 = 269$ is a solution of

$$101y_2^* = 372m_1 + 13.$$

So, the general solution of $101y_2^* = 372m_1 + 13$ is $y_2^* = 372m_2 + 269$.
$m_1 = 101m_2 + 73$, when m_2 is an arbitrary integer.
But, $x = 62m_1 + 35 = 62(101m_2 + 73) + 35$.
So, $x = 6262m_2 + 4561$ (∗).
Equation (3) is $17y_3 = 8x + 9$.
Substituting the value of x, we get

$$17y_3 = 8(6262m_2 + 4561) + 9$$

i.e.,

$$17y_3 = 50,096m_2 + 36,497$$

$$17(y_3 - 2146) = 50,096m_2 + 15$$

i.e., $17y_3^* = 50,096m_2 + 15$, where $y_3^* = y_3 - 2146$.
We now solve this equation.

(Continued)

Example (Bhaskara II) (cont'd)

```
17)50096(2946
   34
   160
   153
    79
    68
    116
    102
     14)17(1
        14
         3)14(4
           12
            2)3(1
              2
              1
```

The number of quotients is 4 and so even.
We now form the following table:

2946	2946	2946	2946	2946	265215
1	1	1	90	90	90
4	4	75	75	75	
1	15	15	15		
15	15	15			
0	0				

So, $Y = 265{,}215 \equiv 14{,}735 (\text{mod } 50{,}096)$.

$$X = 90 \equiv 5 (\text{mod } 17)$$

So, $m_2 = 5$ and $y_3^* = 14,735$ is a solution of

$$17y_3^* = 50,096m_2 + 15$$

So, the general solution of $17y_3^* = 50,096m_2 + 15$ is

$$m_2 = 17m_3 + 5$$

$$y_3^* = 50,096m_3 + 14,735$$

where m_3 is an arbitrary integer.

Substituting the value of m_2 in (*), we get

$$x = 6262(17m_3 + 5) + 4561$$
$$= 106,454m_3 + 31,310 + 4561$$
$$= 10,645m_5 + 35,871$$

where m_3 is an arbitrary integers. So, the least value of x is 35,871.
Now, the general solution of $x = 106.454m_3 + 35.871$.
Substituting in (1), we get

$$62y_1 = 7x + 3 = 7(106,456m_3 + 35,871) + 3$$
$$62y_1 = 745,192m_3 + 251,097 + 3$$
$$62y_1 = 745,192m_3 + 251,100$$

Least value of $62y_1 = 251{,}100$.
So, the least value of

$$y_1 = \frac{251,100}{62} = 4050$$

Similarly, substituting the value of x in (2), we get

$$101y_2 = 6(106,456m_3 + 35,871) + 5$$

i.e.,

$$101y_2 = 638,736m_3 + 215,226 + 5$$
$$101y_2 = 638,736m_3 + 215,231$$

So, the least value of $101y_2 = 312{,}831$.
Thus, the least value of

$$y_2 = \frac{215,231}{101} = 2131$$

From (3), $17y_3 = 8x + 9$.

$$17y_3 = 8(35,871) + 9 = 286,977$$
$$y_3 \quad = \frac{286,977}{17} = 16,881$$

The least value of $y_3 = 16.881$.

Shadow Problems

The shadow problems by Bhaskara II in "Leelavati" are purely geometrical and they have been modeled by the method of Aryabhata I. The following problem by Bhaskara II is noteworthy:

Lemma (Bhaskara):

छाययोः कर्णयोरन्तते ये तयोवंर्गविश्लेषभक्ताः रसाद्रीषवः।
सैकलब्धेः पदध्नं तु कर्णान्तरं भांन्तरेणोनयुक्तं दले स्तः प्रभे ॥

(Lil. 232)

i.e., "Divide 576 by the difference of the squares of difference of the shadow lengths and the difference of the hypotenuse. The product of the difference between the hypotenuses and the square root of the above quotients plus 1 is added to and decreased by the difference between the shadow lengths. The resulting quantities divided by 2 give the shadow lengths separately."

The figure shown does not show the shadows cast by two equal gnomons at different distances from the source of light instead, it shows a triangle got by juxtaposition of two right triangles *PQR* and *PQS*, as if two sources of light at different distances or at different heights or both were producing on either side of gnomon were giving shadows *RS* and *QR* on either side of the gnomon.

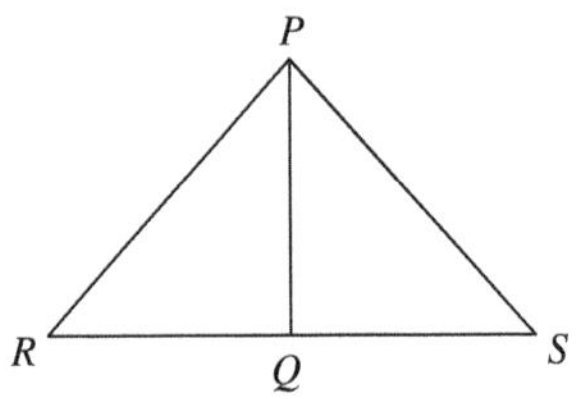

Now, *QR* and *RS* are two "abadhas" of the triangle *PRS*. Now, the difference of the two shadows = $QS - RQ$.

Let

$$QS - RQ = a \text{ (say)} \quad \text{(i)}$$

Let the difference between the hypotenuses

$$= PS - PR = b \text{ (say)} \quad \text{(ii)}$$

Let

$$QS + RQ = x \quad \text{(iii)}$$

Adding (i) and (ii), we get

$$2QS = a + x \Rightarrow QS = \frac{a + x}{2}$$

Subtracting (i) from (iii), we get

$$2RQ = x - a \Rightarrow RQ = \frac{x-a}{2}$$

Now, $PS^2 - PR^2 = (PQ^2 + QS^2) - (PQ^2 + RQ^2) = QS^2 - RQ^2$.
i.e.,

$$PS^2 - PR^2 = \left(\frac{a+x}{2}\right)^2 - \left(\frac{a-x}{2}\right)^2 = \frac{a^2+2ax+x^2}{4} - \frac{a^2-2ax-x^2}{4}$$

$$= \frac{4ax}{4} = ax$$

$$(PS + PR)(PS - PR) = ax$$

$$PS + PR = \frac{ax}{PS - PR} = \frac{ax}{b} \qquad \text{(iv)}$$

Adding (ii) and (iv), we get

$$2PS = \frac{ax}{b} + b \Rightarrow PS = \frac{1}{2}\left(\frac{ax}{b} + b\right)$$

Subtracting (ii) from (iv), we get

$$2PR = \frac{ax}{b} - b \Rightarrow PR = \frac{1}{2}\left(\frac{ax}{b} - b\right)$$

Now,

$$PS^2 = PQ^2 + QS^2$$

$$\frac{1}{4}\left(\frac{ax}{b} + b\right)^2 = PQ^2 + \left(\frac{a+x}{2}\right)^2$$

So, $(ax + b^2)^2 = [4PQ^2 + (a + x)^2]b^2 = b^2[4PQ^2 + a^2 + 2ax + x^2]$.
So, $a^2x^2 + 2axb^2 + b^4 = b^2\ (4PQ^2 + a^2) + 2axb^2 + x^2b^2$.
We have, then, $x^2(a^2 - b^2) = b^2(4PQ^2 + a^2 - b^2)$.

$$x^2 = \left(\frac{4PQ^2}{a^2 - b^2} + 1\right)b^2$$

$$x = b\sqrt{\left\{1 + \frac{4PQ^2}{a^2 - b^2}\right\}}$$

But,

$$QS + QR = x = b\sqrt{\left\{1 + \frac{4PQ^2}{a^2 - b^2}\right\}}$$

and

$$QS = \frac{a+x}{2} = \frac{a + b\sqrt{1 + (4PQ^2/(a^2 - b^2))}}{2}$$

$$QR = \frac{x-a}{2} = \frac{b\sqrt{1 + (4PQ^2/(a^2 - b^2))} - a}{2}$$

This is the result given in the verse in "Leelavati," page 232.
Since PQ is given to be 12, $4PQ^2 = 4(12)^2 = 576$.
That is why in the verse 576 appears in place of $4PQ^2$.

The Chord of a Circle

Bhaskara II rejected the method of finding the chord of a circle in terms of the arc of the circle and the vice versa. He gave a new method which is a much better approximation.

चापोननिघ्नपरिधिः प्र थमाह्वयः स्यात्
पंचाहतः परिधिवर्गचतुर्थभागः ।
त्र्यासाहतं प्रथममाप्तमिह ज्यका स्यात् ॥

(Lil. 210)

i.e., "The circumference decreased and multiplied by the arc is called the first ('prathama'). One-quarter of the square of the circumference multiplied by 5 is to be decreased by 'prathama.' The 'prathama' multiplied by four times the diameter is to be divided by the above difference. The quotient is the chord."

Let c be the chord of the arc a. Let d and C be the diameter and circumference of the circle. Then,

$$c = \frac{4d(C-a)a}{(5/4)C^2 - (C-a)a}$$

This formula is a modification of Bhaskara I's formula given in his "Mahabhaskariya."

The Arc of a Circle

व्यासाब्धिवातयुतमौविंकया विभक्तो
जीवांघ्रिपंचगुणितः परिधेस्तु वगंः ।
लब्धोनितात् परिधिवगं चतुर्थपादात्
त्राप्ते पदे वृतिदलात् पतिते धनुंः स्यात् ॥

(Lil. 212)

"i.e., The square of the circumference multiplied by 5/4 of the chord is divided by the sum of the chord and four times the diameter. If we subtract from the semi-circumference the square root of one-fourth the square of the circumference decreased by the above quotient, we get the arc."

This means

$$a = \frac{C}{2} - \sqrt{\frac{C^2}{4} - \frac{5C^2c}{4(c+4d)}}$$

Proof: The Chord

$$c = \frac{4d(C-a)a}{(5/4)C^2 - (C-a)a}$$

So, we have

$$\frac{5}{4}C^2c - ac(C-a) = 4d\left(Ca - a^2\right)$$

i.e.,

$$\frac{5}{4}C^2c + a^2c - acC = 4dCa - 4da^2$$

$$\Rightarrow \frac{5}{4}C^2c + a^2(c+4d) - aC(c+4d) = 0$$

i.e.,

$$\frac{5}{4}C^2c + (c+4d)(a^2 - aC) = 0$$

$$a^2 - aC + \frac{(5/4)C^2c}{c+4d} = 0$$

$$a = \frac{C - \sqrt{C^2 - ((5/4)C^2c \cdot 4/c + 4d)}}{2}$$

taking the negative sign

$$a = \frac{C}{2} - \sqrt{\frac{C^2}{4} - \frac{5C^2c}{4(c+4d)}}$$

To Construct a Rational Right Triangle, Given the Hypotenuse

Bhaskara II

"Multiply the given hypotenuse by twice an optional number and then divide this by the sum of the square of the optional number and unity. This gives the upright ('Koti'). Multiply this again by the optional number and the difference between this product and the given hypotenuse is the base ('bhuja')."

If c is the given hypotenuse and m is the optional number, then the upright ("Koti")

$$= \frac{2cm}{m^2+1}$$

and the base ("bhuja")

$$= \frac{2m^2c}{m^2+1} - c \text{ or } c - \frac{2m^2c}{m^2+1}$$

Bhaskara II's commentator, Suryadasa, has given a proof of how Bhaskara II arrived at the above result (see page 687 in this book).

To Construct a Rational Triangle, Given a Side

"Multiply the given side (base) by twice an optional number and divide this by the square of the optional number decreased by unity. That gives the upright ('Koti'). Then, Koti multiplied by the optional number and diminished by 'bhuja' (base) is the hypotenuse."

Let a be the given base ("bhuja"). Let m be the optional number.

The "Koti" (upright)

$$= \frac{2am}{m^2-1}$$

Hypotenuse ("Karna")

$$= \frac{2am^2}{m^2-1} - a$$

Bhaskara II's commentator, Suryadasa, commented how Bhaskara II arrived at this aforementioned result (see page 686 in this book).

Bhaskara II also gave another method, which is identically the same that was given by Brahmagupta in this book.

To Construct a Rational Cyclic Quadrilateral

"The 'Kotis' and 'bhujas' of two optional rational right triangles multiplied by each other's hypotenuse are the four sides a rational quadrilateral with unequal sides."

Let *ABC* be a right triangle with sides $m^2 + n^2$, $2mn$, $m^2 - n^2$, and let *DEF* be a right triangle with sides $p^2 + q^2$, $2pq$, $p^2 - q^2$, where m, n, p, and q are rational numbers.

Then, the required rational quadrilateral will have four sides equal to $2mn(p^2 + q^2)$, $(m^2 - n^2)(p^2 + q^2)$, $2pq(m^2 + n^2)$, and $(p^2 - q^2)(m^2 + n^2)$ in order.

Bhaskara II pointed out "the product of the uprights added with product of the sides gives one diagonal; the sum of reciprocal product of the sides and uprights gives the second diagonal." He also pointed out that "the cyclic quadrilateral formed by changing the arrangement of side and upright, then the second diagonal will be equal to the product of the two hypotenuses."

Bhaskara II illustrated in "Leelavati," page 51, by taking two right triangles (3,4,5) and (5,12,13). The resulting cyclic quadrilateral will have sides equal to $5 \times 5 = 25$, $3 \times 13 = 39$, $12 \times 5 = 60$, and $4 \times 13 = 52$. The diagonals of this quadrilateral are $3 \cdot 12 + 4 \cdot 5 = 56$ and $4 \cdot 12 + 3 \cdot 5 = 63$ (see Figure 1).

In Figure 2, the diagonals are

$$3 \cdot 12 + 45 = 56 \quad \text{and} \quad 5 \times 13 = 65$$

By taking right triangle (3,4,5) and (15,8,17). Bhaskara got another cyclic quadrilateral whose sides are $17 \times 4 = 68$; $17 \times 3 = 51$; $5 \times 8 = 40$, and $5 \times 15 = 75$. The diagonals of this cyclic quadrilateral are (see Figure 3)

$$3 \cdot 15 + 8 \cdot 4 = 77 \quad \text{and} \quad 17 \times 5 = 85$$

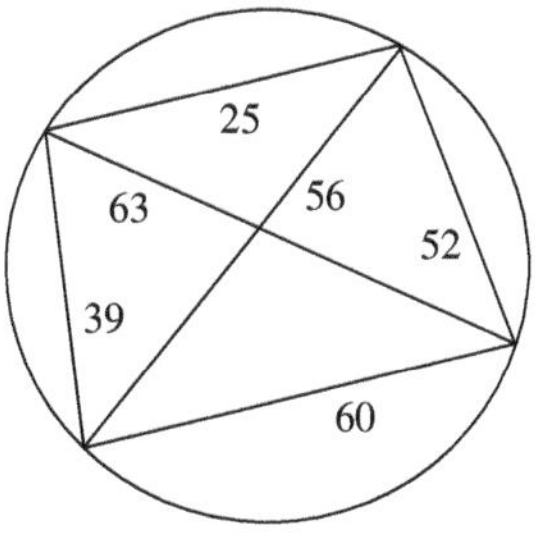

Figure 1

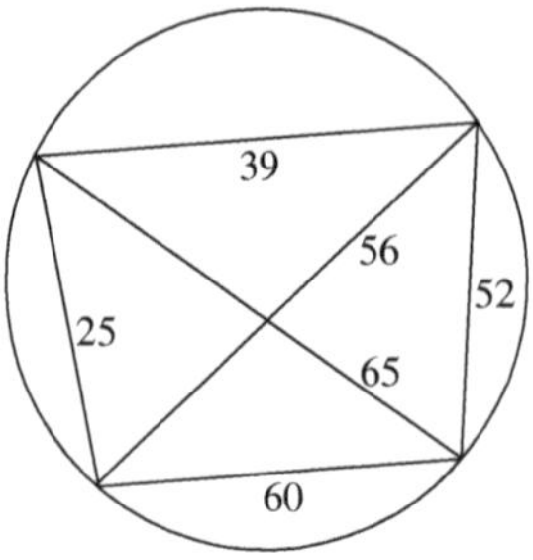

Figure 2

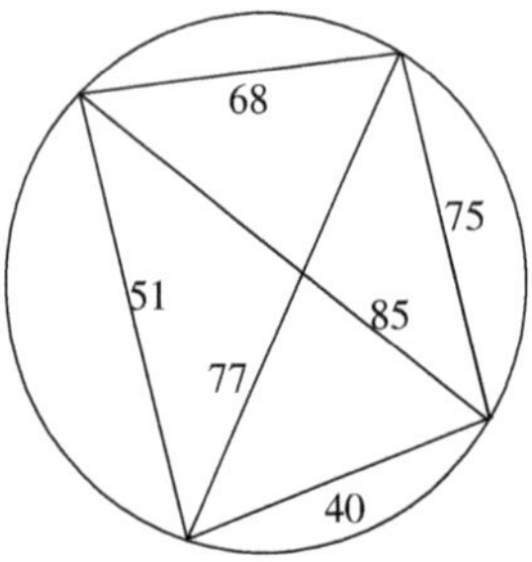

Figure 3

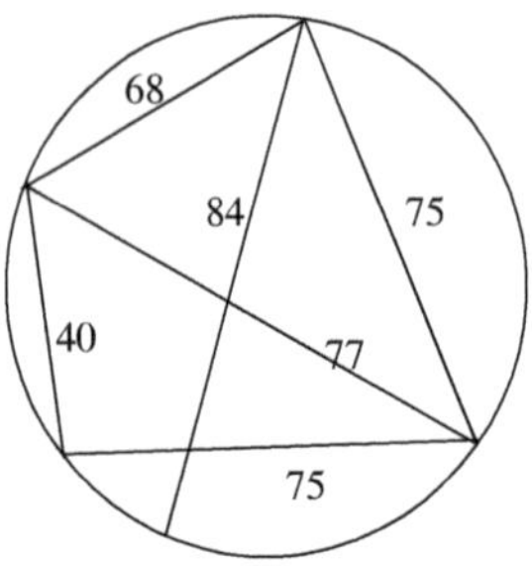

Figure 4

If the sequence of the sides are (68,40,51,75) (see Figure 4), the diagonals are $3 \cdot 15 + 8 \cdot 4 = 77$ and $3 \cdot 8 + 4 \cdot 15 = 84$.

If the sequence of the sides are (68,40,75,51), the diagonals are $3 \cdot 8 + 4 \cdot 15 = 84$ and $17 \cdot 5 = 85$ (see Figure 5 on page 402).

Pythagoras theorem: Bhaskara II's proof of Pythagoras theorem.

Let ABC be a right angled triangle, with right angle at C, with $AB = c$, $AC = b$, and $BC = a$.

Complete the square $ABED$ with each side $= c$. Draw EM_1 perpendicular to CB. Draw DM_2 perpendicular to EM_1. Produce AC to meet DM_2 at M_3.

Now, the four Δ^{les} ABC, ADM_3, DEM_2, and EBM_1 are congruent.

If A = area of Δ^{le} ABC, then sum of the area of these four triangles

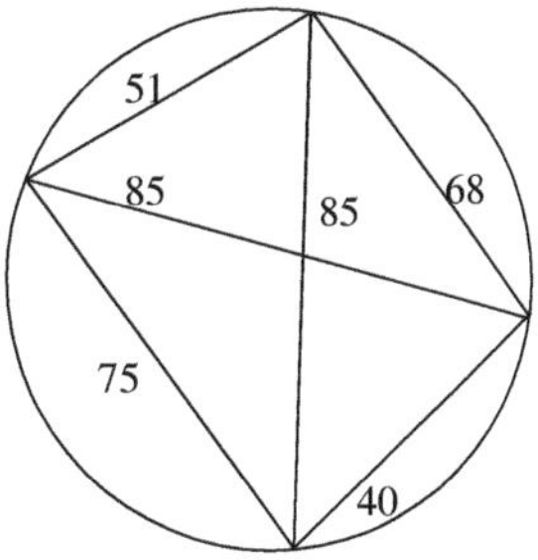

Figure 5

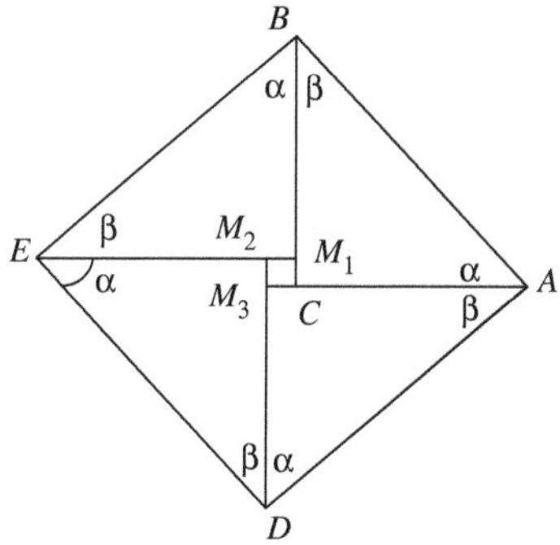

$$= 4A = 4\left(\frac{1}{2}ab\right) = 2ab$$

Since Δ^{les} ABC and DAM_3 are congruent,

$$AM_3 = BC = a$$

$$CM_3 = AM_3 - AC = a - b$$

Since Δ^{les} BEM_1 and ABC are congruent,

$$BM_1 = AC = b$$

$$M_1C = BC - BM_1 = a - b$$

Area of $CM_3M_2M = (a-b)(a-b) = (a-b)^2$.

Thus, the areas of sum of Δ^{les} ABC, ADM_3, DEM_2, and EBM_1 + area of $CM_3M_2M = 2ab + (a-b)^2$

$$= a^2 + b^2 = \text{area of } ABED = c^2$$

That completes the proof of Pythagoras theorem.

Bhaskara II: In the triangle ABC, $BC = \sqrt{18} - 1$, $CA = \sqrt{6}$ and $AB = \sqrt{10} - \sqrt{5}$. Find p, the altitude of A from BC.

Let AD be the altitude from A on BC.

Then, $AD = p$.

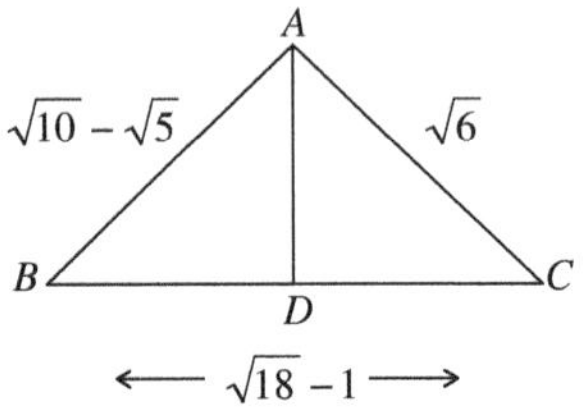

Let $BC = BD + DC = x + y = \sqrt{18} - 1$

i.e., $x + y = 3\sqrt{2} - 1$.

In triangle ADB, $AB^2 = BD^2 + AD^2$.

So, $(\sqrt{10} - \sqrt{5})^2 = x^2 + p^2$.

i.e., $10 + 5 - 2\sqrt{50} = x^2 + p^2$.

Hence,

$$x^2 + p^2 = 15 - 10\sqrt{2} \tag{1}$$

In triangle ABC, $AC^2 = DC^2 + AD^2$.

i.e.,

$$\left(\sqrt{6}\right)^2 = y^2 + p^2$$

i.e.,

$$y^2 + p^2 = 6 \tag{2}$$

Subtracting (2) from (1), we get

$$x^2 - y^2 = 15 - 10\sqrt{2} - 6 = 9 - 10\sqrt{2}$$

i.e.,

$$(x + y)(x - y) = 9 - 10\sqrt{2}$$

So,

$$x - y = \frac{9 - 10\sqrt{2}}{x + y} = \frac{9 - 10\sqrt{2}}{3\sqrt{2} - 1}$$

i.e.,

$$x - y = \frac{9 - 10\sqrt{2}}{3\sqrt{2} - 1} \cdot \frac{3\sqrt{2} + 1}{3\sqrt{2} + 1} = \frac{27\sqrt{2} - 60 + 9 - 10\sqrt{2}}{\left(3\sqrt{2}\right)^2 - 1}$$

Hence,

$$x - y = \frac{17\sqrt{2} - 51}{18 - 1} = \frac{17(\sqrt{2} - 3)}{17} = \sqrt{2} - 3 \qquad (3)$$

But,

$$x + y = 3\sqrt{2} - 1 \qquad (4)$$

Subtracting (3) from (4), we get

$$2y = \left(3\sqrt{2} - 1\right) - \left(\sqrt{2} - 3\right)$$

i.e.,

$$2y = 2\sqrt{2} + 2 \Rightarrow y = \sqrt{2} + 1$$

From (2),

$$p^2 = 6 - y^2 = 6 - \left(\sqrt{2} + 1\right)^2 = 6 - \left(2 + 1 + 2\sqrt{2}\right)$$

So,

$$p^2 = 3 - 2\sqrt{2}$$

Thus,

$$p = \sqrt{3 - 2\sqrt{2}} = \sqrt{\left(\sqrt{2} - 1\right)^2} = \sqrt{2} - 1$$

Example

Find a right triangle whose area equals the hypotenuse.

His method is as follows.

(Continued)

Example (cont'd)

He starts with a rational right triangle whose sides are $3x$, $4x$, and $5x$, where x is a rational number.

$$\text{Area of the triangle} = \frac{1}{2}\text{the base} \times \text{height}$$
$$= \frac{1}{2}3x \times 4x = 6x^2$$

The hypotenuse $= 5x$.
Now, area of the triangle = the hypotenuse

$$6x^2 = 5x$$

Since $x \neq 0$, $6x = 5$ or $x = 5/6$.
So, the sides are

$$3x = 3\left(\frac{5}{6}\right) = \frac{5}{2}$$

$$4x = 4\left(\frac{5}{6}\right) = \frac{10}{3}$$

$$5x = 5\left(\frac{5}{6}\right) = \frac{25}{6}$$

Example

The sides of a triangle are $\sqrt{13}$ and $\sqrt{5}$. The area of the triangle is 4. Find the base of the triangle.

Solution

Let a be base and the sides $b = \sqrt{13}$and $c = \sqrt{5}$.

$$A = \text{area of a triangle} = \sqrt{s(s-a)(s-b)(s-c)} \quad (1)$$

where

$$2s = a + b + c$$

$$s = \frac{a+b+c}{2} = \frac{a+\sqrt{13}+\sqrt{5}}{2}$$

$$s-a = \frac{b+c-a}{2} = \frac{\sqrt{13}+\sqrt{5}-a}{2}$$

$$s-b = \frac{a+c-b}{2} = \frac{a+\sqrt{5}-\sqrt{13}}{2}$$

$$s-c = \frac{a+b+c}{2} = \frac{a+\sqrt{13}-\sqrt{5}}{2}$$

Squaring (1), we get

$$A^2 = s(s-a)(s-b)(s-c)$$

$$16 = \left(\frac{\sqrt{13}+\sqrt{5}+a}{2}\right)\left(\frac{\sqrt{13}+\sqrt{5}-a}{2}\right)\left(\frac{a+(\sqrt{13}-\sqrt{5})}{2}\right)\left(\frac{a-(\sqrt{13}-\sqrt{5})}{2}\right)$$

$$16 = \frac{1}{16}\left[\left(\sqrt{13}+\sqrt{5}\right)^2 - a^2\right]\left[a^2 - \left(\sqrt{13}-\sqrt{5}\right)^2\right]$$

$$256 = \left[13+5+2\sqrt{65}-a^2\right]\left\{a^2-[13+5-2\sqrt{65}]\right\}$$

$$256 = \left[2\sqrt{65}+(18-a^2)\right]\left[2\sqrt{65}-(18-a^2)\right]$$

$$256 = \left(2\sqrt{65}\right)^2 - (18-a^2)^2$$

So, $(18-a^2)^2 = 4(65) - 256$.
i.e., $(18-a^2)^2 = 260 - 256 = 4$.
So, $18 - a^2 = \pm 2$, taking the square root.
i.e., $a^2 = 18 \pm 2 = 20$ or 16.
So, $a = \sqrt{20}$ or $\sqrt{16}$.
i.e., $a = 2\sqrt{5}$ or 4.
Bhaskara II gave the solution as 4 and he did not give the other solution.

"Vargaprakriti" Equation and the "Chakravala" Method

Bhaskara II discovered a method of determining integral solutions of "Vargaprakriti" $Nx^2 + 1 = y^2$. He named his method "चक्रवाल" ("chakravala") or the cyclic method.

Bhaskara II's Method of Finding Integral Solutions of $Nx^2 + 1 = y^2$

First, find the integral solution $(a_0\ b_0)$ of the equation $Nx^2 + k = y^2$ for a suitable choice of k.

We now have the identity $N \cdot 1^2 + (m^2 - N) = m^2$, where m is an integer.

Then,

$$(a_0\ \ b_0\ \ k_0) \circ (1\ \ m\ \ m^2 - N)$$

$$\Rightarrow (a_0 m + b_0 \quad b_0 m + N a_0 \quad k(m^2 - N))$$

which means that $(am_0 + b_0\ b_0 m + Na_0)$ is a solution of $Nx^2 + k(m^2 - N) = y^2$.

We next choose m such that $am_0 + b_0$ is divisible by k such that $m^2 - N$ is small.

Bhaskara II's Theorem 1: Let

$$a_1 = \frac{a_0 m + b_0}{k}, \quad b_1 = \frac{b_0 m + N a_0}{k}, \quad \text{and } k_1 = \frac{m^2 - N}{k}$$

If a_1 is an integer, so are b_1 and k_1.

Bhaskara II did not provide any proof of this theorem.

The following proof was given by Drs. Bibhutibhushan Datta and Avadesh Narayana Singh.

Proof: We have

$$a_1 k = a_0 m + b_0 \tag{1}$$

and

$$b_1 k = b_0 m + N a_0 \tag{2}$$

Multiplying (1) by m, we get

$$a_1 km = a_0 m^2 + b_0 m \tag{3}$$

i.e., $a_1 km = a_0 m^2 + (b_1 k - N a_0)$ from (2).

So, $a_1 km - b_1 k = a_0 m^2 - N a_0$.

i.e., $k(a_1 m - b_1) = a_0 m^2 - N a_0 = a_0(m^2 - N)$.

We can conveniently take k prime to a_0.
In that case, $a_1m - b_1$ is divisible by a_0.
If

$$\frac{a_1m - b_1}{a_0} = k_1$$

then

$$\frac{a_1m - b_1}{a_0} = \frac{m^2 - N}{k} = k_1$$

Since $a_1m - b_1$ is divisible by a_0, k_1 is an integer.
Now,

$$b_1 = \frac{b_0m + Na_0}{k} = \frac{m(a_1k - a_0m) + Na_0}{k} \quad \text{from (1)}$$

$$= a_1m - a_0\left(\frac{m^2 - N}{k}\right)$$

$$= a_1m - a_0k_1 = \text{an integer}$$

since a_0, a_1, m, and k_1 are all integers.

Another proof of Bhaskara II's Theorem 1 has been given by German mathematician and historian Hermann Hankel.

Hankel's proof: Now, $(a_0\, b_0)$ is a solution of $Nx^2 + k = y^2$.
So,

$$Na_0^2 + k = b_0^2 \;\text{ or }\; k = b_0^2 - Na_0^2 \tag{1}$$

We have

$$a_1k = a_0m + b_0 \tag{2}$$

So, $a_1(b_0^2 - Na_0^2) = a_0m + b_0$ from (1).
i.e.,

$$a_1b_0^2 - b_0 = a_0m + Na_0^2a_1$$

$$b_0(a_1b_0 - 1) = a_0(m + Na_0a_1)$$

Since a_0 and b_0 can be chosen to be prime to each others, $a_1b_0 - 1$ is divisible by a_0.

Also,

$$\frac{a_1b_0 - 1}{a_0} = \frac{m + Na_0a_1}{b_0} = \text{an integer}$$

Since $a_1b_1 - 1$ is divisible by a_0, $a_1k = a_0m + b_0$ from (2) and

$$b_1k = b_0m + Na_0 \tag{3}$$

Multiplying (2) by b_0 and (3) by a_0, we get

$$a_1kb_0 = a_0b_0m + b_0^2 \text{ and } b_1ka_0 = a_0b_0m + Na_0^2$$

Subtracting, we get $a_1kb_0 - b_1ka_0 = b_0^2 - Na_0^2 = k$ from (1).
So, $k(a_1b_0 - b_1a_0) = k$.
Hence, $a_1b_0 - b_1a_0 = 1$ or $b_1a_0 = a_1b_0 - 1$.
i.e.,

$$b_1 = \frac{a_1b_0 - 1}{a_0} = \text{an integer}$$

Also from (2),

$$m = \frac{a_1k - b_0}{a_0}$$

So,

$$m^2 - N = \left(\frac{a_1k - b_0}{a_0}\right)^2 - N = \frac{(a_1k - b_0)^2 - Na_0^2}{a_0^2}$$

$$= \frac{(a_1k - b_0)^2 + k - b_0^2}{a_0^2} = \frac{a_1^2k^2 - 2a_1kb_0 + b_0^2 + k - b_0^2}{a_0^2}$$

$$= \frac{a_1^2k^2 - 2a_1kb_0 + k}{a_0^2} = \frac{k(a_1^2k - 2a_1b_0 + 1)}{a_0^2} = \text{an integer}$$

since m and N are integers.

Since a_0 and k are prime to each other, $a_1^2k - 2a_1b_0 + 1$ is divisible by a_0^2.

$$\frac{a_1^2k - 2a_1b_0 + 1}{a_0^2} \text{ is an integer}$$

So,

$$\frac{m^2 - N}{k} = \frac{a_1^2 k - 2a_1 b_0 + 1}{a_0^2} = k_1 = \text{an integer}$$

Bhaskara II's Theorem 2: If $(a_0\ b_0)$ is an integral solution of $Nx^2 + k = y^2$, after a finite number of repeated application of "chakravala" (cyclic process), we will arrive at the equation

$$Nx^2 + c = y^2$$

where

$$c = \pm 1 \text{ or } \pm 2 \text{ or } \pm 4$$

Bhaskara II did not provide the proof of this theorem either. A mathematician of his caliber certainly knew the truth of this theorem. A.A. Krishnaswamy Iyengar [77] has given a tedious and very lengthy proof of this second theorem of Bhaskara II. This long proof had not been given here for the sake of brevity. To illustrate the truth of the theorem of Bhaskara II, we will now give five examples of "Vargaprakriti," solved by this author and then give two examples, solved by Bhaskara II himself. Two examples solved by Bhaskara II's commentator. Narayana (AD 1350) will be given later in the book.

Example 1

Find an integral solution of $179x^2 + 1 = y^2$.

We note that $179 \cdot 1^2 - 10 = 169 = (13)^2$.

So, $(a_0\ b_0) = (1\ 13)$ is a solution of $179x^2 - 10 = y^2$.

Here, $N = 179$, $k_0 = -10$.

Now, we choose an integer m_0 such that

$$a_1 = \frac{a_0 m_0 + b_0}{k_0} = \frac{m_0 + 13}{-10}$$

is an integer such that

$$k_1 = \frac{m_0^2 - N}{k_0} = \frac{m_0^2 - 179}{-10}$$

is numerically small.

(*Continued*)

Example 1 (cont'd)

We choose $m_0 = 17$.
Then,

$$k_1 = \frac{(17)^2 - 179}{-10} = \frac{289 - 179}{-10} = \frac{110}{-10} = -11$$

$$a_1 = \frac{17 + 13}{-10} = \frac{30}{-10} = -3$$

$$b_1 = \frac{b_0 m_0 + a_0 N}{k_0} = \frac{13(17) + 179}{-10} = \frac{221 + 179}{-10} = \frac{400}{-10} = -40$$

So, (3 40) is a solution of $179x^2 - 11 = y^2$.
We choose an integer m_1 such that

$$a_2 = \frac{a_1 m_1 + b_1}{k_1} = \frac{3m_1 + 40}{-11}$$

is an integer such that

$$k_2 = \frac{m_1^2 - N}{k_1} = \frac{m_1^2 - 179}{-11}$$

is numerically small.
We choose $m_1 = 16$.
Then,

$$a_2 = \frac{3(16) + 40}{-11} = \frac{48 + 40}{-11} = \frac{88}{-11} = -8$$

$$b_2 = \frac{b_1 m_1 + a_1 N}{k_1} = \frac{40(16) + 3(179)}{-11} = \frac{640 + 537}{-11} = \frac{1177}{-11} = -107$$

$$k_2 = \frac{(16)^2 - 179}{-11} = \frac{256 - 179}{-11} = \frac{77}{-11} = -7$$

which means $(a_2\ b_2) = (8\ 107)$ is a solution of $179x^2 - 7 = y^2$.
We next choose an integer m_2 such that

$$a_3 = \frac{a_2 m_2 + b_2}{k_2} = \frac{8m_2 + 107}{-7}$$

is an integer such that

$$k_3 = \frac{m_2^2 - N}{k_2} = \frac{m_2^2 - 179}{-7}$$

is numerically small.

We choose $m_2 = 12$.

Then,

$$a_3 = \frac{8(12) + 107}{-7} = \frac{96 + 107}{-7} = \frac{-203}{7} = -29$$

$$b_3 = \frac{b_2 m_2 + a_2 N}{-7} = \frac{(107)(12) + 8(179)}{-7}$$

$$= \frac{1284 + 1432}{-7} = \frac{-2716}{7} = -388$$

$$k_3 = \frac{(12)^2 - 179}{-7} = \frac{144 - 179}{-7} = \frac{-35}{-7} = 5$$

So, $(a_3\ b_3) = (29\ 388)$ is a solution of $179x^2 + 5 = y^2$.

We next choose an integer m_3 such that

$$a_4 = \frac{a_3 m_3 + b_3}{k_3} = \frac{29m_3 + 388}{5}$$

is an integer such that

$$k_4 = \frac{m_3^2 - N}{k_3} = \frac{m_3^2 - 179}{5}$$

is numerically small.

We choose $m_3 = 13$.

Then,

$$k_4 = \frac{169 - 179}{5} = \frac{-10}{5} = -2$$

$$a_4 = \frac{29(13) + 388}{5} = \frac{377 + 388}{5} = \frac{765}{5} = 153$$

(*Continued*)

Example 1 (cont'd)

$$b_4 = \frac{b_3 m_3 + a_3 N}{k_3} = \frac{388(13) + 29(179)}{5} = \frac{5044 + 5191}{5}$$

$$= \frac{10235}{5} = 2047$$

So, $(\alpha\ \beta) = (a_4\ b_4) = (153\ 2047)$ is a solution of $179x^2 - 2 = y^2$.

Then, by Brahmagupta's Theorem 2, $(a\ b)$ is a solution of $179x^2 + 1 = y^2$, where

$$a = \alpha\beta = (153)(2047) = 313,191$$

$$b = \frac{\beta^2 + N\alpha^2}{2} = \frac{(2047)^2 + (179)(153)^2}{2}$$

$$= \frac{4,190,209 + 4,190,211}{2} = 4,190,210$$

Thus, $(a\ b) = (313{,}191\ 4{,}190{,}210)$ is an integral solution of $179x^2 + 1 = y^2$.

Infinite solutions of $179x^2 + 1 = y^2$ can be found by repeated samasa of (313,191 4,190,210 1) on itself.

Example 2

Determine the integral solution of $131x^2 + 1 = y^2$.

We have the identity $131 \cdot 1^2 - 10 = 121 = (11)^2$.

So, $(a_0\ b_0) = (1\ 11)$ is a solution of $131x^2 - 10 = y^2$.

Here, $N = 131$, $K_0 = -10$.

We choose an integer m_0 such that

$$a_1 = \frac{a_0 m + b_0}{K_0} = \frac{m_0 + 11}{-10}$$

is an integer such that

$$K_1 = \frac{m_0^2 - N}{K_0} = \frac{m_0^2 - 131}{-10}$$

is numerically small.

We choose $m_0 = 9$.
Then,

$$a_1 = \frac{9+11}{-10} = \frac{20}{-10} - 2$$

$$b_1 = \frac{b_0 m_0 + N a_0}{K_0} = \frac{(11)(9)+(131)(1)}{-10} = \frac{99+131}{-10} = \frac{230}{-10} = -23$$

$$K_1 = \frac{m_0^2 - N}{K_0} = \frac{(9)^2 - 131}{-10} = \frac{81-131}{-10} = \frac{-50}{-10} = 5$$

Then, (2 23) is a solution of $131x^2 + 5 = y^2$.
We choose an integer m_1 such that

$$a_2 = \frac{a_1 m_1 + b_1}{K_1} = \frac{2m_1 + 23}{5}$$

is an integer such that

$$K_2 = \frac{m_1^2 - N}{K_1} = \frac{m_1^2 - 131}{5}$$

is numerically small.
We choose $m_1 = 11$.
Then,

$$a_2 = \frac{2(11)+23}{5} = \frac{22+23}{5} = \frac{45}{5} = 9$$

$$b_2 = \frac{b_1 m_1 + N a_1}{K_1} = \frac{(23)(11)+(131)(2)}{5} = \frac{253+262}{5} = \frac{515}{5} = 103$$

$$K_2 = \frac{m_1^2 - 131}{5} = \frac{121-131}{5} = \frac{-10}{5} = -2$$

Then, $(\alpha\ \beta) = (a_2\ b_2) = (9\ 103)$ is an integral solution of the equation $131x^2 - 2 = y^2$.
Then, by Brahmagupta's Theorem 2, $(a\ b)$ is an integral solution of $131x^2 + 1 = y^2$, where

$$a = \alpha\beta = 9(103) = 927$$

(*Continued*)

Example 2 (cont'd)

$$b = \frac{\beta^2 + N\alpha^2}{2} = \frac{(103)^2 + (131)(81)}{2} = \frac{10,609 + 10,611}{2} = 10,610$$

Thus, $(a\ b) = (927\ 10{,}610)$ is an integral solution of $131x^2 + 1 = y^2$.

As before, infinite solutions of $131x^2 + 1 = y^2$ can be found by repeated samasa of (927 10,610 1) on itself.

Example 3

Find an integral solution of the equation $71x^2 + 1 = y^2$.

We observe that $71 \cdot 1^2 - 7 = 64 = (8)^2$.

So, $(a_0\ b_0) = (1\ 8)$ is a solution of $71x^2 - 7 = y^2$.

Here $N = 71$, $K_0 = -7$.

Choose an integer m_0 such that

$$a_1 = \frac{a_0 m_0 + b_0}{K_0} = \frac{m_0 + 8}{-7}$$

is an integer such that

$$K_1 = \frac{m_0^2 - N}{K_0} = \frac{m_0^2 - 71}{-7}$$

is numerically small.

We choose $m_0 = 6$.

Then,

$$a_1 = \frac{6+8}{-7} = \frac{14}{-7} = -2$$

$$b_1 = \frac{b_0 m_0 + a_0 N}{K_0} = \frac{(8)6 + 1.71}{-7} = \frac{48 + 71}{-7} = \frac{119}{-7} = -17$$

$$K_1 = \frac{(6)^2 - 71}{-7} = \frac{36 - 71}{-7} = \frac{-35}{-7} = 5$$

So, $(a_1\ b_1) = (2\ 17)$ is a solution of $71x^2 + 5 = y^2$. Choose an integer m_1 such that

$$a_2 = \frac{a_1 m_1 + b_1}{K_1} = \frac{2m_1 + 17}{5}$$

is an integer such that

$$K_2 = \frac{m_1^2 - N}{K_1} = \frac{m_1^2 - 71}{5}$$

is numerically small.

We choose $m_1 = 9$.

Then,

$$a_2 = \frac{2(9) + 17}{5} = \frac{18 + 17}{5} = \frac{35}{5} = 7$$

$$b_2 = \frac{b_1 m_1 + a_1 N}{K_1} = \frac{17(9) + 2(71)}{5} = \frac{153 + 142}{5} = \frac{295}{5} = 59$$

$$K_2 = \frac{(9)^2 - 71}{5} = \frac{81 - 71}{5} = \frac{10}{5} = 2$$

So, $(\alpha\ \beta) = (a_2\ b_2) = (7\ 59)$ is an integral solution of $71x^2 + 2 = y^2$.

Then, by Brahmagupta's Theorem 2, we get $(a\ b)$ is a solution of $71x^2 + 1 = y^2$, where $a = \alpha\beta = 7(59) = 413$ and

$$b = \frac{\beta^2 + N\alpha^2}{2} = \frac{(59)^2 + 71(7)^2}{2} = \frac{3481 + (71)(49)}{2}$$

$$= \frac{3481 + 3479}{2} = 3480$$

Thus, $(a\ b) = (413\ 3480)$ is an integral solution of $71x^2 + 1 = y^2$.

As before, infinite solutions of $71x^2 + 1 = y^2$ can be found by repeated samasa of (413 3480 1) on itself.

Example 4

Find an integral solution of $31x^2 + 1 = y^2$.

We observe that $31(3)^2 + 10 = (31)9 + 10 = 279 + 10 = 289 = (17^2)$.

So, $(a_0\ b_0) = (3\ 17)$ is a solution of $31x^2 + 10 = y^2$.

Here, $N = 31$, $K_0 = 10$.

(*Continued*)

Example 4 (cont'd)

Choose an integer m_0 such that

$$a_1 = \frac{a_0 m_0 + b_0}{K_0} = \frac{3m_0 + 17}{10}$$

is an integer such that

$$K_1 = \frac{m_0^2 - N}{K_0} = \frac{m_0^2 - 31}{10}$$

is numerically small.

We choose $m_0 = 1$.

Then,

$$a_1 = \frac{3 + 17}{10} = \frac{20}{10} = 2$$

$$b_1 = \frac{b_0 m_0 + N a_0}{K_0} = \frac{(17)(1) + (31)(3)}{10}$$

$$= \frac{17 + 93}{10} = \frac{110}{10} = 11$$

$$K_1 = \frac{(1)^2 - 31}{10} = \frac{-30}{10} = -3$$

So, $(a_1\ b_1) = (2\ 11)$ is a solution of $31x^2 - 3 = y^2$.

Choose an integer m_1 such that

$$a_2 = \frac{a_1 m_1 + b_1}{K_1} = \frac{2m_1 + 11}{-3}$$

is an integer such that

$$K_2 = \frac{m_1^2 - N}{K_1} = \frac{m_1^2 - 31}{-3}$$

is numerically small.

We choose $m_1 = 5$.

Then,

$$a_2 = \frac{2(5) + 11}{-3} = \frac{10 + 11}{-3} = \frac{21}{-3} = -7$$

$$b_2 = \frac{b_1 m_1 + N a_1}{K_1} = \frac{11(5) + 31(2)}{-3} = \frac{55 + 62}{-3} = \frac{117}{-3} = -39$$

$$K_2 = \frac{m_1^2 - 31}{-3} = \frac{25 - 31}{-3} = \frac{-6}{-3} = 2$$

Then, $(\alpha\ \beta) = (a_2\ b_2) = (7\ 39)$ is a solution of $31x^2 + 2 = y^2$.

So, by Brahmagupta's Theorem 2, $(a\ b)$ is a solution of $31x^2 + 1 = y^2$, where

$$a = \alpha\beta = 7(39) = 273$$

$$b = \frac{\beta^2 + N\alpha^2}{2} = \frac{(39)^2 + 31(7)^2}{2} = \frac{1521 + (31)(49)}{2}$$

$$= \frac{1521 + 1519}{2} = 1520$$

Thus, $(a\ b) = (273\ 1520)$ is an integral solution of $31x^2 + 1 = y^2$.

As before, infinite solutions of $31x^2 + 1 = y^2$ can be found by repeated samasa of (273 1520 1) on itself.

Example 5

Find an integral solution of $231x^2 + 1 = y^2$. We have the identity $231 \cdot 1^2 - 6 = 225 = (15)^2$.

So, $(a_0\ b_0) = (1\ 15)$ is a solution of $231x^2 - 6 = y^2$.

Here, $N = 231$, $K_0 = -6$.

We choose an integer m_0 such that

$$a_1 = \frac{a_0 m_0 + b_0}{K_0} = \frac{m_0 + 15}{-6}$$

is an integer such that

$$K_1 = \frac{m_0^2 - N}{K_0} = \frac{m_0^2 - 231}{K_0}$$

is numerically small.

(*Continued*)

Example 5 (cont'd)

We choose $m_0 = 15$.
Then,

$$a_1 = \frac{a_0 m_0 + b_0}{K_0} = \frac{15 + 15}{-6} = \frac{30}{-6} = -5$$

$$b_1 = \frac{b_0 m_0 + N a_0}{K_0} = \frac{(15)(15) + (231)}{-6}$$

$$= \frac{225 + 231}{-6} = \frac{456}{-6} = -76$$

$$K_1 = \frac{m_0^2 - N}{K_0} = \frac{(15)^2 - 231}{-6} = \frac{225 - 231}{-6} = \frac{-6}{-6} = 1$$

So, $(a_1\ b_1) = (5\ 76)$ is a solution of $231x^2 + 1 = y^2$.

As before, infinite solution of $231x^2 + 1 = y^2$ can be found by repeated samasa of (5 76 1) on itself.

Example 1 of Bhaskara II

Find an integral solution of $61x^2 + 1 = y^2$.

This is one of the examples given by Bhaskara II in his "Bijaganitha" (algebra) in AD 1150, which is of great historical interest. This problem was proposed by French mathematician Pierre De Fermat (AD 1601–1605) to Frenicle in a letter of February AD 1657 as a challenge to Frenicle and other fellow mathematicians of his time. Euler solved the problem in AD 1732.

We have the identity $61 \cdot 1^2 + 3 = 64 = (8)^2$.

So, $(a_0\ b_0) = (1\ 8)$ is a solution of $61x^2 + 3 = y^2$.

Here, $N = 61$, $K_0 = 3$.

We choose an integer m_0 such that

$$a_1 = \frac{a_0 m_0 + b_0}{K_0} = \frac{m_0 + 8}{3}$$

is an integer such that

$$K_1 = \frac{m_0^2 - N}{K_0} = \frac{m_0^2 - 61}{3}$$

is numerically small.

We choose $m_0 = 7$.

Then,

$$a_1 = \frac{7+8}{3} = \frac{15}{3} = 5$$

$$b_1 = \frac{b_0 m_0 + N a_0}{K_0} = \frac{(8)(7)+(61)(1)}{3} = \frac{56+61}{3} = \frac{117}{3} = 39$$

$$K_1 = \frac{m_0^2 - N}{K_0} = \frac{(7)^2 - 61}{3} = \frac{49-61}{3} = \frac{-12}{3} = -4$$

Then, $(\alpha\ \beta) = (a_1\ b_1) = (5\ 39)$ is a solution of the equation $61x^2 - 4 = y^2$.

Then, by Brahmagupta's Theorem 4, $(a\ b)$ is an integral solution of $61x^2 + 4 = y^2$, where

$$a = \frac{\alpha\beta}{2}(\beta^2+1)(\beta^2+3) = \frac{5(39)}{2}\left[\left(39^2+1\right)\left(39^2+3\right)\right]$$

$$= \frac{195}{2}[(1521+1)(1521+3)]$$

$$= \frac{195}{2}(1522)(1524) = 195(761)(1524)$$

$$= 195(1,161,764) = 226,543,980$$

$$b = \frac{(\beta^2+2)}{2}\left[(\beta^2+1)(\beta^2+3) - 2\right]$$

$$= \left(\frac{(39)^2+2}{2}\right)\left[(39^2+1)(39^2+3) - 2\right]$$

$$= \frac{1521+2}{2}[(1521+1)(1521+3) - 2]$$

$$= \frac{1523}{2}[(1522)(1524) - 2]$$

$$= 1523[(761)(1524) - 1] = (1523)(1,161,764 - 1)$$

$$= (1523)(1,161,763) = 1,766,319,049$$

Thus, $(a\ b) = (226{,}543{,}980\ 1{,}766{,}319{,}049)$ is an integral solution.

As before, infinite solutions of $61x^2 + 1 = y^2$ can be found by repeated samasa of (226,543,980 1,766,319,049 1) on itself.

Example 2 of Bhaskara II

Find an integral solution of $67x^2 + 1 = y^2$. We have the identity $67 \cdot 1^2 - 3 = 64 = (8)^2$. Hence, $(a_0\ b_0) = (1\ 8)$ is a solution of $67x^2 - 3 = y^2$.

Here, $N = 67$, $K_0 = -3$.

We choose an integer m_0 such that

$$a_1 = \frac{a_0 m_0 + b_0}{K_0} = \frac{m_0 + 8}{-3}$$

is an integer such that

$$K_1 = \frac{m_0^2 - N}{K_0} = \frac{m_0^2 - 67}{-3}$$

is numerically small.

We choose $m_0 = 7$.

Then,

$$a_1 = \frac{a_0 m_0 + b_0}{K_0} = \frac{7 + 8}{-3} = \frac{15}{-3} = -5$$

$$b_1 = \frac{b_0 m_0 + N a_0}{K_0} = \frac{(8)(7) + (67)(1)}{-3} = \frac{56 + 67}{-3} = \frac{123}{-3} = -41$$

$$K_1 = \frac{m_0^2 - N}{K_0} = \frac{49 - 67}{-3} = \frac{-18}{-3} = 6$$

So, $(a_1\ b_1) = (5\ 41)$ is a solution of $67x^2 + 6 = y^2$.

We choose an integer m_1 such that

$$a_2 = \frac{a_1 m_1 + b_1}{K_1} = \frac{5m_1 + 41}{6}$$

is an integer such that

$$K_2 = \frac{m_1^2 - N}{K_1} = \frac{m_1^2 - 67}{6}$$

is numerically small.

We choose $m_1 = 5$.

Then,

$$a_2 = \frac{5(5) + 41}{6} = \frac{25 + 41}{6} = \frac{66}{6} = 11$$

$$b_2 = \frac{b_1 m_1 + Na_1}{K_1} = \frac{(41)(5) + (67)(5)}{6} = \frac{205 + 335}{6} = \frac{540}{6} = 90$$

$$K_2 = \frac{m_1^2 - 67}{6} = \frac{25 - 67}{6} = \frac{-42}{6} = -7$$

So, $(a_2\, b_2) = (11\ 90)$ is a solution of $61x^2 - 7 = y^2$.

We next choose an integer m_2 such that

$$a_3 = \frac{a_2 m_2 + b_2}{K_2} = \frac{11m_2 + 90}{-7}$$

is an integer such that

$$K_3 = \frac{m_2^2 - N}{K_2} = \frac{m_2^2 - 67}{-7}$$

is numerically small.

We choose $m_2 = 9$.

Then,

$$a_3 = \frac{11(9) + 90}{-7} = \frac{99 + 90}{-7} = \frac{189}{-7} = -27$$

$$b_3 = \frac{b_2 m_2 + Na_2}{K_2} = \frac{(90)(9) + (67)(11)}{-7} = \frac{810 + 737}{-7} = \frac{1547}{-7} = -221$$

$$K_3 = \frac{m_2^2 - 67}{-7} = \frac{(9)^2 - 67}{-7} = \frac{81 - 67}{-7} = \frac{14}{-7} = -2$$

So, $(\alpha\ \beta) = (a_3\, b_3) = (27\ 221)$ is an integral solution of the equation $67x^2 - 2 = y^2$. Then, by Brahmagupta's Theorem 2, $(a\ b)$ is a solution of the equation $67x^2 + 1 = y^2$, where

$$a = \alpha\beta = 27(221) = 5967$$

$$b = \frac{\beta^2 + N\alpha^2}{2} = \frac{(221)^2 + 67(27)^2}{2} = \frac{48,841 + (67)(729)}{2}$$

$$= \frac{48,841 + 48,843}{2} = 48,842$$

Thus, $(a\ b) = (5967\ 48,842)$ is an integral solution of $67x^2 + 1 = y^2$.

As before, infinite solutions of $67x^2 + 1 = y^2$ can be found by repeated samasa of (5767 48,862 1) on itself.

Remark: A.A. Krishnaswamy Iyengar has shown in his earlier mentioned paper that Bhaskara II's "chakravala" (cyclic method) is much shorter than the Modern Euler–Lagrange method of employing continued fractions to solve "Vargaprakriti" equation.

Rational Solutions of "Vargaprakriti" Equation

Bhaskara II has pointed out the equation $Nx^2 - 1 = y^2$ has no rational solution, unless N is the sum of squares of rational numbers. In fact, his assertion is true more generally for the equation $Nx^2 - k^2 = y^2$.

Theorem: A rational solution of the equation $Nx^2 - K^2 = y^2$ is impossible unless N is the sum of squares of rational numbers.

Proof: Suppose

$$x = \frac{a_1}{b_1} \text{ and } y_2 = \frac{a_2}{b_2}$$

is a rational solution of the equation $Nx^2 - K^2 = y^2$.

Then,

$$N\left(\frac{a_1}{b_1}\right)^2 - K^2 = \left(\frac{a_2}{b_2}\right)^2$$

$$N\left(\frac{a_1}{b_1}\right)^2 = K^2 + \left(\frac{a_2}{b_2}\right)^2$$

Dividing by $(a_1/b_1)^2$, we get

$$N = K^2\left(\frac{b_1}{a_1}\right)^2 + \left(\frac{a_2}{b_2}\right)^2\left(\frac{b_1}{a_1}\right)^2$$

i.e.,

$$N = \left(\frac{b_1K}{a_1}\right)^2 + \left(\frac{a_2b_1}{b_2a_1}\right)^2 = p^2 + q^2$$

where p and q are rational numbers.

Theorem (Bhaskara II): Let $(\alpha\ \beta)$ be a rational solution of the equation $Nx^2 \pm c = y^2$. If $(\alpha^1\ \beta^1)$ is a rational solution of the equation $Nx^2 + 1 = y^2$, then $(\alpha\beta^1 \pm \beta\alpha^1\ \beta\beta^1 \pm N\alpha\alpha^1)$ is a rational solution of $Nx^2 \pm c = y^2$.

Proof: $(\alpha\ \beta)$ is a solution of $Nx^2 \pm c = y^2 \Rightarrow (\alpha\ \beta \pm c)$.
$(\alpha^1\ \beta^1)$ is a solution of $Nx^1 \pm 1 = y^2 \Rightarrow (\alpha^1\ \beta^1\ 1)$.
Then,

$$(\alpha\ \beta\ \pm c)\circ(\alpha^1\ \beta^1\ 1) = (\alpha\alpha^1\ \pm \beta\alpha^1\ \beta\beta^1 \pm N\alpha\alpha^1\ \pm c)$$

$$\Rightarrow (\alpha\beta^1 \pm \beta\alpha^1\ \beta\beta^1 \pm N\alpha\alpha^1) \text{ is a solution of } Nx^2 \pm c = y^2.$$

Theorem (Bhaskara II): Two rational solutions of $Nx^2 - 1^2 = y^2$, where $N = (m^2 + n^2)$ will be

$$\left(\frac{1}{m}\ \frac{n}{m}\right) \text{ and } \left(\frac{1}{n}\ \frac{m}{n}\right)$$

Proof: Let

$$\left(\frac{k}{m}\ \beta\right)$$

be a rational solution of $Nx^2 - k^2 = y^2$, where $N = m^2 + n^2$.
Then,

$$(m^2 + n^2)\left(\frac{k}{m}\right)^2 - k^2 = \beta^2 \Rightarrow k^2 + \frac{n^2k^2}{m^2} - k^2 = \beta^2$$

So,

$$\beta^2 = \frac{n^2k^2}{m^2} \Rightarrow \beta = \frac{nk}{m}$$

$$\Rightarrow \left(\frac{k}{m}\ \frac{nk}{m}\right)$$

is a solution of

$$Nx^2 - k^2 = y^2 \Rightarrow N\frac{k^2}{m^2} - k^2 = \frac{n^2k^2}{m^2}$$

i.e.,

$$\frac{N}{m^2} - 1 = \frac{n^2}{m^2}$$

Hence,

$$\left(\frac{1}{m}\ \frac{n}{m}\right)$$

is a solution of $Nx^2 - 1 = y^2$. Similarly, if

$$\left(\frac{k}{n}\ \beta\right)$$

be a rational solution of $Nx^2 - k^2 = y^2$, where $N = (m^2 + n^2)$, we get

$$(m^2 + n^2)\frac{k^2}{n^2} - k^2 = \beta^2$$

i.e.,

$$\frac{k^2m^2}{n^2} + \frac{k^2n^2}{n^2} - k^2 = \beta^2 \Rightarrow \beta^2 = \frac{k^2m^2}{n^2}$$

$$\Rightarrow \beta = \frac{km}{n}$$

So,

$$\left(\frac{k}{n}\ \frac{km}{n}\right)$$

is a solution of $Nx^2 - k^2 = y^2$.

$$N\frac{k^2}{n^2} - k^2 = \frac{k^2m^2}{n^2} \Rightarrow N \cdot \frac{1}{n^2} - 1 = \frac{m^2}{n^2}$$

$$\left(\frac{1}{n}\ \frac{m}{n}\right)$$

is a rational solution of $Nx^2 - 1 = y^2$, if $N = m^2 + n^2$.

Related Equations of Brahmagupta–Bhaskara Equation

Theorem (Bhaskara II): If $(\alpha\ \beta)$ is a solution of $a^2x + bx + c = y^2$, then

$$\alpha = \frac{-b}{2a} + \frac{1}{a}(mq \pm alp)$$

and $\beta = lq \pm mp$, where $ap^2 + 1 = q^2$ and $al^2 + k = m^2$, where

$$k = \frac{1}{4}(b^2 - 4ac)$$

Proof: Now, $ax^2 + bx + c = y^2$.

Multiplying by a, we get

$$a^2x^2 + abx + ac = ay^2$$

$$\left(ax + \frac{b}{2}\right)^2 - \frac{b^2}{4} + ac = ay^2$$

So,

$$\left(ax + \frac{b}{2}\right)^2 = ay^2 + \frac{1}{4}(b^2 - 4ac)$$

Let

$$ax + \frac{b}{2} = z \text{ and } \frac{1}{4}(b^2 - 4ac) = k$$

Then, we get

$$z^2 = ay^2 + k \tag{1}$$

First, we will prove the following lemma.

Lemma: If $(y\ z) = (l\ m)$ be a solution of the (1) namely, $ay^2 + k = z^2$, then another solution of (1) is $(y\ z) = (lq \pm mp\ mq \pm alp)$, where $ap^2 + 1 = q^2$.

Proof: If $(l\ m)$ be a solution of $ay^2 + k = z^2$, then

$$al^2 + k = m^2 \Rightarrow k = m^2 - al^2$$

Let $(lq \pm mp\ \beta_1)$ be a solution of $ay^2 + k = z^2$, then

$$a(lq \pm mp)^2 + k = \beta_1^2$$

$$a(l^2q^2 \pm 2lqmp + m^2p^2) + k = \beta_1^2$$

i.e., $a(l^2q^2 \pm 2lqmp + m^2p^2) + m^2 - al^2 = \beta_1^2$
i.e., $al^2(q^2 - 1) \pm 2alqmp + m^2(ap^2 + 1) = \beta_1^2$
i.e., $al^2(ap^2) \pm 2alqmp + m^2q^2 = \beta_1^2$

$$\Rightarrow a^2l^2p^2 \pm 2alqmp + m^2q^2 = \beta_1^2$$

So, $\beta_1^2 = (alp \pm mq)^2 \Rightarrow \beta_1 = alp \pm mq \Rightarrow (lq \pm mp \;\; mq \pm alp)$ is a solution of $ay^2 + k = z^2$. This proves the lemma.

Now,

$$z = ax + \frac{b}{2} = mq \pm alp \Rightarrow ax = mq \pm alp - \frac{b}{2}$$

So,

$$x = \frac{-b}{2a} + \frac{1}{a}(mq \pm alp) \text{ and } y = lq \pm mp$$

Thus, if $(\alpha \; \beta)$ is a solution of $ax^2 + bx + c = y^2$, then

$$\alpha = \frac{-b}{2a} + \frac{1}{a}(mq \pm alp) \text{ and } \beta = lq \pm mp$$

Corollary:

$$\alpha^1 = \frac{1}{2a}(bq - b) + qr \pm lp$$

$$\beta^1 = lq \pm \left(apr + \frac{1}{2}bp\right)$$

is also a solution of $ax^2 + bx + c = y^2$, where $ap^2 + 1 = q^2$ and $l^2 = ar^2 + br + c$.

Proof: Suppose $x = r$ and $z = m$.

Since

$$z = ax + \frac{b}{2} \Rightarrow ar + \frac{b}{2} = m$$

From the previous theorem if $(\alpha \; \beta)$ be a solution of $ax^2 + bx + c = y^2$, then we have

$$\alpha = -\frac{b}{2a} + \frac{1}{a}(mq \pm alp)$$

$$\beta = lq \pm mp \text{ and } m = ar + \frac{b}{2}$$

So,

$$\beta = lq \pm \left(apr + \frac{1}{2}bp\right)$$

So,

$$\alpha = -\frac{b}{2a} + \frac{1}{a}\left(q\left(ar + \frac{b}{2}\right) \pm alp\right)$$

$$\alpha = -\frac{b}{2a} + rq + \frac{bq}{2a} \pm lp = \frac{1}{2a}(bq - b) + qr \pm lp$$

and

$$\beta = lq \pm \left(apr + \frac{1}{2}bp\right)$$

Also since, $y = l$, $l^2 = ar^2 + br + c$, and $x = r$.

Remark: Having one solution of $ax^2 + bx + c = y^2$, we can find infinite number of solutions, by finding the infinite solutions of $ap^2 + 1 = q^2$.

Theorem (Bhaskara II): The solution $(x\, y)$ of the equation $ax^2 + bx + c = a'y^2 + b'y + c'$ is given by

$$x = \frac{-b}{2a}(lq \pm mp)$$

$$y = \frac{-b'}{2a'} + \frac{1}{aa'}(mq \pm aa'lp)$$

where $a'ap^2 + 1 = q^2$ and $aa'l^2 - \beta = m^2$, if

$$\beta = a^2a'c' + \left(\frac{1}{4}b^2 - ac\right)aa' - \frac{1}{2}(ab')^2$$

Proof:

$$ax^2 + bx + c = a'y^2 + b'y + c'$$

Multiplying by a, we get

$$a^2x^2 + abx + ac = aa'y^2 + ab'y + ac'$$

$$\left(ax + \frac{b}{2}\right)^2 - \frac{b^2}{4} + ac = aa'y^2 + ab'y + ac'$$

$$\left(ax + \frac{b}{2}\right)^2 = aa'y^2 + ab'y + ac' + \left(\frac{b^2}{4} - ac\right)$$

Let

$$z = ax + \frac{b}{2}$$

This becomes

$$z^2 = aa'y^2 + ab'y + ac' + \left(\frac{b^2}{4} - ac\right)$$

Multiplying by aa', we get

$$aa'z^2 = a^2a'^2y^2 + a^2a'b'y + a^2c'a' + aa'\left(\frac{b^2}{4} - ac\right)$$

$$aa'z^2 = \left(aa'y + \frac{1}{2}ab'\right)^2 - \frac{a^2b'^2}{4} + a^2c'a' + aa'\left(\frac{b^2}{4} - ac\right)$$

$$aa'z^2 = w^2 + \beta, \quad \text{where } w = aa'y + \frac{1}{2}ab' \text{ and}$$

$$\beta = a^2a'c' + aa'\left(\frac{h^2}{4} - ac\right) - \left(\frac{1}{2}ab'\right)^2$$

So,

$$w^2 = aa^1z^2 - \beta \tag{1}$$

If $z = l$, $w = m$ be a solution of (1), another solution of (1) will be

$$z = lq \pm mp = ax + \frac{b}{2} \tag{2}$$

$$w = mq \pm aa'lp = aa'y + \frac{1}{2}ab' \tag{3}$$

where $aa'p^2 + 1 = q^2$.

So,

$$ax = lq \pm mp - \frac{b}{2} \Rightarrow x = \frac{-b}{2a} + \frac{1}{a}(lq \pm mp) \quad \text{from (2)}$$

and

$$aa'y = w - \frac{1}{2}ab' = mq \pm aa'lp - \frac{1}{2}ab' \quad \text{from (3)}$$

So,

$$y = \frac{1}{aa'}[mq + aa'lp] - \frac{1}{2}\frac{b'}{a'}$$

Corollary: If $x = r$, $y = s$, then

$$z = ax + \frac{b}{2} \Rightarrow ar + \frac{b}{2} = z = l \quad \text{(4) from (2)}$$

$$w = aa'y + \frac{1}{2}ab' = aa's + \frac{1}{2}ab' = m \quad \text{(5) from (3)}$$

But

$$x = -\frac{b}{2a} + \frac{1}{a}(lq + mp)$$

Then,

$$x = \frac{-b}{2a} + \frac{1}{a}\left[\left(ar + \frac{b}{2}\right)q \pm p\left(aa's + \frac{1}{2}ab'\right)\right]$$

So,

$$x = \frac{1}{2a}[bq \pm pb'a - b] + rq \pm pa's \tag{6}$$

Also,

$$y = \frac{1}{aa'}\left[w - \frac{ab'}{2}\right] = \frac{1}{aa'}[mq \pm aa'lp] - \frac{b'}{2a'}$$

$$y = \frac{1}{aa'}\left[q\left(aa's + \frac{1}{2}ab'\right) \pm aa'p\left(ar + \frac{b}{2}\right)\right] - \frac{1}{2}\frac{b'}{a'} \quad \text{from (3)}$$

So,

$$y = \frac{1}{2a'}(qb' \pm pba' - b') + qs \pm par \tag{7}$$

where

$$aa'p^2 + 1 = q^2$$

Since $x = r$, $y = s$, we get $ar^2 + br + c = a's^2 + b's^2 + b's + c$ from given equation.

Equations (6) and (7) demonstrate that, having found one solution, Bhaskara II could find an infinite number of solutions of the given equation, $ax^2 + bx + c = a'y^2 + b'y + c'$, by finding infinite solutions of $aa'p^2 + 1 = q^2$.

Example 1 (Bhaskara II)

Find a rational solution of $6x^2 + 2x = y^2$.

$$6x^2 + 2x = y^2$$

Multiplying by 6, we get $36x^2 + 12x = 6y^2$, i.e.,

$$(6x+1)^2 - 1 = 6y^2$$

Let $z = 6x + 1$, then we have

$$6y^2 + 1 = z^2 \tag{1}$$

We use Brahmagupta–Bhaskara method to solve (1).

We have the identity $6 \cdot 1^2 - 2 = 4 = 2^2$.

So, $(\alpha\ \beta) = (1\ 2)$ is a solution of $6y^2 - 2 = z^2$.

Then, by Brahmagupta's Theorem 2, $(y\ z) = (l\ m)$ is a solution of $6y^2 + 1 = z^2$, where $l = \alpha\beta = (1)(2) = 2$.

$$m = \frac{\beta^2 + N\alpha^2}{2} = \frac{2^2 + 6 \cdot 1^2}{2} = \frac{4+6}{2} = 5$$

So, $(l\ m) = (2\ 5)$ is a solution of $6y^2 + 1 = z^2$.

Then, we have from page 428

$$x = \frac{-b}{2a} + \frac{1}{a}(mq \pm alp)$$

$$y = +lq \pm mp$$

where x, y are solutions of $ax^2 + bx + c = y^2$, if l and m are solutions of $ay^2 + k = z^2$ and $ap^2 + 1 = q^2$.

Here,

$$a = 6,\ b = 2,\ c = 0$$

Thus,

$$x = \frac{-2}{12} + \frac{1}{6}(5q \pm 6 \cdot 2p) \text{ and } y = +2q \pm 5p$$

Thus,

$$x = \frac{1}{6}[-1 + 5q \pm 12p] \text{ and } y = 2q \pm 5p$$

where $6p^2 + 1 = q^2$ are the general solutions of the given equation.
If $p = 0$, $q = 1$, then

$$x = \frac{1}{6}[-1 + 5] = \frac{4}{6} = \frac{2}{3} \text{ and } y = 2(1) + 0 = 2$$

So,

$$(x\ y) = \left(\frac{2}{3}\ 2\right)$$

is one rational solutions of the given equation, the infinite solutions can be obtained by different values of p and q satisfying $6p^2 + 1 = q^2$. We observe that if $p = 2$, $q = 5$, infinite solutions of $6p^2 + 1 = q^2$ can be found by repeated samasa of (2 5 1) on itself.

Example 2

Find the general solution of the equation $16x^2 + 80x + 69 = 9y^2 - 24y$.

$$4(4x^2 + 20x) + 69 = (3y-4)^2 - 16$$

$$4[(2x+5)^2 - 25] + 69 = (3y-4)^2 - 16$$

Let $2x + 5 = z$ and $3y - 4 = w$.
Then, we have, $4z^2 - 100 + 69 + 16 = w^2$, i.e.,

$$4z^2 - 15 = w^2 \qquad (1)$$

Now, we solve (1) by Brahmagupta–Bhaskara II method.

(Continued)

Example 2 (cont'd)

We observe that $4 \cdot 1^2 - 3 = 1^2$.
So, (1 1) is a solution of $4z^2 - 3 = w^2 \Rightarrow (1\ 1\ -3)$.
Again we observe that

$$4 \cdot 1^2 + 5 = 3^2$$

Hence, (1 3) is a solution of $4z^2 + 5 = w^2 \Rightarrow (1\ 3\ 5)$.

$$(1\ 1\ -3) \circ (1\ 3\ 5)$$

$$= [\,1 \cdot 3 + 1 \cdot 1 \qquad 1 \cdot 3 + 4 \cdot 1 \cdot 1 \qquad (-3)\,(5)\,] = (4\ 7\ -15)$$

which means that (4 7) is a solution of $4z^2 - 15 = w^2$.

$$z = 4 = 2x + 5 = 4 \Rightarrow 2x = -1 \Rightarrow x = -\frac{1}{2} = r$$

$$w = 7 \Rightarrow 3y - 4 = 7 \Rightarrow 3y = 11 \Rightarrow y = \frac{11}{3} = s$$

We know from the previous theorem,

$$x = \frac{1}{2a}(bq \pm pab' - b) + rq \pm pa's$$

Here,

$$a = 16,\ b = 80,\ c = 69,\ a' = 9,\ b' = -24,\ c' = 0$$

So,

$$x = \frac{1}{32}[80q \pm p \cdot 16(-24) - 80] - \frac{1}{2}q \pm p(9)\left(\frac{11}{3}\right)$$

$$= \frac{1}{32}[80q \pm 384p - 80] - \frac{1}{2}q \pm 33p$$

$$x = \frac{5}{2}q \pm 12p - \frac{5}{2} - \frac{1}{2}q \pm 33p$$

$$x = 2q \pm 45p - \frac{5}{2} \qquad (2)$$

Also,

$$y = \frac{1}{2a'}[qb' \pm pba' - b'] + qs \pm par$$

$$= \frac{1}{2(9)}[q(-24) \pm p \cdot 80.9 + 24] + q \cdot \left(\frac{11}{3}\right) \pm p(16)\left(-\frac{1}{2}\right)$$

$$y = -\frac{4}{3}q \pm 40p + \frac{4}{3} + \frac{11q}{3} \pm 8p = \frac{7q}{3} \pm 48p + \frac{4}{3} \tag{3}$$

where $aa'p^2 + 1 = q^2 \Rightarrow (16)9p^2 + 1 = q^2$, i.e., $144p^2 + 1 = q^2$, if

$$p^2 = \frac{1}{16}$$

Here,

$$ar^2 + br + c = 16\left(\frac{1}{4}\right) + 80\left(-\frac{1}{2}\right) + 69 = 4 - 40 + 69 = 33$$

$$a's^2 + b's + c' = 9\left(\frac{121}{9}\right) - 24\left(\frac{11}{3}\right) + 0 = 121 - 88 = 33$$

We observe that if $p = 0$, $q = 1$, then

$$x = 2 - \frac{5}{2} = -\frac{1}{2} \text{ and } y = \frac{7}{3} + \frac{4}{3} = \frac{11}{3}$$

If

$$p^2 = \frac{1}{16}, \quad q^2 = 1 + 144p^2 = 1 + \frac{144}{256} = 1 + \frac{9}{16} = \frac{25}{16} \Rightarrow q = \frac{5}{4}$$

Then,

$$x = 2 \cdot \frac{5}{4} \pm 45 \cdot \frac{1}{16} - \frac{5}{2} = \frac{40 \pm 45 - 40}{16} = \pm \frac{45}{16}$$

$$y = \frac{7}{3} \cdot \frac{5}{4} \pm 48\frac{1}{16} + \frac{4}{3} = \frac{140 \pm 48 \pm 64}{48} = \frac{21}{4} \text{ or } \frac{13}{4}$$

Hence, repeated samasa of $(\alpha\ \beta\ 1)$ over itself gives the infinite solutions of $144p^2 + 1 = q^2$, where $144\alpha^2 + 1 = \beta^2$.

Theorem (Bhaskara II): Find the rational solutions of the equation

$$ax^2 + by^2 + c = z^2 \tag{1}$$

Bhaskara II gave four different methods of solving (1).

First method:

Let $x = my$.

Then, we have

$$am^2y^2 + by^2 + c = z^2.$$

$$(am^2 + b)y^2 + c = z^2$$

$$Ay^2 + c = z^2$$

where $A = am^2 + b$.

If $(l'\ m')$ is a solution of $Ay^2 + c = z^2$, then another solution of this equation $Ay^2 + c = z^2$ is given by $x = l'q \pm m'p$; $y = m'q \pm Al'p$, where $Ap^2 + 1 = q^2$ and $Al'^2 + c = m'^2$.

Second method:

Let $x = my \pm n$.

Then, (1) becomes

$$a(my \pm n)^2 + by^2 + c = z^2$$

$$a(m^2y^2 \pm 2myn + n^2) + by^2 + c = z^2$$

i.e.,

$$(am^2 + b)y^2 \pm 2amyn + an^2 + c = z^2$$

$$Ay^2 \pm 2amyn + B = z^2$$

where

$$A = am^2 + b; \quad B = an^2 + c$$

Multiplying by A, we get

$$A^2y^2 \pm 2amynA + AB = AZ^2$$

$$(Ay \pm amn)^2 - a^2m^2n^2 + AB = Az^2$$

Let $w = Ay \pm amn$. Then, we have

$Az^2 - c = w^2$, where $c = -a^2m^2n^2 + AB$.

If $(r\ s)$ is a solution of $Az^2 - c = w^2$, another solution of the equation $Az^2 - c = w^2$ will be

$$z = sq \pm rp; \quad w = rq \pm Asp$$

where $q^2 = Ap^2 + 1$.

Hence, the solution $(x\ y)$ of $ax^2 + by^2 + c = z^2$ is given as follows.

Now, $w = rq \pm Asp = Ay \pm amn$.

So,

$$Ay = rq \pm Asp \pm amn \Rightarrow y = \frac{1}{A[rq \pm Asp \pm amn]} \tag{2}$$

$$x = my \pm n = \frac{m}{A}[rq \pm Asp \pm amn] \pm n \tag{3}$$

where

$$q^2 = Ap^2 + 1, \;\; s^2 = Ar^2 - c, \;\; A = am^2 + b \text{ and } c = AB - a^2m^2n^2$$

So, $c = A(an^2 + c) - a^2m^2n^2$.

Now, $c = (am^2 + b)(an^2 + c) - a^2m^2n^2$

$$= a^2m^2n^2 + a(m^2c + n^2b) + bc - a^2m^2n^2$$

$$= a(m^2c + n^2b) + bc$$

Now, (2) and (3) give the general solution of (1).

Third method:

Let $w^2 = by^2 + c$.

Then, the equation $ax^2 + by^2 + c = z^2$ becomes

$$ax^2 + w^2 = z^2 \text{ or } ax^2 = z^2 - w^2 = (z + w)(z - w)$$

Let

$$z + w = \frac{ax}{m}$$

Then,

$$z - w = \frac{z^2 - w^2}{z + w} = \frac{ax^2}{(ax/m)} = mx$$

Adding, we get

$$2z = \frac{ax}{m} + mx \Rightarrow z = \frac{1}{2}\left(\frac{ax}{m} + mx\right)$$

Subtracting the first equation from the second equation, we get

$$2w = \frac{ax}{m} - mx = x\left(\frac{a}{m} - m\right)$$

Thus,

$$x = \frac{2w}{(a/m) - m} = \frac{2wm}{a - m^2}$$

and

$$z = \frac{1}{2}\left(\frac{ax}{m} + mx\right) = \frac{x}{2}\left(\frac{a}{m} + m\right) = \frac{x(a + m^2)}{2m}$$

So,

$$z = \frac{a + m^2}{2m} \cdot \frac{2wm}{a - m^2} = \frac{a + m^2}{a - m^2}w$$

If $(r\ s)$ be a solution of $w^2 = by^2 + c$ (*), then another solution of (*) will be $y = rq \pm sp$; $w = sq \pm brp$, where $q^2 = bp^2 + 1$.

So,

$$x = \frac{2wm}{a - m^2} = \frac{2m}{a - m^2}(sq \pm brp)$$

$$z = \left(\frac{a + m^2}{a - m^2}\right)(sq \pm brp)$$

where $q^2 = bp^2 + 1$ and $br^2 + c = s^2$.

Fourth method:

Suppose $c = 0$.

Then, the equation $ax^2 + by^2 + c = z^2$ becomes $ax^2 + by^2 = z^2$.

Let $x = uy$ and $z = vy$.

We get $au^2y^2 + by^2 = v^2y^2$.

So, we have $au^2 + b = v^2$, which is solved by the Brahmagupta–Bhaskara II method.

Example 1 (Bhaskara II)

Find the rational solutions of $7x^2 + 8y^2 + z^2$.

Let $x = uy$ and $z = vy$.

Then we have $7(u^2\ y^2) + 8y^2 = v^2y^2$

Or $7u^2 + 8 = v^2$, which is solved by Brahmagupta–Bhaskara II method.

We observe that $7 \cdot 2^2 + 8 = 28 + 8 = 36 = 6^2$.

So, $(l \;\; m) = (2 \;\; 6)$ is a solution of $7u^2 + 8 = v^2$. So, another solution of $7u^2 + 8 = v^2$ is

$$u = lq \pm mp \quad \text{and} \quad v = mp \pm 7lp$$

where $7p^2 + 1 = q^2$.

Now,

$$x = uy = (lq \pm mp)y = (2q \pm 6p)y$$

$$z = vy = (mp \pm 7lp)y = (6q \pm 14p)y$$

So,

$$x{:}y{:}z = 2q \pm 6p{:}1{:}6q \pm 14p$$

where $7p^2 + 1 = q^2$.

If $p = 0$, $q = 1$, then $x{:}y{:}z = 2{:}1{:}6$.

If $p = 3$, $q^2 = 7 \cdot 3^2 + 1 = 63 + 1 = 64 \Rightarrow q = 8$.

Hence, $x{:}y{:}z = 2(8) \pm 6(3){:}1{:}6(8) \pm 14(3) = 16 \pm 18{:}1{:}48 \pm 42$.

$$x{:}y{:}z = 34{:}1{:}90 \quad \text{or} \quad -2{:}1{:}6 \text{ and so on}$$

Example 2

Solve $7x^2 - 8y^2 + 1 = z^2$.

Let

$$x = 2y$$

$$7(2y)^2 - 8y^2 + 1 = z^2$$

$$28y^2 - 8y^2 + 1 = z^2$$

$$20y^2 + 1 = z^2$$

We now solve this equation by Brahmagupta–Bhaskara II method.

We observe that $20 \cdot 2^2 + 1 = 80 + 1 = 9^2$.

So, $(l \; m) = (2 \; 9)$ is a solution of $20y^2 + 1 = z^2$ (∗).

Another solution of (∗) is

$$y = lq \pm mp \quad \text{and} \quad z = mq \pm 20lp$$

(Continued)

Example 2 (cont'd)

i.e.,

$$y = 2q \pm 9p \text{ and } z = 9q \pm 40p$$

where $20p^2 + 1 = q^2$.

Now, $x = 2y = 2(2q \pm 9p)$.

$$x{:}y{:}z = 2(2q \pm 9p){:}(2q \pm 9p){:}(9q \pm 40p)$$

If $p = 0$, $q^2 = 1 + 20p = 1 + 0 = 1$, then, $x{:}y{:}z = 2(2 \pm 0){:}(2 \pm 0){:}(9 \pm 0)$.

i.e., $x{:}y{:}z = 4{:}2{:}9$.

If $p = 2$, $q^2 = 1 + 20 \cdot 2^2 = 1 + 80 = 81 = 9^2 \Rightarrow q = 9$.

So,

$$x{:}y{:}z = 2(2 \cdot 9 \pm 9.2){:}(2 \cdot 9 \pm 9 \cdot 2){:}9(9) \pm 40(2).$$

$$= 2(18 \pm 18){:}(18 \pm 18){:}81 \pm 80$$

i.e., $x{:}y{:}z = 72{:}36{:}161$ and so on.

Example 3

Find the rational solutions of the equation

$$45x^2 + 120x + 200y^2 - 160y - 145 = 441z^2 - 252z \quad \text{(A)}$$

$$(45x^2 + 120x) + (200y^2 - 160y) - 145 = 9(49z^2 - 28z)$$

$$5(9x^2 + 24x) + 8(25y^2 - 20y) - 145 = 9(49z^2 - 28z)$$

$$5[(3x+4)^2 - 16] + 8[(5y-2)^2 - 4] - 145 = 9\{[7z-2]^2 - 4\}$$

$$5(3x+4)^2 - 80 + 8(5y-2)^2 - 32 - 145 + 36 = 9(7z-2)^2$$

Let $u = 3x + 4$, $v = 5y - 2$, and $w = 3(7z - 2)$

$$5u^2 + 8v^2 - 221 = w^2$$

Let $u = 3v$.

Then,

$$5(3v)^2 + 8v^2 - 221 = w^2.$$

$$45v^2 + 8v^2 - 221 = w^2$$

$$53v^2 - 221 = w^2 \tag{1}$$

We will now solve this equation by Brahmagupta–Bhaskara II method.
We observe that $53 \cdot 1^2 - 17 = 36 = 6^2$.
So, $(1 \cdot 6)$ is a solution of $53v^2 - 17 = w^2 \Rightarrow (1 \cdot 6 - 17)$.
Again we observe that, $53.2^2 + 13 = 212 + 13 = 225 = 15^2$.
So, $(2\ 15)$ is a solution of $53v^2 + 13 = w^2 \Rightarrow (2\ \ 15\ 13)$.
Then,

$$(1\ 6 - 17) \circ (2\ 15\ 13)$$

$$\Rightarrow [\,1 \cdot (15) + (6) \cdot (2) \quad 6(15) + 1 \cdot 2 \cdot (53) \quad (-17) \cdot (13)\,]$$

$$= (27 \quad 196 \quad -221)$$

which means that $(v\ w) = (l\ m) = (27\ 196)$ is a solution of $53v^2 - 221 = w^2$.
Another solution of (1) is given by

$$v = lq \pm mp \ \text{ and } \ w = mq \pm 53lp$$

where $53p^2 + 1 = q^2$. But $u = 3v$.
So,

$$u = 3(lq \pm mp = 3(27q \pm 196p)$$

$$v = (lq \pm mp) = (27q \pm 196p)$$

$$w = mq \pm 53lp = (196q \pm (53)(27)p) = (196q \pm 1431p)$$

$$3x + 4 = u = 3(27q \pm 196p) \Rightarrow 3x = 3(27q \pm 196p) - 4$$

So,

$$x = (27q \pm 196p) - \frac{4}{3} \tag{2}$$

$$5y - 2 = v = (27q \pm 196p) \Rightarrow 5y = (27q \pm 196p) + 2$$

(*Continued*)

Example 3 (cont'd)

So,

$$y=\frac{1}{5}(27q\pm 196p)+\frac{2}{5} \tag{3}$$

$$3(7z-2)=w=(196q\pm 1431p)\Rightarrow 7z=\frac{1}{3}(196q\pm 1431p)+2$$

So,

$$z=\frac{1}{21}(196q\pm 1431p)+\frac{2}{7} \tag{4}$$

Equations (2), (3), and (4) give the general solution of the given equation (A) For example, if $p=0$, $q^2=53p^2+1=0+1=1=q=1$, then

$$x=27\cdot 1-\frac{4}{3}=\frac{81-4}{3}=\frac{77}{3}$$

$$y=\frac{1}{5}(27)+\frac{2}{5}=\frac{29}{5}$$

$$z=\frac{1}{21}(196)+\frac{2}{7}=\frac{202}{21}$$

Solve $a^2x^2+by^2+c=z^2$ (Bhaskara II).

Solution

Now, $by^2+c=z^2-a^2x^2=(z+ax)(z-ax)$.

Let

$$z-ax=m \tag{1}$$

where m is a nonzero number.

Then,

$$z+ax=\frac{z^2-a^2x^2}{z-ax}=\frac{by^2+c}{m} \tag{2}$$

Adding (1) and (2), we get

$$2z=m+\frac{by^2+c}{m}\Rightarrow z=\frac{1}{2}\left[m+\frac{by^2+c}{m}\right]$$

Subtracting (1) from (2), we get

$$2ax = \frac{by^2 + c}{m} - m$$

$$x = \frac{1}{2a}\left[\frac{by^2 + c}{m} - m\right]$$

For different values of m, we get different solutions.
If $c = 0$, $a = 1$, Bhaskara II's solution reduces to

$$x = \frac{1}{2}\left[\frac{by^2}{m} - m\right] \text{ and } z = \frac{1}{2}\left[\frac{by^2}{m} + m\right]$$

where m is an arbitrary number.

Observation: 1

If $b = \alpha\beta$, $m = 2\beta d^2$, and $y = 2de$, where d and e are arbitrary integers, we get

$$x = \frac{1}{2}\left[\frac{\alpha\beta(2de)^2}{2\beta d^2} - 2\beta d^2\right] = \frac{1}{2}\left[\alpha(2e^2) - 2\beta d^2\right]$$

i.e.,

$$x = \alpha e^2 - \beta d^2$$

$$z = \frac{1}{2}\left[\frac{\alpha\beta(2de)^2}{2\beta d} + 2\beta d^2\right] = \frac{1}{2}\left[\alpha(2e^2) + 2\beta d^2\right]$$

i.e.

$$z = \alpha e^2 + \beta d^2$$

Thus,

$$(x\ y\ z) = (\alpha e^2 - \beta d^2\ 2de\ \alpha e^2 + \beta d^2)$$

is a solution of $x^2 + by^2 = z^2$.

This solution was given by A. Desboves (1879) (Dickson's book 1999, Number II, page 405).

Observation: 2

If $m = 2u^2$ and $y = 2\mu u$.

$$x = \frac{1}{2}\left[\frac{by^2}{m} - m\right] = \frac{1}{2}\left[\frac{b \cdot 4\mu^2 u^2}{2u^2} - 2u^2\right]$$

$$x = b\mu^2 - u^2$$

$$z = \frac{1}{2}\left[\frac{by^2}{m} + m\right]$$

$$= \frac{1}{2}\left[\frac{b \cdot 4\mu^2 u^2}{2u^2} + 2u^2\right]$$

$$= \frac{1}{2}\left[2b\mu^2 + 2u^2\right]$$

$$= [b\mu^2 + u^2]$$

where μ and u are arbitrary integers.

Thus,

$$(x\ y\ z) = (b\mu^2 - u^2\ 2\mu u\ b\mu^2 + u^2)$$

is a solution of $x^2 + by^2 = z^2$.

This solution was given by Matsungao (1735).

Theorem (Bhaskara II)

Find the solutions of the equation $ax^2 + bxy + cy^2 = z^2$.

Two cases arise:

Case (i): Suppose $a = p^2$.

Then, we have $(p^2x^2 + bxy) + cy^2 = z^2$.

i.e.,

$$\left(px + \frac{by}{2p}\right)^2 - \frac{b^2y^2}{4p^2} + cy^2 = z^2$$

$$\left(px + \frac{by}{2p}\right)^2 + y^2\left(c - \frac{b^2}{4p^2}\right) = z^2$$

Let

$$w = px + \frac{by}{2p}$$

Then, we have

$$z^2 - w^2 = y^2\left(c^1 - \frac{b^2}{4p^2}\right)$$

Let (1) $z - w = \lambda$, where λ is a rational number $\neq 0$.
Then, (2)

$$z + w = \frac{z^2 - w^2}{z - w} = \frac{y^2(c^1 - (b^2/4p^2))}{\lambda}$$

Thus, subtracting (1) from (2), we get

$$2w = \frac{y^2}{\lambda}\left(c^2 - \frac{b^2}{4p^2}\right) - \lambda$$

$$w = \frac{1}{2}\frac{y^2}{\lambda}\left(c^1 - \frac{b^2}{4p^2}\right) - \frac{\lambda}{2}$$

But

$$px + \frac{by}{2p} = w$$

$$px = w - \frac{by}{2p} = \frac{1}{2}\frac{y^2}{\lambda}\left(c^2 - \frac{b^2}{4p^2}\right) - \frac{by}{2p} - \frac{\lambda}{2}$$

So,

$$x = \frac{1}{2p}\left[\frac{y^2}{\lambda}\left(c^1 - \frac{b^2}{4p^2}\right)\right] - \frac{by}{2p^2} - \frac{\lambda}{2p}$$

Adding (1) and (2), we get

$$2z = \lambda + \frac{y^2}{\lambda}\left(c^1 - \frac{b^2}{4p^2}\right)$$

So,

$$z = \frac{1}{2}\left[\lambda + \frac{y^2}{\lambda}\left(c^1 - \frac{b^2}{4p^2}\right)\right]$$

Suppose $y = m/n$, where m and n are arbitrary integers and $n \neq 0$.
Then, $y = m/n$ and

$$x = \frac{1}{2p}\left[\frac{m^2}{n^2\lambda}\left(c - \frac{b^2}{4p^2}\right) - \lambda\right] - \frac{bm}{2p^2n}$$

$$x = \frac{1}{8p^3n^2\lambda}\left[m^2(4p^2c - b^2) - 4\lambda^2n^2p^2 - 4bpmn\lambda\right]$$

$$z = \frac{1}{2}\left[\frac{m^2}{n^2\lambda}\left(c - \frac{b^2}{4p^2}\right) + \lambda\right]$$

$$= \frac{1}{8n^2\lambda p^2}\left[m^2(4p^2c - b^2) + 4p^2n^2\lambda^2\right]$$

Since the given equation is homogenous, any multiplies of x, y, and z are also solutions.

Multiplying by $8n^2\lambda p^2$, we get

$$x = m^2(4p^2c - b^2) - 4\lambda p^2n^2 - 4pbmn\lambda$$

$$y = 8n\lambda p^2m$$

$$z = m^2(4p^2c - b^2) + 4p^2n^2\lambda^2$$

where m and n are arbitrary integers.

In particular, if $a = b = c = 1$ and $\lambda = p = 1$, we get

$$x = m^2(4 - 1) - 4n^2 - 4mn = 3m^2 - 4n^2 - 4mn$$

$$y = 8mn$$

$$z = m^2(4 - 1) + 4n^2 = 3m^2 + 4n^2$$

as the solution of $x^2 + xy + y^2 = z^2$.

Case (ii): Bhaskara II: Neither a nor c is square.

We have

$$ax^2 + bxy + cy^2 = z^2 \tag{1}$$

Multiplying by a, we get $a^2x^2 + abxy + acy^2 = az^2$.
i.e.,

$$\left(ax+\frac{by}{2}\right)^2 - \frac{b^2}{4}y^2 + acy^2 = az^2 \tag{2}$$

Let

$$w = ax + \frac{b}{a}y$$

and

$$\beta = \frac{1}{4}(b^2 - 4ac)$$

Then, (2) becomes $w^2 = az^2 + \beta y^2$.
Let $w = vy$ and $z = uy$.
Then,

$$v^2y^2 = au^2y^2 + \beta y^2 \Rightarrow v^2 = au^2 + \beta \tag{3}$$

which is a "Vargaprakriti" equation.
If $u = m$ and $v = n$ be a solution of (3), another solution of (3) is

$$u = mq \pm np \text{ and } v = nq \pm map$$

where $ap^2 + 1 = y^2$.
Hence,

$$ax + \frac{b}{2}y = w = vy = y[nq \pm map]$$

$$ax = y\left[nq \pm map - \frac{b}{2}\right]$$

$$x = \frac{y}{a}\left[nq \pm map - \frac{b}{2}\right] = \frac{y}{2a}[2nq \pm 2map - b]$$

$$z = uy = (mq \pm np)y$$

where $ap^2 + 1 = q^2$ and $n^2 = am^2 + \beta$ (from (3)).

Putting $n = ar + (1/2)b$ and $y = s/t$, we get

$$y = \frac{s}{t}$$

$$x = \frac{s}{2at}[q(2ar + b) \pm 2map - b]$$

$$z = \left[mq \pm p\left(ar + \frac{b}{2}\right)\right]\frac{s}{t} \Rightarrow z = \frac{s}{2t}[2mq \pm p(2ar + b)]$$

Since the given equation is homogenous, multiplying by $2at$, we get

$$x = s[q(2ar + p) \pm 2amp - b]$$

$$y = 2as$$

$$z = as[2mq \pm p(2ar + b)]$$

where

$$ap^2 + 1 = q^2$$

and

$$am^2 + \beta = \left(ar + \frac{1}{2}b\right)^2 \text{ from (3)}$$

Now,

$$am^2 = \left(ar + \frac{1}{2}b\right)^2 - \beta = \left(ar + \frac{1}{2}b\right)^2 - \frac{1}{4}(b^2 - 4ac)$$

$$am^2 = a^2r^2 + arb + \frac{b^2}{4} - \frac{b^2}{4} + ac$$

$$am^2 = a^2r^2 + arb + ac$$

So,

$$m^2 = ar^2 + rb + c$$

Example 1 (Bhaskara 1)

Find the solutions of $x^2 + xy + y^2 = z^2$.

Multiplying by 36, we get

$$36x^2 + 36xy + 36y^2 = 36z^2$$

$$(6x+3y)^2 - 9y^2 + 36y^2 = 36z^2$$

$$(6x+3y)^2 + 27y^2 = 36z^2$$

i.e.,

$$w^2 + 27y^2 = 36z^2$$

where $w = 6x + 3y$.

Let $w = uy$, $z = vy$.

Then, we have $u^2y^2 + 27y^2 = 36v^2y^2 \Rightarrow u^2 + 27 = 36v^2$.

Let $Q = 6v$.

Then, we have

$$u^2 + 27 = Q^2 \tag{1}$$

We solve (1) by Brahmagupta–Bhaskara II method.

We observe that $3^2 + 27 = 9 + 27 = 36 = 6^2$.

So, $(l\ m) = (3\ 6)$ is a solution of (1), namely

$$u^2 + 27 = Q^2$$

Another solution of (1) is

$$u = lq \pm mp \ \text{ and } \ Q = mq \pm lp$$

i.e.,

$$u = 3q \pm 6p \ \text{ and } \ Q = 6q \pm 3p$$

where $q^2 = p^2 + 1$.

Now, $w = 6x + 3y = uy = (3q \pm 6p)y$.

So,

$$6x = \{(3q \pm 6p) - 3\}y$$

(Continued)

Example 1 (Bhaskara 1) (cont'd)

$$x = \frac{1}{2}\{q \pm 2p - 1\}y$$

$$z = vy = \frac{Q}{6}y = \frac{1}{6}(6q \pm 3p)y = \frac{1}{2}(2q \pm p)y,$$

where $q^2 = p^2 + 1$.

If

$$p = \frac{3}{4}, \quad q^2 = 1 + p^2 = 1 + \frac{9}{16} = \frac{25}{16} \Rightarrow q = \frac{5}{4}$$

Then,

$$x = \frac{1}{2}\left[\frac{5}{4} \pm \frac{3}{2} - 1\right]y = \frac{7y}{8}$$

taking the positive sign.

$$z = \frac{1}{2}\left[2 \cdot \frac{5}{4} \pm \frac{3}{4}\right]y = \frac{13}{8}y$$

taking the positive sign.

If $y = 8$, $x = 7$, $z = 13$.

Then, $(x\ y\ z) = (8\ 7\ 13)$ is one solution of the equation

$$x^2 + xy + y^2 = z^2$$

Example 2

Solve $3x^2 - 5xy + 4y^2 - 15 = 9z^2$.

Let $x = 2y$.

Then, we have

$$3(2y)^2 - 5(2y^2) + 4y^2 - 15 = 9z^2$$

i.e.,

$$12y^2 - 10y^2 + 4y^2 - 15 = 9z^2$$

i.e., (1) $6y^2 - 15 = 9z^2 = w^2$, where $w = 3z$.
We will solve (1) by Brahmagupta–Bhaskara II method.
We observe that $6 \cdot 1^2 + 3 = 9 = 3^2$.
So, (1 3) is a solution of $6y^2 + 3 = w^2 \Rightarrow (1\ 3\ 3)$.
Again we observe that $6 \cdot 1^2 - 5 = 6 - 5 = 1^2$.
So, (1 1) is a solution of $6y^2 - 5 = w^2 \Rightarrow (1\ 1\ -5)$.
Hence,

$$(1\ 3\ 3) \circ (1\ 1\ -5) = [\,1 \cdot 1 + 3 \cdot 1 \quad 3 \cdot 1 + 1 \cdot 1 \cdot 6 \quad 3(-15)\,] = (4 \quad 9 \quad -15),$$

which means that $(y\ w) = (l\ m) = (4\ 9)$ is a solution of (1), $6y^2 - 15 = w^2$.
Another solution of (1) is given by

$$y = lq \pm mp, \quad w = mq \pm l.6p = mq \pm 6lp$$

So,

$$y = 4q \pm 9p, \quad w = 9q \pm 24p \tag{2}$$

$$x = 2y = 2(4q \pm 9p) = 8q \pm 18p \tag{3}$$

Also,

$$z = \frac{w}{3} = \frac{1}{3}(9q \pm 24p) = 3q \pm 8p \tag{4}$$

where $6p^2 = 1 + q^2$ and (2), (3), and (4) is the general solution of the given equation.
In particular, if $p = 0, \quad q = 1$, then $y = 4, \quad x = 8, \quad z = 3$.
If $p = 2, \quad q^2 = 6p^2 + 1 = 6(4) + 1 = 5^2 \Rightarrow q = 5$.
Then,

$$y = 4(5) \pm 9(2) = 20 \pm 18 = 38$$

taking the positive sign.

$$x = 2y = 76$$

$$z = 3(5) \pm 8(2) = 15 + 16 = 31$$

taking the positive sign.
So, if $p = 2$, $y = 38$, $x = 76$, $z = 31$, and so on.
We observe that repeated samasa of (2 5 1) on itself will give infinite solutions of $6p^2 + 1 = q^2$.

Example 3

Solve the equation $72x^2 + 18xy + 28y^2 + 69x - 72y + 302 = 64z^2 - 96z$ (∗).

Equation (∗) can be written as

$$(72x^2 + 96x + 32) + (28y^2 - 84y + 63) + (18xy - 27x + 12y - 18) + 261$$
$$= 64z^2 - 96z + 36$$

$$8(9x^2 + 12x + 4) + 7(4y^2 - 12y + 9) + [9x(2y - 3) + 6(2y - 3)] + 261$$

$$= 4[16z^2 - 24z + 9]$$

$$8(3x+2)^2 + 7(2y-3)^2 + 3(2y - 3)(3x + 2) + 261 = 4(4z-3)^2$$

Let $3x + 2 = u$, $2y - 3 = v$, and $4z - 3 = w$.
Then, we have $8u^2 + 7v^2 + 3uv + 261 = 4w^2$.
Let $u = 2v$.
Then, we have $8(2v)^2 + 7v^2 + 3(2v)v + 261 = 4w^2$.

$$32v^2 + 7v^2 + 6v^2 + 261 = 4w^2$$

$$45v^2 + 261 = Q^2$$

where

$$Q = 2w \tag{1}$$

We will now solve (1) by Brahmagupta–Bhaskara II method.
We have the identity $45 \cdot 1^2 - 9 = 36 = 6^2$.
So, (1 6) is a solution $45v^2 - 9 = Q^2 \Rightarrow (1\ 6 - 9)$.
Again, we observe that $45 \cdot 1^2 - 29 = 16 = 4^2$.
So, (1 4) is a solution of $45v^2 - 29 = Q^2 \Rightarrow (1\ 4 - 29)$.
where

$$(1\ 6 - 9) \circ (1\ 4 - 29)$$

$$= [\ 1 \cdot 4 + 6 \cdot 1 \quad 6 \cdot 4 + 1 \cdot 1 \cdot (45) \quad (-9) \cdot (-29)\ ]$$

$$= (10 \quad 69 \quad 261)$$

which means that $(v\ Q) = (l\ m) = (10\ 69)$ is a solution of $45v^2 + 261 = Q^2$. So, another solution of (1), namely $45v^2 + 261 = Q^2$, is $v = lq \pm mp$ and $Q = mq \pm l \cdot 45 \cdot p = mq \pm 45\,lp$, where $45(p + 1)^2 = q^2$.

i.e., $v = 10q \pm 69p$ and $Q = 69q \pm 450p$, where $45(p+1)^2 = q^2$.
Now, $3x + 2 = u = 2v = 2(10q \pm 69p) = 20q \pm 138p$.

$$3x = 20q \pm 138p - 2$$

$$x = \frac{1}{3}(20q \pm 138p - 2) \tag{2}$$

$$2y - 3 = v = 10q \pm 69p$$

So,

$$2y = 10q \pm 69p + 3$$

$$y = \frac{1}{2}(10q \pm 69p + 3) \tag{3}$$

Also,

$$4z - 3 = w = \frac{Q}{2} = \frac{1}{2}(69q \pm 450p)$$

$$4z = \frac{1}{2}(69q \pm 450p) + 3$$

$$z = \frac{1}{8}(69q \pm 450p) + \frac{3}{4} \tag{4}$$

where $45p^2 + 1 = q^2$.
Then, (2), (3), and (4) is the general solution of the given equation (*).
Now, in particular, if $p = 0$, $q = 1$, then

$$x = \frac{1}{3}(20 - 2) = \frac{18}{3} = 6$$

$$y = \frac{1}{2}(10 \cdot 1 + 3) = \frac{13}{2}$$

$$z = \frac{1}{8}(69) + \frac{3}{4} = \frac{75}{8}$$

So,

$$(x\ y\ z) = \left(6\ \frac{13}{2}\ \frac{75}{8}\right)$$

is one solution of $(*)$.

(*Continued*)

Example 3 (cont'd)

Other solutions can be found for different values of p and q satisfying $45p^2 + 1 = q^2$.

We observe that if

$$p = \frac{1}{2}, \quad q^2 = 45p^2 + 1 = 45 \cdot \frac{1}{4} + 1 = \frac{45+4}{4} = \frac{49}{4}$$

So,

$$q = \frac{7}{2}$$

If

$$p = \frac{1}{2}, \quad q = \frac{7}{2}$$

we get

$$x = \frac{1}{3}\left[20 \cdot \frac{7}{2} \pm 138\frac{1}{2} - 2\right] = \frac{1}{6}[140 \pm 138 - 4]$$

So,

$$x = \frac{137}{3}$$

or

$$-\frac{1}{3}$$

Similarly,

$$y = \frac{1}{2}\left[10 \cdot \frac{7}{2} \pm 69\frac{1}{2} + 3\right] = \frac{1}{2}\left(\frac{70 \pm 69 + 6}{2}\right) = \frac{145}{4} \text{ or } \frac{45}{16}$$

and

$$z = \frac{1}{8}\left(69 \cdot \frac{7}{2} \pm 450 \cdot \frac{1}{2}\right) + \frac{3}{4} = \frac{1}{16}[483 \pm 450 + 12] = \frac{945}{16} \text{ or } \frac{45}{16}$$

We observe that infinite solutions of $q^2 = 45p^2 + 1$ can be found by repeated samasa of $\left(\frac{1}{2}\ \frac{7}{2}\ 1\right)$ over itself.

Example 4

Solve the equation

$$9x^2 + 8xy + 2y^2 + 20 = 5z^2 \quad (*)$$

Multiplying (*) by 5, we get

$$45x^2 + 40xy + 10y^2 + 100 = 25z^2$$

Let $y = x$, then we have

$$45x^2 + 40x^2 + 10x^2 + 100 = 25z^2 \qquad (1)$$

$$95x^2 + 100 = w^2$$

where $w = 5z$.

We will now solve (1) by Brahmagupta–Bhaskara II method.

We observe that $95 \cdot 1^2 + 5 = 100 = 10^2$.

So, (1 10) is a solution of $95x^2 + 5 = w^2 \Rightarrow (1\ 10\ 5)$.

We also observe that $95(2)^2 + 20 = 95(4) + 20 = 400 = (20)^2$.

So, (2 20) is a solution of the equation $95x^2 + 20 = w^2$.

$$\Rightarrow (2\ 20\ 20)$$

$$(1\ 10\ 5) \circ (2\ 20\ 20)$$

$$= [\, 1 \cdot (20) + 2 \cdot (10) \quad (10) \cdot (20) + 1 \cdot 2 \cdot (95) \quad (5)\,(20)\,]$$

$$= (40 \quad 390 \quad 100)$$

which means that $(x\ w) = (l\ m) = (40\ 390)$ is a solution of $95x^2 + 100 = w^2$. Another solution of (1) is

$$x = lq \pm mp \text{ and } w = mq \pm 95lp$$

i.e.,

$$x = 40q \pm 390p \quad (**)$$

and

$$w = 390q \pm 95(40)p$$

i.e.,

$$w = 390q \pm 3800p$$

where $q^2 = 95p^2 + 1$.

(*Continued*)

Example 4 (cont'd)

Now,

$$y = x = 40q \pm 390p \tag{2}$$

$$z = \frac{w}{5} = \frac{1}{5}(390q \pm 3800p) = 78q \pm 760p \tag{3}$$

where $q^2 = 95p^2 + 1$.

Equations (2) and (3) will give the general solution of the given equation (*).

In particular, if $p = 0$, $q = 1$,

$$y = x = 40 \text{ and } w = 78$$

$(x\ \ y\ z) = (40\ 40\ 78)$ is a solution of the given equation $(*)$.

We observe that if $p = 4$, $q^2 = 95p^2 + 1 = 95(16) + 1$

$$= 1520 + 1 = 1521 = 39^2$$

Hence, $q = 39$.

Then,

$$x = y = 40q \pm 390p = 40(39) \pm 390(4) = 3120$$

taking the positive sign.

$$z = 78q \pm 760p = 78(39) \pm 760(4) = 6082$$

taking the positive sign.

We observe that infinite solution of $95p^2 + 1 = q^2$ can be found by repeated samasa of (4 39 1) over itself.

Solve

$$ax^{2n+2} + bx^{2n} = y^2 \quad \text{(Bhaskara II)}$$

We have $x^{2n}(ax^2 + b) = y^2$.

Let $z = ax^2 + b$.

Then, we have $x^2z^2 = y^2$.

So, $y = x^n z$, where $z^2 = ax^2 + b$.

We note that the equation $z^2 = ax^2 + b$ is a Vargaprakriti equation, which can be solved by Brahmagupta–Bhaskara II method.

Examples

Solve

$$5x^4 - 100x^2 = y^2 \quad \text{(BBI, 103)}$$

$$x^2(5x^2 - 100) = y^2$$

Let

$$z^2 = 5x^2 - 100$$

Then, we have $x^2z^2 = y^2 \Rightarrow y = xz$.where

$$z^2 = 5x^2 - 100 \quad (*)$$

which is "Vargaprakriti" equation.

We now solve this equation.

We observe that $5 \cdot 5^2 - 100 = 5^2$.

So, $(x\ z) = (l\ m) = (5\ 5)$ is a solution of $z^2 = 5x^2 - 100$.

Another solution of $(*)$ is given by

$$x = lq \pm mp = 5p \pm 5q = 5(p \pm q) \quad (1)$$

$$z = mq \pm l \cdot 5p, \quad \text{where } 5p^2 + 1 = q^2 \quad (2)$$

$$= 5q \pm 25p = 5(q \pm 5p), \quad \text{where } 5p^2 + 1 = q^2$$

Now,

$$y = xz = 5(p \pm q) \cdot 5(q \pm 5p)$$

So,

$$y = 25 \cdot (p \pm q)(q \pm 5p) \quad (3)$$

So, (1), (2), and (3) is the general solution of the given equation $5x^4 - 100x^2 = y^2$.

In particular, if $p = 0$, $q = 1$.

Then, $x = 5q = 5$, taking the positive sign.

$$y = (5q \pm 0)(5q \pm 0) = 25q^2 = 25$$

(*Continued*)

Examples (cont'd)

$(x\quad y) = (5\ 25)$ is a solution of the given equation. Other solutions can be found for different values of p and q satisfying $5p^2 + 1 = q^2$. For example, if $p = 4$, $q = 9$, and so on.

If $p = 4$, $q = 9$, we get

$$x = 5(p \pm q) = 5(4 \pm 9) = 65$$

taking the positive sign.

$$y = 25(p \pm q)(q \pm 5p) = 25(4 \pm 9)(9 \pm 20)$$

$$= (25)(13)(29) = 9425$$

We observe that we get the infinite solutions of $5p^2 + 1 = q^2$ can be found by repeated samasa of (4 9 1) over itself.

Example 2

Solve

$$8x^6 + 49x^4 = y^2 \quad (*) \quad \text{(BBI, 107)}$$

$$x^4(8x^2 + 49) = y^2$$

Let

$$z^2 = 8x^2 + 49$$

Then, we have $x^4z^2 = y^2 \Rightarrow y = x^2z$, where

$$z^2 = 8x^2 + 49 \quad (*)$$

which is a "Vargaprakriti" equation.

We now solve (∗).

We observe that $8 \cdot 2^2 + 49 = 81 = 9^2$.

So, $(x\ z) = (l\ m) = (2\ 9)$ is a solution of $8x^2 + 49 = z^2$.

Another solution of (∗) is

$$x = lq \pm mp, \quad z = mq \pm 8lp$$

$$x = 2q \pm 9p, \quad z = 9q \pm 8(2)p = 9q \pm 16p$$

where $8p^2 + 1 = q^2$. Now, $y = x^2z = (9q \pm 16p)(2q \pm 9p)^2$.

So,

$$x = 2q \pm 9p$$

$$y = (9q \pm 16p)(2q \pm 9p)^2$$

is the general solution of (∗), where $8p^2 + 1 = q^2$. If $p = 0$, $q = 1$; and if $p = 1$, $q = 3$ and if $p = 6$, $q = 17$, and so on. These give infinitely many solutions. If $p = 0$, $q = 1$, $x = 2$, $y = 36$; if $p = 1$, $q = 3$, we get $x = (6 \pm 9)$, $y = (6 \pm 9)^2(27 \pm 16) \Rightarrow x = 15$, $y = 9675$, $x = -3$, $y = 99$, and so on. Infinite solutions of (*) can be found by repeated samasa of ($\alpha\ \beta$ 1), where $8\alpha^2 + 1 = \beta^2$.

Solve

$$(a+x^2)^2 + a^2 = u^3 \quad (*) \quad \text{(BBI, 50)}$$

So, $x^4 + 2ax^2 + 2a^2 = u^3$.

Let $u = x^2$.

We have that $x^4 + 2ax^2 + 2a^2 = x^6$.

$$x^6 - x^4 = 2a^2 + 2ax^2$$

Multiplying by 2, we get

$$2x^6 - 2x^4 = 4a^2 + 4ax^2$$

$$2x^6 - x^4 = 4a^2 + 4ax^2 + x^4$$

i.e.,

$$x^4(2x^2 - 1) = (x^2 + 2a)^2$$

Let $2x^2 - 1 = v^2$.

So,

$$x^4v^2 = (x^2 + 2a)^2$$

$$x^2v = x^2 + 2a, \quad \text{where } v^2 = 2x^2 - 1 \quad (*)$$

We now solve (∗), which is a "Vargaprakriti" equation.

We observe that $2 \cdot 1^2 - 1 = 1^2$.

(Continued)

Example 2 (cont'd)

So, $(x\ v) = (l\ m) = (1\ 1)$ is a solution of $2x^2 - 1 = v^2$ (∗).
Another solution of (∗) is given by

$$x = ql \pm mp \text{ and } v = mq \pm 2l \cdot p$$

where $q^2 = 2p^2 + 1$.
So, $x = q \pm p$ and $v = q \pm 2p$, where $q^2 = 2p^2 + 1$.
Now,

$$u = x^2 = (q \pm p)^2$$

$x = q \pm p$ is the general solution of (∗), where $q^2 = 2p^2 + 1$.
If $p = 0$, $q = 1$, $x = 1$, $u = 1$, $v = 1$ and $a = 0$.
If $p = 2$, $q = 3$,

$$x = q \pm p = 3 \pm 2 = 5 \text{ or } -1$$

$$u = x^2 = 25 \text{ or } 1,\ v = 7 \text{ or } -1 \text{ and } a = 75$$

infinite solutions of $q^2 = 2p^2 + 1$ can be found by repeated samasa of (2 3 1) over itself.

Varga-Kuttaka Equation

The equation $bx + c = y^2$ is called Varga-kuttaka (square pulvarizer).
Solve

$$bx + c = y^2 \quad \text{(BBI, 122)}$$

Case 1: Let c be a perfect square $= \beta^2$.
Then, we have $bx + \beta^2 = y^2$.
Bhaskara II's rule states the following.
Find p such that $p^2 = bq$ and $2p\beta = br$ and if $y = pu + \beta$, then $x = qu^2 + ru$.

Proof: Let $y = pu + \beta$.
Then, $bx + \beta^2 = y^2$ becomes

$$bx + \beta^2 = (pu+\beta)^2 = p^2u^2 + 2pu\beta + \beta^2$$

$$bx = p^2u^2 + 2pu\beta = bqu^2 + ubr$$

So, $x = qu^2 + ur$.

Corollary:
Let $p = b$ and $u = v$, then $br = 2p\beta = 2b\beta \Rightarrow r = 2\beta$.
Then, $y = vb + \beta$, since $y = pu + \beta$.

$$x = qv^2 + vr = qv^2 + 2\beta v$$

Case 2: Suppose c is not a square.
Let $c = bm + n$.
Let $n + sb = r^2$ and $y = bu \pm r$.
Substituting in $bx + c = y^2$, we get

$$bx + bm + n = (bu \pm r)^2 = b^2u^2 \pm 2bur + r^2$$

$$bx + bm + n = b^2u^2 \pm 2bur + (n + sb)$$

$$bx + bm - sb = b^2u^2 \pm 2bur \Rightarrow x + m - s = bu^2 \pm 2ur$$

i.e.,

$$x = bu^2 \pm 2ur + s - m$$

Solve

$$7x + 30 = y^2 \quad (*) \quad \text{(BBI, 120, 121)}$$

If $bx + c = y^2$, here $b = 7$, $c = 30$.
Let

$$c = bm + n$$

i.e.,

$$30 = 7m + n$$

if

$$m = 4, \quad n = 2$$

Now,

$$r^2 = n + bs$$

i.e.,

$$r^2 = 2 + 7s$$

if

$$s = 2, \quad r = 4$$

and

$$y = bu \pm r = 7u \pm 4$$

$$x = bu^2 \pm 2ur + s - m$$

$$= 7u^2 \pm 2u(4) + 2 - 4$$

$$x = 7u^2 \pm 8u - 2$$

Hence, $x = 7u^2 \pm 8u - 2$ and $y = 7u \pm 4$ give the general solution of (∗).

"Ghana-Kuttaka" Equation

The equation $bx + c = y^3$ is called the "Ghana-kuttaka" (cubic pulvarizer) equation.
Solve

$$bx + c = y^3 \quad \text{(BBI, 122)}$$

Two cases arise:

Case 1: Let c be a cube and let $c = \beta^3$.
Bhaskara II's rule states the following.
Find p such that $p^3 = bq$, $3p\beta = br$ and if $y = pv + \beta$, then $x = b^2v^3 + 3pv(bv + \beta)$.

Proof: Substituting in the equation $bx + c = y^3$, we get

$$bx + \beta^3 = (pv + \beta)^3 = p^3v^3 + 3p^2v^2\beta + 3pv\beta^2 + \beta^3$$

$$bx = p^3v^3 + 3p^2v^2\beta + 3pv\beta^2 = p^3v^3 + 3pv\beta(pv + \beta)$$

$$bx = bqv^3 + vbr(pv + \beta)$$

$$x = qv^3 + vr(pv + \beta)$$

In particular, if $y = bv + \beta$ instead of $pv + \beta$, we get

$$bx + \beta^3 = (bv+\beta)^3 = b^3v^3 + 3bv\beta(bv + \beta) + \beta^3$$

$$bx = b^3v^3 + 3bv\beta(bv + \beta)$$

$$x = b^2v^3 + 3v\beta(bv + \beta)$$

Case 2: Suppose c is not a cube. Suppose $c = bm + n$.
Find s such that $n + sb = r^3$.
Assume $y = bv + r$.
Substituting in $bx + c = y^3$, we get

$$bx + bm + n = (bv+r)^3 = b^3v^3 + 3bvr(bv + r) + r^3$$

$$bx + bm + n = b^3v^3 + 3bvr(bv + r) + n + sb$$

$$bx = b^3v^3 + 3bvr(bv + r) + sb - bm$$

$$x = b^2v^3 + 3vr(bv + r) + s - m$$

and $y = bv + r$ is the general solution.

Example

Solve

$$5x + 6 = y^3 \quad \text{(BBI, 121)}$$

If $bx + c = 5x + 6 = y^3$, then $b = 5,\ \ c = 6$.

$$6 = c = bm + n = 5m + n$$

Let $m = 1$, $n = 1$.
Also $n + sb = r^3$. Hence, we have

$$1 + 5s = r^3$$

If

$$s = 43,\ \ 1 + 5s = 1 + 5(43) = 1 + 215 = 216 = (6)^3 = r^3$$

$$\Rightarrow r = 6$$

(*Continued*)

Example (cont'd)

So, the general solution is $y = bv + r = 5v + 6$ and

$$x = b^2v^3 + 3vr(v + r) + s - m$$
$$= 25v^3 + 18v(v + 6) + 43 - 1$$
$$= 25v^3 + 18v(v + 6) + 42$$

Solve

$$bx \pm c = ay^2 \quad \text{(Bhaskara II)}$$

Multiplying both sides by a, we get

$$abx \pm ac = a^2y^2 \tag{1}$$

Let $u = ay$ and $v = ax$.
Then, (1) becomes

$$bv \pm ac = u^2$$

which can be solved as we solved $y^2 = bx + c$.

Example

Solve $5y^2 + 3 = 16x$.
Multiplying both sides by 5, we get

$$25y^2 + 15 = 80x \tag{1}$$

Let $u = 5y$, $v = 5x$.
Equation (1) becomes

$$u^2 + 15 = 16v \Rightarrow u^2 = 16v - 15$$

If $bv + c = 16v - 15$,

$$b = 16, \quad c = -15$$

$$-15 = c = bm + n = 16m + n$$

Let $m = -1, n = 1$.
Also, $n + sb = r^2$.
i.e!., $1 + 16s = r^2$.
If $s = 5, \quad r = 9$.
Let $u = bw + r = 16w + 9$.

$$v = bw^2 + 2rw - (m - s)$$

$$= 16w^2 + 18w - (-1 - 5)$$

$$= 16w^2 + 18w + 6$$

$$x = \frac{v}{5} = \frac{1}{5}(16w^2 + 18w + 6)$$

$$y = \frac{u}{5} = \frac{1}{5}(16w + 9)$$

Solve

$$bx \pm c = ay^n \quad \text{(BBI, 121)}$$

Here again, two cases arise.

Case 1: Let $a = 1$.
Then, we have $bx \pm c = y^n$.
So, $bx = y^n \pm c$.
or

$$x = \frac{y^n \pm c}{b} \quad (*)$$

Let $y = mz \pm k$.
Then,

$$\frac{y^n \pm c}{b} = \frac{1}{b}[(mz \pm k)^n \pm c]$$

$$= \frac{1}{b}\left[m^n z^n \pm nm^{n-1}z^{n-1}k \pm \cdots \pm nmzk^{n-1} \pm k^n + c\right]$$

$$= \frac{1}{b}\left[m^n z^n \pm nm^{n-1}z^{n-1}k \pm \cdots \pm nmzk^{n-1}\right] + \frac{\pm k^n \pm c}{b}$$

Now, if $(k^n \pm c)/b = a_n$ integral number, then $(y^n \pm c)/b$ will be an integral number, when either (1) $m = b$ or (2) b is a factor of m^n, $nm^{n-1}k, \ldots, nmk^{n-1}$. This means, knowing one integral solution of (*), an infinite number of other solutions can be found.

Case 2: Solve

$$x = \frac{ay^n \pm c}{b} \quad \text{(Bhaskara II)}$$

Multiplying by a^{n-1}, we get

$$xa^{n-1} = \frac{a^n y^n \pm ca^{n-1}}{b}$$

Let $u = ay$, $v = xa^{n-1}$, and $d = ca^{n-1}$, we get

$$v = \frac{u^n \pm d}{b}$$

which is of the form of case 1.

Simultaneous Indeterminate Quadratic Equations

Solve

$$ax + c = u^2 \tag{1}$$

$$bx + d = v^2 \quad \text{(BBI, } 177 - 179\text{)} \tag{2}$$

Bhaskara II provided two methods to solve this equation.

First method:

Let $u = mv + \beta$.

Substituting in (1), we have

$$ax = u^2 - c = (mv + \beta)^2 - c = m^2v^2 + 2mv\beta + \beta^2 - c$$

or

$$x = \frac{1}{a}(m^2v^2 + 2mv\beta + \beta^2 - c)$$

Substituting this value of x in (2), we get

$$\frac{b}{a}(m^2v^2 + 2mv\beta + \beta^2 - c) + d = v^2 \tag{3}$$

If $w^2 = (m^2v^2 + 2mv\beta + \beta^2 - c)$ and $A = (b/a)$, then (3) becomes $Aw^2 + d = v^2$, which is "Vargaprakriti," which can be solved by Brahmagupta–Bhaskara II method.

Second method: Multiplying (1) by b and (2) by a and subtracting the second from the first, we get $bc - ad = bu^2 - av^2$.

So, we have $av^2 = bu^2 + (ad - bc)$.

Multiplying both sides by a, we get

$$a^2v^2 = abu^2 + a(ad - bc)$$

Let $w = av$ and $k = a(ad - bc)$.

Then, we have

$$w^2 = abu^2 + k \quad (*)$$

which is a Vargaprakriti equation.

Now, if $u = r$, $w = s$ is a solution of (*), another solution of (*) is

$$u = rq \pm ps \ \text{ and } \ w = qs \pm abrp$$

where $abp^2 + 1 = q^2$ and $a^2v^2 = w^2 = s^2 = abr^2 + a^2d - abc$.

Then,

$$x = \frac{1}{a}(u^2 - c) = \frac{1}{a}(rq \pm ps)^2 - \frac{c}{a}$$

where $abp^2 + 1 = q^2$ is the general solution of (1) and (2).

Now, a rational solution of $abp^2 + 1 = q^2$ is as given by Sripati is follows.

Let

$$p = \frac{2t}{t^2 - ab} \ \text{ and } \ q = \frac{t^2 + ab}{t^2 - ab}$$

where t is any rational number. Therefore, the general solution of (1) and (2) is

$$x = \frac{1}{a}\frac{1}{(t^2 - ab)^2}\{r(t^2 + ab) \pm 2st\}^2 - \frac{c}{a}$$

$$u = \frac{1}{t^2 - ab}\{r(t^2 + ab) \pm 2st\}$$

and

$$v = \frac{1}{a(t^2 - ab)}\{s(t^2 + ab) \pm 2abrt\},$$

where $abr^2 + a^2d - abc = s^2$.

Special case: Suppose in (1) and (2), $c = \alpha^2$, $d = \beta^2$.
Then, (1) and (2) become

$$ax + \alpha^2 = u^2$$

$$bx + \beta^2 = v^2 \qquad (4)$$

The auxiliary equation $abr^2 + a^2d - abc = s^2$ becomes

$$abr^2 + (a^2\beta^2 - ab\alpha^2) = s^2 \quad (**)$$

An obvious solution of (**) is $r = \alpha$ and $s = a\beta$. Hence, in this case, the general solution of (4) will be

$$x = \frac{1}{a(t^2 - ab)^2}\{\alpha(t^2 + ab) \pm 2a\beta t\}^2 - \frac{\alpha^2}{a}$$

$$u = \frac{1}{(t^2 - ab)}\left\{\alpha(t^2 + ab) \pm 2a\beta t\right\}$$

$$v = \frac{1}{(t^2 - ab)}\left\{\beta(t^2 + ab) \pm 2b\alpha t\right\}$$

where t is any rational number.

If in particular, we take $\alpha = \beta = 1$, $t = a$ and taking only the positive sign, we get a particular solution of $ax + 1 = u^2$ and $bx + 1 = v^2$ as

$$x = \frac{1}{a(a^2 - ab)^2}\{a^2 + ab + 2a^2\}^2 - \frac{1}{a}$$

$$= \frac{a^2}{a^3(a-b)^2}\{3a + b\}^2 - \frac{1}{a} = \frac{1}{a}\left\{\frac{(3a+b)^2}{(a-b)^2} - 1\right\}$$

$$= \frac{1}{a}\left\{\frac{9a^2 + 6ab + b^2 - a^2 + 2ab - b^2}{(a-b)^2}\right\} = \frac{8a^2 + 8ab}{a(a-b)^2} = \frac{8(a+b)}{(a-b)^2}$$

$$u = \frac{a}{a^2 - ab}\{3a + b\} = \frac{3a + b}{a - b}$$

and

$$v = \frac{a}{a^2 - ab}\{a + 3b\} = \frac{a + 3b}{a - b}$$

This solution has also been given by Brahmagupta and Narayana (later in this book).

Example (Bhaskara II)

Solve

$$3x + 1 = u^2 \tag{1}$$

$$5x + 1 = v^2 \quad \text{in positive integers (BBI, 118)} \tag{2}$$

First method:
Let $u = 3y + 1$.
Then, from (1), $3x + 1 = (3y + 1)^2 = 9y^2 + 6y + 1$.

$$\Rightarrow x = \frac{9y^2 + 6y}{3} = 3y^2 + 2y$$

Substituting in (2), we get

$$5(3y^2 + 2y) + 1 = v^2$$

$$15y^2 + 10y + 1 = v^2$$

Multiplying by 15, we get

$$225y^2 + 150y + 15 = 15v^2$$

$$(15y+5)^2 - 25 + 15 = 15v^2$$

$$(15y+5)^2 = 15v^2 + 10$$

Let $w = 15y + 5$.
Then, we have

$$15v^2 + 10 = w^2 \quad (*)$$

which is a Vargaprakriti equation.
We observe that $(v,w) = (l\ m) = (1\ 5)$ is a solution of (*).
Then, another solution of (*) is

$$v = lq + mp = q \pm 5p$$

$$w = mq + 15pl = 5q + 15p$$

(*Continued*)

Example (Bhaskara II) (cont'd)

where $15p^2 + 1 = q^2$.

$$15y + 5 = w = 5q \pm 15p$$

$$y = \frac{1}{15}[5q \pm 15p - 5] = \frac{1}{3}[q \pm 3p - 1]$$

$$u = 3y + 1 = q \pm 3p - 1 + 1 = q \pm 3p$$

Hence,

$$x = \frac{1}{3}(u^2 - 1) = \frac{1}{3}\left[(q \pm 3p)^2 - 1\right]$$

$$u = q \pm 3p$$

$$v = q \pm 5p$$

is the general solution of (1) and (2), where $15p^2 + 1 = q^2$.

In particular, if $p = 0$, $q = 1$, then $x = 0$, $u = 1$, $v = 1$.

If $p = 1$, $q = 4$, then $x = 0, 16$.

$u = 1$ or 7; $v = -1$ or 9 respectively discarding the negative values, we have $x = 16$, $u = 7$, $v = 9$.

To find another solution of $15p^2 + 1 = q^2$, we have seen (1 4) is a solution of $15p^2 + 1 = q^2 \Rightarrow (1\ 4\ 1)$.

$$(1\ 4\ 1) \circ (1\ 4\ 1) = (8\ 16 + 1 \cdot 15\ 1 \cdot 1) = (8\ 31\ 1)$$

which means that (8 31) is a solution of $15p^2 + 1 = q^2$.

So, substituting $p = 8$, $q = 31$, we get

$$u = q \pm 3p = 31 \pm 3(8) = 55 \text{ or } 7$$

$$v = q \pm 5p = 31 \pm 40 = 71 \text{ or } -9$$

$$x = \frac{1}{3}\left[\{31 \pm 318\}^2 - 1\right] = \frac{1}{3}\left[(55)^2 - 1\right]$$

or 16 respectively discarding the negative values, we have

$$u = 55, \quad v = 71, \text{ and } x = \frac{1}{3}[3025 - 1] = \frac{1}{3}(3024) = 1008$$

We observe that infinite solutions of $15p^2 + 1 = q^2$ can be found by repeated samasa of ($\alpha\ \beta\ 1$), where $15\alpha^2 + 1 = \beta^2$.

Second method:

From (1),

$$x = \frac{1}{3}(u^2 - 1)$$

Substituting in (2), we get

$$\frac{5}{3}(u^2 - 1) + 1 = v^2$$

i.e., $5u^2 - 5 + 3 = 3v^2$.

i.e., $5u^2 = 3v^2 + 2$, which is a Vargaprakriti equation.

Multiplying by 5, we get

$$25u^2 = 15v^2 + 10$$

Let $w = 5u$.

Then, we have $15v^2 + 10 = w^2$.

As before $(v\ w) = (9\ 35), (71\ 275), \ldots$

$$v = 9 \Rightarrow 5x + 1 = 81 \Rightarrow 5x = 80 \Rightarrow x = 16$$

$$v = 71 \Rightarrow 5x + 1 = (71)^2 = 5041 \Rightarrow 5x = 5040 \Rightarrow x = 1008$$

as before.

Bhaskara II gave the following two examples to illustrate the solution simultaneous indeterminate equations of second degree.

Example 1

Solve

$$x^2 + y^2 + 1 = u^2 \qquad (1)$$

$$x^2 - y^2 + 1 = v^2 \quad \text{(BBI, 108)} \qquad (2)$$

Let $x^2 = 5z^2 - 1$ and $y^2 = 4z^2$.

Then, we have $u^2 = 5z^2 - 1 + 4z^2 + 1 = 9z^2 \Rightarrow u = 3z$ and $v^2 = 5z^2 - 1 - 4z^2 + 1 = z^2 \Rightarrow v = z$ from (1) and (2).

(Continued)

Example 1 (cont'd)

We now have $x^2 = 5z^2 - 1$ (*), which is "Vargaprakriti" equation.
We observe that $(z\ x) = (l\ m) = (1\ 2)$ is a solution of (*).
Then, another solution of (*) is

$$z = ql \pm mp = q \pm 2p \Rightarrow y = 2z = 2(q \pm 2p)$$

$$x = mq \pm 5lp = 2q \pm 5p$$

where $5p^2 + 1 = q^2$.
We observe that if $p = 0$, $q = 1$ also, if $p = 4$, $q = 9$.
If $p = 0$, $q = 1$,

$$x = 2(1) \pm 0 = 2, \quad y = 2(1 \pm 0) = 2, \text{ and } z = 1$$

Thus, if $p = 0$, $q = 1$, $x = 2$, $y = 2$, $u = 3z = 3$, or $v = z = 1$.
If $p = 4$, $q = 9$, $x = 2(9) \pm 5(4) = 18 \pm 20 = 38$, leaving the negative sign

$$y = 2(9) \pm 4(4) = 18 \pm 16 = 34$$

leaving the negative sign

$$z = \frac{y}{2} = 17, \ u = 3z = 51, \text{ and } v = z = 17$$

and so on.
Infinite solutions of $5p^2 + 1 = q^2$ can be found by repeated samasa of (4 9 1) over itself.

Example 2 (Bhaskara II)

Solve

$$x^2 + y^2 - 1 = u^2 \quad (1)$$

$$x^2 - y^2 - 1 = v^2 \quad \text{(BBI, 108)} \quad (2)$$

Let $x^2 = 5z^2 + 1$, $y^2 = 4z^2$.
Then, from (1) and (2), we have

$$5z^2 + 1 + 4z^2 - 1 = u^2 \Rightarrow u^2 = 9z^2 \Rightarrow u = 3z$$

$$5z^2 + 1 - 4z^2 - 1 = v^2 \Rightarrow z^2 \Rightarrow v = z$$

Also, we have $x^2 = 5z^2 + 1$ (*), which is Vargaprakriti equation.

We observe that $(z\ x) = (l\ m) = (4\ 9)$ is a solution of (*). Then, another solution of (*) is

$$z = lq \pm mp = 4q \pm 9p \Rightarrow y = 2z = 8q \pm 18p \tag{3}$$

Also, $v = z = 4q \pm 9p$, $u = 3z = 12q \pm 27p$.

$$x = mq \pm 5pl = 9q \pm 20p \tag{4}$$

where $5p^2 + 1 = q^2$. Here (3) and (4) are the general solution of (1); we observe that if $p = 0$, $q = 1$; $x = 9(1) \pm 0 = 9$.

$$z = 4(1) \pm 0 = 4 \Rightarrow y = 2z = 8$$

Also, $u = 3z = 12$ and $v = z = 4$.

We also observe that if $p = 4$, $q = 9$.

Then, $x = 9(9) \pm 20(4) = 81 \pm 80 = 161$ or 1.

$$z = 4(9) \pm 9(4) = 72 \text{ or } 0$$

$$y = 2z = 144 \text{ or } 0$$

$$u = 3z = 3(72) \text{ or } 0, \text{ so } u = 216 \text{ or } 0$$

$$v = z = 72 \text{ or } 0$$

Bhaskara II pointed out that for the solutions of above two examples, a better general assumption will be

$$(*) x^2 = pz^2 \mp 1 \text{ and } y^2 = q^2 z^2$$

where p and q are such that $p \pm q^2 =$ a square, the upper sign taken for the first example and the lower sign taken for the second example.

Suppose

$$p + q^2 = a^2 \tag{i}$$

and

$$p - q^2 = b^2 \tag{ii}$$

Then,

$$a^2 - b^2 = 2q^2 \Rightarrow q^2 = \frac{1}{2}(a^2 - b^2)$$

(3) Let $a + b = m$, where m is any rational number, $m \neq 0$.
Then, $a^2 - b^2 = (a + b)(a - b) = 2q^2$.
i.e.,

$$m(a - b) = 2q^2 \Rightarrow a - b = \frac{2q^2}{m} \tag{4}$$

Adding (3) and (4), we get

$$2a = m + \frac{2q^2}{m} \Rightarrow a = \frac{1}{2}\left(m + \frac{2q^2}{m}\right)$$

Subtracting (4) from (3), we get

$$2b = m - \frac{2q^2}{m} \Rightarrow b = \frac{1}{2}\left(m - \frac{2q^2}{m}\right)$$

Now,

$$p = a^2 - q^2 = \left\{\frac{1}{2}\left(m + \frac{2q^2}{m}\right)\right\}^2 - q^2$$

So,

$$p = \frac{1}{4}\left[m^2 + 2m\frac{2q^2}{m} + \frac{4q^4}{m^4}\right] - q^2 = \frac{1}{4}m^2 + q^2 + \frac{q^4}{m^4} - q^2$$

$$= \frac{1}{4}m^2 + \frac{q^4}{m^4}$$

Bhaskara II stated that m^2 should be chosen so that p will be an integer.
Hence,

$$x^2 = pz^2 \mp 1 = \left(\frac{1}{4}m^2 + \frac{q^4}{m^4}\right)z^2 \mp 1 \quad (*)$$

$$y^2 = q^2z^2$$

the upper sign being taken for the first example and the lower sign taken for the second example, on pages 471 and 472, respectively

For the first example,

$$u^2 = x^2 + y^2 + 1$$

$$v^2 = x^2 - y^2 + 1$$

So,

$$u^2 = \left(\frac{m^2}{4} + \frac{q^4}{m^2}\right)z^2 + q^2z^2 - 1 + 1$$

$$= \frac{m^2z^2}{4} + \frac{q^4z^2}{m^2} + q^2z^2$$

$$= \frac{z^2}{4}\left(m^2 + \frac{4q^4}{m^2} + 4q^2\right)$$

$$u^2 = \frac{z^2}{4}\left(m + \frac{2q^2}{m}\right)^2 \Rightarrow u = \frac{z}{2}\left(m + \frac{2q^2}{m}\right)$$

Also,

$$v^2 = \left(\frac{m^2}{4} + \frac{q^4}{m^2}\right)z^2 - q^2z^2 - 1 + 1$$

$$= \left(\frac{m^2}{4} + \frac{q^4}{m^2} - q^2\right)z^2 = \frac{z^2}{4}\left(m^2 + \frac{4q^4}{m^2} - 4q^2\right)$$

$$v^2 = \frac{z^2}{4}\left(m - \frac{2q^2}{m}\right)^2 \Rightarrow v = \frac{z}{2}\left(m - \frac{2q^2}{m}\right)$$

In the second example,

$$x^2 + y^2 - 1 = u^2$$

$$x^2 - y^2 - 1 = v^2$$

So,

$$u^2 = \left(\frac{m^2}{4} + \frac{q^4}{m^2}\right)z^2 + 1 + q^2z^2 - 1$$

$$= z^2\left[\frac{m^2}{4} + \frac{q^4}{m^2} + q^2\right] = \frac{z^2}{4}\left[m^2 + \frac{4q^4}{m^2} + 4q^2\right]$$

$$= \frac{z^2}{2}\left(m + \frac{2q^2}{m}\right)^2$$

$$u = \frac{z}{2}\left(m + \frac{2q^2}{m}\right)$$

Also,

$$v^2 = \left(\frac{m^2}{4} + \frac{q^4}{m^2}\right)z^2 + 1 - q^2z^2 - 1$$

$$= \left(\frac{m^2}{4} + \frac{q^4}{m^2}\right)z^2 - q^2z^2 = z^2\left[\frac{m^2}{4} + \frac{q^4}{m^2} - q^2\right]$$

$$= \frac{z^2}{4}\left(m^2 + \frac{4q^4}{m^2} - 4q^2\right) = \frac{z^2}{4}\left(m - \frac{2q^2}{m}\right)^2$$

So,

$$v = \frac{z}{2}\left(m - \frac{2q^2}{m}\right)$$

We further observe that the equations $x^2 = pz^2 \mp 1$ being "Vargaprakriti" can be solved by Brahmagupta–Bhaskara II method, the upper sign being taken for the first example and the lower sign taken for the second example. Once we solve these equations, from the values of x and z so determined, values for y, u, and v can be determined. Here, Bhaskara II gave another method of solving the above two examples.

Since $a^2 + b^2 \pm 2ab$ is always a square, we may assume that $x^2 = (a^2 + b^2)z^2 \mp 1$ and $y^2 = 2abz^2$.

From this, we note that if y is to be a rational number, $2ab$ must be a square.

Hence, we take $a = 2m^2$ and $b = n^2$.

So, then, we have $x^2 = (4m^4 + n^4)z^2 \mp 1$ and $y^2 = 4m^2n^2z^2$, the upper sign being taken for the first example and the lower sign for the second example.

From first example, we have

$$x^2 + y^2 + 1 = u^2$$

$$x^2 - y^2 + 1 = v^2$$

Substituting the above values of x^2, y^2, we get

$$u^2 = (4m^4 + n^4)z^2 - 1 + 4m^2n^2z^2 + 1$$

$$= (4m^4 + n^4 + 4m^2n^2)z^2 = (2m^2 + n^2)^2z^2$$

So, $u = z(2m^2 + n^2)$.
Also, $v^2 = (4m^4 + n^4)z^2 - 1 - 4m^2n^2z^2 + 1$

$$= z^2(4m^4 + n^4 - 4m^2n^2) = z^2(2m^2 - n^2)^2$$

So, $v = (2m^2 - n^2)z$.
In the second example, we have

$$x^2 + y^2 - 1 = u^2$$

$$x^2 - y^2 - 1 = v^2$$

So, by substituting the values of x^2 and y^2, we get

$$u^2 = (4m^4 + n^4)z^2 + 1 + 4m^2n^2z^2 - 1$$

$$= z^2(4m^4 + n^4 + 4m^2n^2) = z^2(2m^2 + n^2)^2$$

So, $u = z(2m^2 + n^2)$.
Again, $v^2 = (4m^4 + n^4)z^2 + 1 - 4m^2n^2z^2 - 1$

$$= z^2(4m^4 + n^4 - 4m^2n^2) = z^2(2m^2 - n^2)^2$$

So, $v = z(2m^2 - n^2)$.
So, we have

$$x^2 = (4m^4 + n^4)z^2 \mp 1 \quad (**)$$

$$y^2 = 4m^2n^2z^2$$

the upper sign being taken for the first example and the lower sign taken for the second example.

We observe that the equation (*) on page 474 follow from (**) by replacing z by. Hence, we shall take (**) as our fundamental assumption for the solution of (i) and (ii). Since $(4m^4 + n^4)z^2 \mp 1 = x^2$ is a "Vargaprakriti" equation, it has infinite integral solutions and infinite number of solutions of original equation (1) and (2) can be found.

We will now find a rational solution of the equation $(4m^4 + n^4)z^2 + 1 = x^2$.
Let $a^2 = 4m^4 + n^4$.
Then, we have

$$a^2z^2 + 1 = x^2 \qquad \text{(i)}$$

Let

$$z = \frac{2r}{a^2 - r^2}$$

where r is a rational number such that $0 < r < a$.

Then, substituting the value of z in (i), we get

$$a^2 \cdot \frac{4r^2}{(a^2-r^2)^2} + 1 = x^2$$

$$\frac{4r^2a^2 + (a^2-r^2)^2}{(a^2-r^2)^2} = x^2$$

$$\frac{(a^2+r^2)^2}{(a^2-r^2)^2} = x^2 \Rightarrow x = \frac{a^2+r^2}{a^2-r^2}$$

Hence,

$$x = \frac{4m^4+n^4+r^2}{4m^4+n^4-r^2} \text{ and } z = \frac{2r}{a^2-r^2} = \frac{2r}{4m^4+n^4-r^2}$$

Therefore, the general solution of $x^2+y^2-1=u^2$; $x^2-y^2-1=v^2$ is

$$x = \frac{4m^4+n^4+r^2}{4m^4+n^4-r^2}$$

$$u = z(2m^2+n^2) = \frac{(2m^2+n^2)2r}{4m^4+n^4-r^2}$$

$$v = z(2m^2-n^2) = \frac{(2m^2-n^2)2r}{4m^4+n^4-r^2}$$

and $y^2 = u^2+1-x^2$.

$$y^2 = \frac{4r^2(2m^2+n^2)^2}{(4m^4+n^4-r^2)^2} + 1 - \frac{(4m^4+n^4+r^2)^2}{(4m^4+n^4-r^2)^2}$$

$$= \frac{4r^2(2m^2+n^2)^2 + \{(4m^4+n^4-r^2)^2 - (4m^4+n^4+r^2)^2\}}{(4m^4+n^4-r^2)^2}$$

$$= \frac{4r^2(2m^2+n^2)^2 + \{(4m^4+n^4-r^2+4m^4+n^4+r^2)(-2r^2)\}}{(4m^4+n^4-r^2)^2}$$

$$= \frac{4r^2(2m^2+n^2)^2 + (8m^4+2n^4)(-2r^2)}{(4m^4+n^4-r^2)^2}$$

$$= \frac{4r^2[(2m^2+n^2)^2 - (4m^4+n^4)]}{(4m^4+n^4-r^2)^2}$$

$$y^2 = \frac{4r^2(4m^2n^2)}{(4m^4+n^4-r^2)^2} = \frac{16m^2n^2r^2}{(4m^4+n^4-r^2)^2}$$

So,

$$y = \frac{4mnr}{4m^4 + n^4 - r^2}$$

where, m, n, r are all rational numbers.

If we put

$$r = \frac{s}{t}$$

we get Genocchi's solution (Nouv, Ann. Math x, 1851, pp. 80–85, also Dickson, Theory of Numbers, Vol. II, p. 479).

(a) Let $m = 2t$, $n = 1$, and $r = 8t^2 - 1$.

Then,

$$x = \frac{4(2t)^4 + 1 + (8t^2 - 1)^2}{4(2t)^4 + 1 - (8t^2 - 1)^2}$$

$$= \frac{64t^4 + 1 + 64t^4 - 16t^2 + 1}{64t^4 + 1 - 64t^4 + 16t^2 - 1}$$

$$x = \frac{128t^4 - 16t^2 + 2}{16t^2}$$

$$= \frac{64t^4 - 8t^2 + 1}{8t^2}$$

$$= \frac{64t^4 - 16t^2 + 1 + 8t^2}{8t^2} = \frac{(8t^2 - 1)^2 + 8t^2}{8t^2}$$

$$x = \frac{1}{2}\left(\frac{8t^2 - 1}{2t}\right)^2 + 1$$

$$y = \frac{4(2t)(1)(8t^2 - 1)}{16t^2} = \frac{8t^2 - 1}{2t}$$

$$u = \frac{2(8t^2 - 1)[2 \cdot 4t^2 + 1]}{16t^2}$$

$$= \frac{2(8t^2 - 1)[8t^2 + 1]}{16t^2}$$

$$= \frac{2(64t^4 + 8t^2 - 8t^2 - 1)}{16t^2}$$

$$= \frac{64t^4 - 1}{8t^2}$$

$$v = \frac{2(8t^2 - 1)[8t^2 - 1]}{16t^2} = \frac{(8t^2 - 1)^2}{8t^2} = \frac{1}{2}\left(\frac{8t^2 - 1}{2t}\right)^2$$

(b) Let $m = t$, $n = 1$, and $r = 2t^2 + 2t + 1$.
Then,

$$x = \frac{4t^4 + 4t^2 + 1 + 8t^3 + 4t^2 + 4t + 4t^4 + 1}{4t^4 + 1 - (4t^4 + 4t^2 + 1 + 8t^3 + 4t^2 + 4t)}$$

$$= \frac{8t^4 + 8t^3 + 8t^2 + 4t + 2}{-(8t^3 + 8t^2 + 4t)}$$

$$= -\frac{t(8t^3 + 8t^2 + 4t) + 4t^2 + 4t + 2}{8t^3 + 8t^2 + 4t} = -t - \frac{1}{2t}$$

$$y = \frac{4t(2t^2 + 2t + 1)}{-4t(2t^2 + 2t + 1)} = -1$$

$$u = \frac{2(2t^2 + 2t + 1)(2t^2 + 1)}{-4t(2t^2 + 2t + 1)} = -\frac{2t^2 + 1}{2t} = -t - \frac{1}{2t}$$

$$v = \frac{2(2t^2 + 2t + 1)(2t^2 - 1)}{-4t(2t^2 + 2t + 1)} = -\frac{2t^2 - 1}{2t} = -t + \frac{1}{2t}$$

So,

$$x = t + \frac{1}{2t},\ y = 1,\ u = t + \frac{1}{2t},\ \text{and } v = t - \frac{1}{2t}$$

discarding the negative sign.

(c) Let $m = t$, $n = 1$, and $r = 2t^2$.
Then,

$$x = \frac{4t^4 + 1 + 4t^4}{4t^4 + 1 - 4t^4} = 8t^4 + 1$$

$$y = \frac{4t \cdot 2t^2}{4t^4 + 1 - 4t^4} = 8t^3$$

$$u = \frac{2(2t^2)(2t^2 + 1)}{4t^4 + 1 - 4t^4} = 4t^2(2t^2 + 1)$$

$$v = \frac{2(2t^2)(2t^2 - 1)}{4t^2 + 1 - 4t^4} = 4t^2(2t^2 - 1)$$

Next, we consider the equation $(4m^4 + n^4)z^2 - 1 = x^2$, which is a Vargaprakriti equation.

Let

$$n^4 z^2 = 1 \Rightarrow z^2 = \frac{1}{n^4} \Rightarrow z = \frac{1}{n^2}$$

Then, we have

$$4m^4 \frac{1}{n^4} = x^2$$

So,

$$x = \frac{2m^2}{n^2}$$

So,

$$(z\ x) = \left(\frac{1}{n^2}\ \frac{2m^2}{n^2}\right)$$

is a solution of $(4m^4 + n^4)z^2 - 1 = x^2$, which means

$$\left(\frac{1}{n^2}\ \frac{2m^2}{n^2}\ -1\right)$$

Hence,

$$\left(\frac{1}{n^2}\ \frac{2m^2}{n^2}\ -1\right)\circ\left(\frac{1}{n^2}\ \frac{2m^2}{n^2}\ -1\right)\circ\left(\frac{1}{n^2}\ \frac{2m^2}{n^2}\ -1\right)$$

$$= \left(\left(\frac{1}{n^2}\ \frac{2m^2}{n^2}\cdot 2\right)\left\{\left(\frac{2m^2}{n^2}\right)^2 + \frac{1}{n^4}(4m^4 + n^4)\right\}1\right)\circ\left(\frac{1}{n^2}\ \frac{2m^2}{n^2}\ -1\right)$$

$$= \left(\frac{4m^2}{n^4}\ \frac{8m^4 + n^4}{n^4}\ 1\right)\circ\left(\frac{1}{n^2}\ \frac{2m^2}{n^2}\ -1\right)$$

$$= \left[\frac{4m^2}{n^4}\cdot\frac{2m^2}{n^2} + \frac{1}{n^2}\left(\frac{8m^4 + n^4}{n^4}\right)\left(\frac{8m^4 + n^4}{n^4}\right)\left(\frac{2m^2}{n^2}\right) + \frac{4m^2}{n^6}(4m^4 + n^4) - 1\right]$$

$$= \left(\frac{16m^4 + n^4}{n^6}\ \frac{32m^6 + 6m^2n^4}{n^6}\ -1\right)$$

which mean that

$$(z\ x) = \left(\frac{16m^4 + n^4}{n^6}\ \frac{32m^6 + 6m^2n^4}{n^6}\right)$$

is a solution of $(4m^4 + n^4)z^2 - 1 = x^2$.

We will now find another set of solutions of the equation

$$(4m^4 + n^4)z^2 - 1 = x^2$$

Let

$$4m^4z^2 - 1 = 0 \Rightarrow z^2 = \frac{1}{4m^4} \Rightarrow z = \frac{1}{2m^2}$$

Then,

$$n^4z^2 = x^2 \Rightarrow x = n^2z = \frac{n^2}{2m^2}$$

Then,

$$(z\ x) = \left(\frac{1}{2m^2}\ \frac{n^2}{2m^2}\right)$$

is a solution of $(4m^4 + n^4)z^2 - 1 = x^2$, which means

$$\left(\frac{1}{2m^2}\ \frac{n^2}{2m^2}\ -1\right)$$

Hence,

$$\left(\frac{1}{2m^2}\ \frac{n^2}{2m^2}\ -1\right)\circ\left(\frac{1}{2m^2}\ \frac{n^2}{2m^2}\ -1\right)\circ\left(\frac{1}{2m^2}\ \frac{n^2}{2m^2}\ -1\right)$$

$$= \left(\left(\frac{1}{2m^2}\ \frac{n^2}{2m^2}\cdot 2\right)\left\{\left(\frac{n^2}{2m^2}\right)^2 + \left(\frac{1}{2m^2}\right)^2(4m^4 + n^4)\right\}\ 1\right)\circ\left(\frac{1}{2m^2}\ \frac{n^2}{2m^2}\ -1\right)$$

$$= \left(\frac{n^2}{2m^4}\ \frac{n^4 + 2m^4}{2m^4}\ 1\right)\circ\left(\frac{1}{2m^2}\ \frac{n^2}{2m^2}\ -1\right)$$

$$= \left(\left(\frac{n^2}{2m^4}\frac{n^2}{2m^2} + \frac{n^4 + 2m^4}{2m^4}\cdot\frac{1}{2m^2}\right)\left(\frac{n^4 + 2m^4}{2m^4}\frac{n^2}{2m^2} + \frac{n^2}{2m^4}\frac{1}{2m^2}(4m^4 + n^4)\right)\ -1\right)$$

$$= \left(\frac{m^4 + n^4}{2m^6}\ \frac{n^6 + 3m^4n^2}{2m^6}\ -1\right)$$

which means that

$$(z\ x) = \left(\frac{m^4 + n^4}{2m^6}\ \frac{n^6 + 3m^4n^2}{2m^6}\right)$$

is a solution of $(4m^4 + n^4)z^2 - 1 = x^2$.

(1) If

$$(z\ x) = \left(\frac{1}{n^2}\ \frac{2m^2}{n^2}\right)$$

then from page 477

$$u = (2m^2 + n^2)z = \frac{2m^2 + n^2}{n^2}$$

$$v = (2m^2 - n^2)z = \frac{2m^2 - n^2}{n^2}$$

$$y = 2mnz = 2mn \cdot \frac{1}{n^2} = \frac{2m}{n}$$

(2) If

$$(z\ x) = \left(\frac{16m^4 + n^4}{n^6}\ \frac{32m^6 + 6m^2n^4}{n^6}\right)$$

$$u = (2m^2 + n^2)z = \frac{(2m^2 + n^2)(16m^4 + n^4)}{n^6}$$

$$v = (2m^2 - n^2)z = \frac{(2m^2 - n^2)(16m^4 + n^4)}{n^6}$$

$$y = 2mnz = \frac{2mn(16m^4 + n^4)}{n^6} = \frac{2m(16m^4 + n^4)}{n^5}$$

(3) If

$$(z\ x) = \left(\frac{1}{2m^2}\ \frac{n^2}{2m^2}\right)$$

$$u = (2m^2 + n^2)z = \frac{(2m^2 + n^2)}{2m^2}$$

$$v = (2m^2 - n^2)z = \frac{(2m^2 - n^2)}{2m^2}$$

$$y = 2mnz = 2mn\frac{1}{2m^2} = \frac{n}{m}$$

(4) If

$$(z\,x) = \left(\frac{m^4 + n^4}{2m^6}\ \frac{n^6 + 3n^2m^4}{2m^6}\right)$$

$$u = (2m^2 + n^2)z = \frac{(2m^2 + n^2)(m^4 + n^4)}{2m^6}$$

$$v = \frac{(2m^2 - n^2)(m^4 + n^4)}{2m^6}$$

$$y = 2mnz = 2mn\frac{(m^4 + n^4)}{2m^6} = \frac{n(m^4 + n^4)}{m^5}$$

(a) Let $n = 1$ in (1).
We get $x = 2m^2$, $y = 2m$, $u = 2m^2 + 1$, and $v = 2m^2 - 1$.
(b) Let $n = 1$ in (2).
We get $x = 2m^2(16m^4 + 3)$ and $y = 2m(16m^4 + 1)$.

$$u = (16m^4 + 1)(2m^2 + 1)$$

$$v = (16m^4 + 1)(2m^2 - 1)$$

(c) Let $m = 1$ in (3).
We get

$$x = \frac{1}{2}n^2, \quad y = n$$

$$u = \frac{1}{2}(n^2 + 2), \quad v = \frac{1}{2}(n^2 - 2)$$

(d) Let $m = 1$ in (4).
We get

$$x = \frac{1}{2}n^2(n^4 + 3), \quad y = n(n^4 + 1)$$

$$u = \frac{1}{2}(n^4 + 1)(n^2 + 2) \text{ and } v = \frac{1}{2}(n^4 + 1)(n^2 - 2)$$

Solution (a) was given by Drummond (American Mathematician, Mon. IX, 1902, p. 232).

Solve:

$$ax^2 + by^2 = u^2 \tag{1}$$

$$Ax^2 + By^2 + C = v^2 \quad \text{(BBI, 119)} \tag{2}$$

Let the solution of (1) be $x = my$ and $u = ny$.
Then, substituting in (1), we get

$$a(m^2\ y^2) + by^2 = n^2y^2 \Rightarrow am^2 + b = n^2$$

Substituting in (2), we get

$$Am^2y^2 + By^2 + C = v^2$$

$(Am^2 + B)y^2 + C = v^2$, which is "Vargaprakriti" which can be solved by Brahmagupta–Bhaskara II method.

Example

Solve:

$$7x^2 + 8y^2 = u^2 \tag{1}$$

$$7x^2 - 8y^2 + 1 = v^2 \quad \text{(BBI, 119)} \tag{2}$$

Let $x = 2y$ in (1), we get

$$7(4y^2) + 8y^2 = u^2 \Rightarrow 36y^2 = u^2 \Rightarrow u = 6y$$

Letting $x = 2y$ in (2), we get

$$7(4y^2) - 8y^2 + 1 = v^2$$

(*Continued*)

Example (cont'd)

$20y^2 + 1 = v^2$, which is "Vargaprakriti" equation. Observe that $(y\ v) = (l\ m) = (2\ 9)$ is a solution of $20y^2 + 1 = v^2$.

Another solution of this equation is given by

$$y = lq \pm mp \text{ and } v = mq \pm 20pl, \text{ where } 20p^2 + 1 = q^2$$

i.e., $y = 2q \pm 9p$ and $v = 9q \pm 40p$, where $20p^2 + 1 = q^2$.

If $p = 0$, $q = 1$, then $y = 2$, $x = 2y = 4$, $v = 9$, $u = 6y = 12$.

If $p = 2$, $q = 9$, then $y = 2(9) \pm 9(2) = 36$ or 0.

$$x = 2y = 72 \text{ or } 0, \quad u = 6y = 216 \text{ or } 0$$

$$v = 9(9) \pm 40(2) = 81 \pm 80 = 161 \text{ or } 1$$

If $x = 0$, $y = 0$, $u = 0$, $v = 1$, which is a trivial solution.

If $x = 4$, $y = 2$, $u = 12$, $v = 9$.

If $x = 72$, $y = 36$, $u = 216$, $v = 161$. We observe that we obtain infinite solution of $20p^2 + 1 = q^2$ by taking samasa of (2 9 1) on itself.

Solve:

$$a(x^2 \pm y^2) + b = u^2 \quad \text{(Bharakara II)}$$

$$A(x^2 \pm y^2) + B = v^2$$

Let $x^2 \pm y^2 = w$.

Then, we have

$$aw + b = u^2$$

$$Aw + B = v^2$$

The method of solving this equation has been given on page 466 in this book.

Example

Solve:

$$2(x^2 - y^2) + 3 = u^2$$

$$3(x^2 - y^2) + 3 = v^2 \quad (*) \quad \text{(BBI, 119)}$$

Let $x^2 - y^2 = w$.
Then, we have

$$2w + 3 = u^2 \tag{1}$$

$$3w + 3 = v^2 \tag{2}$$

Multiplying (1) by 3, we get $6w + 9 = 3u^2$.
Multiplying (2) by 2, we get $6w + 6 = 2v^2$.
Subtracting, we get

$$3 = 3u^2 - 2v^2 \text{ or } 2v^2 + 3 = 3u^2 \tag{3}$$

which is "Vargaprakriti" equation. We will now solve (3) by Brahmagupta–Bhaskara II method.

Multiplying (3) by 3, we get $9u^2 = 6v^2 + 9$.
Let $z = 3u$. Then, we have $6v^2 + 9 = z^2$.
We observe that $6 \cdot 1^2 + 3 = 9 = 3^2$.
So, (1,3) is a solution of $6v^2 + 3 = z^2 \Rightarrow (1\ 3\ 3)$.
Now, $(1\ 3\ 3)(1\ 3\ 3) = ((1 \cdot 3 + 3 \cdot 1\ 3 \cdot 3 + 1 \cdot 1 \cdot 6\ 3 \cdot 3)$
$= (6\ 15\ 9) \Rightarrow (v\ z) = (6\ 15)$ is a solution of $6v^2 + 9 = z^2$.
Another solution of $6v^2 + 9 = z^2$ is $v = lq \pm mp$ and $z = mq \pm 6pl$, where $(l\ m) = (6\ 15)$ and $6p^2 + 1 = q^2$.
So, $v = 6q \pm 15p$ and $z = 15q \pm 36p$.
If $p = 0$, $q = 1$, then $v = 6$ and $z = 15$.
Now,

$$u = \frac{z}{3} = \frac{15}{3} = 5$$

(*Continued*)

Example (cont'd)

If $p = 2$, $q^2 = 6(4) + 1 = 25 \Rightarrow q = 5$.
Then, $v = 6(5) \pm 15(2) = 60$ or 0.

$$z = 15(5) \pm 36(2) = 75 \pm 72 = 147 \text{ or } 3$$

$$u = \frac{z}{3} = \frac{147}{3} \text{ or } \frac{3}{3} = 49 \text{ or } 1$$

$$2w + 3 = u^2 \Rightarrow w = \frac{1}{2}(u^2 - 3)$$

If

$$u = 5, \quad w = \frac{1}{2}(25 - 3) = 11$$

If

$$u = 49, \quad w = \frac{1}{2}(49^2 - 3) = \frac{1}{2}[2401 - 3]$$

$$= \frac{1}{2}[2398] = 1199$$

So, $w = 11, 1199, \ldots$
But $w = x^2 - y^2 = (x + y)(x - y)$.
Let $x - y = m$, where m is a rational number $\neq 0$.
Then,

$$x + y = \frac{x^2 - y^2}{x - y} = \frac{w}{m}$$

Adding, we get

$$2x = m + \frac{w}{m} \Rightarrow x = \frac{1}{2}\left(m + \frac{w}{m}\right)$$

Subtracting, we get

$$2y = \frac{w}{m} - m \Rightarrow y = \frac{1}{2}\left(\frac{w}{m} - m\right)$$

where $w = 11, 1199, \ldots$ and w is a rational number $\neq 0$.

Solve:

$$ax^2 + by^2 + c = u^2 \quad (1)$$

$$Ax^2 + By^2 + C = v^2 \quad (2)$$

From (1),

$$ax^2 = u^2 - by^2 - c \text{ or } x^2 = \frac{1}{a}(u^2 - by^2 - c)$$

Substituting in (2), we get

$$\frac{A}{a}(u^2 - by^2 - c) + By^2 + C = v^2$$

$$y^2\left(B - \frac{Ab}{a}\right) + \left(C - \frac{Ac}{a}\right) = v^2 - \frac{A}{a}u^2$$

Let

$$v^2 - \frac{A}{a}u^2 = z^2$$

$$B - \frac{Ab}{a} = A^*$$

and

$$C - \frac{Ac}{a} = B^*$$

We have $A{*}y^2 + B{*} = z^2$ (*), which is "Vargaprakriti," equation that can be solved by Brahmagupta–Bhaskara II method. From, different values of z obtained from (*), we further solve the equation $v^2 = (A/a)u^2 + z^2$ as another "Vargaprakriti" equation. Bhaskara II has not given any example on this equation. We will provide an example.

Find an integral solution of:

$$3x^2 + 2y^2 - 5 = u^2 \tag{1}$$

$$8x^2 + 6y^2 + 14 = v^2 \tag{2}$$

From (1),

$$x^2 = \frac{1}{3}(u^2 - 2y^2 + 5) \tag{3}$$

Then substituting in (2), we get

$$\frac{8}{3}(u^2 - 2y^2 + 5) + 6y^2 + 14 = v^2$$

$$8u^2 - 16y^2 + 40 + 18y^2 + 42 = 3v^2$$

$$2y^2 + 82 = 3v^2 - 8u^2$$

Let $z^2 = 3v^2 - 8u^2$.
Then, we have

$$2y^2 + 82 = z^2 \quad (*)$$

We observe that $(y\ z) = (3\ 10)$ is a solution of $(*)$.
We now solve $z^2 = 3v^2 - 8y^2$.
Let $v = cz$ and $u = dz$.
We get $z^2 = 3c^2z^2 - 8d^2z^2 \Rightarrow 3c^2 - 8d^2 = 1 \Rightarrow 3c^2 = 8d^2 + 1$.
We observe that if

$$d = \frac{1}{2}, \quad 3c^2 = 8 \cdot \frac{1}{4} + 1 = 3 \Rightarrow 3c^2 = 3 \Rightarrow c = 1$$

Hence, $u = 5$, $v = 10$, and $y = 3$.
Substituting in (3), we get

$$x^2 = \frac{1}{3}(5^2 - 2(3)^2 + 5) = \frac{1}{3}(25 - 18 + 5) = \frac{12}{3} = 4$$

Hence, $x = 2$.
Hence, $(x\ y\ u\ v) = (2\ 3\ 5\ 10)$ is a solution of (1) and (2).

Solve:

$$x^2 + xy + y^2 = u^2 \tag{1}$$

$$(x + y)u + 1 = v^2 \quad \text{(BBI, 107)} \tag{2}$$

Multiplying (1) by 36, we get

$$36x^2 + 36xy + 36y^2 = 36u^2$$

$$(6x+3y)^2 + 27y^2 = 36u^2$$

$$27y^2 = 36u^2 - (6x+3y)^2$$

$$= \{6u + (6x + 3y)\}\{6u - (6x + 3y)\}$$

Let

$$6u - (6x + 3y) = \lambda, \text{ where } \lambda \text{ is any rational number} \neq 0 \tag{3}$$

Then,

$$6u + 6x + 3y = \frac{36u^2 - (6x+3y)^2}{6u - (6x + 3y)} = \frac{27y^2}{\lambda} \tag{4}$$

Adding (3) and (4), we get

$$12u = \lambda + \frac{27y^2}{\lambda} \Rightarrow u = \frac{1}{12}\left[\lambda + \frac{27y^2}{\lambda}\right]$$

Subtracting (3) from (4), we get

$$2(6x + 3y) = \frac{27y^2}{\lambda} - \lambda$$

$$\Rightarrow 6x + 3y = \frac{1}{2}\left[\frac{27y^2}{\lambda} - \lambda\right]$$

If we take $\lambda = y$, then

$$6x + 3y = \frac{1}{2}[27y - y] = 13y$$

$$6x = 10y \Rightarrow x = \frac{5}{3}y$$

and

$$u = \frac{1}{12}[y + 27y] = \frac{14}{6}y = \frac{7}{3}y$$

Substituting in (2), we get

$$\left(\frac{5}{3}y + y\right)\frac{7}{3}y + 1 = v^2$$

$$\frac{56y^2}{9} + 1 = v^2 \Rightarrow 56y^2 + 9 = 9v^2 = w^2$$

where $w = 3v$.

We observe that $56y^2 + 9 = w^2$ is a "Vargaprakriti" equation. We now solve this by Brahmagupta–Bhaskara II method.

We observe that

$$56 \cdot 2^2 + 1 = 224 + 1 = 225 = 15^2 \quad (*)$$

So, (2 15) is a solution of $56y^2 + 1 = w^2$.

Multiplying (*) by 3^2, we get

$$56 \cdot 2^2 \cdot 3^2 + 3^2 = 15^2 \cdot 3^2$$

$$\Rightarrow 56(6)^2 + 9 = (45)^2,$$

which means that $(y\ w) = (l\ m) = (6\ 45)$ is a solution of $56y^2 + 9 = w^2$. Then, another solution of $56y^2 + 9 = w^2$ is

$$y = lq \pm mp = 6q \pm 45p$$

$$w = mq \pm 56pl = 45q \pm 336p$$

Since $56p^2 + 1 = q^2$, and if $p = 0$, $q = 1$, then $y = 6$, $w = 45$.

$$v = \frac{w}{3} = \frac{45}{3} = 15, \quad x = \frac{5}{3}y = \frac{5}{3}(6) = 10$$

$$u = \frac{7y}{3} = \frac{7}{3}(6) = 14$$

and if $p = 2$, $q = 15$, then

$$y = 6(15) \pm 45(2) = 180 \text{ or } 0$$

$$x = \frac{5}{3}y = \frac{5}{3}(180) = 300 \text{ or } 0$$

$$u = \frac{7}{3}y = \frac{7}{3}(180) = 420 \text{ or } 0$$

$$w = 45(15) \pm 336(2) = 675 \pm 672 = 1347 \text{ or } 3$$

$$v = \frac{w}{3} = 449 \text{ or } 1$$

So, if $x = 0$, $y = 0$, $u = 0$, $v = 1$. This is a trivial solution.

If $x = 10$, $y = 6$, $u = 14$, $v = 15$.

If $x = 300$, $y = 180$, $u = 420$, $v = 449$, and so on.

We observe that infinite solutions of $56p^2 + 1 = q^2$ can be found by taking repeated samasa of (2 15 1) on itself.

Example

"Tell me quickly those two numbers such that the sum of cube of one and square of the other becomes a square and whose sum is also a square."

(BBI, 107)

Solution

Let x and y be the two numbers.

Then, we have to solve

$$x^3 + y^2 = u^2 \quad (1)$$

$$x + y = v^2 \quad (2)$$

Bhaskara II has solved this problem in two different ways.

Method 1: From (1), $x^3 = u^2 - y^2 = (u - y)(u + y)$.

Let $u - y = \lambda$, where λ is any arbitrary number $\neq 0$.

Then,

$$u + y = \frac{x^3}{\lambda}$$

(*Continued*)

Example (cont'd)

Adding, we get

$$2u = \lambda + \frac{x^3}{\lambda} \Rightarrow u = \frac{1}{2}\left(\lambda + \frac{x^3}{\lambda}\right)$$

Subtracting the first equation from the second, we get

$$2y = \frac{x^3}{\lambda} - \lambda \Rightarrow y = \frac{1}{2}\left(\frac{x^3}{\lambda} - \lambda\right)$$

Let $\lambda = x$.
Then, we get

$$u = \frac{1}{2}(x + x^2) \text{ and } y = \frac{1}{2}(x^2 - x)$$

Then, from (2)

$$x + \frac{1}{2}(x^2 - x) = v^2$$

$$2x + x^2 - x = 2v^2$$

$$x^2 + x = 2v^2$$

Multiplying by 4, we get $4x^2 + 4x = 8v^2$, i.e., $(2x + 1)^2 = 8v^2 + 1$.
Let $w = 2x + 1$.
Then, we have $8v^2 + 1 = w^2$ (*), which is a "Vargaprakriti" equation.
We observe that $v = 1$, $w = 3$ is a solution of (*).
If $(l\ m) = (v\ w) = (1\ 3)$ is a solution of (*), then another solution of (*) is given by $(1\ 3\ 1) \cdot (1\ 3\ 1) = (1 \cdot 3 + 3 \cdot 1 \quad 3 \cdot 3 + 1 \cdot 1 \cdot 8 \quad 1) = (6 \quad 17 \quad 1)$, which means that $(6 \quad 17)$ is also a solution of $8v^2 + 1 = w^2$.
If

$$w = 3, \quad w = 2x + 1 \Rightarrow 2x + 1 = 3 \Rightarrow x = 1$$

$$v = 1$$

If

$$w = 17, \quad 2x + 1 = w = 17 \Rightarrow x = 8$$

$$v = 6$$

Now,

$$y = \frac{1}{2}(x^2 - x), \quad u = \frac{1}{2}(x^2 + x)$$

If

$$x = 1, \ y = \frac{1}{2}(1 - 1) = 0, \ u = \frac{1}{2}(1 + 1) = 1, \ v = 1$$

This is a trivial solution.

If

$$x = 8, \ y = \frac{1}{2}(64 - 8) = \frac{1}{2}(56) = 28, \ u = \frac{1}{2}(64 + 8) = 36$$

Another solution of (*) is

$$[\,1 \cdot (17) + 3(6) \quad 3 \cdot (17) + 1 \cdot 6 \cdot 8 \quad 1 \cdot 1\,] = [\,35 \quad 99 \quad 1\,]$$

which means (35 99) = (v w) is also a solution of (*).

$$w = 99 = 2x + 1 \Rightarrow x = 49$$

$$v = 35$$

$$y = \frac{1}{2}(x^2 - x) = \frac{1}{2}(49^2 - 41) = \frac{1}{2}[2401 - 49] = \frac{2352}{2} = 1176$$

$$u = \frac{1}{2}(x^2 + x) = \frac{1}{2}(2401 + 49) = \frac{2450}{2} = 1225$$

$$(x\ y\ u\ v) = (8\ 28\ 36\ 6), (49\ 1176\ 1225\ 35), \text{ and so on}$$

Method 2: Let $x = 2w^2$ and $y = 7w^2$.

From (2), $x + y = 2w^2 + 7w^2 = 9w^2 = v^2 \Rightarrow v = 3w$.

From (1), we have $(2w^2)^3 + (7w^2)^2 = u^2$.

$$8w^6 + 49w^4 = u^2$$

i.e., $w^4(8w^2 + 149) = u^2$.

i.e., $w^4z^2 = u^2$, where $z^2 = 8w^2 + 49$.

So, $u = w^2z$, where $z^2 = 8w^2 + 49$.

We now have to solve $z^2 = 8w^2 + 49$, which is "Vargaprakriti" equation.

(*Continued*)

Example (cont'd)

We observe that $8 \cdot 1^2 - 7 = 1^2$.
So, (1 1) is a solution of $8w^2 - 7 = z^2$.

$$\Rightarrow (1 \ \ 1 \ \ -7)$$

So, $(1\ 1\ \ -7) \circ (1\ 1\ \ -7) = (2\ \ (1 + 1 \cdot 8)\ \ (-7)(-7)) = (2\ \ 9\ \ 49)$
So, $(w,z) = (l\ m) = (2\ 9)$ is a solution of $8w^2 + 49 = z^2$ (*).
Another solution of (*) is

$$w = ql \pm mp \Rightarrow w = 2q \pm 9p$$

$$z = qm \pm 8pl \Rightarrow z = 9q \pm 16p$$

where $8p^2 + 1 = q^2$.
If $p = 0$, $q = 1$, then $w = 2$ and $z = 9$.

$$x = 2w^2 = 2(4) = 8$$

$$y = 7w^2 = 7(4) = 28$$

$$u = w^2 z = 4(9) = 36$$

$$v = 3w = 6$$

If $p = 1$, $q = 3$, then, $w = 2(3) \pm 9(1) = 6 \pm 9 = 15$.

$$z = 9(3) \pm 16(1) = 27 \pm 16 = 43$$

discarding the negative signs.

$$x = 2w^2 = 2(225) = 450$$

$$y = 7w^2 = 7(225) = 1575$$

$$v = 3w = 3(15) = 45$$

$$u = w^2 z = 225 \cdot 43 = 9675$$

So, $x = 8$, $y = 28$, $u = 36$, $v = 6$ is a solution of (1) and (2) as well as $x = 450$, $y = 1575$, $v = 45$, and $u = 9675$. We observe that infinite solutions of $8p^2 + 1 = q^2$ can be found by repeated samasa of (1 3 1) on itself.

Example (Bhaskara II)

"The sum of cubes of two numbers is a square and the sum of the squares is a cube. If you know them, then I will accept that you are a great algebraist."

(BBI, 56)

Let x and y be the two numbers.
Then, we have

$$x^3 + y^3 = u^2 \tag{1}$$

$$x^2 + y^2 = v^3 \tag{2}$$

Let $x = pz^2$ and $y = qz^2$.
Then, $x^3 + y^3 = p^3z^6 + q^3z^6 = (p^3 + q^3)z^6 = u^2$.

$$\Rightarrow u = (p^3 + q^3)^{\frac{1}{2}}z^3$$

Then from (2),

$$p^2z^4 + q^2z^4 = v^3 \Rightarrow (p^2 + q^2)z^4 = v^3$$

Let $v = rz$.
Then, $(p^2 + q^2)z^4 = r^3z^3$.
i.e.,

$$(p^2 + q^2)z = r^3 \Rightarrow z = \frac{r^3}{p^2 + q^2}$$

$$x = pz^2 = \frac{pr^6}{(p^2+q^2)^2}, \quad y = qz^2 = \frac{qr^6}{(p^2+q^2)^2}$$

$$v = rz = \frac{r^4}{p^2 + q^2}$$

$$u = (p^3 + q^3)^{\frac{1}{2}}z^3 = (p^3 + q^3)^{\frac{1}{2}}\frac{r^9}{(p^2+q^2)^3}$$

Here p and q are integers such that $p^3 + q^3$ is a square. One simple solution is $p = 1$, $q = 2$.

(Continued)

Example (Bhaskara II) (cont'd)

Then,

$$x = \frac{r^6}{(1+4)^2} = \frac{r^6}{25}, \quad y = \frac{2r^6}{(1+4)^2} = \frac{2r^6}{25}$$

$$v = \frac{r^4}{1+4} = \frac{r^4}{5}, \quad u = \frac{(1+8)^{\frac{1}{2}} r^9}{(1+4)^3} = \frac{3r^9}{125}$$

This particular solution has been given by Narayana Pandita (G.K.1). In general, $p^3 + q^3$ can be made a square, if we choose

$$p = (m^3 + n^3)m \quad q = (m^3 + n^3)n$$

Then, $p^3 + q^3 = (m^3 + n^3)^3(m^3 + n^3) = (m^3 + n^3)^4$.
Hence, the solution of (1) and (2) is

$$x = \frac{pr^6}{(p^2+q^2)^2} = \frac{(m^3 + n^3)mr^6}{[(m^3+n^3)^2 m^2 + (m^3+n^3)^2 n^2]^2}$$

$$= \frac{(m^3 + n^3)mr^6}{\{(m^3+n^3)^4(m^2+n^2)\}^2}$$

$$= \frac{mr^6}{(m^3+n^3)^3(m^2+n^2)^2}$$

$$y = \frac{qr^6}{(p^2+q^2)^2} = \frac{(m^3 + n^3)nr^6}{(m^3+n^3)^4(m^2+n^2)^2}$$

$$= \frac{nr^6}{(m^3+n^3)^3(m^2+n^2)^2}$$

$$v = \frac{r^4}{p^2 + q^2} = \frac{r^4}{(m^3+n^3)^2 m^2 + (m^3+n^3)^2 n^2}$$

$$v = \frac{r^4}{(m^3+n^3)^2(m^2+n^2)}$$

$$u = \frac{(p^3+q^3)^{\frac{1}{2}} r^9}{(p^2+q^2)^3} = \frac{\{(m^3+n^3)^4\}^{\frac{1}{2}} r^9}{(m^3+n^3)^6(m^2+n^2)^3}$$

$$= \frac{r^9}{(m^2+n^2)^3} \cdot \frac{1}{(m^3+n^3)^4}$$

Letting $r = (m^3 + n^3)(m^2 + n^2)$, we get

$$x = m(m^3 + n^3)^3(m^2 + n^2)^4$$

$$y = n(m^3 + n^3)^3(m^2 + n^2)^4$$

$$u = (m^3 + n^3)^5(m^2 + n^2)^6$$

$$v = (m^3 + n^3)^2(m^2 + n^2)^3$$

Example (Bhaskara II) (Bijaganitha, page 103)

"Oh most learned algebraist, find the various pairs of integers whose difference is a square and the sum of whose square is a cube."

Solution

Let x and y be the two numbers.

Then,

$$y - x = u^2 \quad (1)$$

$$y^2 + x^2 = v^3 \quad (2)$$

From (1), $x = y - u^2$.

Substituting in (2), we have

$$y^2 + (y - u^2)^2 = v^3$$

$$y^2 + y^2 - 2yu^2 + u^4 = v^3$$

$$2y^2 - 2yu^2 + u^4 = v^3$$

Let $v = u^2$.

Then, this becomes $2y^2 - 2yu^2 = u^6 - u^4$.

Multiplying by 2 and adding u^4 to both sides, we get

$$4y^2 - 4yu^2 + u^4 = 2u^6 - u^4$$

$$(2y - u^2)^2 = 2u^6 - u^4 = u^4(2u^4 - 1)$$

(*Continued*)

Example (Bhaskara II) (Bijaganitha, page 103) (cont'd)

So, we have

$$\left(\frac{2y-u^2}{u^2}\right)^2 = 2u^2 - 1$$

i.e., $w^2 = 2u^2 - 1$, where

$$w = \frac{2y - u^2}{u^2} = \frac{2y}{u^2} - 1$$

We now solve

$$w^2 = 2u^2 - 1 \quad (*)$$

We observe that $2 \cdot 1^2 - 1 = 2 - 1 = 1^2$.
So, $(u\ m) = (1\ 1)$ is a solution of $w^2 = 2u^2 - 1$.
Another solution of (*) is

$$u = lq \pm mp = q \pm p$$

$$w = mq \pm 2pl = q \pm 2p$$

where $2p^2 + 1 = q^2$.

If $p = 0$, $q = 1$, then $u = 1$, $w = 1$.

If $p = 2$, $q = 3$, then $u = 3 \pm 2$ and $w = 3 \pm 4 \Rightarrow u = 5$, $w = 7$, discarding the negative signs.

If $p = 12$, $q = 17$, then $u = 17 \pm 12 = 29$ or 5.

$$w = 17 \pm 24 = 41 \text{ or } -7$$

So, $u = 29$, $w = 41$ discarding the negative sign.
So, $(u,w) = (1\ 1), (5\ 7), (29\ 41)$, and so on.
Now, if $u = 1$, $w = 1$, then

$$w = \frac{2y - u^2}{u^2} \Rightarrow 1 = 2y - 1$$

$$\Rightarrow 2y = 2 \Rightarrow y = 1$$

and $x = y - u^2 = 1 - 1 = 0$.

$$v = u^2 = 1$$

So, $x = 0$, $y = 1$, $u = v = 1$ is a solution of (1) and (2). This is a trivial solution.

If

$$u = 5, \quad w = 7, \quad w = \frac{2y}{u^2} - 1$$

$$\Rightarrow 7 = \frac{2y}{25} - 1 \Rightarrow 8 = \frac{2y}{25} \Rightarrow y = 100$$

$$x = y - u^2 = 100 - 25 = 75$$

$$v = u^2 = 25$$

So, $x = 75$, $y = 100$, $u = 5$, $v = 25$ is another solution (1) and (2).

If

$$u = 29, \quad w = 41, \quad w = \frac{2y}{u^2} - 1$$

So,

$$41 = \frac{2y}{(29)^2} - 1 \Rightarrow 2y = (29)^2 \cdot 42$$

$$y = 21(29)^2 = 21(841) = 17,661$$

$$x = y - u^2 = 17,661 - (29)^2 = 17,661 - 841 = 16,820$$

$$v = u^2 = (29)^2 = 841$$

So, $x = 16{,}820$, $y = 17{,}661$, $u = 29$, $v = 841$ is another solution of (1) and (2) and so on.

Bhaskara II pointed out that he took the following problem from an earlier mathematician. Bhaskara II gave two methods.

Method 1: Let x be the number.

Then,

$$3x + 1 = u^3 \tag{1}$$

$$3u^2 + 1 = v^2 \quad \text{(BBI 119)} \tag{2}$$

(*Continued*)

Example (Bhaskara II) (Bijaganitha, page 103) (cont'd)

We note that (2) is a "Vargaprakriti" equation. We solve (2) by Brahmagupta–Bhaskara II method.

We observe that $3 \cdot 1^2 + 1 = 4 = 2^2$.

So, $(u\ v) = (l\ m) = (1\ 2)$ is a solution of

$$3u^2 + 1 = v^2 \quad (*)$$

So, another solution of (*) is given by

$$u = ql \pm mp = q \pm 2p$$

$$v = mq \pm 3lp = 2q \pm 3p, \quad \text{where } 3p^2 + 1 = q^2$$

If $p = 0$, $q = 1$, then $u = 1$, $v = 2$.

$$3x + 1 = u^3 = 1 \Rightarrow x = 0$$

So, $x = 0$, $u = 1$, $v = 2$ is a solution of (1) and (2). This is a trivial solution.

If $p = 1$, $q = 2$, then $u = 2 \pm 2 = 4$ or 0.

$$v = 4 \pm 3 = 7 \text{ or } 1$$

So, $u = 4$, $v = 7$, discarding the negative signs.

$$3x + 1 = u^3 = 4^3 = 64 \Rightarrow 3x = 63 \Rightarrow x = 21$$

So, $x = 21$, $u = 4$, $v = 7$ is a solution of (1) and (2).

If $p = 4$, $q = 7$.

Then, $u = 7 \pm 8 = 15$ or -1.

$$v = 14 \pm 12 = 26 \text{ or } 2$$

Discarding the negative signs, $u = 15$, $v = 26$.

$$3x + 1 = u^3 = (15)^3 = 3375 \Rightarrow 3x = 3374$$

$$\Rightarrow x = \frac{3374}{3}$$

So,

$$x = \frac{3374}{3}, \quad u = 15, \quad v = 26$$

is a also a solution of (1) and (2) and so on. We observe that infinite solutions of $3u^2 + 1 = v^2$ can be found by repeated samasa of (1 2 1) on itself.

Method 2 (Bijaganitha, page 121): Let $u = 3y + 1$.

Then from (1),

$$3x = u^3 - 1 = (3y+1)^3 - 1 = 27y^3 + 27y^2 + 9y$$

$$x = 9y^3 + 9y^2 + 3y = 3y(3y^2 + 3y + 1)$$

From (2), we have $3(3y + 1)^2 + 1 = v^2$.

$$1 + 3(9y^2 + 6y + 1) = v^2$$

$$27y^2 + 18y + 4 = v^2$$

Let $v = my - 2$.
Then, $27y^2 + 18y + 4 = (my - 2)^2 = m^2y^2 - 4my + 4$.
Or $27y + 18 = m^2y - 4m$.
Hence, $18 + 4m = m^2y - 27y = y(m^2 - 27)$.
Thus,

$$y = \frac{18 + 4m}{m^2 - 27}$$

Since $y > 0$, $m^2 - 27 > 0 \Rightarrow m^2 > 27$.

$$x = 3y(3y^2 + 3y + 1)$$

$$= 3\left(\frac{18 + 4m}{m^2 - 27}\right)\left[3\left(\frac{18+4m}{m^2-27}\right)^2 + 3\left(\frac{18 + 4m}{m^2 - 27}\right) + 1\right]$$

where m is a rational number such that $m^2 > 27$.

We observe that if $m = 9$,

$$y = \frac{18 + 4m}{m^2 - 27} = \frac{18 + 36}{81 - 27} = \frac{54}{54} = 1$$

and $x = 3[3 + 3 + 1] = 21$.

$$v = my - 2 = 9(1) - 2 = 7$$

$$u = 3y + 1 = 3 + 1 = 4$$

So, $x = 21$, $u = 4$, $v = 7$ is a solution of (1) and (2).

Example (Bhaskara II, Bijaganitha, page 56)

"If you are an expert in mathematics, tell me quickly those numbers whose sum and difference are squares and whose product is a cube."

Let x and y be the two numbers.

Then, we have

$$x + y = u^2 \tag{1}$$

$$x - y = v^2 \tag{2}$$

$$xy = w^3 \tag{3}$$

Let $x = (m^2 + n^2)z^2$ and $y = 2mnz^2$.

Then, $x + y = z^2(m^2 + n^2 + 2mn) = z^2(m+n)^2 = u^2$ from (1)

$$\Rightarrow u = z(m + n)$$

Again $x - y = z^2(m^2 + n^2 - 2mn) = z^2(m-n)^2 = v^2$ from (2)

$$\Rightarrow v = z(m - n)$$

Let $w = az$ from (3)

Then, $2mnz^4(m^2 + n^2) = a^3z^3$.

$$z = \frac{a^3}{2mn(m^2 + n^2)}$$

$$x = (m^2 + n^2)z^2 = \frac{(m^2 + n^2)a^6}{4m^2n^2(m^2 + n^2)^2} = \frac{a^6}{4m^2n^2(m^2 + n^2)}$$

$$y = 2mnz^2 = \frac{2mna^6}{4m^2n^2(m^2 + n^2)^2} = \frac{a^6}{2mn(m^2 + n^2)^2}$$

where m, n, a are arbitrary numbers.

Example (Bhaskara II, Bijaganitha, page 68)

"Find four numbers such that each of them separately added to a given number become a perfect square. So also are the products of contiguous pair increased by another given number."

Let α and β be the two given numbers and let x, y z, and w be the four required numbers.

Then,

$$x + \alpha = a^2 \quad xy + \beta = e^2$$

$$y + \alpha = b^2 \quad yz + \beta = f^2$$

$$z + \alpha = c^2 \quad zw + \beta = g^2$$

$$w + \alpha = d^2$$

Bhaskara II gave the following method, which has been used by an earlier mathematician.

Now, $x = a^2 - \alpha, \quad y = b^2 - \alpha$

$$xy + \beta = (a^2 - \alpha)(b^2 - \alpha) + \beta$$

$$= a^2b^2 - \alpha(a^2 + b^2) + \alpha^2 + \beta$$

$$= a^2b^2 - 2ab\alpha + \alpha^2 + 2ab\alpha - \alpha(a^2 + b^2) + \beta$$

$$xy + \beta = (ab - \alpha)^2 + [\beta - \alpha(a^2 + b^2 - 2ab)]$$

$$xy + \beta = (ab - \alpha)^2 + \beta - \alpha(a - b)^2$$

In order that $xy + \beta = e^2$, where e is a rational number, sufficient condition is $\beta - \alpha(a - b)^2 = 0$.

$$\Rightarrow \beta = \alpha(a - b)^2 \Rightarrow (b - a)^2 = \frac{\beta}{\alpha}$$

i.e.,

$$b - a = \pm\sqrt{\frac{\beta}{\alpha}}$$

(*Continued*)

Example (Bhaskara II, Bijaganitha, page 68) (cont'd)

So,

$$b = a \pm \sqrt{\frac{\beta}{\alpha}} = a \pm \gamma, \quad \text{where } \gamma = \sqrt{\frac{\beta}{\alpha}}$$

Then, $ab - \alpha = e$.
Now,

$$f^2 = yz + \beta = (b^2 - \alpha)(c^2 - \alpha) + \beta$$

$$= b^2c^2 - \alpha(b^2 + c^2) + \alpha^2 + \beta$$

$$= b^2c^2 - 2bc\alpha + \alpha^2 + 2bc\alpha - \alpha(b^2 + c^2) + \beta$$

$$= (bc - \alpha)^2 + \beta - \alpha(b^2 + c^2 - 2bc)$$

$$= (bc - \alpha)^2 + \beta - \alpha(b - c)^2$$

If f is to be a rational number, sufficient condition is

$$\beta - \alpha(b - c)^2 = 0 \Rightarrow \alpha(c - b)^2 = \beta$$

$$\Rightarrow (c - b)^2 = \frac{\beta}{\alpha} \Rightarrow c - b = \pm\sqrt{\frac{\beta}{\alpha}} \Rightarrow c = b \pm \gamma \text{ and } f = bc - \alpha$$

Similarly, $d = c \pm \gamma$ and $g = cd - \alpha$.
Now, a, b, c, and d are such that

$$b = a \pm \gamma, \quad c = b \pm \gamma = a \pm 2\gamma, \quad d = c \pm \gamma = a \pm 3$$

So, a, b, c, and d form an arithmetic progressions with common difference $\gamma = \sqrt{(\beta/\alpha)}$.
Moreover,

$$x = a^2 - \alpha$$

$$y = b^2 - \alpha = (a \pm \gamma)^2 - \alpha$$

$$z = c^2 - \alpha = (a \pm 2\gamma)^2 - \alpha$$

$$w = d^2 - \alpha = (a \pm 3\gamma)^2 - \alpha$$

These values of the unknowns satisfy all the condition about their products.

$$e^2 = xy + \beta = (ab - \alpha)^2 = [a(a \pm \gamma) - \alpha]^2$$

$$f^2 = yz + \beta = (bc - \alpha)^2 = [(a \pm \gamma)(a \pm 2\gamma) - \alpha]^2$$

$$g^2 = zw + \beta = (cd - \alpha)^2 = [(a \pm 2\gamma)(a \pm 3\gamma) - \alpha]^2$$

so that $e = a(a \pm \gamma) - \alpha$.

$$f = (a \pm \gamma)(a \pm 2\gamma) - \alpha$$

$$g = (a + 2\gamma)(a \pm 3\gamma) - \alpha$$

We observe that this above principle will hold even when all the β's are different. Suppose in the above example, the second set of conditions is replaced by the following:

$$xy + \beta_1 = e^2, \quad yz + \beta_2 = f^2, \quad \text{and} \quad zw + \beta_3 = g^2$$

then proceeding as before, we will find that

$$b = a \pm \sqrt{\frac{\beta_1}{\alpha}}, \quad c = b \pm \sqrt{\frac{\beta_2}{\alpha}}, \quad \alpha = c \pm \sqrt{\frac{\beta_3}{\alpha}},, \quad \text{and} \quad e = ab - \alpha$$

$f = bc - d$ and $g = cd - \alpha$ as before.

Example (Bhaskara II, BBI, page 67)

"What are those four numbers which when added by 2 form a square. Also the products of two contiguous which when added by 18 form a square and which are such that the square root of the sum of all the square roots added by 11 becomes 13? Tell them to me, Oh algebraist friend."

Let x, y, z, and w be the four numbers.

Then, $x + 2 = a^2 \quad xy + 18 = e^2$.

$$y + 2 = b^2 \quad yz + 18 = f^2$$

$$z + 2 = c^2 \quad zw + 18 = g^2$$

$$w + 2 = d^2$$

(Continued)

Example (Bhaskara II, BBI, page 67) (cont'd)

Also

$$\sqrt{a+b+c+d+e+f+g+11} = 13 \quad (3)$$

In the previous general problem, let $\alpha = 2$, $\beta = 18$.

$$\gamma = \pm\sqrt{\frac{\beta}{\alpha}} = \pm\sqrt{\frac{18}{2}} = \pm 3$$

Hence, $x = a^2 - \alpha = a^2 - 2$.

$$y = b^2 - \alpha = b^2 - 2 = (a \pm \gamma)^2 - 2 = (a \pm 3)^2 - 2$$

$$z = c^2 - \alpha = c^2 - 2 = (a \pm 2\gamma)^2 - 2 = (a \pm 6)^2 - 2$$

$$w = d^2 - \alpha = d^2 - 2 = (a \pm 3\gamma)^2 - 2 = (a \pm 9)^2 - 2$$

$$xy + 18 = e^2 = (ab - \alpha)^2 = \{a(a \pm 3) - 2\}^2$$

$$yz + 18 = f^2 = (bc - \alpha)^2 = \{(a \pm 3)(a \pm 6) - 2\}^2$$

$$zw + 18 = g^2 = (cd - \alpha)^2 = \{(a \pm 6)(a \pm 9) - 2\}^2$$

Also,

$$\sqrt{a+b+c+d+e+f+g+11} = 13$$

Squaring, we get

$$a+b+c+d+e+f+g+11 = 169$$

$$a + a + 3 + a + 6 + a + 9 + [a(a+3) - 2]$$
$$+ [(a+6)(a+3) - 2] + [(a+9)(a+6) - 2]$$

$$= 169 - 11 = 158$$

$$4a + 18 + a^2 + 3a - 2 + a^2 + 9a + 18 - 2 + a^2 + 15a + 54 - 2 = 158$$

$$3a^2 + 31a + 84 = 158$$

$$3a^2 + 31a - 74 = 0$$

$$(3a + 37)(a - 2) = 0$$

$$a = 2 \text{ or } a = \frac{-37}{3}$$

Hence, $a = 2$, discarding the negative value of a,

$$b = a + \gamma = 2 + 3 = 5$$

$$c = a + 2\gamma = 2 + 6 = 8$$

$$d = a + 3\gamma = 2 + 9 = 11$$

$$e = a(a + 3) - 2 = 2(2 + 3) - 2 = 10 - 2 = 8$$

$$f = (a + 3)(a + 6) - 2 = (2 + 3)(2 + 6) - 2 = 16 - 2 = 38$$

$$g = (a + 6)(a + 9) - 2 = (2 + 6)(2 + 9) - 2 = 22 - 2 = 86$$

and

$$x = a^2 - 2 = 2^2 - 2 = 4 - 2 = 2$$

$$y = (a+3)^2 - 2 = (2+3)^2 - 2 = 25 - 2 = 23$$

$$z = (a+6)^2 - 2 = (2+6)^2 - 2 = 64 - 2 = 62$$

$$w = (a+9)^2 - 2 = (2+9)^2 - 2 = 121 - 2 = 119$$

Example (Bhaskara II, BBI, page 111)

Find two numbers x and y such that

$$x - y + k = a^2 \quad (1)$$

$$x + y + k = b^2 \quad (2)$$

(Continued)

Example (Bhaskara II, BBI, page 111) (cont'd)

$$x^2 - y^2 + k' = c^2 \tag{3}$$

$$x^2 + y^2 + k'' = d^2 \tag{4}$$

We assume that $a = w \pm \alpha$, where w is any rational number. Then, we will first prove that

$$b = a + \sqrt{\frac{k'}{k}}$$

$$= w \pm \alpha + \sqrt{\frac{k'}{k}} = w + \alpha + \gamma$$

where

$$\gamma = \sqrt{\frac{k'}{k}}$$

Solution

$$x - y = a^2 - k \quad \text{from (1)}$$

$$x + y = b^2 - k \quad \text{from (2)}$$

$$x^2 - y^2 + k' = (x + y)(x - y) + k' = (a^2 - k)(b^2 - k) + k'$$

$$= a^2b^2 - k(a^2 + b^2) + k^2 + k^1$$

$$= a^2b^2 - 2abk + k^2 + 2abk - k(a^2 + b^2) + k'$$

$$= (ab - k)^2 + [k' - k(a^2 + b^2 - 2ab)]$$

$$= (ab - k)^2 + [k' - k(a - b)^2] = c^2$$

If c is a rational number, it is sufficient that if $k' - k(a-b)^2 = 0 \Rightarrow$ $K(a-b)^2 = k'$.

$$\Rightarrow (b-a)^2 = \frac{k'}{k} \Rightarrow (b-a) = \pm\sqrt{\frac{k'}{k}} = \pm\gamma$$

where

$$\gamma = \sqrt{\frac{k'}{k}}$$

So,

$$b = a \pm \sqrt{\frac{k'}{k}} = a \pm \gamma = w \pm \alpha \pm \gamma$$

From (1), $x - y = a^2 - k = (w \pm \alpha)^2 - k$.
From (2), $x + y = b^2 - k = (w \pm \alpha + \gamma)^2 - k$.
Adding, we get $2x = (w \pm \alpha)^2 - k + (w \pm \alpha + \gamma)^2 - k$.
So,

$$x = \frac{1}{2}\left[(w \pm \alpha)^2 + (w \pm \alpha \pm \gamma)^2 - 2k\right]$$

Subtracting (1) from (2), we get $2y = [(w \pm \alpha \pm \gamma)^2 - (w \pm \alpha)^2]$.

$$y = \frac{1}{2}\left[(w \pm \alpha \pm \gamma)^2 - (w \pm \alpha)^2\right]$$

If we replace a by u, we have
Let $a = u = w \pm \alpha$, then

$$x = \frac{1}{2}\left[u^2 \pm (u+\gamma)^2 - 2k\right]$$

$$y = \frac{1}{2}\left[(u \pm \gamma)^2 - u^2\right]$$

(Continued)

Example (Bhaskara II, BBI, page 111) (cont'd)

Taking the positive signs, we get

$$x = \frac{1}{2}\left[u^2 + u^2 + 2u\gamma + \gamma^2 - 2k\right]$$

$$= \frac{1}{2}\left[2u^2 + 2u\gamma + \gamma^2 - 2k\right] = u^2 + u\gamma - k + \frac{\gamma^2}{2}$$

$$y = \frac{1}{2}\left[u^2 + 2u\gamma + \gamma^2 - u^2\right] = u\gamma + \frac{\gamma^2}{2}$$

So,

$$x^2 + y^2 = \left(u^2 + u\gamma - k + \frac{\gamma^2}{2}\right)^2 + \left(u\gamma + \frac{\gamma^2}{2}\right)^2$$

$$= u^4 + 2u^3\gamma - 2u^2k + u^2\gamma^2 + u^2\gamma^2 - 2u\gamma k + u\gamma^3 + k^2 - k\gamma^2$$

$$+ \frac{\gamma^4}{4} + u^2\gamma^2 + u\gamma^3 + \frac{\gamma^4}{4}$$

So,

$$x^2 + y^2 = u^4 + 2u^3\gamma + u^2(3\gamma^2 - 2k) + 2u\gamma(\gamma^2 - k) + \frac{\gamma^2}{2} - k\gamma^2 + k^2$$

$$= u^4 + 2u^3\gamma + u^2(3\gamma^2 - 2k) + 2\gamma u(\gamma^2 - k) + \frac{1}{2}(\gamma^2 - k) + \frac{k^2}{2}$$

Now, $d^2 = x^2 + y^2 + k'$.
So,

$$d^2 = u^4 + 2u^3v + u^2(3\gamma^2 - 2k) + 2\gamma u(\gamma^2 - k) + \frac{1}{2}(\gamma^2 - k)^2 + \frac{k^2}{2} + k' \quad (*)$$

We have to solve (*) which is an indeterminate equation in u.

Example (Bhaskara II, Bijaganitha, page 115)

Solve

$$x - y + 3 = a^2 \tag{1}$$

$$x + y + 3 = b^2 \tag{2}$$

$$x^2 - y^2 + 12 = c^2 \tag{3}$$

$$x^2 + y^2 - 4 = d^2 \tag{4}$$

$$\frac{1}{2}xy + y = e^3 \tag{5}$$

Also,

$$a + b + c + d + e + 2 = f^2 \tag{6}$$

Bhaskara II solved this problem in two different methods.

Method 1: Let $a = w - 1$.

From (1), $(x - y) = a^2 - 3 = (w - 1)^2 - 3 = w^2 - 2w - 2$.

Here $k = 3$, $k' = 12$.

So,

$$\gamma = \sqrt{\frac{k'}{k}} = \sqrt{\frac{12}{3}} = 2$$

So, $b = a + \gamma = a + 2 = w - 1 + 2 = w + 1$.

From (2), $x + y = b^2 - 3 = (w + 1)^2 - 3 = w^2 + 2w - 2$.

So, we have $x - y = w^2 - 2w - 2$.

$$x + y = w^2 + 2w - 2$$

Adding, we get $2x = 2(w^2 - 2) \Rightarrow x = w^2 - 2$.

Subtracting, we get $2y = 4w \Rightarrow y = 2w$.

Now, $c^2 = x^2 - y^2 + 12 = (w^2 - 2)^2 - (2w)^2 + 12$.

$$= w^4 - 8w^2 + 16 = (w^2 - 4)^2 \Rightarrow c = w^2 - 4$$

$$d^2 = x^2 + y^2 - 4 = (w^2 - 2)^2 + (2w)^2 - 4$$

(*Continued*)

Example (Bhaskara II, Bijaganitha, page 115) (cont'd)

$$= w^4 - 4w^2 + 4 + 4w^2 - 4 = w^4 \Rightarrow d = w^2$$

$$\frac{1}{2}xy + y = e^3 \quad \text{from (5)}$$

So,

$$e^3 = \frac{1}{2}(w^2 - 2)2w + 2w = (w^2 - 2)w + 2w = w^3$$

$$\Rightarrow e = w$$

Also, $a + b + c + d + e + 2 = f^2$.

$$w - 1 + w + 1 + w^2 - 4 + w^2 + w + 2 = f^2$$

$$2w^2 + 3w - 2 = f^2$$

Multiplying by 8, we get

$$16w^2 + 24w - 16 = 8f^2$$

$$(4w+3)^2 - 9 - 16 = 8f^2$$

(*) i.e., $z^2 = 8f^2 + 25$, where $z = 4w + 3$ we observe that (*) is a "Vargaprakriti" equation. We solve this by Brahmagupta–Bhaskara II method.

We observe that $8 \cdot 5^2 + 25 = 225 = 15^2$.

So, $(l\ m) = (5\ 15)$ is a solution of $8f^2 + 25 = z^2$.

Another solution of (*) is given by

$$f = ql \pm pm \Rightarrow f = 5q \pm 15p$$

$$z = qm \pm 8pl \Rightarrow z = 15q \pm 40p$$

where $8p^2 + 1 = q^2$.

If $p = 0$, $q = 1$, then $f = 5$ and $z = 15$.

Now, $4w + 3 = z = 15 \Rightarrow 4w = 12 \Rightarrow w = 3$.

Now, $x = w^2 - 2 = 9 - 2 = 7.$

$$y = 2w = 2(3) = 6$$

$$a = w - 1 = 3 - 1 = 2$$

$$b = w + 1 = 3 + 1 = 4$$

$$c = w^2 - 4 = 9 - 4 = 5$$

$$d = w^2 = 3^2 = 9; \quad e = w = 3$$

If $p = 1$, $q = 3$, then $f = 5q \pm 15p = 15 \pm 15 = 30$ or 0.
$z = 15q \pm 40p = 45 \pm 40 = 85$ or 0.
Discarding the negative signs, we get

$$4w + 3 = z = 85 \Rightarrow 4w = 82 \Rightarrow w = \frac{41}{2}$$

Then,

$$x = w^2 - 2 = \left(\frac{41}{2}\right)^2 - 2 = \frac{1681}{4} - 2 = \frac{1673}{4}$$

$$y = 2w = 2\left(\frac{41}{2}\right) = 41$$

$$a = w - 1 = \frac{41}{2} - 1 = \frac{39}{2}$$

$$b = w + 1 = \frac{41}{2} + 1 = \frac{43}{2}$$

$$c = w^2 - 1 = \left(\frac{41}{2}\right)^2 - 1 = \frac{1677}{4}$$

$$d = w^2 = \left(\frac{41}{2}\right)^2 = \frac{1681}{4}$$

$$e = w = \frac{41}{2}$$

and so on. We observe that infinite solutions of $8p^2 + 1 = q^2$ can be obtained by repeated samasa of (1 3 1) on itself.

(*Continued*)

Example (Bhaskara II, Bijaganitha, page 115) (cont'd)

Method 2: As before, $b = a + \gamma = a + 2$.

From (1) we have $x - y + 3 = a^2$.

From (2), we have $x + y + 3 = b^2 = (a + 2)^2 = a^2 + 4a + 4$.

Adding, we get $2x + 6 = 2a^2 + 4a + 4 \Rightarrow x = a^2 + 2a - 1$.

Subtracting, we get $2y = 4a + 4 \Rightarrow y = 2a + 2$.

Now, $c^2 = x^2 - y^2 + 12 = (a^2 + 2a - 1)^2 - (2a + 2)^2 + 12$

$$= a^4 + 4a^3 - 2a^2 - 12a + 9 = (a^2+2a-3)^2$$

So, $c = a^2 + 2a - 3$.

$$d^2 = x^2 + y^2 - 4 = (a^2+2a-1)^2 + (2a+2)^2 - 4$$

$$d^2 = a^4 + 4a^3 + 6a^2 + 4a + 1 = (a^2+2a+1)^2$$

So, $d = a^2 + 2a + 1$.

$$e^3 = \frac{1}{2}xy + y = y\left[\frac{x}{2} + 1\right] = (2a + 2)\left[\frac{a^2 + 2a - 1}{2} + 1\right]$$

$$e^3 = (a + 1)[a^2 + 2a + 1] = (a+1)^3$$

$$e = a + 1$$

Also, $a + b + c + d + e + 2 = f^2$.

i.e., $a + a + 2 + a^2 + 2a - 3 + a^2 + 2a + 1 + a + 1 + 2 = f^2$.

So, we have $2a^2 + 7a + 3 = f^2$.

Multiplying by 8, we get

$$16a^2 + 56a + 24 = 8f^2$$

$$(4a+7)^2 - 49 + 24 = 8f^2$$

$$w^2 = 8f^2 + 25, \text{ where } w = 4a + 7$$

This equation has already been solved.

One solution is $f = 5$, $w = 15$. Another solution is $f = 30$, $w = 85$.

(a) Let $f = 5$, $w = 15$.

If $w = 15$, $4a + 7 = w = 15 \Rightarrow a = 2$.

$$b = a + 2 = 2 + 2 = 4$$

$$c = a^2 + 2a - 3 = 4 + 4 - 3 = 5$$

$$d = a^2 + 2a + 1 = 4 + 4 + 1 = 9$$

$$e = a + 1 = 2 + 1 = 3$$

$$x = a^2 + 2a - 1 = 4 + 4 - 1 = 7$$

$$y = 2a + 2 = 4 + 2 = 6$$

(b) Let $f = 30$, $w = 85$.
If

$$w = 85,\quad 4a + 7 = w = 85 \Rightarrow 4a = 78 \Rightarrow a = \frac{39}{2}$$

$$b = a + 2 = \frac{39}{2} + 2 = \frac{43}{2},\ c = a^2 + 2a - 3$$

So,

$$c = \left(\frac{39}{2}\right)^2 + 2\left(\frac{39}{2}\right) - 3 = \frac{1677}{4}$$

$$d = a^2 + 2a + 1 = (a+1)^2 = \left(\frac{39}{2} + 1\right)^2 = \left(\frac{41}{2}\right)^2 = \frac{1681}{4}$$

$$e = a + 1 = \frac{39}{2} + 1 = \frac{41}{2}$$

$$x = a^2 + 2a - 1 = \left(\frac{39}{2}\right)^2 + 2 \cdot \frac{39}{2} - 1 = \frac{1673}{4}$$

$$y = 2a + 2 = 2 \cdot \frac{39}{2} + 2 = 39 + 2 = 41$$

The following example was borrowed by Bhaskara II from on earlier mathematician.

Example (Bhaskara II, BBI, page 110)

"Tell me quickly, Oh algebraist, two numbers excepting 6 and 8 which are such that the cube root of half of the sum of their product and the smaller one, the square root of the sum of their squares, the square roots of the sum and difference of them (each) increased by 2 and of the difference of their squares plus 8, all being added together is capable of giving a square root."

Let x and y be two numbers such that $x > y$.

Then, we have

$$\sqrt[3]{\left(\frac{xy+y}{2}\right)} + \sqrt{x^2+y^2} + \sqrt{x+y+2} + \sqrt{x-y+2} + \sqrt{x^2-y^2-8} = f^2$$

So, we have to solve

$$x+y+2=a^2 \tag{1}$$

$$x-y+2=b^2 \tag{2}$$

$$x^2+y^2=c^2 \tag{3}$$

$$x^2-y^2+8=d^2 \tag{4}$$

$$\frac{1}{2}(xy+y)=e^3 \tag{5}$$

Also,

$$a+b+c+d+e=f^2 \tag{6}$$

Bhaskara II gave four different methods of solving this system of equations:

First method:

Let $x = w^2 - 1$ and $y = 2w$.
Now, $a^2 = x + y + 2 = w^2 - 1 + 2w + 2 = w^2 + 2w + 1 = (w+1)^2$.
So, $a = w + 1$.

$$b^2 = x - y + 2 = w^2 - 1 - 2w + 2 = w^2 - 2w + 1 = (w-1)^2$$

So, $b = w - 1$.

$$c^2 = x^2 + y^2 = (w^2-1)^2 + (2w)^2 = w^4 - 2w^2 + 1 + 4w^2 = (w^2+1)^2$$

So, $c = w^2 + 1$.

$$d^2 = x^2 - y^2 + 8 = (w^2 - 1)^2 - (2w)^2 + 8 = w^4 - 2w^2 + 1 - 4w^2 + 8$$

$$= w^4 - 6w^2 + 9 = (w^2 - 3)^2$$

So, $d = w^2 - 3$.

$$e^3 = \frac{1}{2}(xy + y) = \frac{1}{2}(x + 1)y = \frac{1}{2}(w^2 - 1 + 1)2w = w^3$$

So, $e = w$.
Also $a + b + c + d + e = f^2$.
i.e., $w + 1 + w - 1 + w^2 + 1 + w^2 - 3 + w = f^2$.

$$2w^2 + 3w - 2 = f^2$$

Multiplying by 8, we get

$$16w^2 + 24w - 16 = 8f^2$$

$$(4w+3)^2 - 9 - 16 = 8f^2$$

i.e., $z^2 = 8f^2 + 25$, where $z = 4w + 3$.
This "Vargaprakriti" equation has been solved earlier and two solutions we found were
(i) $f = 5$, $z = 15$; (ii) f = 30, $z = 85$.

$$f = 5, \quad z = 15$$

$$4w + 3 = 15 \Rightarrow 4w = 12 \Rightarrow w = 3$$

$$a = w + 1 = 3 + 1 = 4, \quad b = w - 1 = 3 - 1 = 2$$

$$c = w^2 + 1 = 3^2 + 1 = 10$$

$$d = w^2 - 3 = 3^2 - 3 = 9 - 3 = 6$$

$$e = w = 3$$

$$x = w^2 - 1 = 9 - 1 = 8$$

$$y = 2w = 6$$

(*Continued*)

Example (Bhaskara II, BBI, page 110) (cont'd)

(ii) $f = 30$, $z = 85$.

$$4w + 3 = z = 85 \Rightarrow 4w = 82 \Rightarrow w = \frac{41}{2}$$

$$x = w - 1 = \left(\frac{41}{2}\right)^2 - 1 = \frac{1681}{4} - 1 = \frac{1677}{4}$$

$$y = 2w = 2 \cdot \frac{41}{2} = 41$$

$$a = w + 1 = \frac{41}{2} + 1 = \frac{43}{2}$$

$$b = w - 1 = \frac{41}{2} - 1 = \frac{39}{2}, \quad c = w^2 + 1 = \left(\frac{41}{2}\right)^2 + 1 = \frac{1685}{4}$$

$$d = w^2 - 3 = \left(\frac{41}{2}\right)^2 - 3 = \frac{1681}{4} - 3 = \frac{1669}{4}$$

$$e = w = \frac{41}{2}$$

Second method:
Let $x = w^2 + 2w$ and $y = 2w + 2$.

$$a^2 = x + y + 2 = w^2 + 2w + 2w + 4 = w^2 + 4w + 4 = (w+2)^2$$

So, $a = w + 2$.

$$b^2 = x - y + 2 = w^2 + 2w - (2w + 2) + 2 = w^2$$

$$b = w$$

$$c^2 = x^2 + y^2 = (w^2+2w)^2 + (2w+2)^2$$

$$= w^2(w^2 + 4w + 4) + 4(w^2 + 2w + 1)$$

$$= w^4 + 4w^3 + 8w^2 + 8w + 1$$

$$c^2 = (w^2+2w+2)^2 \Rightarrow c = w^2 + 2w + 2$$

$$d^2 = x^2 - y^2 + 8 = (w^2+2w)^2 - (2w+2)^2 + 8$$

$$= w^2(w^2 + 4w + 4) - 4(w^2 + 2w + 1) + 8$$

$$= w^4 + 4w^3 - 8w + 4 = (w^2+2w-2)^2$$

So, $d = w^2 + 2w - 2$.

$$e^3 = \frac{1}{2}(xy + y) = \frac{1}{2}y(x + 1) = \frac{1}{2}(2w + 2)\left[w^2 + 2w + 1\right]$$

$$e^3 = (w+1)^3 \Rightarrow e = w + 1$$

Now, $a + b + c + d + e = f^2$.

$$w + 2 + w + w^2 + 2w + 2 + w^2 + 2w - 2 + w + 1 = f^2$$

$$2w^2 + 7w + 3 = f^2$$

Multiplying by 8, we get

$$16w^2 + 56w + 24 = 8f^2$$

$$(4w+7)^2 - 49 + 24 = 8f^2$$

$$z^2 = 8f^2 + 25, \quad \text{where } z = 4w + 7$$

which we have already solved.

We will consider solutions (a) $f = 5$, $z = 15$; (b) $f = 30$, $z = 85$.

(a) $f = 5$, $z = 15$.

$$4w + 7 = z = 15 \Rightarrow 4w = 8 \Rightarrow w = 2$$

$$y = 2w + 2 = 2(2) + 2 = 6$$

$$a = w + 2 = 2 + 2 = 4; \quad b = w = 2$$

$$c = w^2 + 2w + 2 = 4 + 2(2) + 2 = 10$$

$$d = w^2 + 2w - 2 = 2^2 + 2(2) - 2 = 6$$

$$e = w + 1 = 2 + 1 = 3$$

(*Continued*)

Example (Bhaskara II, BBI, page 110) (cont'd)

(b) $f = 30,\ z = 85 \Rightarrow 4w + 7 = 85 \Rightarrow 4w = 78 \Rightarrow w = \frac{39}{2}$.

$$x = w^2 + 2w = \left(\frac{39}{2}\right)^2 + 2\left(\frac{39}{2}\right) = \frac{1677}{4}$$

$$y = 2w + 2 = 2\left(\frac{39}{2}\right) + 2 = 39 + 2 = 41$$

$$a = w + 2 = \frac{39}{2} + 2 = \frac{43}{2}, \quad b = w = \frac{39}{2}$$

$$c = w^2 + 2w + 2 = \left(\frac{39}{2}\right)^2 + 2\left(\frac{39}{2}\right) + 2 = \frac{1685}{4}$$

$$d = w^2 + 2w - 2 = \left(\frac{39}{2}\right)^2 + 2\left(\frac{39}{2}\right) - 2 = \frac{1669}{4}$$

$$e = w + 1 = \frac{39}{2} + 1 = \frac{41}{2}$$

Third method:
Let $x = w^2 + 4w + 3$ and $y = 2w + 4$.

$$a^2 = x + y + 2 = w^2 + 4w + 3 + 2w + 4 + 2 = w^2 + 6w + 9 = (w+3)^2$$

So, $a = w + 3$.

$$b^2 = x - y + 2 = w^2 + 4w + 3 - (2w + 4) + 2 = w^2 + 2w + 1 = (w+1)^2$$

So, $b = w + 1$.

$$c^2 = x^2 + y^2 = (w^2 + 4w + 3)^2 + (2w + 4)^2$$

$$= w^4 + 16w^2 + 9 + 8w^3 + 6w^2 + 24w + 4w^2 + 16w + 16$$

$$c^2 = w^4 + 8w^3 + 26w^2 + 40w + 25 = (w^2 + 4w + 5)^2$$

So, $c = w^2 + 4w + 5$.

$$d^2 = x^2 - y^2 + 8 = (w^2+4w+3)^2 - (2w+4)^2 + 8$$

$$= w^4 + 16w^2 + 9 + 8w^3 + 6w^2 + 24w - 4w^2 - 16w - 16 + 8$$

$$d^2 = w^4 + 8w^3 + 18w^2 + 8w + 1 = (w^2+4w+1)^2$$

So, $d = w^2 + 4w + 1$.

$$e^3 = \frac{1}{2}y(x+1) = \frac{1}{2}(2w+4)(w^2+4w+3+1) = (w+2)^3$$

So, $e = w + 2$.
Now, $a + b + c + d + e = f^2$.

$$w + 3 + w + 1 + w^2 + 4w + 5 + w^2 + 4w + 1 + w + 2 = f^2$$

$$2w^2 + 11w + 12 = f^2$$

Multiplying by 8, we get

$$16w^2 + 88w + 96 = 8f^2$$

$$(4w+11)^2 - 121 + 96 = 8f^2$$

$$(4w+11)^2 = 8f^2 + 25 \Rightarrow z^2 = 8f^2 + 25, \quad \text{where } z = 4w + 11$$

This equation has already been solved. We will consider the two solutions: (a) $f = 5$, $z = 15$; (b) $f = 30$, $z = 85$.

(a) $f = 5$, $z = 15$.

$$4w + 11 = z = 15 \Rightarrow 4w = 4 \Rightarrow w = 1$$

So, $x = w^2 + 4w + 3 = 1 + 4 + 3 = 8$.

$$y = 2w + 4 = 2 + 4 = 6$$

$$a = w + 3 = 1 + 3 = 4$$

$$b = w + 1 = 1 + 1 = 2$$

(*Continued*)

Example (Bhaskara II, BBI, page 110) (cont'd)

$$c = w^2 + 4w + 5 = 1 + 4 + 5 = 10$$

$$d = w^2 + 4w + 1 = 1 + 4 + 1 = 6$$

$$e = w + 2 = 1 + 2 = 3$$

(b) $f = 30,\ \ z = 85 \Rightarrow 4w + 11 = z = 85 \Rightarrow 4w = 74 \Rightarrow w = \dfrac{37}{2}.$

$$x = w^2 + 4w + 3 = \left(\frac{37}{2}\right)^2 + 4\left(\frac{37}{2}\right) + 3 = \frac{1677}{4}$$

$$y = 2w + 4 = 2\left(\frac{37}{2}\right) + 4 = 37 + 4 = 41$$

$$a = w + 3 = \frac{37}{2} + 3 = \frac{43}{2}$$

$$b = w + 1 = \frac{37}{2} + 1 = \frac{39}{2}$$

$$c = w^2 + 4w + 5 = \left(\frac{37}{2}\right)^2 + 4\left(\frac{37}{2}\right) + 5 = \frac{1685}{4}$$

$$d = w^2 + 4w + 1 = \left(\frac{37}{2}\right)^2 + 4\left(\frac{37}{2}\right) + 1 = \frac{1669}{4}$$

$$e = w + 2 = \frac{37}{2} + 2 = \frac{41}{2}$$

Fourth method:
Let

$$x = w^2 - 2w \text{ and } y = 2w - 2$$

$$a^2 = x + y + 2 = w^2 - 2w + 2w - 2 + 2 = w^2 \Rightarrow a = w$$

$$b^2 = x - y + 2 = w^2 - 2w - (2w - 2) + 2$$

$$= w^2 - 4w + 4 = (w-2)^2 \Rightarrow b = w - 2$$

$$c^2 = x^2 + y^2 = (w^2 - 2w)^2 + (2w-2)^2 = w^2(w-2)^2 + 4(w-1)^2$$

$$c^2 = w^2(w^2 - 4w + 4) + 4(w^2 - 2w + 1)$$

$$= w^4 - 4w^3 + 8w^2 - 8w + 4 = (w^2 - 2w + 2)^2$$

So, $c = w^2 - 2w + 2$.

$$d^2 = x^2 - y^2 + 8 = (w^2 - 2w)^2 - (2w-2)^2 + 8$$

$$= w^2(w-2)^2 - 4(w-1)^2 + 8$$

$$= w^4 - 4w^3 + 4w^2 - 4w^2 + 8w - 4 + 8 = w^4 - 4w^3 + 8w + 4$$

$$d^2 = (w^2 - 2w - 2)^2 \Rightarrow d = w^2 - 2w - 2$$

$$e^3 = \frac{y}{2}[x + 1] = \frac{1}{2}(2w - 2)(w^2 - 2w + 1)$$

$$e^3 = (w-1)^3 \Rightarrow e = w - 1$$

Also, $a + b + c + d + e = f^2$.
So,

$$w + w - 2 + w^2 - 2w + 2 + w^2 - 2w - 2 + w - 1 = f^2$$

$$2w^2 - w - 3 = f^2$$

Multiplying by 8, we get $16w^2 - 8w - 24 = 8f^2$.

$$(4w-1)^2 - 1 - 24 = 8f^2$$

$$(4w-1)^2 = 8f^2 + 25 \Rightarrow z^2 = 8f^2 + 25, \quad \text{where } z = 4w - 1$$

We have solved the equation $z^2 = 8y^2 + 25$ already. We will consider the two solutions: (a) $f = 5$, $z = 15$; (b) $f = 30$, $z = 85$.

(a) $f = 5$, $z = 15$.

$$4w - 1 = z = 15 \Rightarrow 4w = 16 \Rightarrow w = 4$$

(Continued)

Example (Bhaskara II, BBI, page 110) (cont'd)

So, $x = w^2 - 2w = 4^2 - 2 \cdot 4 = 16 - 8 = 8$

$$y = 2w - 2 = 2(4) - 2 = 8 - 2 = 6$$

$$a = w = 4,\ b = w - 2 = 4 - 2 = 2$$

$$c = w^2 - 2w + 2 = 4^2 - 2(4) + 2 = 16 - 8 + 2 = 10$$

$$d = w^2 - 2w - 2 = 4^2 - 2(4) - 2 = 16 - 8 - 2 = 6$$

$$e = w - 1 = 4 - 1 = 3$$

(b) $f = 30$, $z = 85$.

$$4w - 1 = z = 85 \Rightarrow 4w = 85 + 1 = 86 \Rightarrow w = \frac{86}{4} = \frac{43}{2}$$

$$x = w^2 - 2w = \left(\frac{43}{2}\right)^2 - 2\left(\frac{43}{2}\right) = \frac{1849}{4} - 43 = \frac{1849 - 172}{4} = \frac{1677}{4}$$

$$y = 2w - 2 = 2\left(\frac{43}{2}\right) - 2 = 43 - 2 = 41$$

$$a = w = \frac{43}{2}, \quad b = w - 2 = \frac{43}{2} - 2 = \frac{39}{2}$$

$$c = w^2 - 2w + 2 = \left(\frac{43}{2}\right)^2 - 3\left(\frac{43}{2}\right) + 2 = \frac{1849}{4} - 41$$

$$= \frac{1849 - 164}{4} = \frac{1685}{4}$$

$$d = w^2 - 2w - 2 = \left(\frac{43}{2}\right)^2 - 2\left(\frac{43}{2}\right) - 2 = \frac{1849}{4} - 45$$

$$= \frac{1849 - 180}{4} = \frac{1669}{4}$$

$$e = 4 - 1 = \frac{43}{2} - 1 = \frac{41}{2}$$

Bakshali Equation
Solve

$$axy = bx + cy + d \qquad \text{(Bhaskara II, Bijaganitha, page 124)}$$

Dividing by a, we get

$$xy - \frac{b}{a}x - \frac{c}{a}y = \frac{d}{a}$$

$$xy - \frac{b}{a}x - \frac{c}{a}y + \frac{bc}{a^2} = \frac{d}{a} + \frac{bc}{a^2}$$

So,

$$x\left(y - \frac{b}{a}\right) - \frac{c}{a}\left(y - \frac{b}{a}\right) = \frac{d}{a} + \frac{bc}{a^2}$$

i.e.,

$$\left(x - \frac{c}{a}\right)\left(y - \frac{b}{a}\right) = \frac{d}{a} + \frac{bc}{a^2}$$

Let

$$\frac{d}{a} + \frac{bc}{a^2} = mn$$

where m and n are rational numbers.
Then, we have

$$\left(x - \frac{c}{a}\right)\left(y - \frac{b}{a}\right) = mn$$

Hence, if

$$x - \frac{c}{a} = \pm m, \quad \text{then } y - \frac{c}{a} = \pm n$$

and if

$$x - \frac{c}{a} = \pm n, \quad \text{then } y - \frac{b}{a} = \pm m$$

Bhaskara II gave the following example:

Example

"The sum of two numbers multiplied by 4 and 3, added by 2 is equal to the product of these numbers. Tell me, if you know those two numbers."

Let x and y be the two numbers.

Then, we have $4x + 3y + 2 = xy$.

Comparing this equation with $axy = bx + cy + d$, we have

$$a = 1,\ b = 4,\ c = 3,\ d = 2,\ \text{and } mn = \frac{d}{a} + \frac{bc}{a^2}$$

Let $m = 1$.

Then,

$$n = \frac{d}{a} + \frac{bc}{a^2} = \frac{2}{1} + \frac{4 \cdot 3}{1} = 2 + 12 = 14$$

$$\text{(a) } x - \frac{c}{a} = \pm m \Rightarrow x = \frac{3}{1} \pm 1 = 4 \text{ or } 2$$

$$y - \frac{b}{a} = \pm n \Rightarrow y = \frac{4}{1} \pm 14 = 18 \text{ or } -10$$

Since $x > 0$, $y > 0$, discarding the negative values, we have $x = 4$ and $y = 18$.

$$\text{(b) } x - \frac{c}{a} = \pm n \Rightarrow x = \frac{3}{1} \pm 14 = 17 \text{ or } -11$$

$$y - \frac{b}{a} = \pm m \Rightarrow y = \frac{4}{1} \pm 1 = 5 \text{ or } 3$$

Since $x > 0$, $y > 0$, discarding the negative values, we have $x = 17$ and $y = 5$.

A different geometrical method of solution of $axy = bx + cy + d$ is given by Bhaskara II when $a = 1$.

Consider the rectangle $ABCD$ with side $AB = y$ and $BC = x$. Take a point M on AB such that $AM = b$. Draw MN perpendicular to DC. Take a point P on MN such that $MP = c$. Draw PQ perpendicular to BC.

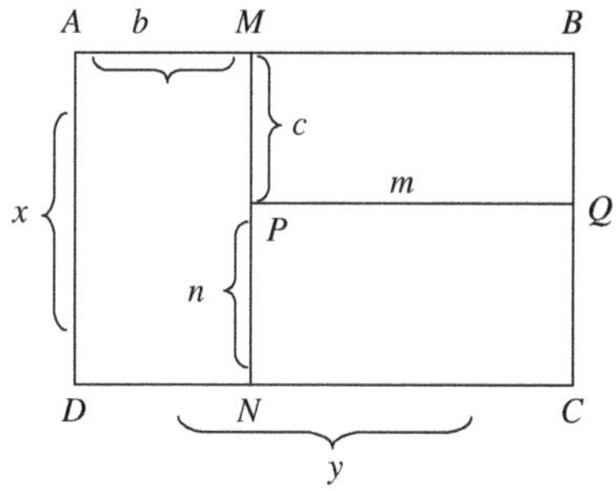

Let $PQ = m$ and $PN = n$.
Now, area of rectangle $PQCN = mn$.
Also, area of rectangle

$$PQCN = \text{area of the rectangle } ABCD$$
$$- \text{area of the rectangle } AMND$$
$$- \text{area of the rectangle } MBQP$$

So, $mn = xy - bx - (y - b)c = xy - bx - cy + bc$.

$$\Rightarrow xy = bx + cy + mn - bc$$

But $xy = bx + cy + d$.
So, $mn = d + bc$.
And $x = n + c$ and $y = m + b$.

Differential Calculus

Munjala (AD 932) is the world's first mathematician who conceived of the differential calculus. B.B. Dutta (Roc. Cit) pointed out that the works of Munjala and his commentator, Prashastidhara (AD 958) demonstrated that they knew the formula $\delta(\sin\theta) = \cos\theta\,\delta\theta$. Bhaskara II is the world's third mathematician to exhibit the knowledge of differential calculus. He has introduced in the chapter of positions of planets ("spastādhikāra") in his work "Siddhantasiromani," the concept of instantaneous ("Tatkaligati") motion by dividing time into a large number of small intervals and comparing the positions of a planet at the end of these successive small intervals. He brought out clearly the distinction between average velocity ("Sthulagati") and correct velocity ("Suksmagati").

In this connection, Bhaskara II enunciated the following:

बिम्बार्धस्य कोटिज्यागुणस्त्रिज्याहरः फलं दोर्ज्यायोरन्तरम् ॥

i.e., "The product of the cosine of the radius and the element of the radius is the difference of the two sines."

Let x and x^1 be the mean anomalies of a planet at the end of two consecutive intervals. Then, Bhaskara II stated that

$\sin x - \sin x^1 = (x - x^1)\cos x$, which is equivalent to stating that $\delta(\sin x) = \cos x\ \delta x$.

Bhaskara II has further dealt with differential calculus. He stated that the derivative of a function vanishes at a point of extrema. He also mentioned the idea of Rolle's theorem. In the chapter on "Spasthadhikara" of his work, "Graha Ganita," he enunciated the following:

यत्र ग्रहस्य परमंफ्लं तत्रैवगतिफलाभावेन भवितव्यम्

i.e., "When the planet's motion is an extremum, the fruit of the motion at that point is absent." This means that at an extremum point, the motion of the planet is stationary. The above statements of Bhaskara II demonstrated strong evidence for giving credit to him as a pioneer of differential calculus.

Integral Calculus

The concept of integral calculus has been known in a rough way by ancient Greeks and they used this knowledge to determine areas and volumes. We find the method of summation envisaged in the integral calculus from the time of Archimedes to Kepler. Bhaskara II determined the surface area and the volume of a sphere, using similar methods. So, Bhaskara II knew the concept of integral calculus as much as known at that time.

Bhaskara II gave correct formulae for the surface area as well as the volume of the sphere.

वृत्तक्षे त्ने परिधिगुणितव्यासपादः फलं य–
तक्षु ण्णं वेदैरुपरिपरितः कन्दुकस्येव जालम् ।
गोलस्यैवं तदपि च फलं पृष्ठजं व्यासनिघ्नम्
षड्भिर्भक्तं भवति नियतं गोलगर्भे घनाख्यम् ॥

(Līl. 201)

i.e., "In a circle, the product of the circumference and one-fourth of the diameter gives its area. This area of the circle multiplied by 4 is the surface area of the sphere. When the product of this surface area and the diameter of the circle are divided by 6, we get the volume of the sphere."

Now, area of the circle

$$= A = c \cdot \frac{d}{4}$$

where c is the circumference and d is the diameter of the circle.

So,

$$A = \pi d \times \frac{d}{4} = \frac{\pi d^2}{4} = \frac{\pi (2r)^2}{4} = \pi r^2$$

where r is the radius of the circle.

$$S = \text{surface area of the sphere} = A \cdot 4 = 4\pi r^2$$

$$V = \text{volume of the sphere} = \frac{S \times d}{6} = \frac{4\pi r^2 \times 2r}{6} = \frac{8\pi r^3}{6} = \frac{4}{3}\pi r^3$$

Bhaskara II has provided a method of finding the surface area of a sphere, by dividing the surface into elementary lines by drawing circles passing through two diametrically opposite points on the sphere. Then, each line is again divided into a large number of elementary quadrilaterals by circles parallel to the horizontal great circle. Now, the area of line is the sum of the areas of the quadrilaterals. The surface area of the sphere is the sum of the areas of all the lines. For another proof, see "Yuktibhasa" in this book.

To find the volume of a sphere, Bhaskara II took a system of pyramids with vertex at the center and standing on the surface on curvilinear areas of unity. The volume of the sphere is the sum of the volumes of the pyramids.

An alternate method and proof is provided in "Yuktibhasa" on page 648 in this book.

Bhaskara II's Contributions to Trigonometry

Bhaskara II has given the following results in his work "Siddhantasiromani," part III ("Goladyaya")

(1) चापयोरष्टियोः दोर्ज्ये मिथः कोटि ज्यकाहते
त्रिज्या भक्ते तयोरक्यं स्यात् चापकस्य दोर्ज़्यका
चापांतरस्य जीवास्यात् तयोरतरसमिता

(Sl. 21–22)

i.e.

$$(1) \quad H \sin(A \pm B) = \frac{H \sin AH \cos B \pm H \cos AH \sin B}{R}$$

Here $H \sin A = R \sin A$ and $H \cos A = R \cos A$.

The proof of this result has been given in Jayasthedeva's "yuktibhasa" of this book.

$$(2) \quad H \sin \frac{A - B}{2} = \frac{1}{2}\{(H \sin A - H \sin B)^2 + (H \cos A - H \cos B)^2\}^{\frac{1}{2}}$$

Proof:

$$\text{R.H.S.} = \frac{1}{2}\{(H\sin^2 A + H\cos^2 A) + (H\sin^2 B + H\cos^2 B) - 2(H\cos A \cdot H\cos B + H\sin A \cdot H\sin B)\}^{\frac{1}{2}}$$

$$= \frac{1}{2}\{2R^2 - 2(H\cos A \cdot H\cos B + H\sin A \cdot H\sin B)\}^{\frac{1}{2}} \quad (*)$$

Since $H\sin^2\theta + H\cos^2\theta = R^2$.
Now, we know that $H\sin(90 - A) = H\cos A$ and

$$H\cos(90 - A) = H\sin A$$

$$H\cos(A - B) = H\sin[90 - (A - B)] = H\sin[(90 - A) + B]$$

$$= \frac{H\sin(90 - A)H\cos B + H\cos(90 - A)H\sin B}{R} \quad \text{from (1)}$$

$$H\cos(A - B) = \frac{H\cos A\, H\cos B + H\sin A\, H\sin B}{R} \quad (**)$$

$$\text{RHS} = \frac{1}{2}[2R^2 - 2RH\cos(A - B)]^{\frac{1}{2}} \quad (***) \text{ by substituting } (**) \text{ in } (*),$$

Now, $H\cos(A + B) = H\sin[90 - (A + B] = H\sin(\overline{90 - A} - B)$

$$H\cos(A + B) = \frac{H\sin(90 - A)H\cos B - H\cos(90 - A)H\sin B}{R} \quad \text{from (1)}$$

So,

$$H\cos(A + B) = \frac{H\cos A\, H\cos B - H\sin A\, H\sin B}{R}$$

Let $A = B = \theta$.
Then,

$$H\cos 2\theta = \frac{H\cos\theta \cdot H\cos\theta - H\sin\theta \cdot H\sin\theta}{R}$$

$$H\cos 2\theta = \frac{H\cos^2\theta - H\sin^2\theta}{R}$$

$$H\cos 2\theta = \frac{R^2 - H\sin^2\theta - H\sin^2\theta}{R}$$

$$H\cos 2\theta = \frac{R^2 - 2\,H\sin^2\theta}{R}$$

Let

$$\theta = \frac{A-B}{2}$$

Then,

$$RH\cos(A-B) = R^2 - 2H\sin^2\left(\frac{A-B}{2}\right) \quad (* * **)$$

Substituting (****) in (***), we get

$$\text{RHS} = \frac{1}{2}\left[2\cdot 2H\sin^2\frac{A-B}{2}\right]^{\frac{1}{2}} = \frac{1}{2}\cdot 2H\sin\frac{A-B}{2}$$

Thus,

$$\text{RHS} = H\sin\frac{A-B}{2} = \text{LHS}$$

(3) $H\sin\dfrac{\theta}{2} = \dfrac{1}{2}[H\sin^2\theta + H\,\text{versin}^2\,\theta]^{\frac{1}{2}}$

Proof: We have shown in (2) that

$$H\sin\frac{A-B}{2} = \frac{1}{2}[(H\sin A - H\sin B)^2 + (H\cos A - H\cos B)^2]^{\frac{1}{2}} \quad (*)$$

Let $A = \theta$ and $B = 0$.
Then, $H\sin B = H\sin 0 = R\sin 0 = 0$.

$$H\cos B = H\cos 0 = R\cos 0 = R$$

Then, (*) becomes

$$H\sin\frac{\theta}{2} = \frac{1}{2}[H\sin^2\theta + (H\cos\theta - R)^2]^{\frac{1}{2}}$$

But H versin $\theta = R(1 - \cos\theta) = R - R\cos\theta = R - H\cos\theta$.
Hence, we get

$$H\sin\frac{\theta}{2} = \frac{1}{2}[H\sin^2\theta + H\text{ versin}^2\theta]^{\frac{1}{2}}$$

(4) $$H\sin 18^\circ = \frac{\sqrt{5}-1}{4}R \text{ and } H\sin 36^\circ = \frac{\sqrt{5R^2 - \sqrt{5}R^4}}{8}$$

Proof: Let $\theta = 18^\circ$, then $5\theta = 90^\circ$ and $H\cos 90^\circ = 0$.
Now, $H\sin 2\theta = H\sin(\theta+\theta)$.

$$H\sin 2\theta = \frac{H\sin\theta\cos\theta + H\cos\theta\, H\sin\theta}{R} = \frac{2H\sin\theta H\cos\theta}{R} \quad \text{from (1)}$$

$$H\sin 3\theta = H\sin(2\theta+\theta) = \frac{H\sin 2\theta H\cos\theta + H\cos 2\theta H\sin\theta}{R}$$

$$= \frac{2H\sin\theta H\cos^2\theta + H\sin\theta(R^2 - 2H\sin^2\theta)}{R^2}$$

$$= \frac{H\sin\theta(2H\cos^2\theta + R^2 - 2H\sin^2\theta)}{R^2}$$

$$= \frac{H\sin\theta[2(R^2 - H\sin^2\theta) + R^2 - 2H\sin^2\theta]}{R^2}$$

$$H\sin 3\theta = \frac{H\sin\theta[3R^2 - 4H\sin^2\theta]}{R^2}$$

$$H\cos 3\theta = H(\cos 2\theta + \theta)$$

$$= \frac{H\cos 2\theta H\cos\theta - H\sin 2\theta H\sin\theta}{R}$$

$$= \frac{H\cos\theta(R^2 - 2H\sin^2\theta) - (2H\sin\theta H\cos\theta)H\sin\theta}{R^2}$$

$$= \frac{H\cos\theta[R^2 - 2H\sin^2\theta - 2H\sin^2\theta]}{R^2}$$

$$= \frac{H \cos \theta[R^2 - 4H \sin^2 \theta]}{R^2}$$

$$R.H \cos 5\theta = H \cos(3\theta + 2\theta)$$

$$= \frac{H \cos 3\theta \cos 2\theta - H \sin 3\theta H \sin 2\theta}{R}$$

$$= H \cos \theta \left(\frac{R^2 - 4H \sin^2 \theta}{R^2}\right) \left(\frac{R^2 - 2H \sin^2 \theta}{R}\right)$$

$$-H \sin \theta \left(\frac{3R^2 - 4H \sin^2 \theta}{R^2}\right) \frac{H \sin\theta \, H \cos \theta}{R}$$

$$R.H \cos 5\theta = \frac{H \cos \theta}{R^3} \left\{(R^2 - 4H \sin^2 \theta)(R^2 - 2H \sin^2 \theta) - 2H \sin^2 \theta(3R^2 - 4H \sin^2 \theta)\right\}$$

Let $\theta = 18^\circ$, then $\cos 5\theta = \cos 90^\circ = 0$.
Hence, we get

$$0 = H \cos 18^\circ\{(R^2 - 4H \sin^2 18^\circ)(R^2 - 2H \sin^2 18^\circ) - 2H \sin^2 18^\circ(3R^2 - 4H \sin^2 18^\circ)\}$$

Since $H \cos 18 \neq 0$, we get

$$(R^2 - 4H \sin^2 18^0)(R^2 - 2H \sin^2 18^0) - 2H \sin^2 18^\circ(3R^2 - 4H \sin^2 18^\circ) = 0$$

Let $x = H \sin 18^\circ$, then we get

$$(R^2 - 4x^2)(R^2 - 2x^2) - 2x^2(3R^2 - 4x^2) = 0$$

i.e., $R^4 - 4x^2R^2 - 2R^2x^2 - 6R^2x^2 + 8x^4 + 8x^4 = 0$.
i.e., $16x^4 - 12x^2R^2 + R^4 = 0$.
Let $x^2 = y$.
Then, we have $16y^2 - 12yR^2 + R^4 = 0$.
So,

$$y = \frac{12R^2 \pm \sqrt{144R^4 - 64R^4}}{32} = R^2 \cdot \frac{12 \pm \sqrt{80}}{32} = R^2 \cdot \frac{12 \pm 4\sqrt{5}}{32}$$

$$y = R^2 \cdot \frac{3 \pm \sqrt{5}}{8}$$

i.e.,

$$x^2 = \frac{R^2}{8}\left(3 \pm \sqrt{5}\right)$$

$$x = \frac{R}{2\sqrt{2}}\sqrt{3 \pm \sqrt{5}}$$

Let

$$\sqrt{3 \pm \sqrt{5}} = \sqrt{a} \mp \sqrt{b}$$

Squaring, we get

$$3 \pm \sqrt{5} = a + b \pm 2\sqrt{ab}$$

$$a + b = 3$$

$$\pm 2\sqrt{ab} = \sqrt{5} \Rightarrow 4ab = 5$$

$$(a-b)^2 = (a+b)^2 - 4ab = 9 - 5 = 4$$

So, $a - b = 2$.
Also, we have $a + b = 3$.
Adding, we get

$$2a = 5 \text{ or } a = 5/2$$

Subtracting the first equation from the second, we get

$$2b = 1 \text{ or } b = 1/2$$

So,

$$\sqrt{a} \pm \sqrt{b} = \sqrt{\frac{5}{2}} \pm \sqrt{\frac{1}{2}}$$

So,

$$x = H \sin 18^\circ = \frac{R}{2\sqrt{2}}\left[\sqrt{\left(\frac{5}{2}\right)} \pm \sqrt{\frac{1}{2}}\right]$$

Thus,

$$H\sin 18^\circ = \frac{R}{4}\left[\sqrt{5}\pm 1\right]$$

We know that $H\sin\theta$ increases as θ increases in the first quadrant,

$$H\sin 18^\circ < H\sin 30^\circ = R\sin 30^\circ = \frac{R}{2}$$

Since

$$\frac{\sqrt{5}+1}{4}R > \frac{R}{2}$$

So, we discard the value

$$\frac{R}{4}\left(\sqrt{5}+1\right)$$

Thus,

$$H\sin 18^\circ = \frac{R}{4}\left[\sqrt{5}-1\right]$$

So,

$$H\sin 18^\circ = \frac{\sqrt{5}-1}{4}R$$

$$(H\cos 18^\circ)^2 = R^2 - \left(\frac{\sqrt{5}-1}{4}\right)^2 R^2$$

$$= R^2\left[\frac{16-(\sqrt{5}-1)^2}{16}\right] = \frac{R^2}{16}\left[16-\left(6-2\sqrt{5}\right)\right]$$

$$(H\cos 18^\circ)^2 = \frac{R^2}{16}\left(10+2\sqrt{5}\right)$$

$$(H\cos 18^\circ\ H\sin 18^\circ)^2 = \left(\frac{R^2}{16}\right)^2\left[\left(10+2\sqrt{5}\right)\left(6-2\sqrt{5}\right)\right]$$

$$\left(\frac{R}{2}H\sin 36^\circ\right)^2 = \left(\frac{R^2}{16}\right)^2\left[60+12\sqrt{5}-20\sqrt{5}-20\right]$$

$$= \left(\frac{R^2}{16}\right)^2 \left[40 - 8\sqrt{5}\right]$$

$$(H \sin 36°)^2 = \left(\frac{R}{8}\right)^2 \left(40 - 8\sqrt{5}\right)$$

$$H \sin 36° = \frac{R}{8}\sqrt{40 - 8\sqrt{5}} = R\sqrt{\frac{40 - 8\sqrt{5}}{64}}$$

So,

$$H \sin 36° = R\sqrt{\frac{5 - \sqrt{5}}{8}} = \sqrt{\left\{\frac{5R^2 - \sqrt{5R^4}}{8}\right\}}$$

Dr. D.A. Somyaji's Comments on the Hindu Sine Table

Let

$$\alpha = 3\frac{3°}{4}$$

Then, Hindu sine table consists of $H \sin(K\alpha)$, $K = 1, 2, 3, \ldots, 24$, where $H \sin \theta$ denotes the Hindu sine of $\theta = R \sin \theta$, R being the radius of the circle.

In his Ph.D. thesis to the Karnataka University, Dr. D.A. Somyaji drew attention to Bhaskara II's comments in the verses 1–25 of "Jyoti Vasana" in his "Golaladya." Bhaskara II was of the opinion that the 24 Hindu sine table was constructed with the help of the two formulae.

$$H \sin^2\theta + H \cos^2 \theta = R^2 \tag{1}$$

$$H \sin\frac{\theta}{2} = \frac{1}{2}[H \sin^2 \theta + H \operatorname{versin}^2 \theta]^{\frac{1}{2}} \tag{2}$$

Let

$$\alpha = 3\frac{3°}{4}$$

Then, the Hindu sine table consists of $H \sin(K\alpha)$, $K = 1, 2, 3, \ldots, 24$.

If $H_k(\alpha) = H_k = H\sin(K\alpha)$, $K = 1, 2, 3,\ldots, 24$ and taking $R = 3438$.

$$H_{12+i} = H\sin(12+i)\alpha, \quad \text{for } i = 1, 2, 3, \ldots, 11$$

$$= H\sin(12\alpha + i\alpha) = H\sin[24\alpha - (12\alpha - i\alpha)]$$

$$= H\sin[90^\circ - (12\alpha - i\alpha)]$$

So,

$$H_{12+i} = H\cos(12-i)\alpha$$

$$= \sqrt{R^2 - H\sin^2(12-i)\alpha}$$

Hence,

$$H_{12+i} = \sqrt{R^2 - H_{12-i}^2}, \quad \text{for } i = 1, 2, 3, \ldots, 11 \tag{3}$$

We know that

$$H_8 = H\sin(8\alpha) = H\sin 30^\circ = R\sin 30^\circ = \frac{R}{2} = \frac{3438}{2} = 1719$$

$$H_{12} = H\sin(12\alpha) = H\sin 45^\circ = R\sin 45^\circ \frac{R}{\sqrt{2}} = \frac{3438}{\sqrt{2}} = 2431$$

$$H_{16} = H\sin(16\alpha) = H\sin 60^\circ = R\sin 60^\circ = \frac{\sqrt{3}}{2}R = \frac{\sqrt{3}}{2}(3438) = 2977$$

$$H_{24} = H\sin(24\alpha) = H\sin 90^\circ = R\sin 90^\circ = R = 3438$$

Now, with the help of (2), $H_4 = H\sin(4\alpha) = H\sin 15^\circ$ can be obtained from $H_8 = H(\sin 8\alpha) = H\sin 30^\circ$, $H_2 = H(\sin 2\alpha) = H\sin(7\frac{1}{2}^\circ)$ can be obtained from $H_4 = H\sin 15^\circ$ and $H_1 = H\sin\alpha = H\sin(3\frac{3}{4}^\circ)$ can be obtained from $H_2 = H\sin(7\frac{1}{2})^\circ$. Similarly, $H_6 = H\sin(6\alpha) = H\sin(22\frac{1}{2}^\circ)$ can be obtained from $H_{12} = H\sin(12\alpha) = H\sin(45^\circ)$ and $H_3 = H(\sin 3\alpha) = H\sin(11\frac{1}{4}^\circ)$ can be obtained from $H_6 = H\sin(22\frac{1}{2}^\circ)$. With the help of (3), $H_1 = H\sin(3\frac{3}{4}^\circ)$, $H_2 = H\sin(7\frac{1}{2}^\circ)$, $H_3 = H\sin(11\frac{1}{4}^\circ)$, $H_4 = H\sin 15^\circ$ and $H_6 = H\sin(22\frac{1}{2}^\circ)$ lead us to find the values of $H_{23} = H\sin(23\alpha) = H\sin(86\frac{1}{4}^\circ)$, $H_{22} = H(\sin 22\alpha) = H\sin(82\frac{1}{2}^\circ)$, $H_{21} = H\sin(21\alpha) = H\sin(78\frac{3}{4}^\circ)$, $H_{20} = H\sin(20\alpha) = H\sin 75^\circ$ and $H_{18} = H\sin(18\alpha) = H\sin(67\frac{1}{2}^\circ)$, respectively. Again using (2), from H_{20}, H_{22}, and H_{18}, we can find the values of $H_{10} = H\sin(10\alpha) = H\sin(37\frac{1}{2}^\circ)$, $H_{11} = H\sin(11\alpha) = H\sin(41\frac{1}{4}^\circ)$, and

$H_9 = H\sin(9\alpha) = H\sin(33\frac{3}{4}^\circ)$, respectively. Once again, we can find the value of $H_5 = H\sin(5\alpha) = H\sin(18\frac{3}{4})^\circ$ from H_{10} using (2) and from (3), H_5, H_{10}, H_{11}, and H_9 lead us to find the values of $H_{19} = H\sin(19\alpha) = H\sin(71\frac{1}{4}^\circ)$, $H_{14} = H\sin(14\alpha) = H\sin(52\frac{1}{2}^\circ)$, $H_{13} = H\sin(13\alpha) = H\sin(48\frac{3}{4}^\circ)$, and $H_{15} = H\sin(15\alpha) = H\sin(56\frac{1}{4}^\circ)$, respectively. Then from (2), we can find $H_7 = H\sin(7\alpha) = H\sin(26\frac{1}{4}^\circ)$ from H_{14} and (3) leads to find $H_{17} = H\sin(17\alpha) = H\sin(63\frac{3}{4}^\circ)$ from H_7. This way, entire table of 24 Hindu sines can be computed.

Dr. Somyaji opined that it is by this above Bhaskara II's method, Aryabhata I and his predecessors constructed the Hindu sine table. It is not yet known whether Aryabhata I and his predecessors knew this method of Bhaskara II for the construction of the Hindu sine table.

Bhaskara II's "Leelavati" has been translated into Persian by Abul Fazl, Minister of Mughal emperor, Akbar in 1587. His "Bijaganitha" was translated into Persian by Utta ulla Rushdee in 1634. Many commentaries have been written on the works of Bhaskara II. These include "Bijaganitha" (AD 1350) "Ganitakaumadi" (AD 1356) by Narayana Pandita. "Ganithamrita" (AD 1538) and "Suryaprakasa" (AD 1541) by Suryadasa, "Buddivilasini" by Ganesha Daivjna in AD 1545 and "Vasana Bhasya" by Ranganatha in the early part of the seventeenth century.

Bhaskara II concluded his favorite work, "Leelavati" with the following verse (sloka):

येषां सुजातिगुणवर्ग विभूषिताङ्गी
शुद्धाखिलव्यवहृतिः खलुकण्ठसक्ता।
लीलावतीहसरसोक्तमुदाहरन्ती
तषां सदैवसुखसम्पदुपैतिवृद्धिम्॥

i.e., "If Leelavati a gentle lady of good virtues throws her arms around a person, that person will always be immensely happy, rich, and prosperous."

12 Narayana Pandita

Narayana Pandita (fourteenth century) is the son of Nrsimha. He composed his marvelous work "Ganita Kaumudi" in AD 1356. Narayana Pandita is often confused with Narayana the commentator of Bhaskara II's "Lilavati." He is the first mathematician to enunciate certain theorems on cyclic quadrilateral; given the sides, three and only three diagonals are possible for a cyclic quadrilateral, the area of a cyclic quadrilateral is the product of the three diagonals divided by four times the radius. Narayana Pandita has taken the study of the cyclic quadrilateral much deeper than Brahmagupta did. There is no evidence whether he belonged to Aryabhata I School or not. Ganita Kaumudi overshadows the accomplishments of Aryabhata I School. The entire knowledge of triangles contained in the work of his predecessors has been found in Ganita Kaumudi, in fact, in a much precise and elegant form. Like Mahavira, Narayana Pandita presented accurate formulae for the area enclosed by three or more equal, mutually touching circles. He reproduced Bhaskara II's formula for the chord from the arc in a slightly different form.

The Hindu name for the science of algebra is "*bijaganita*." Here, "*bija*" means elements and "*ganita*" means the Science of Calculation (mathematics). The ancient Hindus revered algebra as a Science of utmost importance, besides being a great working tool. Narayana, the commentator of Bhaskara II, was an eminent algebraist. He made the following statement in his work "*Bijaganitha*" (AD 1350):

> *"I adore Brahma, the creator of the universe, as well as the science of computation with the unknown, which is the unique invisible root cause of the visible and multiple qualities of this universe."*
>
> (NBI I. R. 1)

He continues:

> *"Just like this entire universe, visible and endless is created by Brahma, the entire arithmetic with endless rules follows out of Algebra."*
>
> (NBI II. R. 1)

He further adds:

> *"People raise questions, whose solutions arithmetic cannot answer. Algebra can generally provide answer for such questions. People with less intelligence cannot find solutions for such questions, with the help of Arithmetic. I must talk loudly about the elegant and intelligent rules of Algebra."*
>
> (NBI I. R. 5–6)

Mathematical Achievements of Pre-modern Indian Mathematicians. DOI: http://dx.doi.org/10.1016/B978-0-12-397913-1.00012-0

Narayana's Lemma on "Chakravala" Method of Bhaskara II

According to Bhaskara II's "Chakravala" method, if we have $Na_0^2 + k = b_0^2$, we have to choose an integer m_0 such that

$$a_1 = \frac{a_0 m_1 + b_0}{k_0}$$

is an integer such that

$$k_1 = \frac{m_0^2 - N}{k_0}$$

is small, where

$$b_1 = \frac{b_0 m_0 + N a_0}{k_0}$$

Narayana does not clearly state

$$\frac{m_0^2 - N}{k_0}$$

should be small. It is very likely that he had realized that $b_1 = a_1 m_0 - a_0 k_1$, also,

$$a_0 = \frac{a_1 m_0 - b_1}{k_1}$$

and

$$b_0 = \frac{b_1 m_0 - N a_1}{k_1}$$

Proof: We have

$$a_1 = \frac{a_0 m_0 + b_0}{k_0} \tag{1}$$

$$b_1 = \frac{b_0 m_0 + N a_0}{k_0} \tag{2}$$

$$k_1 = \frac{m_0^2 - N}{k_0} \tag{3}$$

Now,

$$b_1 = \frac{m_0(a_1k_0 - a_0m_0) + Na_0}{k_0} \quad \text{since } b_0 = a_1k_0 - a_0m_0 \quad \text{from (1)}$$

i.e.,

$$b_1 = a_1m_0 - \frac{a_0(m_0^2 - N)}{k_0} = a_1m_0 - a_0k_1 \quad \text{from (3)}$$

i.e.,

$$b_1 = a_1m_0 - a_0k_1 \tag{4}$$

Also from (4),

$$a_0 = \frac{a_1m_0 - b_1}{k_1}$$

Multiplying (4) by m_0, we get

$$\begin{aligned} b_1m_0 &= a_1m_0^2 - a_0k_1m_0 \\ &= a_1(k_1k_0 + N) - a_0k_1m_0 \end{aligned}$$

since $k_1k_0 = m^2 - N$ from (3)

$$b_1m_0 = a_1N + a_1k_1k_0 - a_0k_1m_0 = a_1N + k_1(a_1k_0 - a_0m_0)$$

But from (1), $b_0 = a_1k_0 - a_0m_0$
So,

$$b_1m_0 = a_1N + b_0k_1 \Rightarrow b_0 = \frac{b_1m_0 - Na_1}{k_1}$$

Example (Narayana)

"Tell me the number whose square is multiplied by 13 and then increased or decreased by 17 is capable of having a square root."

Solution

Let x be the number, then we have two equations, each of which have to be solved.

$$13x^2 + 17 = y^2 \tag{1}$$

$$13x^2 - 17 = y^2 \tag{2}$$

(*Continued*)

Example (Narayana) (cont'd)

Solution of (1)

We observe that $13 \cdot 1^2 + 3 = 13 + 3 = 16 = 4^2$.

So, (1 4) is a solution of $13x^2 + 3 = y^2$.

Again, we observe that $13 \cdot 1^2 + 51 = 13 + 51 = 64 = 8^2$.

So, (1 8) is a solution of $13x^2 + 51 = y^2$.

So,

$$(1\ 4\ 3) \cdot (1\ 8\ 51) = (1.8 + 4.1 \quad 4.8 + 1.1.13 \quad 3.51) = (12\ 45\ 153)$$

which means that (12 45) is a solution of $13x^2 + 153 = y^2$, i.e., (12 45) is a solution of $13x^2 + 3^2 \cdot 17 = y^2$.

Then, by Brahmagupta's lemma 2 on page 165, we have

$$\left(\frac{12}{3}\frac{45}{3}\right) = (4\ 15)$$

is a solution of $13x^2 + 17 = y^2$.

The required answer is $x = 4$ and $y = 15$.

Solution of (2)

To solve $13x^2 - 17 = y^2$.

We have already shown that (4 15) is a solution of $13x^2 + 17 = y^2$.

We will find a solution of $13x^2 - 1 = y^2$.

We observe that $13 \cdot 1^2 - 4 = 13 - 4 = 9^2 = (3)^2$.

So, (1, 3) is a solution of $13x^2 - 4 = y^2$, i.e., (1, 3) is a solution of $13x^2 - 2^2 \cdot 1 = y^2$.

Then, by Brahmagupta's lemma 2 on page 165, ((1/2) (3/2)) is a solution of $13x^2 - 1 = y^2$

$$\begin{aligned}
&(4\ 15\ 17)\circ\left(\frac{1}{2}\ \frac{3}{2}\ \ -1\right)\\
&= \left(4 \cdot \frac{3}{2} + 15 \cdot \frac{1}{2} \quad 15 \cdot \frac{3}{2} + 4 \cdot \frac{1}{2} \cdot 13 \quad (17)(-1)\right)\\
&= \left(6 + \frac{15}{2}\ \frac{45}{2} + 26\ \ -17\right)\\
&= \left(\frac{27}{2}\ \frac{97}{2}\ \ -17\right)
\end{aligned}$$

which means that

$$\left(\frac{27}{2}\ \frac{97}{2}\right)$$

is a solution of $13x^2 - 17 = y^2$.

Narayana has given the following two examples of solving "vargaprakrti" equations.

Example 1

Find an integral solution of $97x^2 + 1 = y^2$.

Solution

We have the identity $97 \cdot 1^2 + 3 = 100 = 10^2$.

So, $(a_0\ b_0) = (1\ 10)$ is a solution of $97x^2 + 3 = y^2$.

Here $N = 97$, $K_0 = 3$.

We choose an integer m_0 such that

$$a_1 = \frac{a_0 m_0 + b_0}{K_0} = \frac{m_0 + 10}{3}$$

is an integer and

$$k_1 = \frac{m_0^2 - N}{K_0} = \frac{m_0^2 - 97}{3}$$

is numerically small.

We choose $m_0 = 11$

Then,

$$a_1 = \frac{11 + 10}{3} = \frac{21}{3} = 7$$

$$b_1 = \frac{b_0 m_0 + N a_0}{K_0} = \frac{10(11) + 97(1)}{3} = \frac{110 + 97}{3} = \frac{207}{3} = 69$$

$$K_1 = \frac{(11)^2 - 97}{3} = \frac{121 - 97}{3} = \frac{24}{3} = 8$$

which means that $(a_1\ b_1) = (7\ 69)$ is a solution of $97x^2 + 8 = y^2$.

We choose an integer m_1 such that

$$a_2 = \frac{a_1 m_1 + b_1}{K_1} = \frac{7m_1 + 69}{8}$$

is an integer and

$$K_2 = \frac{m_1^2 - N}{K_1} = \frac{m_1^2 - 97}{8}$$

is numerically small.

(*Continued*)

Example 1 (cont'd)

We take $m_1 = 13$.
Then,

$$a_2 = \frac{7(13) + 69}{8} = \frac{91 + 69}{8} = \frac{160}{8} = 20$$

$$b_2 = \frac{b_1 m_1 + N a_1}{K_1}$$

$$= \frac{(69)(13) + (97)7}{8} = \frac{897 + 679}{8} = \frac{1576}{8} = 197$$

$$K_2 = \frac{m_1^2 - 97}{8} = \frac{(13)^2 - 97}{8} = \frac{169 - 97}{8} = \frac{72}{8} = 9$$

So, $(a_2\ b_2) = (20\ 197)$ is a solution of $97x^2 + 9 = y^2$.
We choose an integer m_2 such that

$$a_3 = \frac{a_2 m_2 + b_2}{K_2} = \frac{20m_2 + 197}{9}$$

is an integer and

$$K_3 = \frac{m_2^2 - N}{K_2} = \frac{m_2^2 - 97}{9}$$

is numerically small.
We let $m_2 = 14$.
Then,

$$a_3 = \frac{20(14) + 197}{9} = \frac{280 + 197}{9} = \frac{477}{9} = 53$$

$$b_3 = \frac{b_2 m_2 + N a_2}{K_2} = \frac{(197)(14) + (97)(20)}{9}$$

$$= \frac{2758 + 1940}{9} = \frac{4698}{9} = 522$$

$$K_3 = \frac{m_2^2 - 97}{9} = \frac{(14)^2 - 97}{9} = \frac{196 - 97}{9} = \frac{99}{9} = 11$$

Thus, $(a_3\ b_3) = (53\ 522)$ is a solution of $97x^2 + 11 = y^2$.

We select m_3 such that

$$a_4 = \frac{a_3 m_3 + b_3}{k_3} = \frac{53m_3 + 522}{11}$$

is an integer such that

$$K_4 = \frac{m_3^2 - N}{K_3} = \frac{m_3^2 - 97}{11}$$

is numerically small.

We choose $m_3 = 8$.

Then,

$$a_4 = \frac{(53)8 + 522}{11} = \frac{424 + 522}{11} = \frac{946}{11} = 86$$

$$b_4 = \frac{b_3 m_3 + N a_3}{K_3} = \frac{(522)8 + (97)(53)}{11}$$

$$= \frac{4176 + 5141}{11} = \frac{9317}{11} = 847$$

$$K_4 = \frac{(8)^2 - 97}{11} = \frac{64 - 97}{11} = \frac{-33}{11} = -3$$

So, $(a_4\ b_4) = (86\ 847)$ is a solution of $97x^2 - 3 = y^2$.

We choose an integer m_4 such that

$$a_5 = \frac{a_4 m_4 + b_4}{K_4} = \frac{86m_4 + 847}{-3}$$

$\therefore$ an integer and

$$K_5 = \frac{m_4^2 - N}{K_4} = \frac{m_4^2 - 97}{-3}$$

is numerically small.

We choose $m_4 = 10$. Then,

$$a_5 = \frac{86(10) + 847}{-3} = \frac{860 + 847}{-3} = \frac{1707}{-3} = -569$$

(*Continued*)

Example 1 (cont'd)

$$b_5 = \frac{b_4 m_4 + N a_4}{K_4} = \frac{(847)(10) + (97)(86)}{-3} = \frac{8470 + 8342}{-3}$$

$$= \frac{16812}{-3} = -5604$$

$$K_5 = \frac{m_4^2 - N}{K_4} = \frac{100 - 97}{-3} = \frac{3}{-3} = -1$$

So, $(\alpha\ \beta) = (a_5\ b_5) = (569\ 5604)$ is a solution of $97x^2 - 1 = y^2$.

Then, from Brahmagupta's theorem on page 166, $(a\ b)$ is a solution of $97x^2 + 1 = y^2$, where

$$a = 2\alpha\beta = 2(569)(5604) = 2(3,188,676) = 6,377,352$$

$$b = \beta^2 + N\alpha^2 = (5604)^2 + 97(569)^2 = 31,404,816 + 97(323,761)$$

$$= 31,404,816 + 31,404,817 = 62,809,633$$

So, (6,377,352 62,809,633) is an integral solution of $97x^2 + 1 = y^2$.

Example 2

Solve in integers $103x^2 + 1 = y^2$.

Solution

We have the identity $103 \cdot 1^2 - 3 = 100 = 10^2$.

So, $(a_0\ b_0) = (1\ 10)$ is a solution of $103x^2 - 3 = y^2$.

Here $N = 103$, $K_0 = -3$.

We choose an integer m_0 such that

$$a_1 = \frac{a_0 m_0 + b_0}{K_0} = \frac{m_0 + 10}{-3}$$

is an integer and

$$K_1 = \frac{m_0^2 - N}{K_0} = \frac{m_0^2 - 103}{-3}$$

is numerically small.

We choose $m_0 = 11$.

Thus,

$$a_1 = \frac{11 + 10}{-3} = \frac{21}{-3} = -7$$

$$b_1 = \frac{b_0 m_0 + N a_0}{K_0} = \frac{(10)(11) + 103(1)}{-3} = \frac{110 + 103}{-3} = \frac{213}{-3} = -71$$

$$K_1 = \frac{m_1^2 - N}{K_0} = \frac{(11)^2 - 103}{-3} = \frac{121 - 103}{-3} = \frac{18}{-3} = -6$$

So, $(a_1\ b_1) = (7\ 71)$ is a solution of $103x^2 - 6 = y^2$.
We next choose an integer m_1 such that

$$a_2 = \frac{a_1 m_1 + b_1}{K_1} = \frac{7m_1 + 71}{-6}$$

is an integer and

$$K_2 = \frac{m_1^2 - N}{K_1} = \frac{m_1^2 - 103}{-6}$$

is numerically small.
We choose $m_1 = 7$.
Then,

$$a_2 = \frac{7(7) + 71}{-6} = \frac{49 + 71}{-6} = \frac{120}{-6} = -20$$

$$b_2 = \frac{b_1 m_1 + N a_1}{K_1} = \frac{(71)(7) + 103(7)}{-6} = \frac{497 + 721}{-6} = \frac{1218}{-6} = -203$$

$$K_2 = \frac{7^2 - 103}{-6} = \frac{49 - 103}{-6} = \frac{-54}{-6} = 9$$

Thus, we have

$$(a_2\ b_2) = (20\ 203)$$

is a solution at $103x^2 + 9 = y^2$.
We next choose an integer m_2 such that

$$a_3 = \frac{a_2 m_2 + b_2}{K_2} = \frac{20m_2 + 203}{9}$$

is an integer and

(*Continued*)

Example 2 (cont'd)

$$K_3 = \frac{m_2^2 - N}{K_2} = \frac{m_2^2 - 103}{9}$$

is numerically small.

We choose $m_2 = 11$.

Then,

$$a_3 = \frac{20(11) + 203}{9} = \frac{220 + 203}{9} = \frac{423}{9} = 47$$

$$b_3 = \frac{b_2 m_2 + N a_2}{K_2} = \frac{(203)(11) + 103(20)}{9} = \frac{2233 + 2060}{9} = \frac{4293}{9} = 477$$

$$K_3 = \frac{(11)^2 - 103}{9} = \frac{121 - 103}{9} = \frac{18}{9} = 2$$

So, $(\alpha\ \beta) = (a_3\ b_3) = (47\ 477)$ is a solution of $103x^2 + 2 = y^2$.

Then, by Brahmagupta's theorem 2 on page 166, $(a\ b)$ is an integral solution of $103x^2 + 1 = y^2$, where

$$a = \alpha\beta = 47(477) = 2249$$

$$b = \frac{\beta^2 + N\alpha^2}{2} = \frac{(477)^2 + 103(47)^2}{2} = \frac{227,529 + (103)(2109)}{2}$$

$$= \frac{227,529 + 227,527}{2} = 227,528$$

Thus, (2249 227,528) is an integral solution of $103x^2 + 1 = y$.

Example

Solve $13x^2 + 1 = y^2$.

Solution

As we have seen in an earlier example, $(\alpha\ \beta) = (1\ 3)$ is a solution of $13x^2 - 4 = y^2$.

Then, by Brahmagupta's theorem on page 170, we get

$$x = \frac{1}{2}\alpha\,\beta\,(\beta^2 + 3)(\beta^2 + 1) = \frac{1}{2}\cdot 1 \cdot 3(3^2 + 3)(3^2 + 1)$$

i.e.,

$$x = \frac{1}{2} \cdot 3(12)(10) = 180$$

$$y = (\beta^2 + 2)\left[\frac{1}{2}(\beta^2 + 3)(\beta^2 + 1) - 1\right]$$

$$= (3^2 + 2)\left[\frac{1}{2}(3^2 + 3)(3^2 + 1) - 1\right] = 11 \cdot \left[\frac{1}{2} \cdot 12 \cdot 9\right]$$

$$= 11(6 \cdot 10 - 1) = 649$$

So, $(x\ y) = (180\ 649)$ is a solution of $13x^2 + 1 = y^2$.

Simultaneous Quadratic Equations

(i) Solve

$$x + y = a \quad \text{(G. K. i. 35)}$$

$$xy = b$$

Narayana Pandita provided the following solution:

$$x = \frac{1}{2}[(x + y) + (x - y)] = \frac{1}{2}\left[(x + y) + \sqrt{x^2 - 2xy + y^2}\right]$$

$$x = \frac{1}{2}\left[(x + y) + \sqrt{(x+y)^2 - 4xy}\right]$$

Thus,

$$x = \frac{1}{2}\left[a + \sqrt{(a^2 - 4b)}\right]$$

$$y = \frac{1}{2}[(x + y) - (x - y)] = \frac{1}{2}\left[(x + y) - \sqrt{x^2 - 2xy + y^2}\right]$$

$$y = \frac{1}{2}\left[(x + y) - \sqrt{(x+y)^2 - 4xy}\right]$$

Thus,

$$y = \frac{1}{2}\left[a - \sqrt{a^2 - 4ab}\right]$$

(ii) Solve

$$x - y = a \quad \text{(G. K. i. 35)}$$

$$xy = b$$

(Continued)

Example (cont'd)

Narayana Pandita gave the following solution:

$$x + y = \sqrt{x^2 + 2xy + y^2} = \sqrt{(x^2 - 2xy + y^2) + 4xy} = \sqrt{a^2 + 4b}$$

Now,

$$x = \frac{1}{2}[(x + y) + (x - y)] = \frac{1}{2}\left[\sqrt{a^2 + 4b} + a\right]$$

$$y = \frac{1}{2}[(x + y) - (x - y)] = \frac{1}{2}\left[\sqrt{a^2 + 4b} - a\right]$$

(iii) Solve

$$x^2 + y^2 = a \quad \text{(G. K. i. 33)}$$

$$x - y = b$$

Narayana Pandita gave the following solution:

$$x + y = \sqrt{x^2 + 2xy + y^2} = \sqrt{2x^2 + 2y^2 - (x^2 - 2xy + y^2)} = \sqrt{2a - b^2}$$

Now,

$$x - y = b$$

Thus,

$$x = \frac{1}{2}[(x + y) + (x - y)] = \frac{1}{2}\left[\sqrt{2a - b^2} + b\right]$$

Similarly,

$$y = \frac{1}{2}[(x + y) - (x - y)] = \frac{1}{2}\left[\sqrt{2a - b^2} - b\right]$$

(iv) Solve

$$x^2 - y^2 = a \quad \text{(G. K. i. 34)}$$

$$xy = b$$

Narayana Pandita gave the following solution:

$$x^2 + y^2 = \sqrt{(x^2 - y^2)^2 + 4x^2y^2} = \sqrt{a^2 + 4b^2}$$

Now,

$$x^2 = \frac{1}{2}\left[(x^2 + y^2) + (x^2 - y^2)\right] = \frac{1}{2}\left[\sqrt{a^2 + 4b^2} + a\right]$$

So,

$$x = \sqrt{\frac{1}{2}\left[\sqrt{a^2 + 4b^2} + a\right]}$$

Similarly,

$$y^2 = \frac{1}{2}\left[(x^2 + y^2) - (x^2 - y^2)\right] = \frac{1}{2}\left[\sqrt{a^2 + 4b} - a\right]$$

Thus,

$$y = \sqrt{\frac{1}{2}\left[\sqrt{a^2 + 4b} - a\right]}$$

Single Indeterminate Equations

(i) Solve

$$x^3 + y^3 = x^2 + y^2 \quad (^*) \quad (G.\ \text{K. i. } 58)$$

Solution

Let,

$$\frac{x}{y} = \frac{m}{n}$$

$$x \Rightarrow \frac{m}{n}y$$

Substituting in (*), we get

$$\frac{m^3}{n^3}y^3 + y^3 = \frac{m^2}{n^2}y^2 + y^2$$

Dividing by y^2, we get

$$\frac{m^3}{n^3}y + y = \frac{m^2}{n^2} + 1$$

$$y\left(\frac{m^3}{n^3} + 1\right) = \frac{m^2 + n^2}{n^2}$$

(Continued)

Example (cont'd)

$$y\left(\frac{m^3+n^3}{n^3}\right)=\frac{m^2+n^2}{n^2}$$

$$y=\frac{m^2+n^2}{m^3+n^3}\cdot n$$

$$x=\frac{m}{n}y=\frac{m^2+n^2}{m^3+n^3}\cdot m$$

(ii) Solve

$$x^3+y^3=(x+y)^2 \quad \text{(G. K. i. 58)}$$

Solution

Let

$$\frac{x}{y}=\frac{m}{n}\Rightarrow x=\frac{m}{n}y$$

Thus, we have

$$\frac{m^3}{n^3}y^3+y^3=\left(\frac{m}{n}y+y\right)^2$$

$$y^3\left(\frac{m^3+n^3}{n^3}\right)=y^2\left(\frac{m+n}{n}\right)^2$$

So,

$$y\left(\frac{m^3+n^3}{n^3}\right)=\frac{(m+n)^2}{n^2}$$

Hence,

$$y=\frac{(m+n)^2}{m^3+n^3}\cdot n$$

$$x=\frac{m}{n}y=\frac{(m+n)^2}{m^3+n^3}\cdot m$$

(iii) Solve

$$x^3+y^3=xy \quad (G.\text{ K. i. 58})$$

Solution

Let

$$x = \frac{m}{n}y$$

Then, we have

$$\frac{m^3}{n^3}y^3 + y^3 = \left(\frac{m}{n}y\right)y$$

$$y^3\left[\frac{m^3 + n^3}{n^3}\right] = \frac{m}{n}y^2$$

which gives

$$y\left(\frac{m^3 + n^3}{n^3}\right) = \frac{m}{n}$$

So,

$$y = \frac{mn^2}{m^3 + n^3}$$

$$x = \frac{m}{n}y = \frac{m^2n}{m^3 + n^3}$$

(iv) Solve

$$(x+y)^3 = x^2 + y^2 \quad \text{(G. K. i. 58)}$$

Solution

Let

$$x = \frac{m}{n}y$$

Then, we have

$$\left(\frac{m}{n}y+y\right)^3 = \frac{m^2}{n^2}y^2 + y^2$$

$$y^3\left(\frac{m+n}{n}\right)^3 = y^2\left(\frac{m^2 + n^2}{n^2}\right)$$

(*Continued*)

Example (cont'd)

which gives

$$y\left(\frac{m+n}{n}\right)^3 = \frac{m^2+n^2}{n^2}$$

i.e.,

$$y = \frac{(m^2+n^2)n}{(m+n)^3}$$

$$x = \frac{m}{n}y = \frac{(m^2+n^2)m}{(m+n)^3}$$

(v) Solve

$$(x+y)^3 = (x+y)^2 \quad (G. \text{ K. i. } 58)$$

Solution

Let

$$x = \frac{m}{n}y$$

Then, we have

$$\left(\frac{my}{n}+y\right)^3 = \left(\frac{m}{n}y+y\right)^2$$

i.e.,

$$y^3\left(\frac{m+n}{n}\right)^3 = \left(\frac{m+n}{n}\right)^2 y^2$$

$$y\left(\frac{m+n}{n}\right) = 1$$

i.e.,

$$y = \frac{n}{m+n}$$

$$x = \frac{m}{n}y = \frac{m}{m+n}$$

(vi) Solve

$$(x+y)^3 = xy \quad (\text{G. K. i. } 58)$$

Solution

Let

$$y = \frac{m}{n}y$$

Then, we have

$$\left(\frac{m}{n}y + y\right)^3 = \frac{m}{n}y \cdot y$$

$$y^3\left(\frac{m+n}{n}\right)^3 = \frac{m}{n}y^2$$

i.e.,

$$\frac{y(m+n)^3}{n^3} = \frac{m}{n}$$

$$y = \frac{mn^2}{(m+n)^3}$$

$$x = \frac{m}{n}y, \quad x = \frac{m^2 n}{(m+n)^3}$$

Double Equations of First Degree

(i) Solve

$$x + a = u^2, \quad x + b = v^2 \quad \text{(G. K. i. 53)}$$

Solution

Let $a > b$.

Then,

$$u^2 - v^2 = a - b$$

$$(u + v)(u - v) = a - b$$

Let $u + v = m$, where m is any optional number.

Then,

$$u - v = \frac{a - b}{m}$$

(Continued)

Example (cont'd)

Adding

$$2u = m + \frac{a-b}{m} \Rightarrow u = \frac{1}{2}\left[m + \frac{a-b}{m}\right]$$

Subtracting, we get

$$2v = \left(m - \frac{a-b}{m}\right) \Rightarrow v = \frac{1}{2}\left[m - \frac{a-b}{m}\right]$$

Now,

$$x = u^2 - a = \frac{1}{2}\left[m + \frac{a-b}{m}\right]^2 - a$$

Taking m as 1, Narayana gave the solution as

$$x = \frac{1}{2}[1 + a - b]^2 - a$$

(ii) Solve

$$x - a = u^2, \quad x - b = v^2$$

If we replace a by $-a$ and b by $-b$ in (i), we get

$$x = \frac{1}{2}\left[m + \frac{b-a}{m}\right]^2 + a$$

Taking $m = 1$, Narayana gave the solution as

$$x = \frac{1}{2}[1 + b - a]^2 + a$$

or

$$x = \frac{1}{2}[a - b - 1]^2 + a$$

Example 1

Solve

$$4x - 3 = u^2 \quad \text{(i)}$$

$$7x - 3 = v^2 \quad \text{(G. K. i. 47)} \quad \text{(ii)}$$

Solution

From (i),

$$4x = u^2 + 3 \Rightarrow x = \frac{1}{4}(u^2 + 3)$$

Then from (ii),

$$\frac{7}{4}(u^2 + 3) - 3 = v^2$$

i.e.,

$$7u^2 + 9 = 4v^2$$

Let

$$w = 2v$$

Then, we have $w^2 = 7u^2 + 9$, which is a "vargaprakrti" equation.
We observe that $(u\ w) = (l\ m) = (1\ 4)$ is a solution of $w^2 = 7u^2 + 9$.
So, another solution of $w^2 = 7u^2 + 9$ is

$$u = lq \pm mp = q \pm 4p$$

and

$$v = mq \pm 7lp = 4q \pm 7p$$

where $7p^2 + 1 = q^2$.
If $p = 0$, $q = 1$, then $u = 1$, $v = 4$.
If $p = 3$, $q = 8$, then $u = 20$ or -4, $v = 53$ or 11.
If $u = 1$, $4x = u^2 + 3 = 1 + 3 = 4 \Rightarrow x = 1$.
If $u = 20$,

$$4x = u^2 + 3 = 400 + 3 \Rightarrow x = \frac{403}{4}$$

and so on, discarding the negative signs.

Simplification of Bhaskara II's solution on page 480 (G. K. i. 46). Bhaskara II has shown that $x = 8t^4 + 1$, $y = 8t^3$, $u = 4t^2(2t^2 + 1)$, and $v = 4t^2(2t^2 - 1)$ is a solution of $x^2 + y^2 - 1 = u^2$ and $x^2 - y^2 - 1 = v^2$.

If we let $t = m/2$, we get

$$x = 8\left(\frac{m}{2}\right)^4 + 1 = \frac{1}{2}m^4 + 1$$

$$y = 8t^3 = 8\left(\frac{m}{2}\right)^3 = m^3$$

(*Continued*)

Example 1 (cont'd)

$$u = 4\left(\frac{m}{2}\right)^2\left(2\cdot\left(\frac{m}{2}\right)^2+1\right) = m^2\left(\frac{m^2}{2}+1\right) = \frac{m^2}{2}(m^2+2)$$

$$v = 4\left(\frac{m}{2}\right)^2\left[2\left(\frac{m}{2}\right)^2-1\right] = m^2\left(\frac{m^2}{2}-1\right) = \frac{m^2}{2}(m^2-2)$$

Example 2

Find two numbers such that sum, their difference, and their product plus unity are all perfect squares.

(G. K. i. 47)

Solution

Let x and y be the two n numbers, we have to solve $x+y=u^2$, $x-y=v^2$, and $1+xy=w^2$.

Let $x=2z^2(a^2+b^2)$ and $y=2z^2(a^2-b^2)$, where a and b are rational numbers.

There

$$x+y=2z^2[a^2+b^2+(a^2-b^2)]=4a^2z^2=u^2 \Rightarrow u=2az$$

$$x-y=2z^2[a^2+b^2-(a^2-b^2)]=4b^2z^2=v^2 \Rightarrow v=2bz$$

But,

$$1+xy=w^2$$

$$1+xy=1+2z^2(a^2+b^2)\cdot 2z^2(a^2-b^2)$$

$$1+xy=1+4z^4(a^4-b^4)$$

Now,

$$1+4z^4(a^4-b^4)=(2z^2a^2-1)^2+4z^2(a^2-z^2b^4)$$

In order that $1+xy$ is a perfect square $=w^2$,

$$a^2-z^2b^4=0 \Rightarrow a^2=z^2b^4 \quad \text{and} \quad w=2z^2a^2-1$$

Hence,

$$2z^2 = \frac{2a^2}{b^4} = \frac{(a^2+b^2)+(a^2-b^2)}{\{(1/2)[(a^2+b^2)-(a^2-b^2)]\}^2}$$

Now,

$$x = 2z^2(a^2+b^2) = 2z^2(z^2b^4+b^2)$$

i.e., $x = 2(z^4b^4 + z^2b^2)$.

Let $zb = p$. Then, p is a rational number.

Hence, $x = 2(p^4 + p^2)$.

Similarly,

$$y = 2z^2(a^2 - b^2)$$

$$y = 2z^2(z^2b^4 - b^2)$$

$$y = 2(z^4b^4 - z^2b^2)$$

$$y = 2(p^4 - p^2)$$

Thus,

$$x = 2(p^4 + p^2) \quad \text{and} \quad y = 2(p^4 - p^2)$$

where p is a rational number.

Bhaskara II on page 480 has given the solution of $x^2 + xy + y^2 = z^2$ as

$$x = 3m^2 - 4n(n+m)$$

$$y = 8mn$$

$$z = 3m^2 + 4n^2$$

Narayana Pandita gave alternate form of this solution as follows.

Dividing by $8n$, we get

$$x = \frac{1}{8n}\left[3m^2 - 4n(n+m)\right] = \frac{1}{2}\left[\frac{3m^2}{4n} - (n+m)\right]$$

$$y = m, \quad z = \frac{1}{8n}(3m^2 + 4n^2)$$

Narayana Pandita preferred integral solutions of his equation.

Area of Triangle

Narayana's new expression for the area of a triangle is embodied in

चतुराहतहृदयहतं त्रिभुजभुचानां वधं गणितप् ।

(G. K. Ks. Vya. 134)

i.e., "The area of a triangle is equal to the product of the sides divided by four times the circumradius."

Let *ABC* be a triangle, Let *AD* be the altitude from *A* on *BC*. Let *R* be the circumradius of the triangle *AB*.

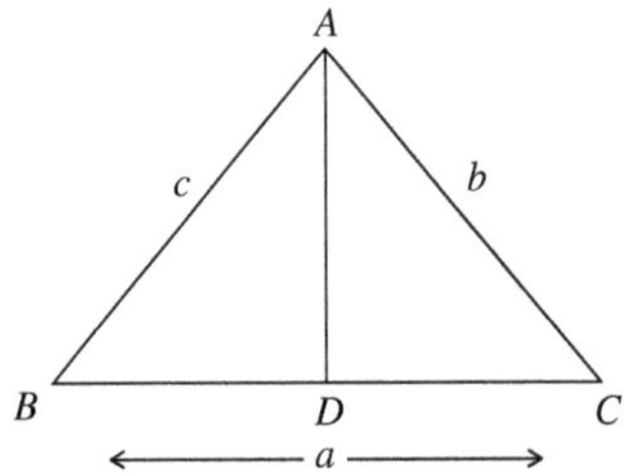

Area of the triangle *ABC*

$$= \frac{1}{2}\text{the base } BC \times \text{altitude } AD$$

$$= \frac{1}{2}a \cdot AD$$

By Brahmagupta's theorem,

$$AD = \frac{\text{Product of the sides } AB \text{ and } AC}{2R}$$

So,

$$AD = \frac{bc}{2R}$$

So, Area of the triangle *ABC*

$$= \frac{1}{2}a \cdot \frac{bc}{2R} = \frac{abc}{4R}$$

"In a triangle, if the altitude is produced to meet the circumcircle, the portion of the altitude beneath the base to the circumcircle is equal to the product of the '*ābādhas*' of the base divided by the altitude. Also the sides of the lower triangle (formed) are equal to the quotients obtained when the product of '*ābādhas*' and nonadjacent sides are divided by the altitude."

Let ABC be a triangle. Let the altitude AD on BC meet the circle at F. Join BF and FC. Then,

$$(i)\ DF = \frac{BD \cdot DC}{AD};\quad \text{(ii)}\ BF = \frac{BD \cdot AC}{AD};\quad \text{(iii)}\ CF = \frac{DC \cdot AB}{AD}$$

Proof: Now triangles ADC and BDF are similar, since

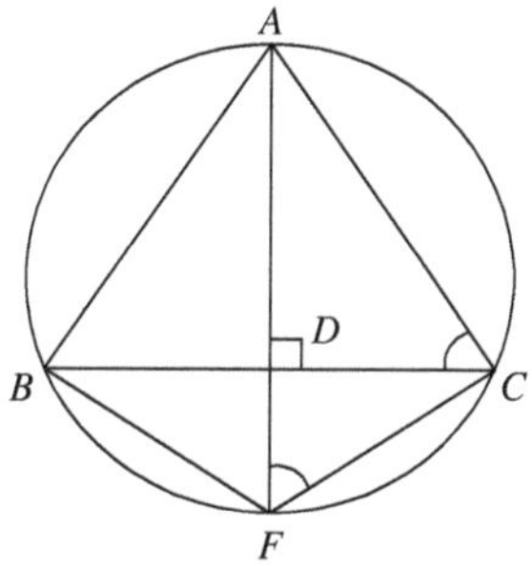

$$A\hat{D}C = B\hat{D}F = 90^\circ$$

$$A\hat{C}D = A\hat{C}B = A\hat{F}B = D\hat{F}B$$

So,

$$\frac{BF}{AC} = \frac{BD}{AD} = \frac{DF}{DC}$$

Hence,

$$BD \cdot DC = AD \cdot DF$$

So,

$$DF = \frac{BD \cdot DC}{AD}$$

Also,

$$BF = \frac{BD \cdot AC}{AD}$$

Again triangles ADB and FDC are similar, since

$$A\hat{D}B = F\hat{D}C = 90^\circ$$

$$A\hat{B}D = A\hat{B}C = A\hat{F}C = D\hat{F}C$$

So,

$$\frac{AB}{CF} = \frac{AD}{DC} = \frac{BD}{DF}$$

So,

$$CF = \frac{AB \cdot DC}{AD}$$

Narayana Pandita in his "*Ganita Kaumudi*" has given new formulae for the circumradius and the area of a triangle.

त्रबधावघेन हीनो लम्बकवर्गोऽवलम्बकविभक्तः ।
तत्कृतिभूकृतियोगान्मूलदलं जायते हृदयम् ॥

(G. K. Ks. Vya. 133)

i.e., "The square of the altitude is decreased by the product of 'ābādhās' and then divided by the altitude. One half of the square root of the sum of the squares of the base and the square of the above quotient gives the circumradius of the triangle."

Let ABC be a triangle. Let AD be the perpendicular drawn from A on BC. Let R be the circumradius of the triangle ABC.

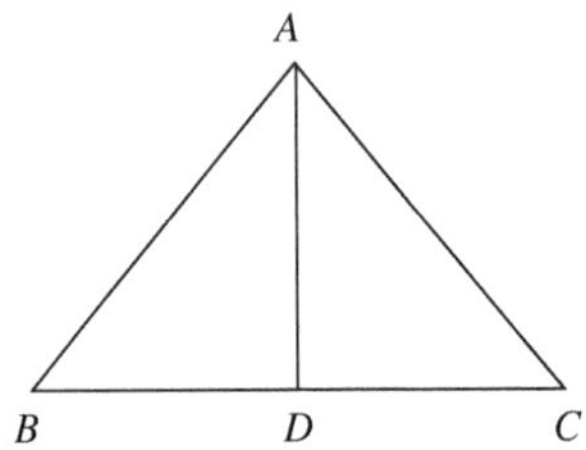

Then,

$$R = \frac{1}{2}\sqrt{BC^2 + \left(\frac{AD^2 - BD \cdot DC}{AD}\right)^2}$$

Proof: From Brahmagupta's theorem

$$R = \frac{AB \cdot AC}{2AD}$$

So,

$$R^2 = \frac{AB^2 AC^2}{4AD^2} \tag{1}$$

In right triangle ADB, $AB^2 = AD^2 + BD^2$.
Again, in right triangle ADC, $AC^2 = AD^2 + DC^2$.
Substituting these in (1), we get

$$R^2 = \frac{(AD^2 + BD^2)(AD^2 + DC^2)}{4AD^2}$$

$$R^2 = \frac{AD^4 + BD^2AD^2 + AD^2DC^2 + BD^2DC^2}{4AD^2}$$

$$R^2 = \frac{AD^4 + AD^2(BD^2 + DC^2) + BD^2DC^2}{4AD^2}$$

$$R^2 = \frac{AD^4 + AD^2[(BD + DC)^2 - 2BD \cdot DC] + BD^2BC^2}{4AD^2}$$

$$R^2 = \frac{AD^4 + AD^2[BC^2 - 2BD \cdot DC] + BD^2DC^2}{4AD^2}$$

$$R^2 = \frac{AD^2BC^2 + (AD^4 - 2AD^2BD \cdot DC + BD^2 \cdot DC^2)}{4AD^2}$$

$$R^2 = \frac{AD^2BC^2 + (AD^2 - BD \cdot DC)^2}{4AD^2}$$

$$= \frac{1}{4}\left[BC^2 + \left(\frac{AD - BD \cdot DC}{AD}\right)^2\right]$$

So,

$$R = \frac{1}{2}\sqrt{(BC)^2 + \left(\frac{AD - BD \cdot DC}{AD}\right)^2}$$

Narayana Pandita has studied the geometry of cyclic quadrilateral much deeper than Brahmagupta did. The theorem on three diagonals is the most significant of the study.

सर्बचतुर्बाहूनां मुखस्य परिवर्तने यदा विहिते ।
कर्णस्तदा तृतीयः पर इति कर्णत्रयं भवति ॥

(G. K. Ks. Vya. 48)

i.e., "When the topside and the flank side of any four-sided figure are interchanged, we obtain a third diagonal called 'para' (the auxiliary diagonal). This interchange of sides without changing the area is possible only in the case of a cyclic quadrilateral."

The area of a cyclic quadrilateral can be expressed in terms of the above three diagonals.

द्विगुणव्यास विभक्ते त्रिकर्णघातेऽथवा गणितम् ।

(G. K. Ks. Vya. 52)

i.e., "The area of a cyclic quadrilateral is equal to the product of the three diagonals divided by twice the circumdiameter."

Proof: Let *ABCD* be a cyclic quadrilateral. Let *E* be the vertex obtained by interchanging the sides *DC* and *BC*.

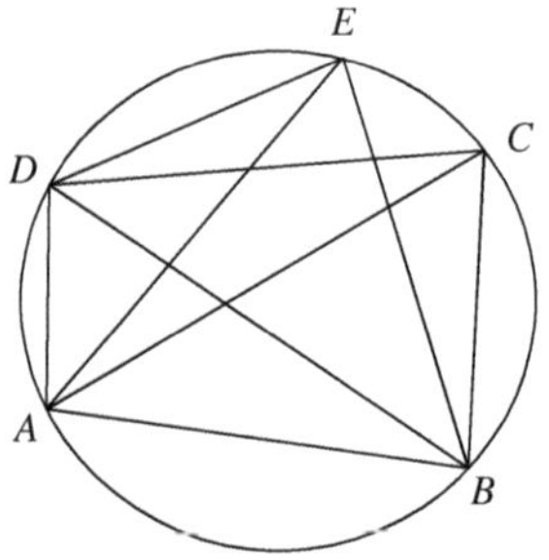

Then, the area of cyclic quadrilateral *ABCD*

$$= \frac{AC \cdot BD \cdot AE}{2d}$$

where *d* is the circumdiameter.

Proof: Area of the quadrilateral *ABCD*

$$= \text{Area of the } \Delta\, ACB + \text{Area of the } \Delta\, ACD$$

We know that area of a triangle

$$= \frac{\text{Product of the three sides}}{\text{Twice the circumdiameter}}$$

So, Area of $\Delta\, ACB$

$$= \frac{AB \cdot BC \cdot AC}{2d}$$

Area of $\Delta\, ACD$

$$= \frac{AC \cdot CD \cdot AD}{2d}$$

Thus, the area of cyclic quadrilateral *ABCD*

$$= \frac{AB \cdot BC \cdot AC + AC \cdot CD \cdot AD}{2d}$$

$$= \frac{AC}{2d}[AB \cdot BC + CD \cdot AD]$$

$$= \frac{AC}{2d}[AB \cdot DE + AD \cdot BE]$$

But by Ptolemy's theorem, the sum of the product of the opposite sides of cyclic quadrilateral is equal to the product of its diagonals. Then, in the cyclic quadrilateral *ADEB*,

$$AB \cdot DE + AD \cdot BE = AE \cdot BD$$

Hence, the area of the cyclic quadrilateral *ABCD*

$$= \frac{AC \cdot AE \cdot BD}{2d}$$

"In the case of cyclic quadrilateral, when the base and the face are separately divided by the third diagonal and the resulting quotients are separately multiplied by the two flanks, we obtain the lower and upper segments of the diagonal."

(G. K. Ks. Vya. 139)

Let *M* be the point of intersection of the diagonal *AC* and *BD* of a cyclic quadrilateral *ABCD*.

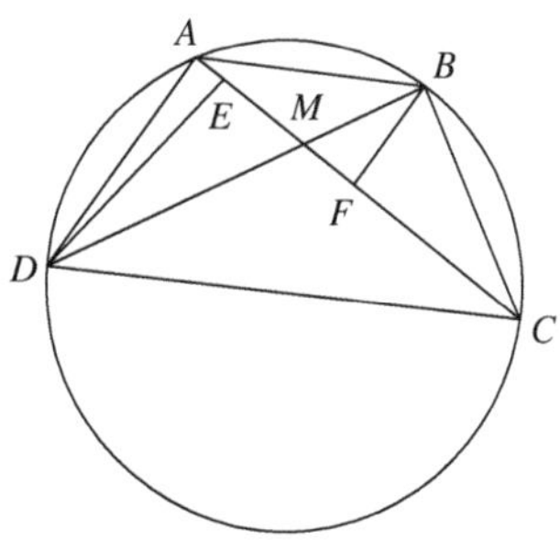

Then,

$$DM = \frac{CD \cdot AD}{d_3}$$

and

$$MB = \frac{BC \cdot AB}{d_3}$$

where d_3 is the third diagonal.

Draw *DE* and *BF* perpendicular from *D* and *B*, respectively, on *AC*.

Proof: In Δ^{le} DEM and MBF,

$$D\hat{E}M = B\hat{F}M = 90^\circ$$

$$E\hat{M}D = B\hat{M}F \text{ (opposite angles)}$$

So, Δ^{les} DEM and MBF are similar

$$\frac{DE}{BF} = \frac{DM}{MB} = \frac{EM}{MF}$$

So,

$$\frac{DM}{MB} = \frac{DE}{BF} \Rightarrow \frac{MB}{DM} = \frac{BF}{DE}$$

$$\frac{MB}{DM} + 1 = \frac{BF}{DE} + 1 \Rightarrow \frac{MB + DM}{DM} = \frac{BF + DE}{DE}$$

$$\frac{BD}{DM} = \frac{DE + BF}{DE}$$

$$DM = \frac{DE \cdot BD}{DE + BF} = \frac{BD}{DE + BF} \frac{AD \cdot CD}{2R} \tag{1}$$

where R is the circumradius.

Now, Area of Δ^{le} ADC

$$= \frac{1}{2} AC \cdot DE$$

Again, Area of Δ^{le} ABC

$$= \frac{1}{2} AC \cdot BF$$

A = Area of cyclic quadrilateral $ABCD$
= Area of Δ^{le} ADC + Area of Δ^{le} ABC

$$= \frac{1}{2} AC(DE + BF)$$

A = Area of cyclic quadrilateral $ABCD$

$$= \frac{AC \cdot BD \cdot d_3}{4R}$$

where d_3 is the third diagonal.

So,

$$\frac{1}{2}AC(DE+BF)=\frac{AC\cdot BD\cdot d_3}{4R}$$

Hence,

$$DE+BF=\frac{2AC\cdot BD\cdot d_3}{4R\cdot AC}=\frac{BD\cdot d_3}{2R} \tag{2}$$

So, from (1), we get

$$DM=\frac{BD}{DE+BF}\frac{AD\cdot CD}{2R}=\frac{2R}{d_3}\frac{AD\cdot CD}{2R}=\frac{AD\cdot CD}{d_3}$$

Hence,

$$DM=\frac{AD\cdot CD}{d_3}$$

Again,

$$\frac{DM}{MB}+1=\frac{DE}{BF}+1$$

$$\frac{DM+MB}{MB}=\frac{DE+BF}{BF}$$

i.e.,

$$\frac{BD}{MB}=\frac{DE+BF}{BF}$$

So,

$$MB=\frac{BD\cdot BF}{DE+BF}=\frac{BD}{DE+BF}\cdot\frac{AB\cdot BC}{2R} \tag{3}$$

Since,

$$BF=\frac{AB\cdot BC}{2R}$$

in triangle ABC.

Then, from (2)

$$MB=\frac{2R}{d_3}\frac{AB\cdot BC}{2R}=\frac{AB\cdot BC}{d_3}$$

Thus,

$$MB = \frac{AB\ BC}{d_3}$$

कर्णाश्रितभुजवधयुतिगणिते त(ऽन्य?) स्मिन् श्रवस्यापि विभक्ते ।
चतुराहतहृदयेन द्विसमादिचतुर्भुजे यणिंतम्॥

(G. K. Ks. Vya. 134)

i.e., "In the case of a cyclic quadrilateral, when the product of a diagonal and the sum of the products of the sides about the other diagonal is divided by four times the circumradius gives the area of the cyclic quadrilateral."

Let *ABCD* be a cyclic quadrilateral of area *A*. Let *R* be the circumradius of this quadrilateral.

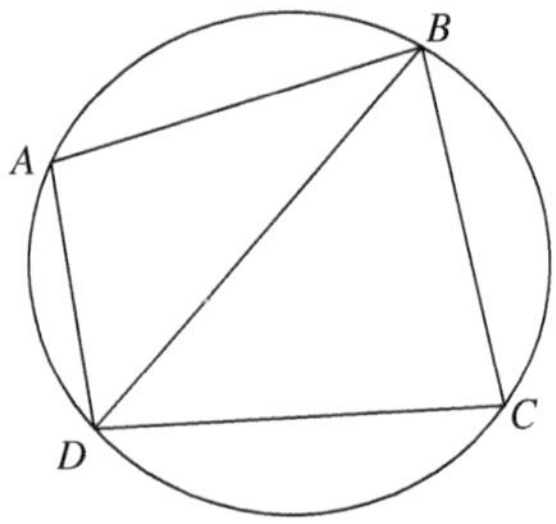

Then,

$$A = \frac{BD[AD \cdot AB + BC \cdot CD]}{4R} \quad \text{(i)}$$

and

$$A = \frac{AC[AB \cdot BC + AD \cdot DC]}{4R} \quad \text{(ii)}$$

Proof: Area of the triangle *ABD*

$$= \frac{1}{2} BD \cdot (\text{Altitude from } C \text{ on } BD)$$

$$= \frac{1}{2} BD \cdot \frac{\text{Product of the sides}}{2R}$$

$$= \frac{1}{2} BD \cdot \frac{AD \cdot AB}{2R} = \frac{BD}{4R}(AD \cdot AB)$$

Area of the triangle BCD

$$= \frac{1}{2}BD \cdot (\text{Altitude from } C \text{ on } BD)$$

$$= \frac{1}{2}BD \cdot \frac{\text{Product of the sides}}{2R}$$

$$= \frac{1}{2}BD \cdot \frac{BC \cdot CD}{2R} = \frac{BD}{4R}(BC \cdot CD)$$

Now, Area of the cyclic quadrilateral $ABCD$

$$= \text{Area of } \Delta^{\text{le}}\ ABD + \text{Area of } \Delta^{\text{le}}\ ACD$$

So,

$$A = \frac{BD}{4R}[AD \cdot AB + BC \cdot CD]$$

This proves (i).
Similarly, Area of $\Delta^{\text{le}}\ ABC$

$$= \frac{1}{2}AC \cdot (\text{Altitude from } B \text{ on } AC)$$

$$= \frac{1}{2}AC \cdot \frac{\text{Product of Sides}}{2R}$$

Area of $\Delta^{\text{le}}\ ABC$

$$= \frac{AC}{4R}[AB \cdot BC]$$

Again, Area of $\Delta^{\text{le}}\ ADC$

$$= \frac{1}{2}AC \cdot \text{Altitude from } D \text{ on } AC$$

$$= \frac{1}{2}AC \cdot \frac{AD \cdot CD}{2R}$$

$$= \frac{AC}{4R}[AD \cdot CD]$$

A = Area of cyclic quadrilateral $ABCD$

= Area of triangle ABC + Area of triangle ACD

A = Area of cyclic quadrilateral $ABCD$

$$= \frac{AC}{4R}[AB \cdot BC + AD \cdot CD]$$

This proves (ii).

Narayana Pandita gave the following formula for the circumradius of a cyclic quadrilateral:

The circumradius is equal to half the square root of the quotient obtained, when the product of the diagonals and the product of flanks are divided by the product of altitudes.

(G. K. Ks. Vya. 138)

Let $ABCD$ be a cyclic quadrilateral. Let R be the circumradius. Let p_1 and p_2 be the altitude from A and B, respectively, on CD.

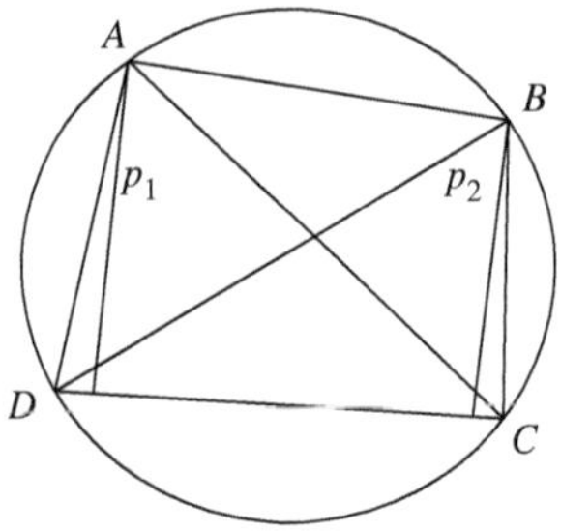

Then,

$$R = \frac{1}{2}\sqrt{\left\{\frac{AC \cdot BD \cdot AD \cdot BC}{p_1 p_2}\right\}}$$

Proof: From Δ^{le} BDC,

Circumradius R

$$= \frac{\text{Product of sides}}{2 \text{ Altitude } p_2}$$

$$R = \frac{BD \cdot BC}{2p_2} \quad (1)$$

Similarly, from Δ^{le} ADC,

Circumradius R

$$= \frac{\text{Product of sides}}{2 \text{ Altitude } p_1}$$

$$R = \frac{AD \cdot AC}{2p_1} \quad (2)$$

Multiplying (1) and (2), we get

$$R^2 = \frac{AC \cdot BD \cdot AD \cdot BC}{4p_1p_2}$$

i.e.,

$$R = \frac{1}{2}\sqrt{\frac{AC \cdot BD \cdot AD \cdot BC}{p_1p_2}}$$

When the quotient obtained by dividing twice the area of the cyclic quadrilateral by the product of the base and the diagonal is multiplied again by the lower segment of the diagonal, we obtain the altitude from the top of the other diagonal.

(G. K. Ks. Vya 140)

Let *AC* and *BD* be the diagonals of a cyclic quadrilateral *ABCD*. Let *M* be the point of interaction of the diagonals *AC* and *BD*. Let *DE*, *MF*, and *CN* be perpendiculars drawn from *D*, *M*, and *C*, respectively, on *AB*. Let *R* be the circumradius of the cyclic quadrilateral *ABCD*.

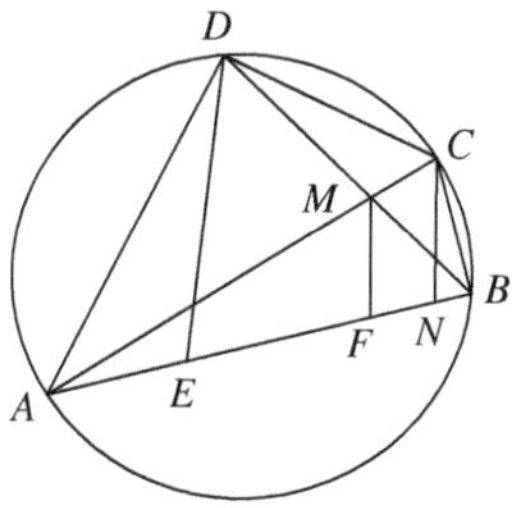

Then,

$$DE = AM \cdot \frac{2 \text{ Area of } ABCD}{AC \cdot AB}$$

and

$$CN = BM \cdot \frac{2 \text{ Area of } ABCD}{BD \cdot AB}$$

Proof: In triangle *ABD*,

$$\text{Altitude } DE = \frac{\text{Product of the sides}}{\text{Twice the circumradius}} = \frac{AD \cdot BD}{2R}$$

But the circumradius

$$R = \frac{\text{Product of the three diagonals}}{4 \text{ Area of } ABCD} \quad (*)$$

Hence, altitude

$$DE = \frac{AD \cdot BD}{(\text{Product of the three diagonals}/2 \text{ Area of } ABCD)}$$

$$= (AD \cdot BD)\frac{2 \text{ Area of } ABCD}{AC \cdot BD\, d_3}$$

if d_3 is the third diagonal.

So,

$$DE = \frac{AD \cdot AB}{d_3} \cdot \frac{2 \text{ Area of } ABCD}{AC \cdot AB}$$

$$DE = AM \cdot \frac{2 \text{ Area of } ABC}{AC \cdot AB}$$

Similarly, in Δ^{le} ABC

$$\text{Altitude } CN = \frac{\text{Product of sides}}{\text{Twice the circumradius}} = \frac{AC \cdot BC}{2R}$$

Hence from (*), we get

$$CN = \frac{AC \cdot BC}{(AC \cdot BD \cdot d_3)/(2 \text{ Area of } ABCD)} = \frac{AC \cdot BC \cdot 2 \text{ Area of } ABCD}{AC \cdot BD\, d_3}$$

where d_3 is the third diagonal.

So,

$$CN = \frac{AB \cdot BC}{d_3} \cdot \frac{2 \text{ Area of } ABCD}{BD \cdot AB} = BM \cdot \frac{2 \text{ Area of } ABCD}{AB \cdot BD}$$

In a cyclic quadrilateral, if the square root of the product of the lower segments is multiplied by the product of the flanks and then divided by the circumdiameter, we obtain the altitude from the point of intersection of the diagonals.

(G. K. Ks. Vya. 142)

Let M be the point of intersection of the diagonal AC and BD of the cyclic quadrilateral $ABCD$. Let DE, MF, and CN be the perpendiculars drawn from D, M, and C, respectively, on the base AB.

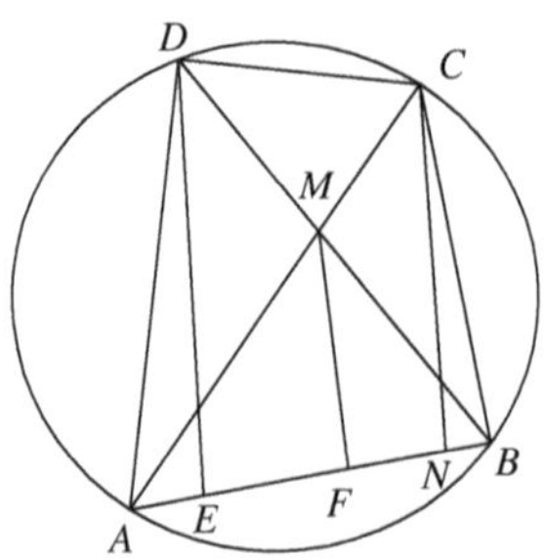

Then,

$$MF = \frac{\sqrt{AM \cdot BM \cdot AD \cdot BC}}{d}$$

where d is the circumdiameter.

Proof: In triangles *BED* and *BMF*

$$B\hat{E}D = B\hat{F}M = 90^\circ$$

$$D\hat{B}E = M\hat{B}F = D\hat{B}A$$

So, Δ^{les} *BED* and *BMF* are similar.
So,

$$\frac{MF}{DE} = \frac{BM}{BD} \Rightarrow MF = \frac{DE \cdot BM}{BD}$$

i.e.,

$$MF = \frac{BM}{BD} \cdot \frac{AD \cdot BD}{d} = \frac{BM \cdot AD}{d} \quad (1)$$

In triangles *CAN* and *MAF*

$$C\hat{N}A = M\hat{F}A = 90^\circ$$

$$C\hat{A}N = M\hat{A}F = C\hat{A}B$$

So, Δ^{les} *CAN* and *MAF* are similar.
Hence,

$$\frac{MF}{CN} = \frac{AM}{AC} \Rightarrow MF = \frac{AM \cdot CN}{AC} \quad (2)$$

So,

$$MF = \frac{AM}{AC} \cdot \frac{AC \cdot BC}{d} = \frac{AM \cdot BC}{d}$$

Multiplying (1) and (2), we get

$$MF^2 = \frac{AM \cdot BM \cdot AD \cdot BC}{d^2}$$

Thus,

$$MF = \frac{\sqrt{AM \cdot BM \cdot AD \cdot BC}}{d}$$

The square of each flank is diminished from the square of the circumdiameter. The square root of this difference divided by the respective flank is called "Sakala." The quotient of the base and the sum of the "Sakalas" give the altitude on the base from the point of intersection of the diagonals. The products of each Sakala and the above altitude will give the segments made by it on the base.

Let M be the point of intersection of the diagonals, AC and BD. Let CE and DF be diameters of circumcircle through C and B, respectively.

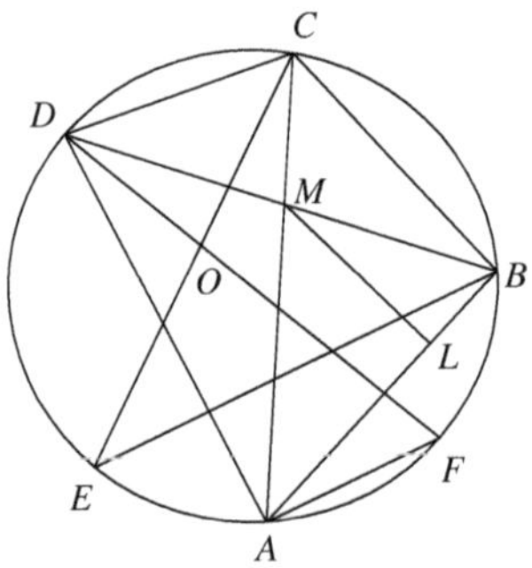

Let ML be the altitude from M on the base AB. Join AF and BE.

Now $C\hat{B}E = 90°$ (angle in a semicircle) and $D\hat{A}F = 90°$ (angle in a semicircle). So, Δ^{les} ECB and DAF are right triangles.

In the right Δ^{le} ECB,

$$BE^2 = EC^2 - BC^2 = d^2 - BC^2$$

In the right Δ^{le} DAF,

$$AF^2 = DF^2 - AD^2 = d^2 - AD^2$$

where d is the circumdiameter.

Now,

$$\frac{\sqrt{BE^2}}{BC} = \frac{BE}{BC}$$

and

$$\frac{\sqrt{AF^2}}{AD} = \frac{AF}{AD}$$

are called "Sakalas."

We have to show

i. $ML = \dfrac{AB}{(BE/BC) + (AF/AD)}$

ii. $AL = \dfrac{BE}{BC} \cdot ML \quad BL = \dfrac{AF}{AD} \cdot ML$

Proof: In Δ^{les} BEC and AML,

$$C\hat{B}E = M\hat{L}A = 90^\circ$$

$$C\hat{E}B = C\hat{A}B = M\hat{A}L \text{ (angles on the same arc)}$$

So, Δ^{les} BEC and AML are similar.
So,

$$\frac{ML}{BC} = \frac{AL}{BE}$$

Hence,

$$AL = \frac{BE}{BC} \cdot ML \tag{1}$$

Again, in Δ^{les} ADF and BLM,

$$D\hat{A}F = M\hat{L}B = 90^\circ$$

$$A\hat{F}D = D\hat{B}A = M\hat{B}L \text{ (angles on the same arc)}$$

So, Δ^{les} ADF and BLM are similar.
So,

$$\frac{BL}{AF} = \frac{ML}{AD}$$

Hence,

$$BL = \frac{AF}{AD} \cdot ML \tag{2}$$

Adding (1) and (2), we get

$$AL + BL = \frac{BE}{BC}ML + \frac{AF}{AD}ML$$

i.e.,

$$AB = ML\left[\frac{BE}{BC} + \frac{AF}{AD}\right]$$

Hence,

$$ML = \frac{AB}{(BE/BC) + (AF/AD)}$$

In connection with a cyclic quadrilateral, "Suci" is the triangle formed by producing the flanks till they meet.

Let *ABCD* be a cyclic quadrilateral. Let *P* be the point of the intersection of the flanks *AD* and *BC* produced.

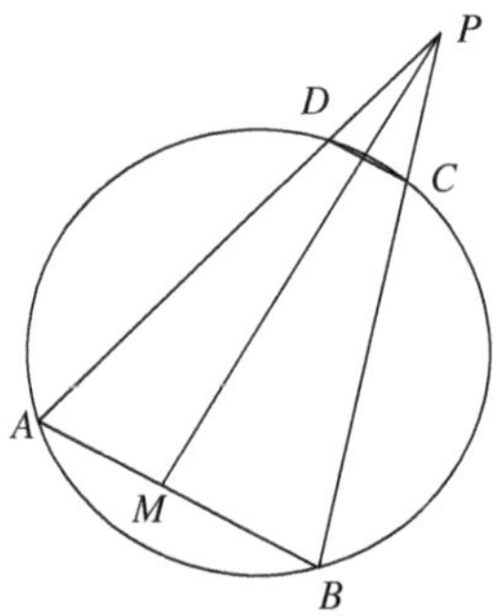

Then the triangle *PAB* is called "Suci" of the cyclic quadrilateral *ABCD*.

Let *PM* be the altitude from *P* on the base *AB* of the cyclic quadrilateral *ABCD*. Then, the altitude *PM* is called "sucilamba of the Suci."

"In a cyclic quadrilateral, if the product of its area and the base is divided by the half the difference between the squares of the base and face, we obtain the altitude 'sucilamba' of the Suci. If the flanks and 'Sandhis' (projections of the flanks on the base) are multiplied by 'Sucilamba' and if these products are divided by the altitude at the end of the respective side, we get the sides and the 'abadhas' of 'suci' respectively."

(G. K. Ks. Vya 143–143½)

Let Δ^{le} *PAB* be the "Suci" and the altitude *PM* from *P* on the base *AB* be the "Sucilamba" of the cyclic quadrilateral *ABCD*. Let *DL* and *CN* be the altitudes drawn from *D* and *C*, respectively, on the base *AB*. Then,

i. $PM = \dfrac{\text{Area of } ABCD \cdot \text{base } AB}{(1/2)[(\text{base } AB)^2 - (\text{face } CD)^2]}$

ii. $AP = \dfrac{ADPM}{DL}$

iii. $BP = \dfrac{BC \cdot PM}{CN}$

iv. $AM = \dfrac{AL \cdot PM}{DL}$

v. $BM = \dfrac{BN}{CN} \cdot PM$

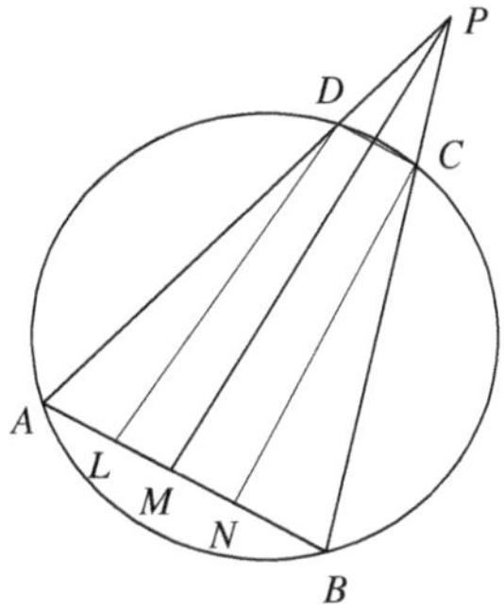

Proof: The following proof is due to Padmakara Dvivedi

In Δ^{le} PDC and PAB,

$P\hat{D}C = A\hat{B}P$ (since $A\hat{D}C$ and $A\hat{B}C$ are supplementary)

$D\hat{P}C = A\hat{P}B$

So, Δ^{les} PDC and PAB are similar.

$$\frac{AB}{CD} = \frac{AP}{PC} = \frac{PB}{PD}$$

So,

$$AP = \frac{AB \cdot PC}{CD} = \frac{AB(PB - BC)}{CD} \tag{1}$$

and

$$PB = \frac{AB \cdot PD}{CD} = \frac{AB(AP - AD)}{CD} \tag{2}$$

Adding (1) and (2), we get

$$AP + PB = \frac{AB(AP - AD)}{CD} + \frac{AB(PB - BC)}{CD}$$

i.e.,

$$AP + PB = \frac{AB}{CD}[AP - AD + PB - BC]$$

Hence,

$$(AP + PB)CD = AB[AP - AD + PB - BC]$$

$$= AB(AP + PB) - AB(AD + BC)$$

So,

$$(AP + PB)[CD - AB] = -AB(AD + BC)$$

Hence,

$$AP + PB = \frac{AB(AD + BC)}{AB - CD} \tag{3}$$

Now subtracting (2) from (1), we get

$$AP - PB = \frac{AB}{CD}(PB - BC) - \frac{AB}{CD}(AP - AD)$$

i.e.,

$$(AP - PB)CD = AB(PB - BC) - AB(AP - AD)$$

$$= -AB(AP - PB) - AB(BC - AD)$$

$$AP - PB = \frac{AB(AD - DC)}{CD + AB} \tag{4}$$

Adding (3) and (4), we get

$$2AP = AB\left[\frac{AD + BC}{AB - CD} + \frac{AD - BC}{CD + AB}\right]$$

$$= AB\left[\frac{AD \cdot CD + AD \cdot AB + BC \cdot CD + BC \cdot AB + AB \cdot AD - AD \cdot CD - BC \cdot AB + BC \cdot CD}{AB^2 - CD^2}\right]$$

$$2AP = \frac{AB}{AB^2 - CD^2}[2AD \cdot AB + 2BC \cdot CD]$$

But the area A of cyclic quadrilateral $ABCD$ is given by

$$A = \frac{BD(AD \cdot AB + CD \cdot BC)}{4R}$$

where R is the circumradius.

So,

$$2AP = \frac{AB}{AB^2 - CD^2} 2 \cdot \frac{4RA}{BD}$$

$$AP = \frac{4RA}{AB^2 - CD^2} \frac{AB}{BD} \tag{5}$$

Now in Δ^{les} PAM and ADL,

$$A\hat{L}D = A\hat{M}P = 90^\circ$$

$$D\hat{A}L = P\hat{A}M = P\hat{A}B$$

So, Δ^{les} PAM and ADL are similar.
Hence,

$$\frac{DL}{PM} = \frac{AD}{AP} = \frac{AL}{AM}$$

So,

$$AM = \frac{AL \cdot PM}{DL} \quad \text{This proves (iv)}$$

$$AP = \frac{AD \cdot PM}{DL} \quad \text{This proves (ii)}$$

Also,

$$PM = \frac{AP \cdot DL}{AD} = \frac{4RA}{AB^2 - CD^2} \frac{AB}{BD} \cdot \frac{DL}{AD} \quad \text{from (5)}$$

$$= \frac{4RA}{AB^2 - CD^2} \frac{AB}{BD} \cdot \frac{1}{AD} \cdot \frac{AD \cdot BD}{2R}$$

$$\left(\text{in } \Delta^{le}\ ABD,\ DL = \frac{AD \cdot BD}{2R}\right)$$

So,

$$PM = \frac{4RA}{AB^2 - CD^2} \frac{AB}{2R} = \frac{2A}{AB^2 - CD^2} AB$$

i.e.,

$$PM = \frac{A}{(1/2)(AB^2 - CD^2)} \cdot AB$$

This completes the proof of (i).
Now in Δ^{les} BMP and CNB,

$$P\hat{M}B = C\hat{N}B = 90^\circ$$

$$P\hat{B}M = C\hat{B}N$$

So, Δ^{les} BMP and CNB are similar.
Hence,

$$\frac{BM}{BN} = \frac{PM}{CN} = \frac{BP}{BC}$$

Then,

$$BM = \frac{PM \cdot BN}{CN} \quad \text{This proves } (v)$$

and

$$BP = \frac{BC \cdot PM}{CN} \quad \text{This proves (iii)}$$

Narayana Pandita obtains Bhaskara II's formula for the chord from the arc in a slightly different form as follows:

On page 400, Bhaskara II's formula for the area a in terms of the circumference C, diameter d, and chord c has been given as

$$a = \frac{C}{2} - \sqrt{\frac{C^2}{4} - \frac{5C^2c}{4(c+4d)}}$$

Narayana Pandita gave the chord c as

$$c = \frac{d\{(C/2)^2 - ((C/2)-a)^2\}}{C^2 + ((C/2)-a)^2/4}$$

Proof:

$$a = \frac{C}{2} - \sqrt{\frac{C^2}{4} - \frac{5C^2c}{4(c+4d)}}$$

So,

$$\left(a - \frac{C}{2}\right) = -\sqrt{\frac{C^2}{4} - \frac{5C^2c}{4(c+4d)}}$$

Squaring both sides, we get

$$a^2 - aC + \frac{C^2}{4} = \frac{C^2}{4} - \frac{5C^2c}{4(c+4d)}$$

We get

$$4(c+4d)(a^2 - aC) = -5C^2c$$

i.e.,

$$4c(a^2 - aC) + 5C^2c = -16d(a^2 - aC)$$

$$c(4a^2 - 4aC + 5C^2) = -16d(a^2 - ac)$$

$$c = \frac{-16d(a^2 - aC)}{4a^2 - 4aC + 5C^2} = \frac{4d[-(a^2 - ac)]}{a^2 - aC + (5/4)C^2}$$

So,

$$c = 4d\frac{[(C^2/4) - ((C^2/4) + a^2 - aC)]}{C^2 + a^2 - aC + (1/4)C^2} = \frac{d[(C/2)^2 - ((C/2) - a)^2]}{[C^2 + ((C/2) - a)^2]/4}$$

13 Kerala Astronomers

Haridatta and His "Parahita" System

The astronomers of Kerala belonged to the Aryabhata I School. But, as time elapsed, they recognized the inaccuracy between the computed and observed positions of planets as well as the occurrence of eclipses. This incorrectness necessitated and demanded the necessary corrections from the then existing adherence to Aryabhata I School. As a result, Kerala astronomer Haridatta (AD 650–700) introduced in AD 683 the so-called "Parahita" system in his works called "Grahacara-nibandana" and "Mahamarga-nibandana." It is believed that the inauguration of this system took place on the holy occasion of "Maha-makha" festival held once in 12 years at Tirunāvāy on the Malabar Coast.

Haridatta still based his new system "Parahita" on the existing "Aryabhatiyam" but with many innovative improvements. Simple multipliers and devisors were used to make calculations easier. Use of Katapayādi system instead of Aryabhata I's letter numerals transformed his work to be simple and elegant. He enunciated graded tables of sines of arcs of anomaly ("Manda-jya") and of conjunction ("Sigra-jya") at intervals of $3°45'$, for the calculation of the correct positions of the planet. One of the corrections injected by Haridatta was called "Bhata-Samskara," after the name of Aryabhata I, since this correction is calculated for the years elapsed since "Sakabda" of Aryabhata I. After nearly seven centuries, when it was discovered that the positions of the planets given by "Parahita" system did not agree with the observed positions, Parameśwara (AD 1360–1433) evolved a new system called DRK system.

Govindaswamin (AD 800–850)

Govindaswamin was an astronomer and a well-known exponent of Aryabhata I School and Bhaskara I. He authored "Mahabhaskariam Bhasya," which expounded new concepts and mathematical reasoning. In this work, he has provided a set of rules to the calculation of intermediary functional values. His formula in modern notation can be written as

$$f(x+nh) = f(x) + n\,\Delta f(x) + \frac{n(n-1)}{2}[\Delta f(x) - \Delta f(x-h)]$$

Mathematical Achievements of Pre-modern Indian Mathematicians. DOI: http://dx.doi.org/10.1016/B978-0-12-397913-1.00013-2

where Δ is finite difference operator. This formula is a special case up to the second order of the general Newton–Gauss formula. Govindaswamin's work has been brought to light by R.C. Gupta ("Fractional parts of Govindaswamin," "Bhasya" on the "Mahabhaskariya." Indian Journal of History of Science, Vol. 6, 1971, pages 51–59). Govindaswamin has also written another work called "Govinda krti" on astronomy and mathematics. Unfortunately, this work has been lost and yet to be traced.

Madhava

Madhava of Saṅgamagrāma (AD 1350–1425) was an astronomer as well as a remarkable first rate mathematician. He belonged to Saṅgamagrāma, which is presently called irinjalakkuda near Cochin in Kerala. Madhava composed a text on astronomy called "Venvaroha." In this work, Madhava has introduced a simple procedure to determine the correct positions of the moon at regular intervals of 36 min.

The power series expansions for sin x and cos x derived in "Yuktibhasa" are two more accomplishments of Madhava. These power series expansions are correct to (1/3600) of a degree. Madhava's verse quoted, extractcd, and explained in the "Tantra-Sangraha," the "Yuktibhasa," the "Karanapaddhati," the "Kriyakramakari," and the "Sadrantamala" present the modern readers to the now well-known powers series of sin x and cos x namely

$$\sin x = x - \frac{x^3}{3!} + \frac{x^5}{5i} + \cdots \text{ and } \cos x = 1 - \frac{x^2}{2!} + \frac{x^4}{4!} + \cdots$$

In the west, the power series was discovered in the seventeenth century, more than three centuries after the great Madhava enunciated these series. Madhava was believed to have composed an elaborate text on astronomy and mathematics, which is yet to be recovered.

Madhava also has given a power series for

$$\tan^{-1} x = x - \frac{x^3}{3} + \frac{x^5}{5} + \cdots$$

which is presently called as Gregory's series.

If $x = 1$, this becomes

$$\frac{\pi}{4} = 1 - \frac{1}{3} + \frac{1}{5} - \cdots$$

which is presently known as Euler series. This series occurs in "Tantra-Sangraha," "Yuktibhasa," and "Karanapaddhati." Madhava went on to demonstrate a rational approximation to the irrational π to any required accuracy.

Since Kerala astronomers belonged to Aryabhata I School, most of the commentaries written are on "Aryabhatiyam" of Aryabhata I and Maha and Laghu Bhaskariyam of Bhaskara I, who himself was a commentator of Aryabhata I. The other texts that were popular in Kerala were "Suryāsiddhantā" and "Laghu mānasam" of Manjula. The most popular text of mathematics in Kerala was Bhaskara II's "Lilavati." At least a dozen commentaries have been written on "Lilavati."

We find that there is a continuous chain of "Guru-śiṣya" (teacher–pupil) tradition among ancient Kerala astronomers. This chain of tradition extends from the thirteenth to seventeenth centuries AD with Govinda Bhattari (AD 1237–1295), grandfather of Parameśwara (AD 1360–1455), Damodara (fifteenth century), son of Parameśvara, Nilakantha Somayaji (AD 1443–1543), Jyesthedeva (AD 1500–1600), and Acyuta Pisārati (AD 1550–1621).

Parameśwara

Parameśwara (AD 1360–1455) was one of the most well-known astronomers of Kerala and enunciator in AD 1430 of the famous DRK system of computation of the positions of planets in his work "Drgganitam." He belonged to Brugu gotra and was a Rgvedin of "Aswalayana Sutra." He was a resident of the village of "Asvatha-grama" (Alattur in Malayalam). His house called "Vatasreni" was the confluence of Nila ("Bharatappuzha" in Malayalam) river with the Arabian Sea. His son Damodara was also a great astronomer.

Parameśwara observed the astronomical events with utmost curiosity and devotion. His methodology included prolonged and repeated observations, recording of readings of observations, examining the values obtained from the observations with those obtained from actual computations, and making postulations of corrections to ensure computations correspond with observations.

Parameśwara was a prolific author and he composed nearly 30 works. Among these writings included are (1) "the Drgganitam," (2) Goldipika I, II, III written in AD 1443, (3) three manuels on improved computations and rationale of the occurrence of eclipses, namely "Grahana," "Grahana mandana," and "Grahananyayadipika," and (4) "Candracchāyā-ganitham" (a manuel on moon's shadow) and "Vakyakarana."

Parameśwara has written superb commentaries on (1) "Aryabhatiyam," (2) "Laghubhaskariam" of Bhaskara I, (3) "Mahabhaskariya-Bhasya" of Govindaswamin, (4) the "Laghu-Manasam" of Munjala, (5) "Lilavati" of Bhaskara II, and (6) "Suryāsiddhantā." Two of his works "Vākyadipikā and Bhādipikā" are yet to be unearthened.

Parameśwara deserved enormous credit for revising the then existing Parahita system of astronomical computation and generating a new system, "DRK" system in his work, "Drgganitam." He stated in the beginning of this work that the positions of planets obtained from the Parahita system differ from the actual positions as observed by the eye. From the DRK system, not only improved multipliers, divisors, and minute corrections for the positions of planets after prolonged periods of time have been obtained, but also the revised values have been given for the position of planets at the commencement of Kaliyuga. The values of the sine of arc of anomaly ("Manda-jya") and of conjunction ("Sigra-jya") have been revised and given for intervals of 6°.

Parameśwara observed the occurrence of the eclipses over a period of 55 years from AD 1393 to 1448. He has recorded the details of those eclipses in his work "Siddhāntadipikā," a superb commentary on "Mahabhaskariya-Bhasya" of Govindaswamin.

Nilakantha Somayaji pays rich tribute to his idol Parameśwara with the following verse:

पञ्चपञ्चाशत् वर्षकालं निरीक्ष्य ग्रहण ग्रहयोगादिषु
परीक्ष्य समदृग्गणितं करणं चकार ॥

Parameśwara provided a formula for the circum radius of a cyclic quadrilateral:

दोष्णं द्वयोर्द्वयोर्धातयुतीनां तिसृणां वधात् ।
एकैकोनेतरऽयैक्यचतुष्क–वध–भाजिते ||
लब्धमूलेन यद्वां विष्कम्भार्धेन निर्मितम् |
सवं चतुर्भुजक्षेज्ञं तस्मिन्नेवावतिष्टते ॥

i.e., "Circum radius of a cyclic quadrilateral is the square root of the quotient obtained when the three sums of the products of the sides taken two at a time are multiplied together and divided by the product of four sums of the sides taken three at a time and decreased by the fourth side."

Proof: Let $ABCD$ be a cyclic quadrilateral with $BC = a$, $CD = b$, $DA = c$, and $AB = d$ and with the chosen diagonal AC ("Istakarna") parallel to x-axis and the center O of the circum circle being the origin.

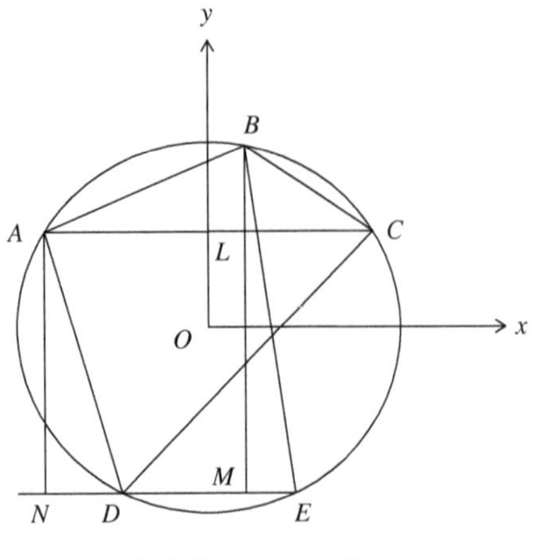

The other diagonal BD is called "Itarakarna."

Let AD and CD be interchanged, AC continues to be the diagonal, but we obtain another diagonal (third diagonal, called "Bhavikarna") different from the second diagonal BD. Let E be the new position of D such that DE is parallel to AC. Let BM be the altitude from B on DE, intersecting AC at L. Let AN be the altitude from A on DE.

Now, $BM = BL + LM = BL + AN$.

Area of the Δ^{le} ABC

$$= \frac{1}{2} AC \cdot BL$$

Area of the Δ^{le} ADC

$$= \frac{1}{2} AC \cdot AN$$

$$\begin{aligned} A &= \text{area of the cyclic quadrilateral } ABCD \\ &= \text{area of the } \Delta^{le} ABC + \text{area of the } \Delta^{le} ADC \end{aligned}$$

$$A = \frac{1}{2} AC(BL + AN) = \frac{1}{2} AC \cdot BM$$

In Δ^{le} BDE,

Altitude BM

$$= \frac{BD \cdot BE}{2R}$$

where R is the circum radius.

So,

$$R = \frac{BD \cdot BE}{2BM} = \frac{BD \cdot BE}{2 \cdot (2A/AC)} = \frac{BD \cdot BE \cdot AC}{4A}$$

$$R = \frac{BD \cdot BE \cdot AC}{4\sqrt{(s-a)(s-b)(s-c)(s-d)}}$$

where $2s = a + b + c + d$, since $A = \sqrt{(s-a)(s-b)(s-d)}$.

From Brahmagupta's theorem on the diagonals of a cyclic quadrilateral,

$$BD^2 = \frac{(AD \cdot BC + AB \cdot CD)(AD \cdot CD + BC \cdot AB)}{AD \cdot AB + CD \cdot BC}$$

i.e.,

$$BD^2 = \frac{(ac + bd)(bc + ad)}{ab + cd}$$

Similarly,

$$AC^2 = \frac{(ab+cd)(ac+bd)}{bc+ad}$$

and

$$BE^2 = \frac{(cd+ab)(ad+bc)}{ac+bd}$$

So, $BD^2 \cdot AC^2 \cdot BE^2 = (ab+cd)(ac+bd)(bc+ad)$.
Now,

$$R = \frac{BD \cdot BE \cdot AC}{\sqrt{(2s-2a)(2s-2b)(2s-2c)(2s-2d)}}$$

i.e.,

$$R = \frac{\sqrt{(ab+cd)(ac+bd)(bc+ad)}}{\sqrt{(b+c+d-a)(a+c+d-b)(a+b+c-d)(a+b+d-c)}}$$

i.e.,

$$R = \sqrt{\frac{(ab+cd)(ac+bd)(bc+ad)}{(b+c+d-a)(a+c+d-b)(a+b+d-c)(a+b+c-d)}}$$

This formula from Parameśwara was rediscovered in Europe more than 200 years later in 1782 by Lhuilier [129] (Vol. II, page 286).

Example (Parameśwara)

Find the least number which when multiplied by 8 and divided by 29 gives 4 as the remainder and when multiplied by 17 and divided by 45 gives remainder 7.

Solution

Let x be the number, then we have

$$8x = 29y + 4 \qquad \text{(i)}$$

$$17x = 45y_2 + 7 \qquad \text{(ii)}$$

We solve (ii) first.

```
17 ) 45 ( 2
     34
   ______
     11 ) 17 ( 1
          11
        -------
          6 ) 11 ( 1
               6
             -----
           5 ) 6 ( 1
               5
             ------
               1
```

Quotients are q_1, q_2, q_3, and q_4, and so the number is 4 and even.
We now form the following table:

2	2	2	2	56	$X = 56 \equiv 11 (\text{mod } 45)$
1	1	1	21	21	$Y = 21 \equiv 4 (\text{mod } 17)$
1	1	14	14	14	$x = 11$, $y_2 = 4$ is the least solution
1	7	7	7		
7	7	7			
0	0				

General solution is

$$x = 45m_1 + 11$$

$$y_2 = 17m_1 + 4$$

where m_1 is an arbitrary integer.
We next solve $8x = 29y + 4$.

```
8)29(3 = q1
  24
  5) 8 (1 = q2
     5
     3) 5 (1 = q3
        3
        2) 3 (1 = q4
           2
           1
```

So, the number of quotients is 4 and so even.
We now form the following table:

3	3	3	3	44	$X = 44$
1	1	1	12	12	$Y = 12$

(*Continued*)

Example (Parameśwara) (cont'd)

1	1	8	8	8	$44 \equiv 15 \pmod{29}$
1	4	4	4		$12 \equiv 4 \pmod{8}$
4	4	4			$y_1 = 4$ and $x = 15$ is the least solution.
0	0				

General solution is

$$y = 8m_2 + 4$$

$$x = 29m_2 + 15$$

where m_2 is an arbitrary integer.

Equating the two general values of x, we get

$$45m_1 + 11 = 29m_2 + 15$$

i.e.,

$$45m_1 = 29m_2 + 4$$

We will now solve this equation

45) 29 (0 $= q_1$

0

29) 45 (1 $= q_2$

29

16) 29 (1 $= q_3$

16

13) 16 (1 $= q_4$

13

3) 13 (4 $= q_5$

12

1

So, the number of quotients is 5 and so odd.

We now form the following table:

0	0	0	0	0	36	$Y = 36$
1	1	1	1	56	56	$X = 56$
1	1	1	36	36	36	$36 \equiv 7 \pmod{29}$
1	1	20	20	20		$56 \equiv 11 \pmod{45}$
4	16	16	16			
4	4	4				
0	0					

Then, $m_2 = 11$ and $m_1 = 7$ is the least solution of

$$45m_1 = 29m_2 - 4$$

So, $m_1 = 29 - 7 = 22$ and $m_2 = 45 - 11 = 34$ is the least solution of

$$45m_1 = 29m_2 + 4$$

General value is

$$m_1 = 29m_3 + 22$$

$$m_2 = 45m_3 + 34$$

where m_3 is an arbitrary integer.

So, general value of $x = 45m_1 + 11$

$$= 45(29m_3 + 22) + 11$$

$$x = 1305m_3 + 990 + 11$$

$$= 1305m_3 + 1001$$

where m_3 is an arbitrary integer.

Least value of $x = 1001$ and the corresponding value of y is obtained by substituting in the equation $8x = 29y + 4$, so, $29y = 8(1001) - 4$.

$29y = 8004 \Rightarrow y = 276$ and y_2 is obtained substituting in the equation $17x = 45y_2 + 7 \Rightarrow 45y_2 = 17(1001) - 7 = 170{,}10 \Rightarrow y_2 = 378$.

Nilakantha Somayaji (AD 1444–1543)

Nilakantha Somayaji was born on Wednesday, June 14, AD 1444. He belonged to "Gargya" gotra and was a resident of Tṛkkaṇtiyur (Sri-Kunda Pura) in South Malabar. His house was called "Kerala-Sadgrama." He received excellent training in Mathematics and astronomy from his teacher, Damodara, son of Parameśwara. Parameśwara was Nilakantha's idol and he tried to follow the footsteps of Parameśwara.

Nilakantha was an accomplished mathematician as well as an astronomer. He wrote six works, "Tantra-Sangraha Bhashya," a commentary on "Aryabhatiyam," "Golasara" (special astronomy), "Siddanta darpana" (laws of astronomy), "Candracchāyā-ganitham" (calculations regarding moon's shadow), and "Sundararajan Prasnotara."

Of the above mentioned six works, the first two are regarded his master pieces. The first one, "Tantra-Sangraha Bhasya," consists of 432 verses, with eight chapters. It deals with mathematics and astronomy. "Sundararajan Prasnotara" contains detailed answers for Sundararajan's many questions on astronomy. Sundararajan was a contemporary of Nilakantha and a well-known Tamil astronomer.

Sankara, a student of Nilakantha, wrote a commentary on his master's work "Tantra-Sangraha." Sankara in his commentary has mentioned that the first and the last verses in "Tantra-Sangraha" contain chronograms which specify the date of commencement and the date of completion respectively of that work. In the very first verse, the words "He visno nihitam kṛtsnam" represent Kaliyuga day 16, 80, 548, which correspond to March 22, AD 1500. In the last verse, the words "Lakṣmiśa nihita dhyāna" represent the Kaliyuga day 1680, 563, which correspond to March 27, AD 1500. In that case, Nilakantha wrote his entire master piece "Tantra-Sangraha" in just five days!

His commentary "Bhasya" on the "Aryabhatiyam" is almost like that on the work of Aryabhata I. In this work, Nilakantha has provided a very clear interpretation of difficult concepts with lot of quotations from contemporary studies as well as from his own personal knowledge. Nilakantha was so pleased with him of about this work, "Bhasya" on the Aryabhatiyam that he called his work "Maha Bhasyam" (great commentary).

Charles M. Whish of East-India Company's civil service at Madras in his memoir of December 15, 1832, quoted the following extracts from the Tantra-Sangraha:

> Ishta vyásè nágavèdavahnyabdhichèndubhih
> Tithyaswivibudhairbhactè susùcshmah paridhirbhavèt
> Paridhèrvyatyayáchaivamsusùcshmam vyásamánayèt
> Tithyaswivibudhairnighnát chachatancèndunètratah
> Sùcshmò vyásòhi vèdagni critòbhrèndu hritòthava
> Trìsaghna chakra liptábhyò vyásòrtthèshwagnibhirhritèh

i.e., "Multiply the given diameter by 104,348 and divide the product by 33,215. The regulating quotients give an accurate value of the circumference. If the circumference is given, multiply the given circumference by 33,215 and divide the product by 104,348, we get the diameter."

Alternatively, multiply the given circumference by 113 and divide by 355. The regulating quotient will give sufficiently correct value for the diameter.

If c is the circumference of a circle of diameter d, then

$$c = \frac{104,348d}{33,215}$$

Since

$$\frac{\text{circumference}}{\text{diameter}} = \frac{c}{d} = \pi$$

This gives the value of

$$\pi = \frac{104,348}{33,215} = 3.14159265392$$

which is correct to nine decimal places.

Also,

$$d = \frac{33,215c}{104,348}$$

Alternatively,

$$d = \frac{113c}{355}$$

Since

$$\frac{c}{d} = \pi$$

here we get the value of

$$\pi = \frac{355}{113} = 3.14159292$$

which is correct to six decimal places.

He proceeded to quote from chapter upon sines, etc., where the author Narayana Pandita introduced a measure called "Anushtubvrittam."

Vyásárdham prathamannitwatatòványátgunannayèt
Sambandhanniyamanchaivam vynèyòvyása vrittayòh

i.e., "Once the radius is known, you may be able to construct the sines. In order to do that, you must first determine the ratio between the diameter and the circumference."

The next verse is of another measure called "Gitivrittam."

Vyásè varidhinihitè rùpa hritè vyásaságarábhihate
Trisarādhi vishama sanchyá bhactamrinamswam prithacramát curyát
Yatsanchyayátra haranè critèrnivrittá hritistujámitayá
Tasyáf ùrdhwa gatáyássamasanchyá taddalamgunòntèsyát
Jadvarggò rùpayutò hárò vyásábdhighátacah prágwat.
Tabhyámáptam swamrinè critèdhanè sòdhananchacaraìyám
Sùcshmah paridhissasyát bahucritwòharaatòti sùcshmascha.

i.e., "Multiply the diameter by 4 and from it, subtract and add alternatively the quotients of four times the diameter divided by the odd numbers 1, 3, 5,... Continue this process to the selected extent. Then, "I will now show how the correction may be made more accurate than the previous rule. Select the even number next greater than the last odd divisor in the selected extent. Square the half of this number and add unity to it. This is the multiplier. The deviser is the product of half of this selected even number and the sum of four times the multiplier plus unity. Multiply this quotient by four times the diameter. Add or subtract this product according to the sign of the last term in the series of selected extent. The resulting sum will give a much more accurate value of the circumference of the circle with the given diameter."

With the above correction, we get an infinite series of the form

$$c = 4d - \frac{4d}{3} + \frac{4d}{5} - \frac{4d}{7} + \cdots \pm \frac{4d[(p/2)^2 + 1]}{(p^2 + 4 + 1)(p/2)}$$

where c is the circumference, d is the diameter, and p being the last odd divisor increased by unity.

Since

$$\frac{c}{d} = \pi$$

this becomes

$$\frac{\pi}{4} = 1 - \frac{1}{3} + \frac{1}{5} - \frac{1}{7} + \cdots \pm \frac{(p/2)^2 + 1}{(p^2 + 4 + 1)(p/2)}$$

Next verse is in "Anushtubvrittam" measure. It reads as follows:

Vyásavargádravihatat padamsyát prathamam phalam
Jatastattat phaláchchápiyávadischántribhirharèt
Rupádyayugmasancyábhirllabdèshwèshuyathácramam
Vishamánám yutètyactè yugmayògè vritirbhavèt

i.e., "Find the square root of the product of 12 and the square of the given diameter. This is the first term. Divide the first term by 3 for the second term. Divide the second term by 3 for the third term, divide the third term by 3 for the fourth term, and so on. Divide the above terms by 1, 3, 5, 7,... in order. Then, add the first, third, fifth, etc. odd new terms, so obtained. Similarly, add the second, fourth, sixth, etc. even new terms so obtained. Then, subtract the sum of the even terms from the sum of the odd terms. The difference obtained will give the circumference of the circle, whose diameter is given."

If we follow the above rule, we get the infinite series:

$$c = \sqrt{12}\left(1 - \frac{1}{3.3} + \frac{1}{5.3^2} - \frac{1}{7.3^3} + \frac{1}{9.3^4} - \frac{1}{11.3^5} + \cdots\right)d$$

Since

$$\frac{c}{d} = \pi$$

we have

$$\pi = \sqrt{12}\left(1 - \frac{1}{3.3} + \frac{1}{5.3^2} - \frac{1}{7.3^3} + \frac{1}{9.3^4} - \frac{1}{11.3^5} + \cdots\right)$$

Then, next verse is of the "Gitivrittam" measure. It follows like this:

Samapancháhatayòyá rupádyayujás chaturghna mùlayutáh
Tabhisshòdasa guitádvyásáft prithagá hritètu vishamayatè
Sama phala yágè tyactè syátishta vyása sambhavah paridhih

i.e., "Divide the product of the diameter and 16 separately by the sum of the fifth power of 1, 3, 5, 7, 9, 11,... and four times of 1, 3, 5, 7, 9, 11,... respectively to form first term, second term, and so on. Subtract the sum of the even terms from the sum of the odd terms. The difference will give the circumference of the circle with a given diameter."

The above rule gives the following infinite series:

$$c = \frac{16d}{1^5 + 4.1} - \frac{16d}{3^5 + 4.3} + \frac{16d}{5^5 + 4.5} - \frac{16d}{7^5 + 4.7} + \cdots$$

Since

$$\frac{c}{d} = \pi$$

this means

$$\frac{\pi}{16} = \frac{1}{1^5 + 4.1} - \frac{1}{3^5 + 4.3} + \frac{1}{5^5 + 4.5} - \frac{1}{7^5 + 4.7} + \cdots$$

The next verse is again of "Gitivrittam" measure and it runs as follows:

Vyásádwaridhi nihatát pritagáptantryáyugvimùld ghanaih
Trighna vyásè swamrinam crasmasah kritwápi
Paridhivánèyah

i.e., "Divide the four times the given diameter by the difference of cubes of 3, 5, 7, 9,... and 3, 5, 7, 9,... in order to form the first term, second term, and so on. Subtract the sum of the even terms from the sum of odd terms. This difference will give the circumference of the circle with the given diameter diminished by three times the diameter."

The above provides the following infinite series:

$$c = 3d + \frac{4d}{3^3 - 3} - \frac{4d}{5^3 - 3} + \frac{4d}{7^3 - 7} - \frac{4d}{9^3 - 9} + \cdots$$

where c is the circumference and d is the diameter of the circle.

Since

$$\frac{c}{d} = \pi$$

we have the following:

$$\pi = 3 + \frac{4}{3^2 - 3} - \frac{4}{5^2 - 5} + \frac{4}{7^2 - 7} - \frac{4}{9^2 - 7} + \cdots$$

$$\frac{\pi - 3}{4} = \frac{1}{3^2 - 3} - \frac{1}{5^2 - 5} + \frac{1}{7^2 - 7} - \frac{1}{9^2 - 7} + \cdots$$

The author Nilakantha Somayaji gave the next verse in the same "Gitivrittam" measure:

Dwyádiyujám vacritayò vyècá hárádwinighna vishcambhe
Dhanamrinamantèntyòrdhwa gataujacritidwi sahitáharòdwighni

i.e., "Divide the product of 4 and the diameter individually the difference of squares of 2, 6, 10, and unity in order to make them the first term, second term, and so on. Subtract the sum of the even terms from the sum of odd terms. Add twice the diameter to the sum obtained. In order to make the new sum more accurate, add or subtract a correction term as required. This correction term is to have four times the diameter as the multiplier and twice the sum of two and the square of the next odd number less than the last even figure squared as the devisor. This will give the circumference accurately."

If the above rule is followed, we get the infinite series

$$c = 2d + \frac{4d}{2^2 - 1} - \frac{4d}{4^2 - 1} + \frac{4d}{6^2 - 1} - \frac{4d}{8^2 - 1} + \cdots + \frac{4d}{2\{(p-1)^2 + 2\}}$$

Since

$$\frac{c}{d} = \pi$$

we get

$$\frac{\pi-2}{4}=\frac{1}{2^2-1}-\frac{1}{4^2-1}+\frac{1}{6^2-1}-\frac{1}{8^2-1}+\cdots+\frac{1}{2\{(p-1)^2+1\}}$$

So,

$$\frac{\pi-2}{4}=\frac{1}{1\cdot 3}-\frac{1}{3\cdot 5}+\frac{1}{5\cdot 7}-\frac{1}{7\cdot 9}+\cdots$$

Two other series are obtained in the next verse of "Giti" measure:

Dwyádèschaturádèrva chaturadhicánánnirèca vargassyuh
Háráh cunjara gunitá vishcambhaswamati calpitòbhájyah
Phalayutirádyè vrittam bhájyadalam phalavihìnamányatra

i.e., "Consider two AP with common difference 4 and with first terms 2 and 4, respectively. Divide eight times the diameter individually by the difference of the squares of 2, 4, 6, 8,... (terms of the first AP) and unity to make them first term, second term, and so on. The sum of these terms forms the first series, which gives the circumference. Next, divide eight times the diameter individually by the difference of squares of 4, 8, 12,... (terms of second AP) and unity, to make them first term, second term, and so on. Subtract the sum of these terms from four times the diameter to form the second series, which also gives the circumference."

If we follow the above rule, we get two infinite series:

$$c=\frac{8d}{2^2-1}+\frac{8d}{6^2-1}+\frac{8d}{10^2-1}+\cdots$$

and

$$c=4d-\left(\frac{8d}{4^2-1}+\frac{8d}{8^2-1}+\frac{8d}{12^2-1}+\cdots\right)$$

where c is the circumference and d is the diameter.

Since

$$\frac{c}{d}=\pi$$

we get

$$\frac{\pi}{8}=\frac{1}{2^2-1}+\frac{1}{6^2-1}+\frac{1}{10^2-1}+\cdots$$

So,

$$\frac{\pi}{8} = \frac{1}{1 \cdot 3} + \frac{1}{5 \cdot 7} + \frac{1}{9 \cdot 11} + \cdots$$

and

$$\frac{4-\pi}{8} = \frac{1}{4^2-1} + \frac{1}{8^2-1} + \frac{1}{12^2-1} + \cdots$$

So,

$$\frac{4-\pi}{8} = \frac{1}{3 \cdot 5} + \frac{1}{7 \cdot 9} + \frac{1}{11 \cdot 13} + \cdots$$

Nilakantha Somayaji's Remarks About π

Nilakantha Somayaji's remarks dated AD 1500 are breathtaking and of monumental historical value. The following is the actual quote from the "Aryabhatiyam" of Aryabhata I with the Bhasya of Nilakantha Somayaji:

> कुतः पुनर्वास्तवीं संख्यामुत्सृज्यासन्नैवेहोक्ता । उच्यते । तस्या वक्तुमशक्यत्वात् । कुतः? येन मानेन मीयमानो व्यासो निरवयवः स्यात् तेनैव मीयमानः परिधिः पुनः सावयव एव स्यात् । येन च मीयमानः परिधिनिरवयवस्तेनैव मीय–मानो व्यासोऽपि सावयव एव, इत्येकनैव मानेन मीयमानयोरुभयोः क्वापि न निरवयवत्वं स्यात् । महान्तमध्वानं गत्वाप्यल्पावयवत्वं एव लभ्यम् । निरवयवत्वं तु क्वापि न लभ्यमिति भावः ॥

i.e., "Why only the approximate value of π is mentioned here, discarding the exact value? This is the answer; because the exact value cannot be given. If the diameter is commensurable with respect to a particular unit of measurement, then with respect to the same unit of measurement, the circumference cannot be exactly measured. On the other hand, if the circumference is commensurable with respect to a particular unit of measurement, then with the respect to the same unit of measurement, the diameter cannot be exactly measured. So, there will never be commensurability between the diameter and the circumference with respect to the same unit of measurement." This is equivalent to saying that the ratio of the circumference to the diameter can never be expressed as the ratio of two integers.

This clearly shows that although the Swiss-German mathematician, physicist, and philosopher Johann Heinrich Lambert (AD 1728–1777) gave formal proof in 1761 that π is an irrational number, he is not the first person who enunciated that π is an irrational number.

Arc in terms of Chord

Nilakantha Somayaji gave the following formula for the arc:

$$a = \sqrt{\left(1 + \frac{1}{3}\right)h^2 + s^2}$$

where $h \Rightarrow$ the height (arrow) and s is the sine chord.

Proof: Let h be the height of the chord PQ. Let P_1, P_2,... be the whole chords of the arc PQ and of the arcs obtained by successively halving it. Let a_1, a_2,... be the corresponding arcs. Let h_1, h_2,... be the corresponding heights of the arcs.

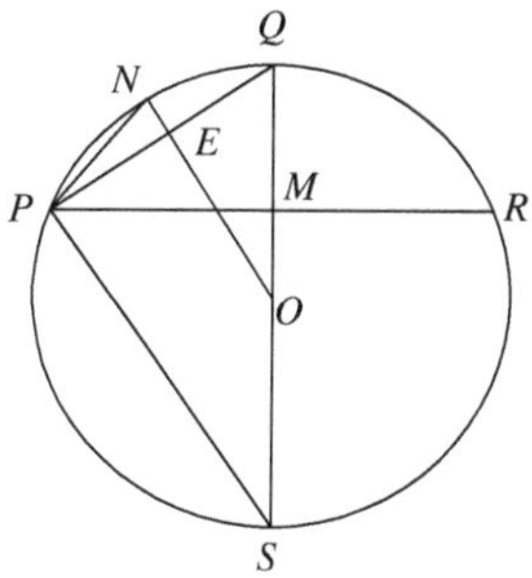

Let arc PR be double that of arc PQ. Now, let the arc PR be halved. We get the arc PQ. Join the chord PQ. Let QS be the diameter through the center O of the circle. Then, the diameter QS bisects PR at right angles at M. Then, PM is the given sine chord ("Ardha Jya") s of the arc PQ.

Now, in right Δ^{le} PQM,

$$P_1^2 = PQ^2 = PM^2 + MQ^2 = s^2 + h^2$$

Let N be the midpoint arc PQ. Join ON. Let ON cut PQ at E.
Now,

$$P_2^2 = PN^2 = PE^2 + NE^2 = \left(\frac{P_1}{2}\right)^2 + h_1^2 = \frac{P_1^2}{4} + h_1^2$$

Similarly,

$$P_3^2 = \frac{P_2^2}{4} + h_2^2 = \frac{(P_1^2/4) + h_1^2}{4} + h_2^2$$

$$= \frac{P_1^2}{4^2} + \frac{h_1^2}{4} + h_2^2$$

$$P_4^2 = \frac{P_3^2}{4} + h_3^2 = \frac{(P_1^2/4^2) + (h_1^2/4) + h_2^2}{4} + h_3^2$$

$$P_4^2 = \frac{P_1^2}{4^3} + \frac{h_1^2}{4^2} + \frac{h_2^2}{4} + h_3^2$$

and so on

So,

$$P_n^2 = \frac{P_1^2}{4^{n-1}} + \frac{h_1^2}{4^{n-2}} + \frac{h_2^2}{4^{n-3}} + \cdots + h_{n-1}^2$$

Let n be large. Then, these arcs will be very small and arc $\sim$ chord. We can write

$$a_n^2 = \frac{P_1^2}{4^{n-1}} + \frac{h_1^2}{4^{n-2}} + \frac{h_2^2}{4^{n-3}} + \cdots + h_{n-1}^2$$

Proceeding inversely the square of twice of this arc,

$$a_{n-1}^2 = \frac{P_1^2}{4^{n-2}} + \frac{h_1^2}{4^{n-3}} + \frac{h_2^2}{4^{n-4}} + \cdots + 4h_{n-1}^2$$

$$a_{n-2}^2 = \frac{P_1^2}{4^{n-3}} + \frac{h_1^2}{4^{n-4}} + \frac{h_2^2}{4^{n-5}} + \cdots + 4^2h_{n-1}^2$$

....

....

$$a_2^2 = \frac{P_1^2}{4} + h_1^2 + 4h_2^2 + \cdots + 4^{n-2}h_{n-1}^2$$

$$a_1^2 = P_1^2 + 4h_1^2 + 4^2h_2^2 + \cdots + 4^{n-1}h_{n-1}^2$$

$$a_1^2 = s^2 + h^2 + 4h_1^2 + 4^2h_2^2 + \cdots + 4^{n-1}h_{n-1}^2 \quad (*)$$

Now, $QM = h$.
In Δ^{les} PQM and QPS,

$$Q\hat{P}S = P\hat{M}Q = 90^\circ$$

$$P\hat{Q}S = P\hat{Q}M$$

So, Δ^{les} PQM and QPS are similar.

$$\frac{PQ}{QS} = \frac{PM}{PS} = \frac{QM}{PQ}$$

So, $PQ^2 = QM \cdot QS$.

Now,

$$h = QM = \frac{PQ^2}{QS} = \frac{P_1^2}{d}$$

where d is the diameter of the circle $PQRS$.

Now,

$$NE = h_1 = \frac{PN^2}{d} = \frac{PE^2 + NE^2}{d}$$

i.e.,

$$h_1 = \frac{(P_1/2)^2 + h_1^2}{d} = \frac{P_1^2}{4d} + \frac{h_1^2}{d}$$

Hence,

$$h_1 = \frac{P_1^2}{4d}$$

neglecting higher powers of h_1, etc.

So,

$$h_1 = \frac{h}{4}$$

Similarly,

$$h_2 = \frac{h_1}{4} = \frac{h^2}{4^2}$$

$$h_3 = \frac{h_2}{4} = \frac{h^2}{4^3}$$

and so on.

Hence,

$$h^2 + 4h_1^2 + 4^2h_2^2 + \cdots = h^2 + \frac{h^2}{4} + \frac{h^2}{4^2} + \cdots$$

$$= h^2\left[1 + \frac{1}{4} + \frac{1}{4^2} + \cdots\right]$$

Now,

$$1 + \frac{1}{4} + \frac{1}{4^2} + \cdots$$

is a geometric series with first term $a = 1$ and common ratio $\Lambda = 1/4$.

Hence,

$$1 + \frac{1}{4} + \frac{1}{4^2} + \cdots = \frac{a}{1 - \Lambda} = \frac{1}{1 - (1/4)} = \frac{1}{(3/4)} = \frac{4}{3}$$

Thus,

$$h^2 + 4h_1^2 + 4^2 h_1^2 + \cdots = h^2 \cdot \frac{4}{3} = \frac{4h^2}{3}$$

Hence, from (*), we get

$$a_1^2 = s^2 + \left(h^2 + \frac{h^2}{4} + \frac{h^2}{4^2} + \cdots \right)$$

$$a_1^2 = s^2 + \frac{4h^2}{3} = s^2 + h^2 \left(1 + \frac{1}{3} \right)$$

$$a_1 = \sqrt{s^2 + h^2 \left(1 + \frac{1}{3} \right)}$$

$$a_1 = \sqrt{s^2 + h^2 \left(1 + \frac{1}{3} \right)}$$

Nilakantha Somayaji observed here the above expressions for the arc is only an approximation, since higher powers of h_1, h_2,... have been neglected. He contented this above formula for arc will be appropriate for small arcs only.

In Tantra-Sangraha, Nilakantha Somayaji has provided the following proof of Aryabhata I's theorem on the common chord and its height with intersecting circles.

Proof: This is derived easily from the expression for the chord c of segment in terms of its diameter d and the height h (see page 76).

We have

$$c = \sqrt{4(d - h)h}$$

Then,

$$\frac{c}{2} = \sqrt{h(d-h)} \text{ or } \frac{c^2}{4} = (d-h)h$$

Since BC is the common chord of the two circles,

$$\frac{\text{chord } BC^2}{4} = (d_1 - h_1)h_1 = (d_2 - h_2)h_2$$

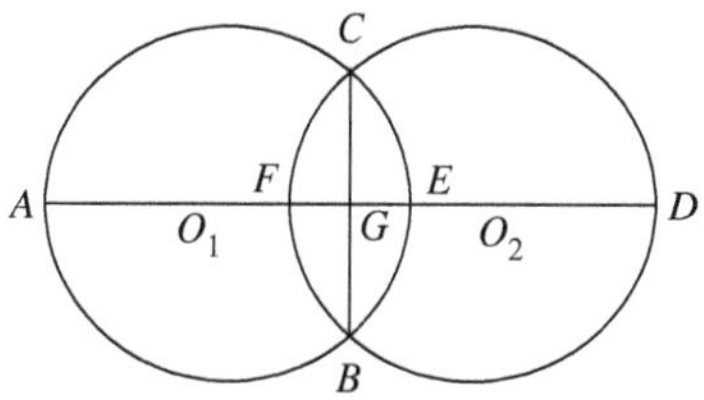

Hence,

$$\frac{h_1}{h_2} = \frac{d_2 - h_2}{d_1 - h_1}$$

$$h_1 d_1 - h_1^2 = h_2 d_2 - h_2^2$$

$$h_1 d_1 - h_1^2 - h_1 h_2 = h_2 d_2 - h_2^2 - h_1 h_2$$

$$h_1 d_1 - h_1(h_1 + h_2) = h_2 d_2 - h_2(h_1 + h_2)$$

$$h_1[d_1 - (h_1 + h_2)] = h_2[d_2 - (h_1 + h_2)]$$

So,

$$\frac{h_1}{h_2} = \frac{d_2 - (h_1 + h_2)}{d_1 - (h_1 + h_2)} \tag{1}$$

Hence,

$$\frac{h_1}{h_2} + 1 = \frac{d_2 - (h_1 + h_2)}{d_1 - (h_1 + h_2)} + 1$$

So,

$$\frac{h_1 + h_2}{h_2} = \frac{d_2 - (h_1 + h_2) + d_1 - (h_1 + h_2)}{d_1 - (h_1 + h_2)}$$

$$\frac{h_1 + h_2}{h_2} = \frac{(d_1 + d_2) - 2(h_1 + h_2)}{d_1 - (h_1 + h_2)}$$

Thus,

$$h_2 = \frac{(h_1 + h_2)[d_1 - (h_1 + h_2)]}{(d_1 + d_2) - 2(h_1 + h_2)}$$

From (1), we have

$$\frac{h_2}{h_1} = \frac{d_1 - (h_1 + h_2)}{d_2 - (h_1 + h_2)}$$

$$\frac{h_2 + h_1}{h_1} = \frac{d_1 - (h_1 + h_2) + d_2 - (h_1 + h_2)}{d_2 - (h_1 + h_2)}$$

$$\frac{h_2 + h_1}{h_1} = \frac{(d_1 + d_2) - 2(h_1 + h_2)}{d_2 - (h_1 + h_2)}$$

Hence,

$$h_1 = \frac{(h_2 + h_1)[d_2 - (h_1 + h_2)]}{(d_1 + d_2) - 2(h_1 + h_2)}$$

Jyesthedeva

"Yuktibhasa" is a Malayalam commentary on Nilakantha Somayaji's "Tantra-Sangraha." The author of "Yuktibhasa," according to Prof. K.V. Sharma (Jyesthedeva and the identification as the author of "Yuktibhasa," Adyar Library Bulletin, XXVI, pages 35–40) is Jyesthedeva, who lived between AD 1475 and 1575. Jyesthedeva was a student of the astronomer, Damodara of Vatasreni, son of Parameśwara. He belonged to parannotta family of Ālattur in south Malabar. Acyuta Pisari (AD 1530–1620), the Kerala astronomer spoke highly about his teacher Jyesthedeva.

"Yuktibhasa" is a remarkable work which has the unique distinction of providing the rationale as well as derivation of all the theorems and formulae that were in use at that time by the astronomers of Kerala. His profound expositions recur (in part) in "Kriyakramakari" and in a Sanskrit commentary (Transcript No. 275(572)) of unknown authorship preserved in the Sanskrit College Library at Trippunithura. At the end of every chapter in this Sanskrit commentary, the following verse occurs:

इत्येष परक्रोडावासाद्विजवरसमीरितो योऽर्थः ।
स तु तन्त्रसंग्रहस्य...ऽध्याये मया कथितः ॥

i.e., "The commentator has received his knowledge from a Brahmin of parannot (parakroda in Sanskrit) family, who is the author of Yuktibhasa."

Charles M. Whish of East-Indian Company's civil service at Madras stated (on the Hindu Quadrature of the Circle and infinite series of the proportion of the circumference in the diameter exhibited in the four shastras, the "Tantra-Sangraha," "Yuktibhasa," "Karaṇapaddhati," and "Sadrantamala," Vol. 3, Transactions of the

Royal Asiatic Society of Great Britain and Ireland, 1835, pages 509–523) that the author of "Yuktibhasa" is a Kelallur Nambudri, which may just mean a Brahmin of Nambudri Sect belonging to Kelallur Village. Whish also pointed out that the author of "Yuktibhasa" also composed "Drkkarana," a comprehensive metrical text in Malayalam on astronomy. The date of "Drkkarana" is mentioned in the latter part of that text as 783d of Malabar era and 4708 of Kaliyuga. These two dates coincidentally correspond to AD 1608.

Sangamagrama Madhava's Theorem on π

Lemma: Let $\{b_n\}$ be a monotonic increasing sequence of positive terms such that $0<b_1<b_2<\cdots<b_n<\cdots$ and $\lim_{n\to\infty} b_n = \infty$. If $\{a_n\}$ is a sequence such that

$$\lim_{n\to\infty} \frac{a_n - a_{n-1}}{b_n - b_{n-1}} = l \tag{1}$$

Then,

$$\lim_{n\to\infty} \frac{a_n}{b_n} = l$$

Proof: Let $c_n = a_n - lb_n$.

Then, $c_n - c_{n-1} = a_n - lb_n - (a_{n-1} - lb_{n-1})$.

$$= (a_n - a_{n-1}) - l(b_n - b_{n-1})$$

So,

$$\frac{c_n - c_{n-1}}{b_n - b_{n-1}} = \frac{a_n - a_{n-1}}{b_n - b_{n-1}} - l$$

Let $n\to\infty$.

In view of (1),

$$\lim_{n\to\infty} \frac{c_n - c_{n-1}}{b_n - b_{n-1}} = \lim_{n\to\infty} \frac{a_n - a_{n-1}}{b_n - b_{n-1}} - l = l - l = 0$$

Therefore, given an $E>0$, there exists an integer N_1 such that

$$\left|\frac{c_n - c_{n-1}}{b_n - b_{n-1}}\right| < \frac{E}{2} \text{ for } n > N_1 \tag{2}$$

Since $b_{n-1}<b_n$, we have $b_n - b_{n-1}>0$.

From (2), we get

$$|c_n - c_{n-1}| < \frac{E}{2}|b_n - b_{n-1}| \text{ for } n > N_1$$

i.e.,

$$|c_n - c_{n-1}| < \frac{E}{2}(b_n - b_{n-1}) \text{ for } n = N_1 + 1, N_1 + 2, \ldots$$

Also, for $n > N_1$, we have

$$c_n - c_{N_1} = (c_n - c_{n-1}) + (c_{n-1} - c_{n-2}) + \cdots + (c_{N_1+1} - c_{N_1})$$

So, for $n > N_1$,

$$|c_n - c_{N_1}| < \frac{E}{2}|[(b_n - b_{n-1}) + (b_{n-1} - b_{n-2})] + \cdots + (b_{N_1+1} - b_{N_1})|$$

i.e.,

$$|c_n - c_{N_1}| < \frac{E}{2}|b_n - b_{N_1}| \text{ for } n > N_1$$

i.e.,

$$\left|\frac{c_n - c_{N_1}}{b_n - b_{N_1}}\right| < \frac{E}{2} \text{ for } n > N_1$$

Now,

$$c_n = c_n - c_{N_1} + c_{N_1}$$

$$\frac{c_n}{b_n} = \frac{c_n - c_{N_1}}{b_n} + \frac{c_{N_1}}{b_n}$$

$$\frac{c_n}{b_n} = \frac{c_n - c_{N_1}}{b_n - b_{N_1}} \frac{b_n - b_{N_1}}{b_n} + \frac{c_{N_1}}{b_n} \tag{3}$$

$$0 < b_{n-1} < b_n \text{ for } n = 1, 2, \ldots$$

For $n > N_1$, $0 < b_{N_1} < b_n$, we have

$$0 < b_n - b_{N_1} < b_n$$

So,

$$0 < \frac{b_n - b_{N_1}}{b_n} < 1$$

Hence, from (3),

$$\left|\frac{c_n}{b_n}\right| < \left|\frac{c_n - c_{N_1}}{b_n - b_{N_1}}\right| + \left|\frac{c_{N_1}}{b_n}\right| \tag{4}$$

Since N_1 is fixed number,

$$\lim_{n\to\infty}\left|\frac{c_{N_1}}{b_n}\right| = 0$$

i.e., for

$$\frac{E}{2} > 0$$

there exists an integer N_2 such that

$$\left|\frac{c_{N_1}}{b_n}\right| < \frac{E}{2} \text{ for } n > N_2 \tag{5}$$

Let $N = \max(N_1, N_2)$.
Then, from, (2), (4), and (5), we have

$$\left|\frac{c_n}{b_n}\right| < \frac{E}{2} + \frac{E}{2} = E \text{ for } n > N$$

i.e.,

$$\lim_{n\to\infty}\frac{c_n}{b_n} = 0$$

$$\lim_{n\to\infty}\frac{a_n - lb_n}{b_n} = 0$$

i.e.,

$$\lim_{n\to\infty}\left(\frac{a_n}{b_n} - l\right) = 0$$

i.e.,

$$\lim_{n\to\infty}\frac{a_n}{b_n} = l$$

This proves the lemma.

Lemma (Madhava of Sangamagrama): If

$$S_p = 1^{p-1} + 2^{p-1} + 3^{p-1} + \cdots + n^{p-1} \quad (p = 1, 2, \ldots)$$

then

$$\lim_{n\to\infty} \frac{S_p}{n^P} = \frac{1}{p}$$

Proof: Let

$$\left.\begin{array}{l} a_n = 1^{p-1} + 2^{p-1} + \cdots + n^{p-1} \\ \text{and } b_n = n^p \end{array}\right\} \text{for } n = 1, 2, \ldots$$

We note that $0 < b_1 < b_2 < \cdots$ and $\lim\limits_{n\to\infty} b_n = \infty$.
Now, $a_n - a_{n-1} = n^{p-1}$ and $b_n - b_{n-1} = n^P - (n-1)^P$.
So,

$$\frac{a_n - a_{n-1}}{b_n - b_{n-1}} = \frac{n^{p-1}}{n^p - (n-1)^p} = \frac{n^{p-1}}{n^P[1 - (1-(1/n))^P]}$$

$$= \frac{1}{n[1 - (1-(1/n))^P]} = \frac{(1/n)}{[1 - (1-(1/n))^P]}$$

If

$$r = 1 - \frac{1}{n}$$

then

$$\frac{1}{n} = 1 - r$$

So, if $n \to \infty$, $r \to 1$.
We have

$$\frac{a_n - a_{n-1}}{b_n - b_{n-1}} = \frac{1-r}{1-r^p} = \frac{1}{(1-r^p)/(1-r)} = \frac{1}{1 + r + r^2 + \cdots + r^{p-1}}$$

$$\lim_{n\to\infty} \frac{a_n - a_{n-1}}{b_n - b_{n-1}} = \lim_{r\to 1} \frac{1}{1 + r + r^2 + \cdots + r^{P-1}} = \frac{1}{1 + 1 + \cdots p \text{ terms}} = \frac{1}{p}$$

So, from previous lemma,

$$\lim_{n\to\infty} \frac{a_n}{b_n} = \frac{1}{p}$$

i.e.,

$$\lim_{n\to\infty} \frac{1^{p-1} + 2^{p-1} + \cdots n^{p-1}}{n^p} = \frac{1}{p}$$

This proves Madhava's lemma.

Madhava's theorem: If

$$|\theta| \leq \frac{\pi}{4}, \text{then } \theta = \tan\theta - \frac{\tan^3\theta}{5} + \frac{\tan^5\theta}{5} - \cdots$$

Or, if $\tan\theta = t$, then

$$\tan^{-1} t = t - \frac{t^3}{3} + \frac{t^5}{5} - \cdots \text{ if } |t| = |\tan\theta| \leq 1$$

Proof: Let C be a circle of unit radius with center O. Let the horizontal and the vertical through O be x- and y-axis, respectively. Let the y-axis cut the circle at the point A.

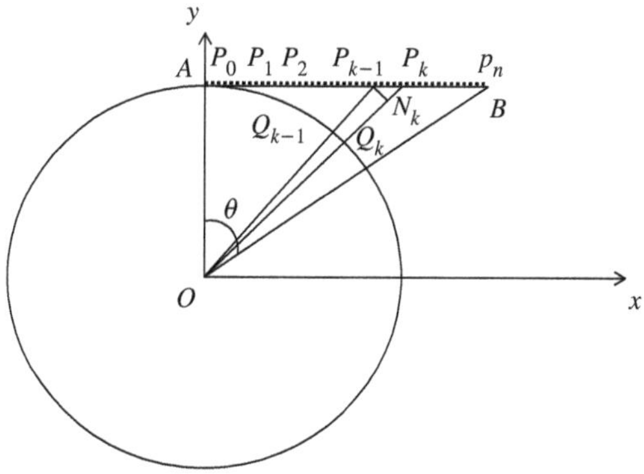

Draw the tangent AB at A to the circle such that $A\hat{O}B = \theta$.
So,

$$t = \tan\theta = \frac{AB}{OA} = AB$$

since radius $OA = 1$.

Divide AB into n equal parts by taking points $P_0, P_1, \ldots, P_n$, with $P_0 = A$ and $P_n = B$.

Then, $AP_1 = P_1P_2 = P_2P_3 = \cdots = P_{n-1}B$

i.e.,

$$P_{k-1}P_k = \frac{AB}{n} = \frac{\tan\theta}{n} \text{ for } k = 1, 2, \ldots, n$$

i.e.,

$$P_{k-1}P_k = \frac{t}{n}$$

where $t = \tan\theta$, for $k = 1, 2, \ldots n$.

Let the line OP_k cut the circle at Q_k and let $OP_k = r_k$ for $k = 0, 1, 2, \ldots, n$.

Now,

$$AP_k = P_0P_1 + P_1P_2 + \cdots P_{k-1}P_k = \frac{kt}{n}$$

In Δ^{le} OAP_k,

$$r_k^2 = OA^2 + AP_k^2 = 1 + \left(\frac{kt}{n}\right)^2 \tag{1}$$

Let $\theta_k = P_{k-1}\hat{O}P_k$ for $k = 1, 2, \ldots, n$

$$= Q_{k-1}OQ_k \quad \text{for } k = 1, 2, \ldots, n$$

Arc $AQ_n =$ arc $Q_0Q_n = \theta$.

Now, area of the Δ^{le} $P_{k-1}OP_k$

$$= \frac{1}{2}OP_k \cdot P_{k-1}N_k$$

where $P_{k-1}N_k$ is the perpendicular drawn from P_{k-1} on OP_k.

But $P_{k-1}N_k = OP_{k-1}\sin\theta_k$, where θ_k is the angle $P_{k-1}OP_k$.

So, area of Δ^{le} $P_{k-1}OP_k$

$$= \frac{1}{2}OP_kOP_{k-1}\sin\theta_k = \frac{1}{2}r_kr_{k-1}\sin\theta_k \tag{2}$$

But area of Δ^{le} $P_{k-1}OP_k$

$$= \frac{1}{2}P_{k-1}P_k \text{ altitude } OA$$

$$= \frac{1}{2}P_{k-1}P_k = \frac{t}{2n} \tag{3}$$

So, from (2) and (3), we get

$$\sin\theta_k \cdot r_kr_{k-1} = \frac{t}{n}$$

i.e.,

$$\sin\theta_k = \frac{t}{n}\frac{1}{r_kr_{k-1}}$$

Let $n \to \infty$.
Then,

$$\frac{r_k^2}{r_{k-1}^2} = \frac{(1+(tk/n))^2}{(1+(t(k-1)/n))^2} \to 1 \text{ as } n \to \infty$$

Also, P_{k-1} and P_k will be very near to each other for large n.
So, $r_{k-1} \to r_k$ and $P_{k-1}\hat{O}P_k = \theta_k \to 0$ as $n \to \infty$.
Now, $\theta = \text{arc } AQ_n = \text{arc } Q_0Q_n$

$$= \sum_{k=1}^{n} \text{arc } Q_{k-1}Q_k = \sum_{k=1}^{n} \theta_k$$

As $n \to \infty$,

$$\theta = \lim_{n\to\infty} \sum_{k=1}^{n} \theta_k = \lim_{n\to\infty} \sum_{k=1}^{n} \sin \theta_k$$

$$\theta = \lim_{n\to\infty} \sum_{k=1}^{n} \frac{t}{n}\frac{1}{r_k^2}$$

$$= \lim_{n\to\infty} \sum_{k=1}^{n} \frac{t}{n}\frac{1}{(1+(tk/n))^2} \tag{4}$$

$$= \lim_{n\to\infty} \frac{t}{n} \sum_{k=1}^{n} \frac{1}{[1+(tk/n)]^2}$$

Here $t = \tan\theta$ and for

$$|\theta| < \frac{\pi}{4}, \quad |t| < 1$$

We also know that if

$$|r| < 1, \quad \frac{1}{1+r} = 1 - r + r^2 - r^3 + \cdots$$

Let

$$\left(\frac{tk}{n}\right)^2 = r, \text{ then } |r| = \frac{tk^2}{n} < 1$$

So,

$$\frac{1}{1+(tk/n)^2} = \left[1 + \left(\frac{tk}{n}\right)^2\right]^{-1}$$

$$= 1 - \left(\frac{tk}{n}\right)^2 + \left(\frac{tk}{n}\right)^4 - \left(\frac{tk}{n}\right)^6 + \cdots$$

Hence, from (4), we have

$$\theta = \lim_{n\to\infty} \left\{ \sum_{k=1}^{n} \frac{t}{n} \left[1 - \left(\frac{tk}{n}\right)^2 + \left(\frac{tk}{n}\right)^4 - \left(\frac{tk}{n}\right)^6 + \cdots \right] \right\}$$

i.e.,

$$\theta = \lim_{n\to\infty} \sum_{k=1}^{n} \left\{ \frac{t}{n} - \frac{t^3k^2}{n^3} + \frac{t^5k^4}{n^5} - \frac{t^7k^6}{n^7} + \cdots \right\}$$

$$\theta = t - \lim_{n\to\infty} \sum_{k=1}^{n} \frac{t^3k^2}{n^3} + \lim_{n\to\infty} \sum_{k=1}^{n} \frac{t^5k^4}{n^5} - \lim_{n\to\infty} \sum \frac{t^7k^6}{n^7} + \cdots$$

i.e.,

$$\theta = t - \lim_{n\to\infty} \frac{t^3}{n^3} \sum_{k=1}^{n} k^2 + \lim_{n\to\infty} \frac{t^5}{n^5} \sum_{k}^{n} k^4 - \lim_{n\to\infty} \frac{t^7}{n^7} \sum_{k=1}^{n} k^6 + \cdots$$

Let

$$\frac{S_p}{n^p} = \frac{\sum_{k=1}^{n} k^{P-1}}{n^p} = \frac{1^{P-1} + 2^{P-1} + 3^{P-1} + \cdots + n^{P-1}}{n^P}$$

So,

$$\theta = t - t^3 \lim_{n\to\infty} \frac{s_3}{n^3} + t^5 \lim_{n\to\infty} \frac{s_5}{n^5} - t^7 \lim_{n\to\infty} \frac{s_7}{n^7} + \cdots$$

But from Madhava's lemma,

$$\lim_{n\to\infty} \frac{S_p}{n^p} = \frac{1}{p}$$

Hence,

$$\theta = t - \frac{t^3}{3} + \frac{t^5}{5} - \frac{t^7}{7} + \cdots \qquad \text{if } |t| \le 1$$

Since $\tan\theta = t$, we have

$$\theta = \tan\theta - \frac{\tan^3\theta}{3} + \frac{\tan^5\theta}{5} - \frac{\tan^7\theta}{7} + \cdots \qquad \text{if } |\theta| \le \frac{\pi}{4}$$

or

$$\tan^{-1} t = t - \frac{t^3}{3} + \frac{t^5}{5} - \frac{t^7}{7} + \cdots \qquad \text{if } |t| \le 1$$

This proves Madhava's theorem. This series is known as Gregory's series, named after the Scottish mathematician and astronomer James Gregory (AD 1638–1675), who rediscovered this series in 1668. This series is also known as Leibnitz series, named after the German philosopher and mathematician Gottfried Wilhelm Leibnitz (AD 1646–1716) who discovered it again in AD 1673.

Corollary 1: Let $t = 1$

Then,

$$\tan^{-1} = 1 - \frac{1}{3} + \frac{1}{5} - \frac{1}{7} + \cdots$$

i.e.,

$$\frac{\pi}{4} = 1 - \frac{1}{3} + \frac{1}{5} - \frac{1}{7} + \cdots$$

This series is known as Euler series, named after the Swiss mathematician Leonhard Euler (AD 1707–1783), who discovered this series in AD 1735.

Remark: Since Sangamagrama Madhava discovered the Gregory–Leibnitz series as well as the Euler series more than two centuries before their rediscovery, to be fair to all these great mathematicians, Gregory–Leibnitz series should be renamed as Madhava–Gregory–Leibnitz series and the Euler series should be renamed as Madhava–Euler series.

Corollary 2: Let

$$t = \frac{1}{\sqrt{3}}$$

Then, we have

$$\tan^{-1}\frac{1}{\sqrt{3}} = \frac{1}{\sqrt{3}} - \frac{(1/\sqrt{3})^3}{3} + \frac{(1/\sqrt{3})^5}{5} - \frac{(1/\sqrt{3})^7}{7} + \cdots$$

i.e.,

$$\frac{\pi}{6} = \frac{1}{\sqrt{3}}\left[1 - \frac{1}{3 \cdot 3} + \frac{1}{5 \cdot 3^2} - \frac{1}{7 \cdot 3^3} + \cdots\right]$$

i.e.,

$$\pi = 2\sqrt{3}\left[1 - \frac{1}{3 \cdot 3} + \frac{1}{5 \cdot 3^2} - \frac{1}{7 \cdot 3^3} + \cdots\right]$$

Series Expansions Given in "Tantra-Sangraha" of Nilakantha Somayaji

We will now describe various transformations which will convert slowly convergent "Euler series" to rapidly convergent series and also obtain some rational approximation for π. We will give a method outlined in "Yuktibhasa" with modern proof.

Now, Euler series is

$$\frac{\pi}{4} = 1 - \frac{1}{3} + \frac{1}{5} - \frac{1}{7} + \cdots \tag{1}$$

Regrouping we get

$$\frac{\pi}{4} = \left(1 - \frac{1}{3}\right) + \left(\frac{1}{5} - \frac{1}{7}\right) + \left(\frac{1}{9} - \frac{1}{11}\right) + \cdots$$

So,

$$\frac{\pi}{4} = \frac{2}{3} + \frac{2}{35} + \frac{2}{99} + \cdots$$

$$\frac{\pi}{8} = \frac{1}{3} + \frac{1}{35} + \frac{1}{99} + \cdots$$

i.e.,

$$\frac{\pi}{8} = \frac{1}{2^2 - 1} + \frac{1}{6^2 - 1} + \frac{1}{10^2 - 1} + \cdots \tag{2}$$

$$= \frac{1}{1 \cdot 3} + \frac{1}{5 \cdot 7} + \frac{1}{9 \cdot 11} + \cdots \tag{2'}$$

(1) can also be regrouped as

$$\frac{\pi}{4} = 1 - \left(\frac{1}{3} - \frac{1}{5}\right) - \left(\frac{1}{7} - \frac{1}{9}\right) - \left(\frac{1}{11} - \frac{1}{13}\right) - \cdots$$

$$= 1 - \frac{2}{15} - \frac{2}{63} - \frac{2}{143} - \cdots$$

So,

$$1 - \frac{\pi}{4} = \frac{2}{15} + \frac{2}{63} + \frac{2}{143} + \cdots$$

i.e.,

$$\frac{4 - \pi}{8} = \frac{1}{15} + \frac{1}{63} + \frac{1}{143} + \cdots$$

$$= \frac{1}{3 \cdot 5} + \frac{1}{7 \cdot 9} + \frac{1}{11 \cdot 13} + \cdots \tag{3}$$

$$\frac{4 - \pi}{8} = \frac{1}{4^2 - 1} + \frac{1}{8^2 - 1} + \frac{1}{12^2 - 1} + \cdots \tag{3'}$$

Subtracting (3) from (2), we get

$$\frac{\pi}{8} - \left(\frac{4 - \pi}{8}\right) = \frac{1}{2^2 - 1} - \frac{1}{4^2 - 1} + \frac{1}{6^2 - 1} - \frac{1}{8^2 - 1} + \frac{1}{10^2 - 1} + \cdots$$

i.e.,

$$\frac{\pi - 2}{4} = \frac{1}{2^2 - 1} - \frac{1}{4^2 - 1} + \frac{1}{6^2 - 1} - \frac{1}{8^2 - 1} + \cdots \tag{4}$$

Series (1) is

$$\frac{\pi}{4} = 1 - \frac{1}{3} + \frac{1}{5} - \frac{1}{7} + \cdots = \sum_{n=1}^{\infty} V_k$$

where

$$V_k = (-1)^{k-1} \frac{1}{2k-1} \quad (*)$$

So,

$$\frac{\pi}{4} = 1 - \frac{1}{3} + \frac{1}{5} - \frac{1}{7} + \cdots = \sum_{k=1}^{\infty} V_k = \sum_{k=1}^{\infty} (-1)^{k-1} \frac{1}{2k-1} \quad (**)$$

Let $S((n+1)/2)$ and $S((n-1)/2)$ be the sums of the first $2m+1$ and $2m$ terms respectively of (**), where $n = 4m+1$. These partial sums will give rational approximation to $\pi/4$, where n is large.

In order to improve the above approximations, we will add correction terms $-f(n+1)$ and $f(n-1)$ respectively to the above partial sums $S((n+1)/2)$ and $S((n-1)/2)$. Let the improved approximation be $S^*((n+1)/2)$ and $S^*((n-1)/2)$.

Then,

$$S^*\left(\frac{n+1}{2}\right) = S\left(\frac{n+1}{2}\right) - f(n+1)$$

and

$$S^*\left(\frac{n-1}{2}\right) = S\left(\frac{n-1}{2}\right) + f(n-1) \quad (5)$$

Let $S^*((n+1)/2)$ and $S^*((n-1)/2)$ be the nth and $(n-1)$th partial sum of a certain series $\sum_{n=1}^{\infty} u_n$.

Then,

$$u_n = S^*\left(\frac{n+1}{2}\right) - S^*\left(\frac{n-1}{2}\right)$$

So,

$$u_n = S\left(\frac{n+1}{2}\right) - S\left(\frac{n-1}{2}\right) - f(n+1) - f(n-1) \quad \text{from (5)}$$

Now,

$$S\left(\frac{n+1}{2}\right) = \sum_{k=1}^{2m+1} V_k$$

$$S\left(\frac{n-1}{2}\right) = \sum_{K=1}^{2m} V_k$$

Then,

$$S\left(\frac{n+1}{2}\right)-S\left(\frac{n-1}{2}\right)=V_{2m+1}=(-1)^{2m}\frac{1}{2(2m+1)-1}\quad \text{from (*)}$$

So,

$$S\left(\frac{n+1}{2}\right)-S\left(\frac{n-1}{2}\right)=\frac{1}{4m+1}=\frac{1}{n} \tag{6}$$

Now, from (5) and (6)

$$u_n=\frac{1}{n}-f(n+1)-f(n-1) \tag{7}$$

Putting $n=3, 5, \ldots, n(4m+1)$, we get

$$u_3=\frac{1}{3}-f(4)-f(2)$$
$$u_5=\frac{1}{5}-f(6)-f(4)$$
$$u_7=\frac{1}{7}-f(8)-f(6)$$
$$\ldots$$
$$\ldots$$
$$u_{n-2}=\frac{1}{n-2}-f(n-1)-f(n-3)$$
$$u_{3+2(2m-1)}=u_{4m+1}=u_n=\frac{1}{n}-f(n+1)-f(n-1)$$

So, we get

$$-u_3+u_5-u_7+\cdots+u_n=-\frac{1}{3}+\frac{1}{5}-\frac{1}{7}+\frac{1}{9}+\cdots+\frac{1}{n}-f(n+1)+f(2)$$

Let $n\to\infty$ and let $f(n+1)\to 0$ as $n\to\infty$.
Then, we have

$$-u_3+u_5-u_7+u_9+\cdots=\frac{\pi}{4}-1+f(2)$$

i.e.

$$\frac{\pi}{4}=1-f(2)-u_3+u_5-u_7+u_9+\cdots \tag{8}$$

Let $f(x)$ be of the form

$$2f(x)=\frac{c_1}{x}+\frac{c_2}{x^2}+\frac{c_3}{x^3}+\cdots \quad (***)$$

By Taylor's theorem,

$$f(x+h)=f(x)+\frac{h}{1!}f'(x)+\frac{h^2}{2!}f''(x)+\cdots+\frac{h^{2p}}{2p!}f^{2p}(x+\theta h)\quad 0<\theta<1$$

First, let $x=n$ and $h=1$.

Then, we get

$$f(n+1) = f(n) + \frac{f'(n)}{1!} + \frac{f''(n)}{2!} + \cdots + \frac{1}{2p!} f^{2p}(n+\theta_1) \quad 0 < \theta_1 < 1 \tag{9}$$

Next, let $x = n$ and $h = -1$.
Then, we have

$$f(n-1) = f(n) - \frac{f'(n)}{1!} + \frac{f''(n)}{2!} - \cdots + \frac{1}{2p!} f^{2p}(n-\theta_2) \quad 0 < \theta_2 < 1 \tag{10}$$

From (9) and (10),

$$f(n+1) + f(n-1) = 2\left[f(n) + \frac{f''(n)}{2!} + \cdots + \frac{1}{2p!}\{f^{2p}(n+\theta_1) - f^{2p}(n-\theta_2\}\right] \tag{11}$$

where $0 < \theta_1 < 1$ and $0 < \theta_2 < 1$.
Then, from (7), we get

$$\frac{1}{n} = u_n + 2\left[f(n) + \frac{f''(n)}{2!} + \cdots + \frac{1}{2p!}\{f^{2p}(n+\theta_1) - f^{2p}(n-\theta_2)\}\right] \tag{12}$$

where $0 < \theta_1 < 1$ and $0 < \theta_2 < 1$.
Now, from (***),

$$f(n) = 0\left(\frac{1}{n}\right)$$

So,

$$f^{2p}(n) = 0\left(\frac{1}{n^{2p+1}}\right) \quad \text{for } p = 1, 2, \ldots$$

If

$$u_n = 0\left(\frac{1}{n^{2p+1}}\right)$$

we obtain from (12) that

$$\frac{1}{n} + 0\left(\frac{1}{n^{2p+1}}\right) = 2\left[f(n) + \frac{f''(n)}{2!} + \cdots + \frac{1}{(2p-2)!} f^{2p-2}(n)\right] \tag{13}$$

Substituting (***) in (13), we can calculate $c_1, c_2, \ldots, c_{2p}$.
We verify that $c_2 = c_4 = \cdots = c_{2p} = 0$ and $c_{2p+1}, c_{2p+2}, \ldots$ arbitrary.
We now employ (7) and (8) to obtain a series for $\pi/4$ in which

$$u_n = 0\left(\frac{1}{n^{2p}+1}\right)$$

Letting $p = 1$ in (13), we get

$$2f(n) = \frac{1}{n} \tag{14}$$

But from (5),

$$S^*\left(\frac{n+1}{2}\right) = S\left(\frac{n+1}{2}\right) - f(n+1)$$

Now,

$$S\left(\frac{n+1}{2}\right) = \sum_{k=1}^{2m+1} V_k = \sum_{k=1}^{2m+1} \frac{(-1)^{k-1}}{2k-1}$$

$$S\left(\frac{n+1}{2}\right) = 1 - \frac{1}{3} + \frac{1}{5} - \dots + \frac{1}{4m+1} = 1 - \frac{1}{3} + \frac{1}{5} - \dots + \frac{1}{n}$$

Also from (14),

$$f(n+1) = \frac{1}{2(n+1)}$$

and

$$S^*\left(\frac{n+1}{2}\right) \approx \frac{\pi}{4}$$

So,

$$\frac{\pi}{4} \approx \left(1 - \frac{1}{3} + \frac{1}{5} - \dots + \frac{1}{n}\right) - \frac{1}{2(n+1)} \tag{15}$$

Also from (7),

$$u_n = \frac{1}{n} - f(n+1) - f(n-1)$$

$$u_n = \frac{1}{n} - \frac{1}{2(n+1)} - \frac{1}{2(n-1)}$$

since

$$f(n) = \frac{1}{2n}$$

from (14),

$$u_n = \frac{2(n^2-1) - n(n-1) - n(n+1)}{2(n+1)(n-1)n}$$

$$= \frac{2n^2 - 2 - n^2 + n - n^2 - n}{2n(n^2-1)}$$

So,

$$u_n = \frac{-2}{2n(n^2-1)} = \frac{-1}{n(n^2-1)}$$

From (8), we have

$$\frac{\pi}{4} = 1 - f(2) - u_3 + u_5 - u_7 + u_9 - \cdots$$

$$\frac{\pi}{4} = 1 - \frac{1}{4} + \frac{1}{3(3^2-1)} - \frac{1}{5(5^2-1)} + \frac{1}{7(7^2-1)} - \frac{1}{9(9^2-1)} + \frac{1}{11(11^2-1)} - \frac{1}{13(13^2-1)} + \cdots$$

$$\frac{\pi-3}{4} = \frac{\pi}{4} - \frac{3}{4} = \frac{1}{3(3^2-1)} - \frac{1}{5(5^2-1)} + \frac{1}{7(7^2-1)} - \frac{1}{9(9^2-1)} + \frac{1}{11(11^2-1)} - \frac{1}{13(13^2-1)} + \cdots$$

i.e.,

$$\frac{\pi-3}{4} = \frac{1}{3^3-3} - \frac{1}{5^3-5} + \frac{1}{7^3-7} - \frac{1}{9^3-9} + \frac{1}{11^3-11} - \frac{1}{13^3-13} + \cdots \quad (16)$$

$$\frac{\pi-3}{4} = \frac{1}{3(3+1)(3-1)} - \frac{1}{5(5+1)(5-1)} + \frac{1}{7(7+1)(7-1)} - \frac{1}{9(9+1)(9-1)} + \frac{1}{11(11+1)(11-1)} - \frac{1}{13(13-1)(13+1)} + \cdots$$

$$\frac{\pi-3}{4} = \frac{1}{2\cdot3\cdot4} - \frac{1}{4\cdot5\cdot6} + \frac{1}{6\cdot7\cdot8} - \frac{1}{8\cdot9\cdot10} + \frac{1}{10\cdot11\cdot12} - \frac{1}{12\cdot13\cdot14} + \cdots + \frac{1}{10\cdot11\cdot12} - \frac{1}{12\cdot13\cdot14} + \cdots \quad (17)$$

Regrouping two by two in the right-hand side of (17), we get

$$\frac{\pi-3}{4} = \left(\frac{1}{2\cdot3\cdot4} - \frac{1}{4\cdot5\cdot6}\right) + \left(\frac{1}{6\cdot7\cdot8} - \frac{1}{9\cdot10\cdot8}\right) + \left(\frac{1}{10\cdot11\cdot12} - \frac{1}{12\cdot13\cdot14}\right) + \cdots$$

$$= \frac{5-1}{120} + \frac{15-7}{5040} + \frac{91-55}{120,120} + \cdots$$

So,

$$\frac{\pi-3}{4} = \frac{4}{120} + \frac{8}{5040} + \frac{36}{120,120} + \cdots$$

$$\frac{\pi-3}{2} = \frac{8}{120} + \frac{16}{5040} + \frac{72}{120,120} + \cdots$$

Hence,

$$\frac{\pi-3}{6} = \frac{8}{360} + \frac{16}{15,120} + \frac{24}{120,120} + \cdots$$

$$\frac{\pi-3}{6} = \frac{1}{45} + \frac{1}{945} + \frac{1}{5005} + \cdots$$

$$\frac{\pi-3}{6} = \frac{1}{9\cdot 5} + \frac{1}{35\cdot 27} + \frac{1}{65\cdot 77} + \cdots$$

$$\frac{\pi-3}{6} = \frac{1}{(7+2)(7-2)} + \frac{1}{(31+4)(31-4)} + \frac{1}{(71+6)(71-6)} + \cdots \quad (17')$$

$$\frac{\pi-3}{6} = \frac{1}{\{(2\cdot 2^2-1)+2\}\{(2\cdot 2^2-1)-2\}} + \frac{1}{\{(2\cdot 4^2-1)+4\}\{(2\cdot 4^2-1)-4\}} + \frac{1}{[(2\cdot 6^2-1)+6][(2\cdot 6^2-1)-6]} + \cdots$$

$$\frac{\pi-3}{6} = \frac{1}{(2\cdot 2^2-1)^2-2^2} + \frac{1}{(2\cdot 4^2-1)^2-4^2} + \frac{1}{[(2\cdot 6^2-1)^2-6^2]} + \cdots \quad (18)$$

From (17′), we also see that

$$\frac{\pi-3}{6} = \frac{1}{1\cdot 3\cdot 3\cdot 5} + \frac{1}{3\cdot 5\cdot 7\cdot 9} + \frac{1}{5\cdot 7\cdot 11\cdot 13} + \frac{1}{7\cdot 9\cdot 15\cdot 17} \quad (18')$$

Letting $p = 2$ in (13), we get

$$\frac{1}{n} + O\left(\frac{1}{n^5}\right) = 2\left[f(n) + \frac{f''(n)}{2!} + \cdots + \frac{1}{(2p-2)!}f^{2p-2}(n)\right]$$

Now,

$$2f(x) = c_1x^{-1} + c_2x^{-2} + c_3x^{-3} + c_4x^{-4} + c_5x^{-5} + \cdots$$

$$2f'(x) = -c_1x^{-2} - 2c_2x^{-3} - 3c_3x^{-4} - 4c_4x^{-5} - 5c_5x^{-6} + \cdots$$

$$2f''(x) = 2c_1x^{-3} + 6c_2x^{-4} + 12c_3x^{-5} + 20c_4x^{-6} + 30c_5x^{-7} + \cdots$$

$$2f^3(x) = -6c_1x^{-4} - 24c_2x^{-5} - 60c_3x^{-6} - 120c_4x^{-7} - 210c_5x^{-8} - \cdots$$

$$2f^4(x) = 24c_1x^{-5} + 120c_2x^{-6} + 360c_3x^{-7} + 840c_4x^{-8} + 1680c_5x^{-9} + \cdots$$

$$\frac{1}{n} + O\left(\frac{1}{n^5}\right) = c_1n^{-1} + c_2n^{-2} + c_3n^{-3} + c_4n^{-4} + c_5n^{-5} + c_6n^{-6} + c_7n^{-7} + \cdots$$
$$+ \frac{1}{2!}\left[2c_1n^{-3} + 6c_2n^{-4} + 12c_3n^{-5} + 20c_4n^{-6} + 30c_5n^{-7} + \cdots\right]$$
$$+ \frac{1}{4!}\left[24c_1n^{-5} + 120c_2n^{-6} + 360c_3n^{-7} + 840c_4n^{-8} + 1680c_5n^{-9} + \cdots\right]$$

$$c_1 = 1,\ \ c_2 = 0,\ \ c_3 + c_1 = 0,\ \ c_4 + 3c_2 = 0$$

$$c_6 + 10c_4 + 5c_2 = 0 \Rightarrow c_6 = 0$$

So,

$$c_1 = 1,\ \ c_2 = 0,\ \ c_3 = -c_1 = -1,\ \ c_4 = -3c_2 = 0,\ \ c_6 = 0 \text{ and so on.}$$

So,

$$2f(x) = \frac{c_1}{x} + \frac{c_2}{x^2} + \frac{c_3}{x^3} + \frac{c_4}{x^4} + \frac{c_5}{x^5} + \cdots$$

$$= \frac{1}{x} - \frac{1}{x^3} + \frac{c_5}{x^5} + \frac{c_7}{x^7} + \cdots$$

Choosing $c_5 = 1$ and $c_7 = -1$, we get

$$2f(n) = \frac{1}{n} - \frac{1}{n^3} + \frac{1}{n^5} - \frac{1}{n^7} + \cdots$$

RHS is an infinite geometric series with first term $a = (1/n)$ and common ratio $r = -(1/n^2)$.

Hence,

$$\frac{1}{n} - \frac{1}{n^3} + \frac{1}{n^5} - \frac{1}{n^7} + \cdots = \frac{a}{1-r} = \frac{(1/n)}{1+(1/n^2)} = \frac{n}{n^2+1}$$

So,

$$2f(n) = \frac{n}{n^2+1} \Rightarrow f(n) = \frac{n}{2(n^2+1)}$$

Hence, from (5), we have

$$S^*\left(\frac{n+1}{2}\right) = S\left(\frac{n+1}{2}\right) - f(n+1)$$

Here,

$$S\left(\frac{n+1}{2}\right) = S\left(\frac{4m+1+1}{2}\right) = S(2m+1) = \sum_{k=1}^{2m+1} V_k$$

$$= \sum_{k=1}^{2m+1} \frac{(-1)^{k-1}}{2k-1} = 1 - \frac{1}{3} + \frac{1}{5} - \cdots + \frac{1}{n}$$

and

$$S^*\left(\frac{n+1}{2}\right) \approx \frac{\pi}{4} \text{ and } f(n+1) = \frac{n+1}{2\{(n+1)^2 - 1\}}$$

So,

$$\frac{\pi}{4} \approx 1 - \frac{1}{3} + \frac{1}{5} - \cdots + \frac{1}{n} - \frac{(n+1/2)}{\{(n+1)^2+1\}} \tag{19}$$

Also,

$$u_n = \frac{1}{n} - f(n+1) - f(n-1) \quad \text{from (7)}$$

So,

$$u_n = \frac{1}{n} - \frac{n+1}{2\{(n+1)^2+1\}} - \frac{n-1}{2\{(n-1)^2+1\}}$$

$$u_n = \frac{2\{(n+1)^2+1\}\cdot\{(n-1)^2+1\} - n\{(n-1)^2+1\}\{n+1\} - n(n-1)\{(n+1)^2+1\}}{2n\{(n+1)^2+1\}\{(n-1)^2+1\}}$$

$$= \frac{2\{n^2+2n+2\}\{n^2-2n+2\} - \{n^2+n\}\{n^2-2n+2\} - \{n^2-n\}\{n^2+2n+2\}}{2n\{(n+1)^2+1\}\{(n-1)^2+1\}}$$

$$= \frac{(n^2-2n+2)\{2n^2+4n+4-n^2-n\} - \{n^4+2n^3+2n^2-n^3-2n^2-2n\}}{2n\{(n+1)^2+1\}\{(n-1)^2+1\}}$$

$$= \frac{(n^2-2n+2)(n^2+3n+4) - (n^4+n^3-2n)}{2n\{(n+1)^2+1\}\{(n-1)^2+1\}}$$

$$= \frac{n^4-2n^3+2n^2+3n^3-6n^2+6n+4n^2-8n+8-n^4-n^3+2n}{2n\{(n+1)^2+1\}\{(n-1)^2+1\}}$$

$$u_n = \frac{8}{2n\{(n+1)^2+1\}\{(n-1)^2+1\}}$$

$$u_n = \frac{4}{n\{(n+1)^2+1\}\{(n-1)^2+1\}} \tag{20}$$

We have from (8),

$$\frac{\pi}{4} = 1 - f(2) - u_3 + u_5 - u_7 + \cdots$$

$$\frac{\pi}{4} - 1 + \frac{2}{2\{5\}} = -u_3 + u_5 - u_7 + \cdots$$

So, from (20), we have

$$\frac{\pi}{4} - \frac{4}{5} = \frac{-4}{3\{4^2+1\}\{2^2+1\}} + \frac{4}{5\{6^2+1\}\{4^2+1\}} - \frac{4}{7\{8^2+1\}\{6^2+1\}} +$$

$$\frac{\pi}{4}-\frac{4}{5}=-\frac{4}{3\cdot 85}+\frac{4}{5\cdot 37\cdot 17}-\frac{4}{7\cdot 65\cdot 37}+\cdots$$

$$\frac{\pi}{16}-\frac{1}{5}=\frac{-1}{3(3^4+4)}+\frac{1}{5(5^4+4)}-\frac{1}{7(7^4+4)}+\cdots$$

So,

$$\frac{\pi}{16}=\frac{1}{5}-\frac{1}{3^5+3\cdot 4}+\frac{1}{5^5+4\cdot 5}-\frac{1}{7^5+4\cdot 7}+\cdots$$

$$\frac{\pi}{16}=\frac{1}{1^5+4\cdot 1}-\frac{1}{3^5+4\cdot 3}+\frac{1}{5^5+4\cdot 5}-\frac{1}{7^5+4\cdot 7}+\cdots \tag{21}$$

Next put $p=3$.
Then, from (13) and (***), we get

$$\begin{aligned}\frac{1}{n}+O\left(\frac{1}{n^7}\right)&=\left[c_1n^{-1}+c_2n^{-2}+c_3n^{-3}+c_4n^{-4}+c_5n^{-5}+c_6n^{-6}+c_7n^{-7}+\cdots\right]\\&+\frac{1}{2!}\left[2c_1n^{-3}+6c_2n^{-4}+12c_3n^{-5}+20c_4n^{-6}+30c_5n^{-7}+\cdots\right]\\&+\frac{1}{4!}\left[24c_1n^{-5}+120c_2n^{-6}+360c_3n^{-7}+840c_4n^{-8}+1680c_5n^{-9}+\cdots\right]+\cdots\end{aligned}$$

Equating like negative powers of n, we get

$$c_1=1,\ c_2=0,\ c_3+c_1=0\Rightarrow c_3=-c_1=-1,\ c_4+3c_2=0\Rightarrow c_4=0$$

$$c_5+6c_3+c_1=0$$

$$c_5=-6c_3-c_1=+6-1=5;\quad c_6+10c_4+5c_2=0\Rightarrow c_6=0\text{ and so on.}$$

Since

$$2f(x)=\frac{c_1}{x}+\frac{c_2}{x^2}+\frac{c_3}{x^4}+\frac{c_5}{x^5}+\frac{c_6}{x^6}+\frac{c_7}{x^7}+\cdots$$

$$2f(x)=\frac{1}{x}-\frac{1}{x^3}+\frac{5}{x^5}+\frac{c_7}{x^7}+\cdots$$

Choosing $c_7=-5^2$, $c_9=+5^3,\ldots$, we get

$$2f(n)=\frac{1}{n}-\frac{1}{n^3}+\frac{5}{n^5}-\frac{5^2}{n^7}+\frac{5^3}{n^9}+\cdots$$

Now,

$$\frac{5}{n^5} - \frac{5^2}{n^7} + \frac{5^3}{n^9} - \cdots$$

is an infinite geometric series with common ratio $r = (-5/n^2)$ and first term $a = (5/n^5)$.

Sum of this infinite series $= a/(1 - r)$

$$= \frac{(5/n^5)}{1 + (5/n^2)} = \frac{5}{n^5} \cdot \frac{n^2}{n^2 + 5} = \frac{5}{n^3(n^2 + 5)}$$

So,

$$2f(n) = \frac{1}{n} - \frac{1}{n^3} + \frac{5}{n^3(n^2 + 5)}$$

$$2f(n) = \frac{n^2 - 1}{n^3} + \frac{5}{n^3(n^2 + 5)} = \frac{(n^2 - 1)(n^2 + 5) + 5}{n^3(n^2 + 5)}$$

So,

$$2f(n) = \frac{n^4 + 4n^2 + 5 - 5}{n^3(n^2 + 5)} = \frac{n^2(n^2 + 4)}{n^3(n^2 + 5)} = \frac{n^2 + 4}{n(n^2 + 5)}$$

Now,

$$S^*\left(\frac{n+1}{2}\right) = S\left(\frac{n+1}{2}\right) - f(n + 1) \quad \text{from (5)}$$

So,

$$\frac{\pi}{4} \approx \sum_{k=1}^{2m+1} V_k - \frac{(n+1)^2 + 4}{2(n + 1)[(n+1)^2 + 5]}$$

i.e.,

$$\frac{\pi}{4} \approx 1 - \frac{1}{3} + \frac{1}{5} - \frac{1}{7} + \cdots + \frac{1}{n} - \frac{(n+1)^2 + 4}{2(n + 1)[(n+1)^2 + 5]} \tag{22}$$

Now,

$$u_n = \frac{1}{n} - f(n + 1) - f(n - 1) \quad \text{from (7)}$$

So,

$$u_n = \frac{1}{n} - \frac{(n+1)^2+4}{2(n+1)[(n+1)^2+5]} - \frac{(n-1)^2+4}{2(n-1)[(n-1)^2+5]}$$

$$u_n = \frac{\begin{array}{l}2(n^2-1)[(n+1)^2+5][(n-1)^2+5] \\ -n\{[(n+1)^2+4][n-1][(n-1)^2+5]-[(n-1)^2+4]n[n+1]\}\{(n+1)^2+5\}\end{array}}{2n(n+1)(n-1)[(n+1)^2+5][(n-1)^2+5]}$$

$$u_n = \frac{\begin{array}{r}2(n^2-1)[(n^2+6)^2-4n^2]-(n^2-n)(n^2+2n+5)(n^2-2n+6) \\ -(n^2+n)(n^2-2n+5)(n^2+2n+6)\end{array}}{2n(n^2-1)[(n+1)^2+5][(n-1)^2+5]}$$

$$u_n = \frac{\begin{array}{r}2(n^2-1)[n^4+12n^2+36-4n^2]-(n^2-n)(n^4+7n^2+2n+30) \\ -(n^2+n)(n^4+7n^2-2n+30)\end{array}}{2n(n^2-1)(n^2+2n+6)(n^2-2n+6)}$$

$$u_n = \frac{\begin{array}{r}2n^6+16n^4+72n^2-2n^4-16n^2-72-n^6-7n^4-2n^3-30n^2 \\ +n^5+7n^3+2n^2+30n-n^6-7n^4+2n^3-30n^2 \\ -n^5-7n^3+2n^2-30n\end{array}}{2n(n^2-1)[n^4+12n^2+36-4n^2]}$$

$$u_n = \frac{-72}{2n(n^2-1)(n^4+8n^2+36)} = \frac{-36}{n(n^2-1)(n^4+8n^2+36)}$$

From (8), we have

$$\frac{\pi}{4} - 1 + f(2) = -u_3 + u_5 - u_7 + u_9 - \cdots$$

i.e.,

$$\frac{\pi}{4} - 1 + \frac{1}{2}\frac{2^2+4}{2(2^2+5)}$$

$$= 36\left[\frac{1}{3(3^2-1)(3^4+8\cdot 3^2+36)} - \frac{1}{5(5^2-1)(5^4+8\cdot 5^2+36)} + \frac{1}{5(7^2-1)(7^4+8\cdot 7^2+36)}\right]$$

i.e.,

$$\frac{\pi}{4} - 1 + \frac{2}{9} = 36\left[\frac{1}{(3^3-3)}\frac{1}{189} - \frac{1}{(5^3-5)}\frac{1}{861} + \frac{1}{(7^3-7)(2829)} - \ldots\right]$$

i.e.,

$$\frac{\pi}{4}-\frac{7}{9}=36\left[\frac{1}{(3^3-3)}\frac{1}{9\cdot 21}-\frac{1}{(5^3-5)}\frac{1}{21\cdot 41}+\frac{1}{(7^3-7)41\cdot 69}-\cdots\right]$$

$$\begin{aligned}\frac{\frac{\pi}{4}-\frac{7}{9}}{36}&=\frac{1}{(3^3-3)}\frac{1}{(2^2+5)(4^2+5)}-\frac{1}{5^3-5}\frac{1}{(4^2+5)(6^2+5)}\\&\quad+\frac{1}{(7^3-7)}\frac{1}{(6^2+5)(8^2+5)}-\cdots\end{aligned} \tag{23}$$

Now, from (4) on page 6, we have

$$\frac{\pi-2}{4}=\frac{1}{2^2-1}-\frac{1}{4^2-1}+\frac{1}{6^2-1}-\cdots=\sum_{k=1}^{\infty}V_k^* \tag{24}$$

where

$$V_k^*=\frac{(-1)^{k-1}}{(2k)^2-1}$$

Let $S(n/2)$ and $S(n-2)/2$ be the sums of the first $2m+1$ and $2m$ terms respectively of (24), when $n=4m+2$. These partial sums as before will give rational approximation to $(\pi-2)/4$, when n is large.

As before, in order to improve the above approximations, we will add correction terms $-f(n+1)$ and $f(n-1)$ respectively to the above partial sums $S(n/2)$ and $S(n-2)/2$. Let the improved approximation be $S^*(n/2)$ and $S^*(n-2)/2$. Then,

$$S^*\left(\frac{n}{2}\right)=S\left(\frac{n}{2}\right)-f(n+1)$$

$$S^*\left(\frac{n-2}{2}\right)=S\left(\frac{n-2}{2}\right)+f(n-1)$$

Let $S^*(n/2)$ and $S^*(n-2)/2$ be the nth and $(n-1)$th partial sum of certain series $\sum_{n=1}^{\infty}u_n^*$, then

$$u_n^*=S^*\left(\frac{n}{2}\right)-S^*\left(\frac{n-2}{2}\right)$$

$$u_n^*=S\left(\frac{n}{2}\right)-S\left(\frac{n-2}{2}\right)-f(n+1)-f(n-1)$$

Now,

$$S\left(\frac{n}{2}\right)=\sum_{k=1}^{2m+1}V_k^* \text{ and } S\left(\frac{n-1}{2}\right)=\sum_{k=1}^{2m}V_k^*$$

Then,

$$S\left(\frac{n}{2}\right) - S\left(\frac{n-1}{2}\right) = V^{*}_{2m+1} = (-1)^{2m}\frac{1}{\{2(2m+1)\}^2 - 1}$$

$$= \frac{1}{(4m+2)^2 - 1} = \frac{1}{n^2 - 1}$$

$$u^{*}_{n} = \frac{1}{n^2 - 1} - f(n+1) - f(n-1)$$

We choose $f(n)$ so that

$$u^{*}_{n} = O\left(\frac{1}{n^{2p+2}}\right)$$

Then,

$$f(n+1) - f(n-1) = 2\left[f(n) + \frac{f''(n)}{2!} + \frac{f^{4}(n)}{4!} + \cdots + \frac{f^{2p-2}(n)}{(2p-2)!}\right]$$

$$= 2\sum_{r=0}^{p-1}\frac{f^{2r}(n)}{2n!}$$

Now,

$$\frac{1}{n^2-1} = \frac{1}{n^2\left(1-\frac{1}{n^2}\right)} = \frac{1}{n^2}\left(1-\frac{1}{n^2}\right)^{-1} = \frac{1}{n^2}\left(1 + \frac{1}{n^2} + \frac{1}{n^4} + \frac{1}{n^6} + \cdots\right)$$

$$= \frac{1}{n^2} + \frac{1}{n^4} + \frac{1}{n^6} + \frac{1}{n^8} + \cdots$$

$$O\left(\frac{1}{n^{2p+2}}\right) = u^{*}_{n} = \left(\frac{1}{n^2} + \frac{1}{n^4} + \frac{1}{n^6} + \frac{1}{n^8} + \cdots\right) - 2\sum_{r=0}^{p-1}\frac{f^{2n}(n)}{(2r)!} \qquad (25)$$

Since,

$$S\left(\frac{n}{2}\right) - S\left(\frac{n-1}{2}\right) = O\left(\frac{1}{n^2}\right),$$

we take

$$2f(x) = \frac{1}{x}\left[\frac{a_1}{x} + \frac{a_2}{x^2} + \frac{a_3}{x^3} + \frac{a_4}{x^4} + \cdots\right]$$

$$2f(x) = a_1x^{-2} + a_2x^{-3} + a_3x^{-4} + a_4x^{-5} + \cdots \quad (*)$$

Then,

$$2f'(x) = -2a_1x^{-3} - 3a_2x^{-4} - 4a_3x^{-5} - 5a_4x^{-6} - \cdots$$

$$2f''(x) = 6a_1x^{-4} + 12a_2x^{-5} + 20a_3x^{-6} + 30a_4x^{-7} + \cdots$$

$$2f^3(x) = -24a_1x^{-5} - 60a_2x^{-6} - 120a_3x^{-7} - 210a_4x^{-8} - \cdots$$

$$2f^4(x) = 120a_1x^{-6} + 360a_2x^{-7} + 840a_3x^{-8} + 1680a_4x^{-9} + \cdots$$

Substituting these in (25), we get

$$O\left(\frac{1}{n^{2P+2}}\right) = \frac{1}{n^2} + \frac{1}{n^4} + \frac{1}{n^6} + \frac{1}{n^8} + \cdots - (a_1n^{-2} + a_2n^{-3} + a_3n^{-4} + a_4n^{-5} + \cdots)$$

$$- \frac{1}{2!}(6a_1n^{-4} + 12a_2n^{-5} + 20a_3n^{-6} + 30a_4n^{-7} + \cdots)$$

$$- \frac{1}{4!}(120a_1n^{-6} + 360a_2n^{-7} + 840a_3n^{-8} + 1680a_4n^{-9} + \cdots)$$

Equating the coefficients of n^{-k}, for $k = 2, 3, \ldots$, we get

$$1 - a_1 = 0 \Rightarrow a_1 = 1$$

$$-a_2 = 0 \Rightarrow a_2 = 0, \quad -a_4 - 6a_2 = 0 \Rightarrow a_4 = 0$$

$$-a_6 - 15a_4 + 15a_2 = 0 \Rightarrow a_6 = 0$$

So,

$$a_2 = a_4 = a_6 = \cdots = a_{2p} = 0$$

If in particular, we choose $p = 1$, then $a_3 = a_5 = \cdots = 0$.
Then,

$$2f(n) = \frac{1}{n^2} \quad \text{from (*)}$$

But,

$$S^*\left(\frac{n}{2}\right) = S\left(\frac{n}{2}\right) - f(n+1) \text{ and } S^*\left(\frac{n}{2}\right) \approx \frac{\pi - 2}{4}$$

Hence,

$$\frac{\pi - 2}{4} = \sum_{k=1}^{2m+1} \frac{(-1)^{k-1}}{(2k)^2 - 1} - \frac{1}{2(n+1)^2}$$

$$\frac{\pi - 2}{4} = \frac{1}{2^2 - 1} - \frac{1}{4^2 - 1} + \frac{1}{6^2 + 1} + \cdots + \frac{1}{(4m+2)^2 - 1} - \frac{1}{2(n+1)^2}$$

i.e.,

$$\frac{\pi - 2}{4} = \frac{1}{2^2 - 1} - \frac{1}{4^2 - 1} + \frac{1}{6^2 - 1} - \cdots + \frac{1}{n^2 - 1} - \frac{1}{2(n+1)^2} \tag{26}$$

Now,

$$u_n^* = \frac{1}{n^2 - 1} - f(n+1) - f(n-1) \tag{26'}$$

$$= \frac{1}{n^2 - 1} - \frac{1}{2(n+1)^2} - \frac{1}{2(n-1)^2}$$

$$= \frac{2(n^2 - 1) - (n-1)^2 - (n+1)^2}{2(n+1)^2(n-1)^2}$$

$$u_n^* = \frac{2n^2 - 2 - n^2 + 2n - 1 - n^2 - 2n - 1}{2(n+1)^2(n-1)^2} = \frac{-4}{2(n+1)^2(n-1)^2}$$

So,

$$u_n^* = O\left(\frac{1}{n^4}\right)$$

Substituting $n = 2, 4, \ldots, n = (4m + 2)$ in (26′), we get

$$u_2^* = \frac{1}{2^2 - 1} - f(3) - f(1)$$

$$u_4^* = \frac{1}{4^2 - 1} - f(5) - f(3)$$

$$u_6^* = \frac{1}{6^2 - 1} - f(7) - f(5)$$

$$u_{2+2(2m)}^* = u_{4m+2}^* = u_n^* = \frac{1}{n^2 - 1} - f(n+1) - f(n-1)$$

Then, $+u_2^* - u_4^* + u_6^* - u_8^* - \cdots + u_n^*$

$$= + \frac{1}{(2^2 - 1)} - \frac{1}{(4^2 - 1)} + \frac{1}{(6^2 - 1)} - \cdots + \frac{1}{(n^2 - 1)} - f(1) - f(n+1) \tag{27}$$

Let $n \to \infty$ and let $f(n+1) \to 0$ as $n \to \infty$,

$$u_2^* - u_4^* + u_6^* - u_8^* + \cdots = \frac{\pi - 2}{4} - f(1) = \frac{\pi - 2}{4} - \frac{1}{2} = \frac{\pi}{4} - 1 = \frac{\pi - 4}{4}$$

Now, substituting

$$u_n^* = \left(\frac{-2}{(n^2 - 1)^2}\right)$$

we get

$$\frac{-2}{(2^2-1)^2}+\frac{2}{(4^2-1)^2}-\frac{2}{(6^2-1)^2}+\cdots=\frac{\pi-4}{4}$$

Hence,

$$\frac{1}{(2^2-1)^2}-\frac{1}{(4^2-1)^2}+\frac{1}{(6^2-1)^3}-\cdots+\cdots=\left(\frac{\pi-4}{4}\right)\left(-\frac{1}{2}\right)$$

i.e.,

$$\frac{1}{(2^2-1)^2}-\frac{1}{(4^2-1)^2}+\frac{1}{(6^2-1)^3}-\cdots=\frac{4-\pi}{8} \tag{28}$$

Now, we take $p=2$, then we get

$$O\left(\frac{1}{n^6}\right)=\text{right-hand side of (25)}$$

As before equating the coefficient of n^{-k} for $k=2,\ 3,\ldots,$ we get $1-a_1=0\Rightarrow a_1=1.$

$$a_2=0,\quad 1-a_3-3a_1=0\Rightarrow a_3=1-3a_1=1-3=-2$$
$$+a_4+6a_2=0\Rightarrow a_4=0$$

So, we take

$$2f(n)=\frac{1}{n^2}-\frac{2}{n^4}+\frac{4}{n^6}-\frac{8}{n^8}+\cdots$$
$$=\frac{1}{n^2}\left[1-\frac{2}{n^2}+\frac{4}{n^4}-\frac{8}{n^6}+\cdots\right]$$

Now, series

$$1-\frac{2}{n^2}+\frac{4}{n^4}-\frac{8}{n^6}+\cdots$$

is an infinite geometric series with first term $a=1$ and common ratio $r=(-2/n^2)$. Its sum

$$=\frac{a}{1-r}=\frac{1}{1+(2/n^2)}=\frac{n^2}{n^2+2}$$

So,

$$2f(n)=\frac{1}{n^2}\left(\frac{n^2}{n^2+2}\right)=\frac{1}{n^2+2}$$
$$f(n)=\frac{1}{2(n^2+2)}$$

Then,

$$S_n^*\left(\frac{n}{2}\right) = S\left(\frac{n}{2}\right) - f(n+1)$$

$$\frac{\pi-2}{4} = \sum_{k=1}^{2m+1} \frac{(-1)^{k-1}}{(2k)^2-1} - \frac{1}{2[(n+1)^2+2]}$$

where

$$n = 4m + 2$$

So,

$$\frac{\pi-2}{4} = \frac{1}{2^2-1} - \frac{1}{4^2-1} + \frac{1}{6^2-1} - \dots + \frac{1}{n^2-1} - \frac{1}{2[(n+1)^2+2]}$$

Now,

$$u_n^* = \frac{1}{n^2-1} - f(n+1) - f(n-1)$$

So,

$$u_n^* = \frac{1}{n^2-1} - \frac{1}{2[(n+1)^2+2]} - \frac{1}{2[(n-1)^2+2]}$$

$$u_n^* = \frac{2[n^2+3+2n][n^2+3-2n] - (n^2-1)[(n^2-2n+3)] - (n^2-1)[n^2+3+2n]}{(n^2-1)2[(n+1)^2+2][(n-1)^2+2]}$$

$$u_n^* = \frac{2[(n^2+3)^2-4n^2] - (n^2-1)[n^2-2n+3+n^2+3+2n]}{2(n^2-1)[(n+1)^2+2][(n-1)^2+2]}$$

So,

$$u_n^* = \frac{2[n^4+6n^2+9-4n^2] - (n^2-1)[2n^2+6]}{2(n^2-1)[(n+1)^2+2][(n-1)^2+2]}$$

$$u_n^* = \frac{2n^4+4n^2+18-2n^4+2n^2-6n^2+6}{2(n^2-1)[(n+1)^2+2][(n-1)^2+2]}$$

$$u_n^* = \frac{24}{2(n^2-1)[(n+1)^2+2][(n-1)^2+2]}$$

$$u_n^* = \frac{12}{(n^2-1)[(n+1)^2+2][(n-1)^2+2]}$$

So,

$$u_n^* = O\left(\frac{1}{n^6}\right)$$

Now, from (27),

$$u_2^* - u_4^* + u_6^* - u_8^* + \cdots = \frac{1}{2^2-1} - \frac{1}{4^2-1} + \frac{1}{6^2-1} - \cdots + \cdots + \frac{1}{n^2-1} - f(1) - f(n+1)$$

Let $n \to \infty$ and let $f(n + 1) \to 0$ as $n \to \infty$, then we get

$$u_2^* - u_4^* + u_6^* - u_8^* + \cdots = \frac{\pi-2}{4} - f(1) = \frac{\pi-2}{4} - \frac{1}{6} = \frac{3\pi-6-2}{12}$$

$$u_2^* - u_4^* + u_6^* - u_8^* + \cdots = \frac{3\pi-8}{12}$$

Substituting for

$$u_n^* = \frac{12}{(n^2-1)[(n+1)^2+2][(n-1)^2+2]}$$

we get

$$12\left[\frac{1}{(2^2-1)(3^2+2)(1^2+2)} - \frac{1}{(4^2-1)(5^2+2)(3^2+2)} + \frac{1}{(6^2-1)(7^2+2)(5^2+2)} - \cdots\right] = \frac{3\pi-8}{12}$$

So,

$$\left[\frac{1}{(2^2-1)(3^2+2)(1^2+2)} - \frac{1}{(4^2-1)(5^2+2)(3^2+2)} + \frac{1}{(6^2-1)(7^2+2)(5^2+2)} - \cdots\right] = \frac{3\pi-8}{144} \quad (29)$$

The following Nilakantha Somayaji's proof of Aryabhata I's theorem that the chord of one-sixth of the circumference of a circle is equal to half its diameter is given in "Yuktibhasa" of Jyesthedeva.

Let AB be a diameter d of a circle with center O. Let C and D be points on the circumference of the circle such that

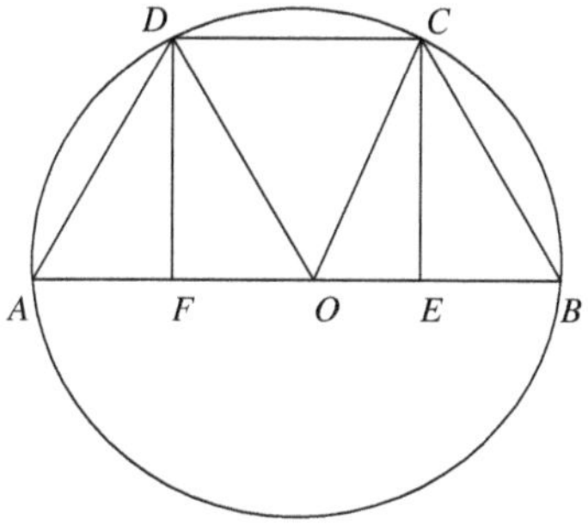

Chord BC = chord $AD = OB = OA = d/2$.

Join OC, DC, and OD. Let CE and DF be the perpendicular drawn from C and D respectively on AB.

Proof: Now, triangles OCB and AOD are both equilateral triangles with side

$$= OB = OA = \frac{d}{2}$$

The perpendicular CE and DF bisects OB and AO, respectively.

Then,

$$OF = \frac{1}{2}OA \text{ and } OE = \frac{1}{2}OB$$

Now,

$$OF + OE = \frac{1}{2}OA + \frac{1}{2}OB = \frac{d}{4} + \frac{d}{4} = \frac{d}{2}$$

i.e.,

$$FE = DC = \frac{d}{2}$$

Hence, the semi circumference $ADCB$ consists of three parts, namely arc $A\widehat{D}$, arc $D\widehat{C}$, and arc $C\widehat{B}$, whose chords AD, DC, and CB are all equal to half the diameter. Then, the chord of one-sixth of the circumference is equal to half the diameter.

In "Yuktibhasa," Jyesthedeva has given the following proof of Brahmagupta's theorem that the circum radius of a circle is equal to the product of the two sides divided by twice the altitude.

Let ABC be a triangle inscribed in its circum circle with center O and radius R^*. Let AD be the altitude from A on BC. Let P be the midpoint of arc AC. Join PO and produce PO to meet the circum circle at Q. Through Q, draw a line parallel to BC meeting the circum circle at R. Join PR.

Then,

$$R^* = \frac{AB \cdot AC}{2AD}$$

Proof: Since QR is drawn parallel to BC,

$$\text{arc } B\widehat{Q} = \text{arc } R\widehat{C} \tag{1}$$

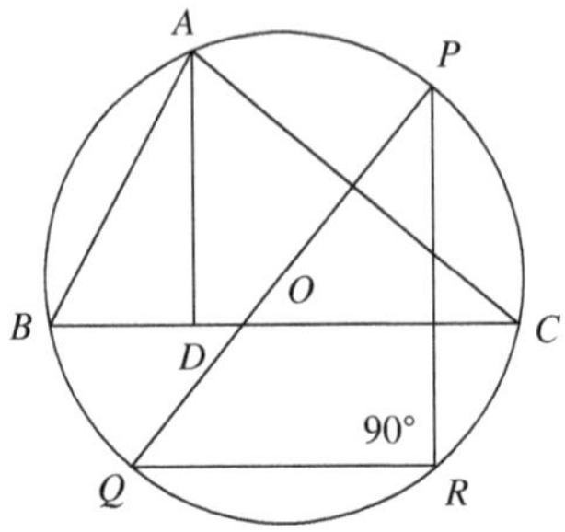

Also, since PQ is the diameter of the circum circle and P is the midpoint of the arc AC, Q is the midpoint of the arc $A\widehat{B}QRC$.

Hence, arc $A\widehat{B}Q$ = arc $Q\widehat{R}C$.

i.e.,

$$\text{arc } A\widehat{B} + \text{arc } Q\widehat{B} = \text{arc } Q\widehat{R} + \text{arc } R\widehat{C} \tag{2}$$

Then, from (1),

$$\text{arc } A\widehat{B} = \text{arc } Q\widehat{R} \Rightarrow \text{chord } AB = \text{chord } QR \tag{3}$$

Then, $A\hat{C}B = Q\hat{P}R$.

Now, in triangles ADC and PQR,

$$ADC = 90^\circ = Q\hat{R}P \text{ (angle in a semicircle)}$$

$$A\hat{C}D = A\hat{C}B = Q\hat{P}R$$

Hence, the triangles ADC and PQR are similar.

So,

$$\frac{PQ}{AC} = \frac{QR}{AD}$$

Thus,

$$PQ = \frac{AC \cdot QR}{AD} = \frac{AB \cdot AC}{AD}$$

But, $PQ = 2R^* =$ diameter of the circum circle.

Hence,

$$2R^* = \frac{AB \cdot AC}{AD} \quad R^* = \frac{AB \cdot AC}{2AD}$$

From this result, we can deduce the following corollary.

Corollary: In triangle *ABC*,

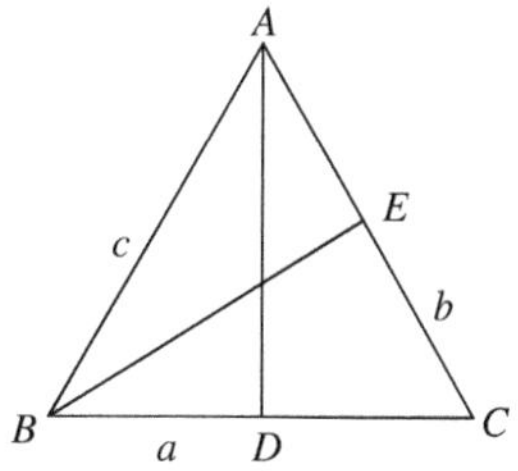

$$\frac{a}{\sin A} = \frac{b}{\sin B} = \frac{c}{\sin C} = 2R^*$$

Proof: In Δ^{le} ABD, $AD = AB \sin B = c \sin B$

In Δ^{le} ADC, $AD = AC \sin C = b \sin C$.

Hence, $c \sin B = b \sin C$.

$$\Rightarrow \frac{b}{\sin B} = \frac{c}{\sin C} \qquad (1)$$

If *BE* is the perpendicular drawn from *B* on *AC*, then from the triangle *ABE*, we have $BE = AB \sin A = c \sin A$.

Again, from the triangle *BEC*, we have

$$BE = BC \sin C = a \sin C$$

So, $c \sin A = a \sin C$.

$$\Rightarrow \frac{a}{\sin A} = \frac{c}{\sin C} \qquad (2)$$

Then, from (1) and (2), we have

$$\frac{a}{\sin A} = \frac{b}{\sin B} = \frac{c}{\sin C}$$

We have

$$R^* = \frac{AB \cdot AC}{2AD}$$

So,

$$2R^* = \frac{AB \cdot AC}{AD} = \frac{AB \cdot AC}{AC \sin C} = \frac{AB}{\sin C} = \frac{c}{\sin C}$$

Hence,

$$\frac{a}{\sin A} = \frac{b}{\sin B} = \frac{c}{\sin C} = 2R^*$$

The following proof of Brahmagupta's theorem, namely if a, b, c, d are the four sides of a cyclic quadrilateral, and if $2s = a + b + c + d$, then the area A of the cyclic quadrilateral $ABCD$ is given by $A = \sqrt{(s-a)(s-b)(s-c)(s-d)}$ is given in "Yuktibhasa," pages 247–257.

Let AE and CF be altitudes from A and C respectively on the diagonal BD of the cyclic quadrilateral $ABCD$. Let $AE = p_1$ and $CF = p_2$. Draw AG perpendicular to CF produced.

Let $BC = a$, $CD = b$, $DA = c$, and $AB = d$.

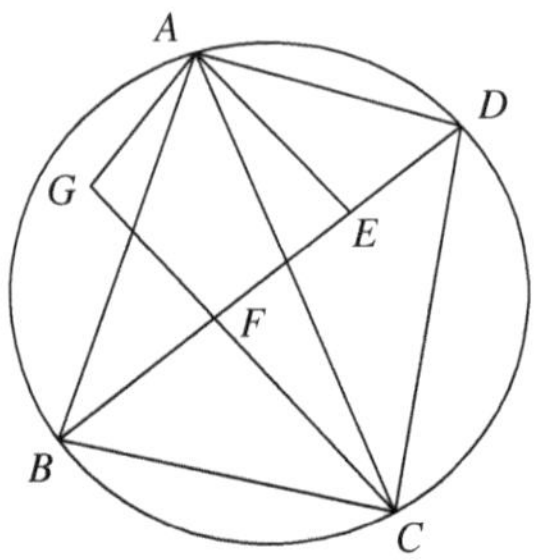

Area of Δ^{le} ABD

$$= \frac{1}{2} BD \cdot AE$$

Area of Δ^{le} BCD

$$= \frac{1}{2} BD \cdot CF$$

Area of cyclic quadrilateral $ABCD$

$$= A = \text{area of } \Delta^{le}\ ABD + \text{area of } \Delta^{le}\ BCA$$

So,

$$A = \frac{1}{2} BD(AE + CF) = \frac{1}{2} BD(p_1 + p_2) \tag{1}$$

In the right Δ^{le} ABE, $AB^2 = BE^2 + AE^2 = BE^2 + p_1^2$.
Again, in the right Δ^{le} ADE, $AD^2 = DE^2 + AE^2 = DE^2 + p_1^2$.
So, $AB^2 - AD^2 = BE^2 - DE^2 = (BE + DE)(BE - DE)$. i.e.,

$$AB^2 - AD^2 = BD \cdot (BE - DE)$$

i.e.,

$$d^2 - c^2 = BD(BE - DE) \tag{2}$$

Again, in the right Δ^{le} BFC, $BC^2 = BF^2 + FC^2$.
And in the right Δ^{le} CFD, $CD^2 = FD^2 + FC^2$.

So, $CD^2 - BC^2 = FD^2 - BF^2 = (FD + BF)(FD - BF)$.
i.e.,

$$b^2 - a^2 = BD(FD - BF) \tag{3}$$

Adding (2) and (3), we get

$$d^2 - c^2 + b^2 - a^2 = BD[BE - DE + FD - BF]$$

$$= BD[(FD - DE) + (BE - BF)]$$

$$= BD[FE + FE] = 2BD \cdot EF$$

So,

$$FE = \frac{d^2 - c^2 + b^2 - a^2}{2BD} \tag{4}$$

In the right triangle AGC,

$$AC^2 = CG^2 + AG^2 = (CF + FG)^2 + FE^2$$

$$AC^2 = (CF + AE)^2 + FE^2 = (p_1 + p_2)^2 + FE^2 \tag{5}$$

So, from (1), (3), and (5), we get

$$A^2 = \frac{1}{4}BD^2(p_1 + p_2)^2 = \frac{1}{4}BD^2\left[AC^2 - FE^2\right]$$

$$= \frac{1}{4}BD^2\left[AC^2 - \left(\frac{d^2 - c^2 + b^2 - a^2}{2BD}\right)^2\right]$$

$$A^2 = \frac{1}{4}\left[BD^2 \cdot AC^2 - \left(\frac{d^2 - c^2 + b^2 - a^2}{2}\right)^2\right]$$

$$= \frac{1}{16}\left[4BD^2 \cdot AC^2 - (d^2 - c^2 + b^2 - a^2)^2\right]$$

But from Ptolemy's theorem, $AC \cdot BD = [AB \cdot CD + BC \cdot AD]$.
i.e., $AC \cdot BD = [ac + bd]$.
So,

$$A^2 = \frac{1}{16}\left[4(ac + bd)^2 - (d^2 - c^2 + b^2 - a^2)^2\right]$$

i.e.,

$$A^2 = \frac{1}{16}\left[\{2(ac+bd)+(d^2-c^2+b^2-a^2)\}\right.$$
$$\left.\{2(ac+bd)-(d^2-c^2+b^2-a^2)\}\right]$$

$$A^2 = \frac{1}{16}\{(b+d)^2-(a-c)^2\}\{(a+c)^2-(b-d)^2\}$$

$$A^2 = \frac{1}{16}(b+c+d-a)(a+b+d-c)(a+b+c-d)(a+c+d-b)$$

$$= \frac{1}{16}(2s-2a)(2s-2c)(2s-2d)(2s-2b)$$

$$A^2 = (s-a)(s-b)(s-c)(s-d)$$

i.e.,

$$A = \sqrt{(s-a)(s-b)(s-c)(s-d)}$$

In Yuktibhasa, pages 228–237, we find the following proof of Brahmagupta's formulae for the diagonals of a cyclic quadrilateral $ABCD$, where

$$BD^2 = \frac{(AD\cdot BC+AB\cdot CD)(AD\cdot CD+BC\cdot AB)}{AD\cdot AB+CD\cdot BC}$$

$$AC^2 = \frac{BC\cdot CD+AB\cdot AD}{AB\cdot BC+CD\cdot AB}(AB\cdot CD+BC\cdot AD)$$

Let $ABCD$ be a cyclic quadrilateral such that $AB > BC > AD > CD$.
Let M be a point on the arc $A\widehat{D}$ such that arc $A\widehat{M}$ = arc $C\widehat{D}$.

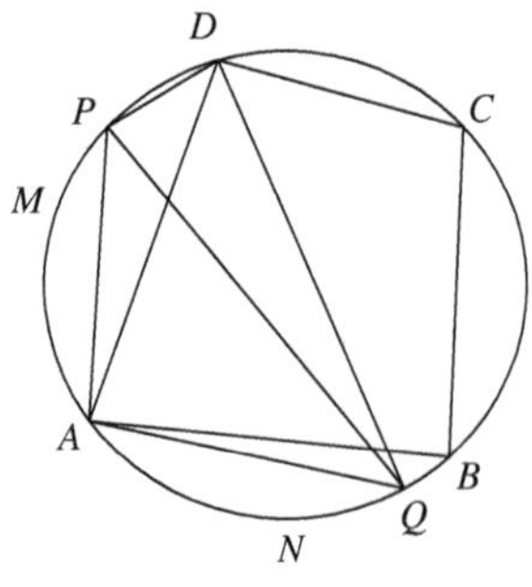

Similarly let N be a point on the arc $A\widehat{B}$ such that arc $A\widehat{N}$ = arc $B\widehat{C}$.

Let P and Q be the midpoints of arcs $D\widehat{M}$ and $B\widehat{N}$. Join PQ.
So, arc $P\widehat{M}$ = arc $P\widehat{D}$ and arc $N\widehat{Q}$ = arc $Q\widehat{B}$.

Now,

$$\begin{aligned}\text{arc } P\widehat{A}Q &= \text{arc } P\widehat{M} + \text{arc } M\widehat{A} + \text{arc } A\widehat{N} + \text{arc } N\widehat{Q} \\ &= \text{arc } P\widehat{D} + \text{arc } C\widehat{D} + \text{arc } B\widehat{C} + \text{arc } B\widehat{Q} \\ &= \text{arc } P\widehat{C}Q\end{aligned}$$

Hence, PQ is a diameter.
Now, $\text{arc } Q\widehat{A} = \text{arc } Q\widehat{N} + \text{arc } A\widehat{N} = \text{arc } B\widehat{Q} + \text{arc } B\widehat{C}$.
So,

$$\begin{aligned}\text{arc } Q\widehat{A} &= \text{arc } Q\widehat{C} \\ \text{arc } P\widehat{A} &= \text{arc } A\widehat{M} + \text{arc } M\widehat{P} \\ &= \text{arc } C\widehat{D} + \text{arc } P\widehat{D} = \text{arc } P\widehat{C}\end{aligned}$$

$$\text{arc } A\widehat{D} + \text{arc } C\widehat{D} = \text{arc } A\widehat{P} + \text{arc } P\widehat{D} + \text{arc } P\widehat{C} - \text{arc } P\widehat{D}$$

$$= \text{arc } A\widehat{P} + \text{arc } A\widehat{P} = 2 \text{ arc } A\widehat{P}$$

$$\text{arc } A\widehat{P} = \frac{1}{2}\left[\text{arc } A\widehat{D} + \text{arc } C\widehat{D}\right]$$

$$\text{chord } AP = \frac{1}{2}[\text{chord } AD + \text{chord } CD]$$

$$\begin{aligned}\text{arc } A\widehat{D} - \text{arc } C\widehat{D} &= \text{arc } A\widehat{P} + \text{arc } P\widehat{D} - [\text{arc } P\widehat{C} - \text{arc } P\widehat{D}] \\ &= \text{arc } A\widehat{P} - \text{arc } A\widehat{P} + 2 \text{ arc } P\widehat{D} = 2 \text{ arc } P\widehat{D}\end{aligned}$$

So,

$$\text{arc } P\widehat{D} = \frac{1}{2}\left[\text{arc } A\widehat{D} - \text{arc } C\widehat{D}\right]$$

Now, $\text{arc } A\widehat{B} + \text{arc } B\widehat{C} = \text{arc } A\widehat{Q} + \text{arc } Q\widehat{B} + \text{arc } B\widehat{C}$

$$\begin{aligned}&= \text{arc } A\widehat{Q} + \text{arc } \widehat{Q}C \\ &= \text{arc } A\widehat{Q} + \text{arc } A\widehat{Q} = 2 \text{ arc } A\widehat{Q}\end{aligned}$$

$$\text{arc } A\widehat{Q} = \frac{1}{2}\left[\text{arc } A\widehat{B} + \text{arc } B\widehat{C}\right]$$

Hence,

$$\text{chord } A\widehat{Q} = \frac{1}{2}[\text{chord } AB + \text{chord } BC]$$

Again, $\text{arc } A\widehat{B} - \text{arc } B\widehat{C} = \text{arc } A\widehat{Q} + \text{arc } Q\widehat{B} - \text{arc } B\widehat{C}$

$$\begin{aligned}&= \text{arc } A\widehat{N} + \text{arc } N\widehat{Q} + \text{arc } Q\widehat{B} - \text{arc } A\widehat{N} \\ &= \text{arc } N\widehat{Q} + \text{arc } Q\widehat{B} \\ &= \text{arc } Q\widehat{B} + \text{arc } Q\widehat{B} = 2 \text{ arc } Q\widehat{B}\end{aligned}$$

So,

$$\text{arc } Q\widehat{B} = \frac{1}{2}\left[\text{arc } A\widehat{B} - \text{arc } B\widehat{C}\right]$$

Hence,

$$\text{chord } QB = \frac{1}{2}[\text{chord } AB - \text{chord } BC]$$

Since

$$pq = \left(\frac{p+q}{2}\right)^2 - \left(\frac{p-q}{2}\right)^2$$

$$\text{chord } AD \cdot \text{chord } CD = \left(\text{chord } \frac{AD+CD}{2}\right)^2 - \left(\text{chord } \frac{AD-CD}{2}\right)^2$$

$$\text{chord } AD \cdot \text{chord } CD = (\text{chord } AP)^2 - (\text{chord } DP)^2$$

So, $AD \cdot CD = AP^2 - DP^2$.
Similarly,

$$AB \cdot BC = \left(\text{chord } \frac{AB+BC}{2}\right)^2 - \left(\text{chord } \frac{AB-BC}{2}\right)^2$$
$$= AQ^2 - QB^2$$

Hence, $AD \cdot CD + AB \cdot BC = AP^2 - DP^2 + AQ^2 - QB^2$.
In the right triangle PAQ (since $P\frown AQ = 90°$, angle in the semicircles)

$$PQ^2 = AP^2 + AQ^2$$

So, $AD \cdot CD + AB \cdot BC = PQ^2 - DP^2 - QB^2$.
Again in the right triangle PDQ,

$$PQ^2 = DP^2 + DQ^2$$

Hence, $AD \cdot CD + AB \cdot BC = DQ^2 - QB^2$

$$= (DQ + QB)(DQ - QB)$$
$$= (DQ + QB)(DQ - QN)$$

i.e., $AD \cdot CD + AB \cdot BC = DB \cdot DN$.

Thus, the sum of the products of the sides about the first diagonal

$BD =$ the first diagonal $BD \cdot$ the third diagonal (1)

Now, consider the cyclic quadrilateral *ANCD*,

$$AN \cdot AD + CD \cdot NC = AC \cdot BD$$

{paracont}from (1), since *BD* is the third diagonal with respect to cyclic quadrilateral *ANCD*.

But $AN = BC$ and $NC = AB$ (by construction), we get

$$BC \cdot AD + CD \cdot AB = AC \cdot BD$$

i.e., the sum of the products of the opposite sides is equal to the product of the diagonals. This is Ptolemy's theorem.

We have already proved that

$$BD \cdot DN = AD \cdot CD + AB \cdot BC \tag{2}$$

$$AC \cdot BD = BC \cdot AD + AB \cdot CD \tag{3}$$

Multiplying (2) and (3), we get

$$BD^2 \cdot DN \cdot AC = (AD \cdot CD + AB \cdot BC)(BC \cdot AB + AB \cdot CD)$$

So,

$$BD^2 = \frac{(AD \cdot CD + AB \cdot BC)(BC \cdot AD + AB \cdot CD)}{AC \cdot DN}$$

But in the cyclic quadrilateral *ANCD*,

$$AC \cdot DN = AN \cdot CD + AD \cdot NC$$

$$= BC \cdot CD + AD \cdot AB$$

So,

$$BD^2 = \frac{(AD \cdot CD + AB \cdot BC)(BC \cdot AD + AB \cdot CD)}{BC \cdot CD + AD \cdot AB}$$

Similarly, we can prove the formulae for the diagonal AC^2. This proves Brahmagupta's formula for the two diagonals in a cyclic quadrilateral.

Corollary: As a corollary, we will prove Narayana Pandita's formula for the area of a cyclic quadrilateral.

Let $ABCD$ be a cyclic quadrilateral. Let N be a point on arc AB such that arc $A\widehat{N}$ = arc $B\widehat{C}$. Join AN, NC, AC, and BN.

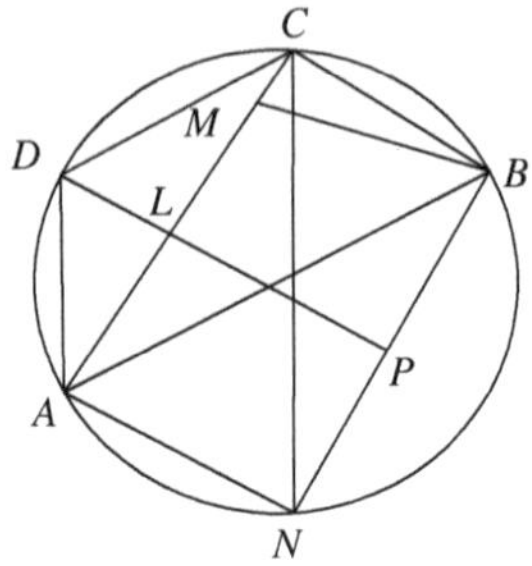

$A\widehat{C}N = B\widehat{A}C$ (angles in equal arcs).

Also $B\widehat{A}C = B\widehat{N}C$ (angles on the same arc).

So, $B\widehat{N}C = A\widehat{C}N$.

Hence, BN is parallel to the diagonal AC.

Since BN is parallel to AC, the altitude DP from D on BN will be equal to the sum of the altitudes DL and BM from D and B respectively on AC, i.e., $DP = DL + BM$.

Now, area of Δ^{le} ADC

$$= \frac{1}{2}AC \cdot DL$$

Area of Δ^{le} ABC

$$= \frac{1}{2}AC \cdot BM$$

Now, area of the cyclic quadrilateral $ABCD$

$$= \text{area of } \Delta^{le} ADC + \text{area of } \Delta^{le} ABC$$

$$= \frac{1}{2}AC(DL + BM) = \frac{1}{2}AC \cdot DP$$

But in Δ^{le} BND,

Altitude DP

$$= \frac{ND \cdot BD}{2R}$$

where R is the circumradius.

Hence, area of the cyclic quadrilateral $ABCD$

$$= \frac{AC \cdot ND \cdot BD}{4R}$$

We observe that ND is the third diagonal for the cyclic quadrilateral $ABCD$.

In "Yuktibhasa," page 214, the following method is given to find the sine chord of the sum and difference of two sine chords.

Let three points A, B, and C lie on a circle.

Let AC and BC be the sine chords of the arcs $A\widehat{C}$ and $B\widehat{C}$, respectively.

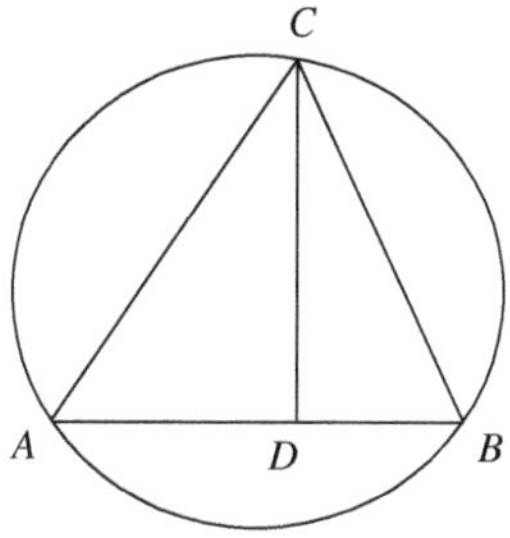

Draw CD perpendicular to AB. Then, the sine chord of the sum of the sine chords AC and BC is $\sqrt{AC^2 - CD^2} + \sqrt{BC^2 - CD^2} = AD + BD = AB$ and the sine chord of the difference of the sine chords AC and BC is $\sqrt{AC^2 - CD^2} \sim \sqrt{BC^2 - CD^2} = AD \sim BD$.

Using the above result, "Yuktibhasa," pages 237–239, proved geometrically the trigonometrical identity.

$$\sin(A \pm B) = \sin A \cos B \pm \cos A \sin B$$

which is given in Nilakantha's "Tantra-Sangraha." This result is attributed to Madhava of Sangamagrama.

जीवे परस्परनिजेतरमौविकाभ्यस्य विस्तृतिदलेन विभज्यमाने ।
यन्नन्योन्ययोगविरहानुगुणे भवेतां।...॥

(T.S.II. 16½)

i.e., "If the sine chords of two arcs reciprocally multiplied by other chord cosine chord and divided by the radius are added or are subtracted from each other will give the sine chords of the sum or the difference of the two arc respectively."

Let MON be a quadrant of a circle of radius R. Let P be a point on the circle such that arc $M\widehat{P} = A$. Let Q be a point on the circle such that arc $PQ = 2B$. Let W be midpoint of arc PQ.

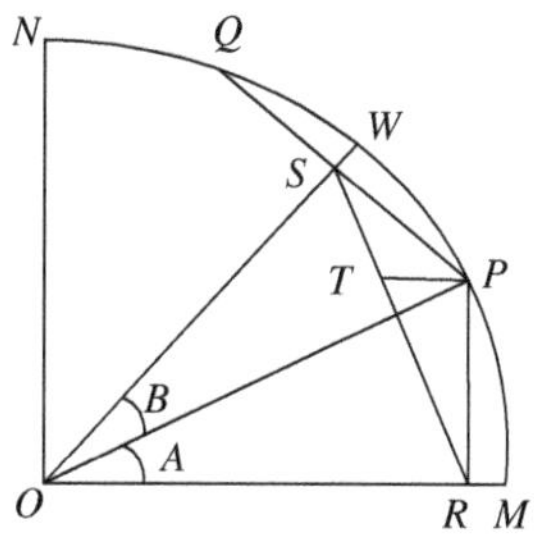

Let OW intersect the line PQ at S. Draw PR perpendicular to OM. Join SR. Draw PT perpendicular to RS.

Now, PR and PS are the sine chords ("bhuja jyas") of the arcs MP and PW, respectively, i.e., $PR = R \sin A$ and $PS = R \sin B$. Now, OR and OS arc cosine chords ("koti jyas") of the arcs MP and PW, respectively, i.e., $OR = R \cos A$ and $OS = R \cos B$. Sine chord of the combined arc MW (sum of arcs MP and PW) of $A + B$ is $R \sin(A + B)$.

Then,

$$R \sin(A+B) = \sqrt{PR^2 - PT^2} + \sqrt{PS^2 - PT^2} \quad (*)$$

and

$$R \sin(A-B) = \sqrt{PR^2 - PT^2} - \sqrt{PS^2 - PT^2} \quad (**)$$

if

$$\sqrt{PR^2 - PT^2} > \sqrt{PS^2 - PT^2}$$

Since W is the midpoint of arc PQ, $SP = SQ$. Since OS bisects PQ, OS must be perpendicular to PQ, So, $O\hat{S}P = 90°$ and $P\hat{R}O = 90°$ (by construction).

Hence, $ORPS$ is a cyclic quadrilateral (since the sum of opposite angles is 180°).

So, $\hat{B} = S\hat{O}P = S\hat{R}P$ (angles in the same arc).

Now, in Δ^{les} PRT and PSO,

$$P\hat{S}O = P\hat{T}R = 90^\circ$$

$$S\hat{O}P = P\hat{R}S = P\hat{R}T = \hat{B}$$

Hence, the Δ^{les} PRT and PSO are similar.

So,

$$\frac{PT}{PS} = \frac{PR}{OP}$$

Hence,

$$PT = \frac{PR \cdot PS}{OP}$$

So,

$$R \sin(A+B) = \left(\sqrt{PR^2 - \frac{PR^2 PS^2}{OP^2}} + \sqrt{PS^2 - \frac{PR^2 PS^2}{OP^2}}\right) \quad \text{from } (*)$$

$$= PR\sqrt{1 - \frac{PS^2}{OP^2}} + PS\sqrt{1 - \frac{PR^2}{OP^2}}$$

$$R\sin(A+B) = R\sin A\sqrt{1-\frac{R^2\sin^2 B}{R^2}} + R\sin B\sqrt{1-\frac{R^2\sin^2 A}{R^2}}$$

$$R\sin(A+B) = R\sin A\sqrt{1-\sin^2 B} + R\sin B\sqrt{1-\sin^2 A}$$

i.e., $R\sin(A+B) = R\sin A\cos B + R\sin B\cos A$.

Thus, $\sin(A+B) = \sin A\cos B + \sin B\cos A$.

Now,

$$\sin(A-B) = \sqrt{PR^2 - \frac{PR^2PS^2}{OP^2}} - \sqrt{PS^2 - \frac{PR^2PS^2}{OP^2}} \quad \text{from(**)}$$

$$\sin(A-B) = \sin A\cos B - \sin B\cos A$$

In, "Yuktibhasa," pages 224–227, the following is the proof geometrically of trigonometric identities:

i. $\sin^2 A - \sin^2 B = \sin(A+B)\cdot\sin(A-B)$.

ii. $\sin^2\frac{A+B}{2} - \sin^2\frac{A-B}{2} = \sin A\sin B$.

Let $ABCD$ be a cyclic quadrilateral in which $AB > CD$. Take a point E on the arc AD such that arc AE = arc CD. Join AE and DE. Draw perpendiculars EM and DN from E and D respectively on the diagonal AC.

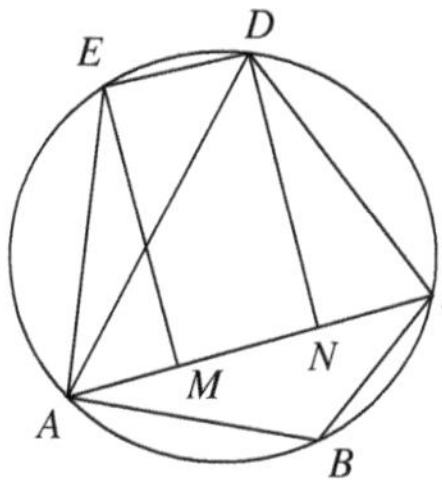

Now, arc DE = arc AD − arc AE = arc AD − arc CD.

Since arc AE = arc CD, we have $EM = DN$.

So, $ED = MN$.

Now, AN = projection of AD on AC and AM is the projection of AE on AC.

So, $MN = AN - AM$

= projection of AD on AC − projection of AE on AC

MN = projection of AD on AC − projection of CD on AC (since $AE = CD$).

So, in Δ^{le} ADC,

MN = difference of projections of AD and CD on the base AC

So, MN = chord of arc $(AD - DC)$.

Similarly, $AC = AN + NC$ = sum of the projections of AD and CD on AC

= chord of arc$(AD + CD)$

Here,

$$AD^2 - DC^2 = (AN^2 + ND^2) - (ND^2 + NC^2)$$

$$= AN^2 - NC^2 = (AN + NC)(AN - NC)$$

$$AD^2 - DC^2 = \text{chord of } (AN + NC) \text{ chord of } (AN - NC) \quad (^*)$$

This is equivalent to

$$\sin^2 A - \sin^2 B = \sin(A + B)\sin(A - B) \quad (^{**})$$

according to the editors of "Yuktibhasa."
Now,

$$\begin{aligned} \text{arc } AC + \text{arc } DE &= \text{arc } AD + \text{arc } CD + \text{arc } AD - \text{arc } AE \\ &= \text{arc } AD + \text{arc } CD + \text{arc } AD - \text{arc } CD \end{aligned}$$

So, arc AC + arc $DE = 2$ arc AD.
Similarly,

$$\text{arc } AC - \text{arc } DE = \text{arc } AD + \text{arc } CD - (\text{arc } AD - CD)$$
$$\text{arc } AC - \text{arc } DE = 2 \text{ arc } CD$$

Let arc $AC = A$ and let $CD = B$.
Then, 2 arc $AD = A + B$ and 2 arc $CD = A-B$.
So,

$$\text{arc } AD = \frac{A + B}{2} \quad \text{and} \quad \text{arc } CD = \frac{A - B}{2}$$

Substituting these two in (*), we get

$$\sin^2 \frac{A + B}{2} - \sin^2 \frac{A - B}{2} = \sin A \sin B$$

In "Yuktibhasa," pages 282–290, the following method has been presented to find the volume of a sphere.

Method: The sphere has to be cut into circular laminae by planes passing through circles drawn parallel to the horizontal great circle of the sphere. These circular laminae should have uniform thickness equal to one unit.

Now, the volume of any circular lamina

$$= \text{area of the circle through its middle} \cdot \text{thickness}$$

$$= \pi r^2 \cdot 1$$

if r is the radius of the circle and is equal to the corresponding sine chord.

Hence, the volume of the sphere V is given by

$$V = \text{the sum of volume of the circular laminae}$$
$$= \pi(r_1^2 + r_2^2 + \cdots + r_n^2)$$

where $r_1, r_2, \ldots, r_n$ are the corresponding sine chords.

Proof: Let h be height of the double arc PQ whose sine chord is r. If d is the diameter of the circle, then

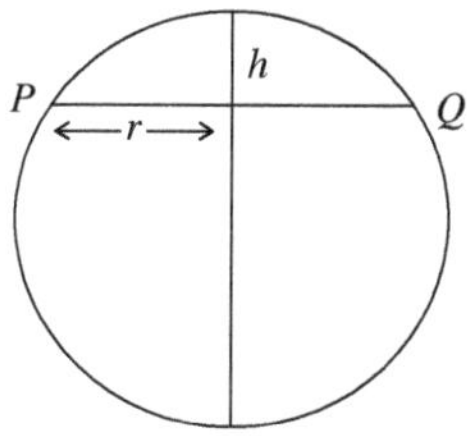

$$r^2 = h(d-h) = \frac{1}{2}\left[(h+[d-h])^2 - (h^2 + (d-h)^2)\right]$$

$$r^2 = \frac{1}{2}\left[d^2 - (h^2 + (d-h)^2)\right]$$

So,

$$r_1^2 + r_2^2 + \cdots + r_n^2$$
$$= \frac{1}{2}\left[d^2 - \left\{h_1^2 + (d-n_1)^2\right\}\right] + \frac{1}{2}\left[d^2 - \left\{h_2^2 + (d-h_2)^2\right\}\right]$$
$$+ \cdots + \frac{1}{2}\left[d^2 - \left\{h_n^2 + (d-h_n)^2\right\}\right]$$

Thus,

$$\sum_{k=1}^{n} r_K^2 = \frac{1}{2}nd^2 - \frac{1}{2}\sum_{k=1}^{n} h_K^2 - \frac{1}{2}\sum_{k=1}^{n}(d-h_K)^2$$

Let the sphere be divided into infinitely large of these circular laminae. Then, the uniform thickness of each lamina is very small.

Now, $d - h_1 = h_{n-1}, d - h_2 = h_{n-2}, \ldots, d - h_{n-1} = h_1$.

i.e.,

$$\sum_{k=1}^{n}(d-h_k)^2 = \sum_{k=1}^{n} h_k$$

Also, $h_1 = \Delta d$, $h_2 = 2\Delta d$, $h_3 = 3\Delta d, \ldots, h_n = n\Delta d$.

$$\sum_{k=1}^{n} h_k^2 = (\Delta d)^2 + (2\Delta d)^2 + \cdots + (n\Delta d)^2$$

$$= \frac{d^3}{3}, \quad \text{if } n = d$$

$$\sum_{k=1}^{n} r_k^2 = \frac{n}{2}d^2 - \frac{1}{2}\frac{d^3}{3} - \frac{1}{2}\frac{d^3}{3}$$

Hence,

$$\sum_{k=1}^{n} r_k^2 = \frac{n}{2}d^2 - \frac{d^3}{3} = \frac{d^3}{2} - \frac{d^3}{3} = \frac{d^3}{6}, \quad \text{if } n = d$$

Hence, the volume V of sphere

$$= \pi \sum_{K=1}^{n} r_K^2$$

$$V = \frac{\pi \cdot d^3}{6} = \frac{4}{3}\pi r^3$$

if r is the radius of the sphere.

The Sine and Cosine Series of Madhava

"Yuktibhasa" has provided the derivation of series attributed to Saṅgamagrāma Madhava for the sine chord ("bhuja jya"). The author quoted the following verse taken from "Tantra-Sangraha":

निहत्य चापवर्गेण चापं तत्तत् फलानि च ।
हरेत् समूलयुगवर्गेस्त्रिज्यावर्गाहतैः क्रमात् ।
चाँप फलानिं चाधोऽधौ न्यस्योपर्युपरि त्यजेत् ।
जीवाप्त्यै.........

(Y.B. 190)

i.e., "The arc is the first term. Obtain a series using the first term to find the succeeding terms by making the square of the arc as multiplier and the product of the square of the radius and the sum of the square of even numbers and their roots as divisor. Subtract the sum of even terms from the sum of odd terms to obtain the sine chord ('bhuja jya')."

Let a be the arc (in radius) and r the radius. Then, a is the first term.
Second term

$$= a \cdot \frac{a^2}{(2^2+2)r^2} = \frac{a^3}{6r^2} = \frac{a^3}{3!r^2}$$

Third term

$$= \frac{a^3}{3!r^2} \cdot \frac{1}{4^2+4} \cdot \frac{a^2}{r^2} = \frac{1}{3!4 \cdot 5} \frac{a^5}{r^4} = \frac{a^5}{5!r^4}$$

Fourth term

$$= \frac{a^5}{5!r^4} \frac{1}{6^2+6} \cdot \frac{a^2}{r^2} = \frac{1}{5!6 \cdot 7} \frac{a^7}{r^6} = \frac{a^7}{7!r^6}$$

and so on.
Hence, the sine chord ("bhuja jya")

$$= a - \frac{a^3}{3!r^2} + \frac{a^5}{5!r^4} - \frac{a^7}{7!r^6} + \cdots$$

The series for the versed sine $a(r\text{-}jya\ (90 - a))$ is given by the following sloka (verse), taken from "Tantra-Sangraha."

निहत्य चापवर्गेण रूपं तत्तत् फलानि च ।
हरेत् विमूलयुग्वर्गैस्त्रिज्यावर्गहतैः क्रमात् ॥
किन्तु व्यासदलेनैव द्विघ्नेनाद्यं विभज्यताम् ।
फलान्यधोऽधःक्रमशो न्यस्योपर्युपरि त्यजेत् ॥
शराप्त्यै........

i.e., "Divide the square of the arc by twice the radius. That is the first term. Using the first term, obtain a series to find the succeeding terms by making the square of the arc as the multiplier and the product of the square of the radius and the difference of the squares of even numbers greater than 2 and their roots as the divisors. Subtract the sum of the even terms from the sum of the odd terms to obtain the verse sine( )."
Again, if r is the radius and a (in radius) is the arc:
First term

$$= \frac{a^2}{2r}$$

Second term

$$= \frac{a^2}{2r} \cdot \frac{1}{4^2-4} \cdot \frac{a^2}{r^2} = \frac{1}{2 \cdot 12} \frac{a^4}{r^3} = \frac{a^4}{4!r^3}$$

Third term

$$= \frac{a^4}{4!r^3}\frac{1}{6^2-6}\frac{a^2}{r^2} = \frac{1}{4!5\cdot 6}\frac{a^2}{r^5} = \frac{a^6}{6!r^5}$$

Fourth term

$$= \frac{a^6}{6!r^5}\frac{1}{8^2-8}\frac{a^2}{r^2} = \frac{1}{6!7\cdot 8}\frac{a^8}{r^7} = \frac{a^8}{8!\cdot 7} \text{ and so on.}$$

Hence, versed sine

$$= \frac{a^2}{2!r} - \frac{a^4}{4!r^3} + \frac{a^6}{6!r^5} - \frac{a^8}{8!r^7} + \cdots$$

The ancient Indians called the versed sine as the "sara" of the sine chord.

In order to derive the series for sine and cosine, we will prove six lemmas. We will first prove the following lemma to find the corresponding formulae for the change in the sine chord ("bhuja jya") and cosine chord ("koti jya") produced by an infinitesimal difference in the arc. This process is equivalent to the differentiation of sine and cosine.

Lemma 1:

$$\mathrm{d}(r \sin a) = \frac{r \cos a \, \mathrm{d}a}{r} \quad (*)$$

$$\mathrm{d}(r \cos a) = \frac{r \sin a \, \mathrm{d}a}{r}$$

in modern notation.

We find the results (*) in "Yuktibhasa," pages 165 and 173, as well as in Nilakantha Somayaji's "Aryabhatiya-bhasya" (section Ganita Pada, pages 48–50).

Proof: Let arc AB be the quadrant of a circle of radius R with center C. Let arc P_0P_1 represents a small increase in the arc AP_0. Let P be middle point of arc P_0P_1.

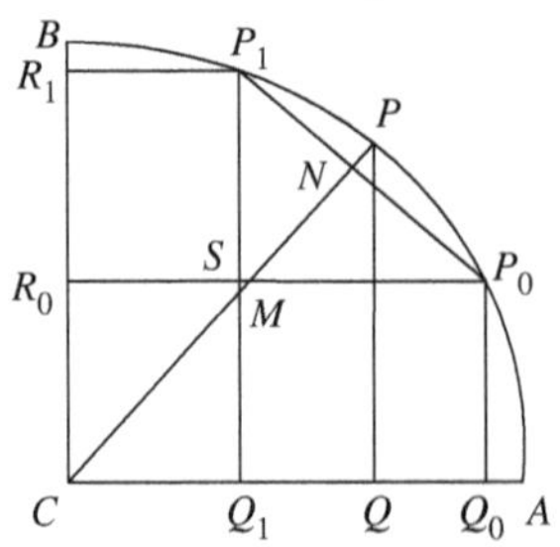

Draw perpendicular P_1Q_1, PQ, and P_0Q_0 from P_1, P, and P_0 respectively on CA. Draw P_0R_0 and P_1R_1 perpendicular to CB, from P_0 and P_1, respectively.

Then, P_1Q_1 and P_1R_1 will be the sine chord ("bhuja jya") and cosine chord ("koti jya") at P_1. Let P_1Q_1 and P_0R_0 intersect at S. Let CP intersect P_1Q_1 at M and P_0P_1 at N, respectively.

Since P is the midpoint of arc P_0P_1, CP bisects P_0P_1 at right angles.

In triangles P_1MN and CMQ_1,

$$P_1\hat{M}N = C\hat{M}Q_1 \text{ (vertically opposite angles)}$$

$$P_1\hat{N}M = MQ_1C = 90^\circ$$

So,

$$M\hat{C}Q_1 = N\hat{P}_1M, \text{ i.e., } P\hat{C}Q = P_\circ\hat{P}_1S$$

Now, in triangles P_0SP_1 and QCP, we have

$$PQC = P_1SP_0 = 90^\circ$$

$$\text{angle } PCQ = \text{angle } P_0P_1S$$

So, triangles P_0SP_1 and QCP are similar.

Hence,

$$\frac{P_0S}{PQ} = \frac{P_0P_1}{PC} = \frac{P_1S}{CQ}$$

Then,

$$P_1S = \frac{CQ \cdot P_0P_1}{PC} \quad (**)$$

and

$$P_0S = \frac{P_0P_1 \cdot PQ}{PC}$$

Now, if the arc P_0P_1 is very small, the sine chord and the cosine chord at P may be regarded to be equal to those at P_0 and also the arc P_0P_1 to coincide with the chord P_0P_1.

Then, from (**), we get

sine difference ("bhuja khanda") $= P_1S$

$$= \frac{\text{cosine chord at } P_0 \cdot \text{the increase in the arc}}{\text{radius } r}$$

Similarly, the cosine difference ("koti khanda") $= P_0S$

$$= \frac{\text{sine chord at } P_0 \cdot \text{the increase in the arc}}{\text{radius } r}$$

If we used modern notations, these become

$$d(r \sin a) = \frac{r \cos a \, da}{r} = \text{cosine chord} \cdot \frac{da}{r}$$

and

$$d(r \cos a) = \frac{r \sin a \, da}{r} = \text{sine chord} \cdot \frac{da}{r}$$

Next, we will prove the following lemma to find the second sine difference [difference between two sine differences ("bhuja Khandas")].

Lemma 2: Second sine difference

$$= r \sin a \frac{(da)^2}{r^2} = \text{``bhuja jya''} \cdot \frac{(da)^2}{r^2}$$

This result is also found in the "Yuktibhasa," page 173, as well as in Nilakantha Somayaji's "Aryabhatiya-bhasya," pages 51–53.

Proof: Second sine difference = difference between two sine difference
= sine difference of the second−sine difference of the first
Then, from Lemma 1,

$$\text{Second sine difference} = \text{cosine chord of the second} \cdot \frac{da}{r} - \text{cosine chord of the first} \cdot \frac{da}{r}$$

$$\begin{aligned}\text{Second sine difference} &= \frac{da}{r}[\text{cosine chord of the second} - \text{cosine chord of the first}] \\ &= \frac{da}{r}\text{``koti khanda''} \\ &= \frac{da}{r}\text{cosine difference} \\ &= \frac{da}{r}\frac{r \sin a}{r}da\end{aligned}$$

Thus, second sine difference

$$= r \sin a \frac{(da)^2}{r^2} = \text{``bhuja jya''} \cdot \frac{(da)^2}{r^2}$$

We will now prove the following lemma to find a formula for the difference between an arc and its sine chord.

Lemma 3: Consider an arc in the first quadrant. Divide this arc into n equal parts Δa. Let S_1, S_2, S_3,... be the sine chords at these points. Let ΔS_1, ΔS_2,... be the corresponding sine differences at these points. Them, the difference between the arc and the sine chord

$$= \{(n-1)S_1 + (n-2)S_2 + \cdots + S_{n-1}\} \frac{(\mathrm{d}a)^2}{r^2}$$

Proof:

$$\Delta S_1 = S_1$$

Then, from Lemma 2,

$$\Delta S_1 - \Delta S_2 = S_1 \frac{(\mathrm{d}a)^2}{r^2} \quad (1)$$

Hence,

$$\Delta S_2 = \Delta S_1 - S_1 \frac{(\mathrm{d}a)^2}{r^2}$$

$$\Delta S_2 = S_1 - S_1 \frac{(\mathrm{d}a)^2}{r^2} \quad (2)$$

Similarly,

$$\Delta S_3 = \Delta S_2 - S_2 \frac{(\mathrm{d}a)^2}{r^2}$$

$$= S_1 - S_1 \frac{(\mathrm{d}a)^2}{r^2} - S_2 \frac{(\mathrm{d}a)^2}{r^2} \quad (3)$$

and so on.

Hence,

$$\Delta S_n = S_1 - S_1 \frac{(\mathrm{d}a)^2}{r^2} - S_2 \frac{(\mathrm{d}a)^2}{r^2} - \cdots - S_{n-1} \frac{(\mathrm{d}a)^2}{r^2} \quad (\mathrm{n})$$

Adding (1), (2), ..., (n), we get

$$\sum_{K=1}^{n} \Delta S_K = nS_1 - \{(n-1)S_1 + (n-2)S_2 + (n-3)S_3 + \cdots + 2S_{n-2} + S_{n-1}\} \frac{(\mathrm{d}a)^2}{r^2}$$

We also have $S_1 = \Delta S_1$,

$$S_2 - S_1 = \Delta S_2$$

$$S_3 - S_2 = \Delta S_3$$

...

...

$$S_n - S_{n-1} = \Delta S_n$$

Hence,

$$\sum_{K=1}^{n} \Delta S_K = S_n$$

Thus we get

$$S_n = nS_1 - \left\{(n-1)S_1 + (n-2)S_2 + (n-3)S_3 + \cdots + 2S_{n-2} + S_n\right\} \frac{(\mathrm{d}a)^2}{r^2}$$

We observe that, if n is sufficiently large, the first sine chord will coincide with its arc.

Hence, nS_1 = whole arc a.

Thus the whole arc $a - S_n$ is the difference between the arc and sine chord and this is equal to

$$\left\{(n-1)S_1 + (n-2)S_2 + (n-3)S_3 + \cdots + 2S_{n-2} + S_{n-1}\right\} \frac{(\mathrm{d}a)^2}{r^2}$$

This completes the proof of Lemma 3.

Lemma 4: The height of the arc

$$a = \frac{1}{r}\frac{a^2}{2}$$

where r is the radius.

Proof: Now, the height of the given arc a is the sum of the cosine differences.

Then, from Lemma 1,

Sum of the cosine differences

$$= S_1 \frac{\mathrm{d}a}{r} + S_2 \frac{\mathrm{d}a}{r} + S_3 \frac{\mathrm{d}a}{r} + \cdots$$

$$= \frac{\mathrm{d}a}{r}[S_1 + S_2 + S_3 + \cdots]$$

If we take that these unknown chords are equal to the respective arc themselves, and if we take $da = 1$, we get

The height of the arc = sum of the cosine differences

$$= \frac{1}{r}[a + (a-1) + (a-2) + \cdots]$$

If n is large, in modern notation,

$$\frac{1}{r}[a + (a-1) + (a-1) + \cdots] = \frac{1}{r}\int a\,\mathrm{d}a = \frac{1}{r^2}a^2$$

Thus the height of the arc

$$= \frac{1}{2r}a^2$$

This proves Lemma 4.

We now prove the following lemma.

Lemma 5: If $\Delta S_1 - \Delta S_2, \ldots$ are sine differences, then

$$\Delta S_1 - \Delta S_n = \frac{\mathrm{d}a}{r} h_n$$

where h_n is the height of the nth arc.

Proof: From Lemma 1,

$$\Delta S_1 = C_{m_1} \frac{\mathrm{d}a}{r}$$

where C_m is the cosine chord at the middle point m_1 of the first arc.
So,

$$\Delta S_1 = \left[r - h_{m_1}\right] \frac{\mathrm{d}a}{r}$$

where h_{m_1} is the height of the middle point of the m_1 of the first arc.
Similarly,

$$\Delta S_2 = \frac{\mathrm{d}a}{r}\left[r - h_{m_2}\right]$$

where h_{m_2} is the height of the middle point of the second arc and so on.
So,

$$\Delta S_n = \frac{\mathrm{d}a}{r}\left[r - h_{m_n}\right]$$

where h_{m_n} is the height of the middle point of the nth arc.
Hence,

$$\Delta S_1 - \Delta S_n = \frac{\mathrm{d}a}{r}\left[r - h_{m_1} - (r - h_{m_n})\right]$$

$$= \frac{\mathrm{d}a}{r}\left[h_{m_n} - h_{m_1}\right]$$

Since da can be made very small, h_m can be made equal to $h_0 = 0$ and $h_{m_n} = h_n$.

Thus, we have

$$\Delta S_1 - \Delta S_n = \frac{\mathrm{d}a}{r} h_n$$

Lemma 6: Arc−sine chord

$$= \frac{a^3}{r^2 3!}$$

where a is the arc and r is the radius.

Proof: From Lemma 5,

$$\Delta S_1 - \Delta S_n = \frac{\mathrm{d}a}{r} h_n$$

But from Lemma 4, the heights of the arc

$$= \frac{1}{r}\frac{a^2}{2}$$

Hence,

$$\Delta S_1 - \Delta S_n = \frac{\mathrm{d}a}{r} \cdot \frac{1}{2}\frac{a^2}{r} = \frac{1}{2}\mathrm{d}a\frac{a^2}{r^2}$$

If $\mathrm{d}a = 1$, then

$$\Delta S_1 - \Delta S_n = \frac{a^2}{2r^2}$$

Similarly,

$$\Delta S_1 - \Delta S_{n-1} = \frac{(a-1)^2}{2r^2}$$

$$\Delta S_1 - \Delta S_{n-2} = \frac{(a-2)^2}{2r^2}$$

and so on.

Summing these is the sum of the second differences, which is the difference between the arc and its sine chord.

Hence, arc−sine chord

$$= \frac{a^2}{2a^2} + \frac{(a-1)^2}{2r^2} + \cdots$$

$$= \frac{1}{2r^2}\int a^2 \,\mathrm{d}r$$

in modern notation

$$= \frac{1}{2r^2}\frac{a^3}{3} = \frac{1}{r^2}\frac{a^3}{3!}$$

This completes the proof of Lemma 6.

In deriving on expression for the height of the arc, the sine chords have been equated to the arcs to which they are not equal. A correction has to be made for the height of the arc. The above same difference between the arc and its sine chord has to be made as correction for the expression of the arc, namely

$$\frac{1}{r}\{a + (a-1) + (a-2) + \cdots\}$$

when this correction is made, the expression for the height of the arc

$$= \frac{1}{r}\left\{\left(a - \frac{a^3}{3!r^2}\right) + \left[(a-1) - \frac{(a-1)^3}{3!r^2}\right] + \left[(a-2) - \frac{(a-2)^3}{3!r^2}\right]\right\} + \cdots$$

$$= \frac{1}{r}[a + (a-1) + \cdots] - \frac{1}{r}\left[\frac{a^3}{3!r^2} + \frac{(a-1)^3}{3!r^2} - \frac{(a-2)^3}{3!r^2} + \cdots\right]$$

Hence, correction to the height of the arc

$$= -\frac{1}{r^3 3!}\left[a^3 + (a-1)^3 + \cdots\right]$$

In modern notation, when n is large, correction to the height of the arc

$$= \frac{-1}{3!r^3}\int a^3\,da = -\frac{1}{3!r^3}\frac{a^4}{4} = -\frac{a^4}{4!r^3}$$

If this corrected value of the height of the arc is used, in Lemma 6, we obtain arc$-$sine chord

$$= \left\{\left(\frac{a^2}{2r^2} - \frac{a^4}{4!r^3}\right) + \left(\frac{(a-1)^2}{2r^2} - \frac{(a-1)^4}{4!r^3}\right) + \cdots\right\}$$

$$= \frac{1}{2r^2}\left[a^2 + (a-1)^2 + (a-2)^2 + \cdots\right] - \frac{1}{4!r^4}\left\{a^4 + (a-1)^4 + \cdots\right\}$$

when n is large, in modern notation, this becomes
arc$-$sine chord

$$= \frac{1}{2r^2}\int a^2\,da - \frac{1}{4!r^4}\int a^4\,da$$

$$= \frac{1}{2r^2}\frac{a^3}{3} - \frac{1}{4!r^4}\frac{a^5}{5}$$

arc−sine chord

$$= \frac{1}{3!}\frac{a^3}{r^2} - \frac{1}{5!}\frac{a^5}{r^4}$$

when this correction is made "ad infinitum" to the formula for the arc−sine chord, we get

$a -$ sine chord

$$= \frac{1}{3!}\frac{a^3}{r^2} - \frac{1}{5!}\frac{a^5}{r^4} + \frac{1}{7!}\frac{a^7}{r^6} - \cdots$$

Thus, sine chord

$$= a - \frac{a^3}{3!r^2} + \frac{1}{5!}\frac{a^5}{r^4} - \frac{1}{7!}\frac{a^7}{r^6} + \cdots$$

Now, cosine chord = r−height of the arc.

When the correction $\frac{-a^4}{4!r^3}$ is made "ad infinitum" to the formula for the height of the arc, we get

cosine chord

$$= r - \left[\frac{a^2}{2r} - \frac{a^4}{4!r^3} + \frac{a^6}{6!r^5} + \cdots\right]$$

cosine chord

$$= r - \frac{a^2}{2r} + \frac{a^4}{4!r^3} - \frac{a^6}{6!r^5} + \cdots$$

arc a = radius × (angle in radius).

$$a = r\theta$$

Substituting this value of a, remembering that sine chord = $r \sin\theta$ and cosine chord = $r\cos\theta$, we get

$$r\sin\theta = r\theta - \frac{r^3\theta^3}{3!r^2} + \frac{1}{5!}\frac{r^5\theta^5}{r^4} - \frac{1}{7!}\frac{r^7\theta^7}{r^6} + \cdots$$

i.e.,

$$\sin\theta = \theta - \frac{\theta^3}{3!} + \frac{\theta^5}{5!} - \frac{\theta^7}{7!} + \cdots$$

Similarly,

$$r\cos\theta = r - \frac{r^2\theta^2}{2r} + \frac{r^4\theta^4}{4!r^3} - \frac{r^6\theta^4}{6!r^5} + \cdots$$

i.e.,

$$\cos\theta = 1 - \frac{\theta^2}{2} + \frac{\theta^4}{4!} - \frac{\theta^4}{6!} + \cdots$$

It is noteworthy to point out these series for sin θ and cos θ were known in India in the fourteenth century.

The same series were rediscovered in Europe in the seventeenth century, three centuries after they were discovered in India.

Carana Paddhati

"Karaṇapaddhati" consists of 10 chapters and is an extensive treatise on astronomy. It has been published by Trivandrum Sanskrit series No. 20. Tamil and Telugu scripts of this work are available and indicate clearly its popularity in South India. This work also carries commentary in Malayalam, Sanskrit, and Tamil.

According to Charles M. Whish, the author of "Karaṇapaddhati" prefers to be an anonymous except that he is a Pathumana Somayaji, a nambudri Brahmin of Tirusivapura Trichur in Malabar. Whish quoted the first verse of the commentary on his work, which is as follows:

nutuna griha sama sutarachitayah carana padhatervidusha
bhasham vilichati caschit balanam bhodhanarthamalpadhiyam

i.e., "Karaṇapaddhati" is a work on arithmetic and astronomy and is composed by a Puthumana Somayaji, belonging to "nutana Griha" or "Navina Vipana" family.

The last verse of his work contains the date of composition of this work. This verse as quoted by Whish is as follows:

iti sivapura nam agramajah capiyajwa
cimapi carana padhatyahwayam tantra rupam
vyadhita ganita metutsamyagalocyasantah
cathitamihavidantassuntu santosha pantah

The date of its composition is given in some chronogram. The words "Ganita Metutsamyai" refers to 176563 Kaliyuga days, which correspond to Saturday, April 4, AD 1733.

The author of "Karaṇapaddhati" is believed to have written some other works in astronomy called "Nyayarathna," "venvārohāstaka" (a booklet for correctly determining the position of the moon at small intervals of time), "Grahana Staka," and

"GrahanaGanitha" (manuals dealing with mathematics involved in determining the occurrences of eclipses).

Charles M. Whish in his memoir; dated December 15, 1832 presents the following verse from "carana-paddhati":

Anùanutnánananunnanutyaih
Samáhataschacra kálávibhactah
Chandámschandrádhana cumbhipálaih
Vyásastadardhantribhamaurvicásyát

i.e., "If the product of the circumference of the circle in minutes and 10000000000 is divided by 31415926536, the quotient obtained will be the diameter of the circle in terms of minutes of circumference and its half will be the radius."

This means, if c is the circumference and d the diameter of the circle respectively, then

$$\frac{c \times 10000000000}{31415926536} = d$$

i.e.,

$$\frac{c}{d} = \frac{31415926536}{10000000000} = 3.1415926536$$

Since,

$$\frac{c}{d} = \pi$$

we have

$\pi = 3.1415926536$, which is correct to nine decimal places.

He proceeded to quote the first verse in Chapter VI of "Karaṇapaddhati" as follows:

vyásácchaturghnát bahusah prithac sthát
tripanja saptadyayu gáhritáni
vyasè chatuoghnè cramasastwrinam swam
curyáttadásyát paridhissusùcshmah

i.e., "Divide the product of the diameter and four individually by the odd numbers 1, 3, 5, 7, 9,... to make them the first term, second term, and so on. Subtract the sum of the even terms from the sum of the odd terms. The difference obtained will precisely be the circumference."

If the above rule is followed, we obtain the following infinite series:

$$c = \frac{4d}{1} - \frac{4d}{3} + \frac{4d}{5} - \frac{4d}{7} + \cdots$$

It must be pointed out that this series was first mentioned by Nilakantha Somayaji, the author of "Tantra-Sangraha."

Since,

$$\frac{c}{d} = \pi$$

we have

$$\frac{\pi}{4} = 1 - \frac{1}{3} + \frac{1}{5} - \frac{1}{7} + \cdots$$

The next verse in Chapter VI of "Karaṇapaddhati" is of measure "Gitivrittam" and it runs as follows:

vyásádwana samgunitát prithgáptantryádyay
ugvimùlaghanaih
irigunita vyáse swamrinam cramasah
critwápi paridhiránèyah

i.e., "Divide the product of the diameter and four individually by the cubes of the odd numbers 3, 5, 7, . . . , subtracting from each cube its corresponding root to form the first term, second term, and so on. Subtract the sum of the even terms from the sum of odd terms. To the difference obtained, add three times the diameter. The resulting sum will give the circumference of the circle with the given diameter."

If we follow this rule, we get the following infinite series:

$$c = 3d + \frac{4d}{3^3 - 3} - \frac{4d}{5^3 - 5} + \frac{4d}{7^3 - 7} - \frac{1}{9^3 - 9} + \cdots$$

Since,

$$\frac{c}{d} = \pi$$

we get

$$\frac{\pi - 3}{4} = \frac{1}{3^3 - 3} - \frac{1}{5^3 - 5} + \frac{1}{7^3 - 7} - \frac{1}{9^3 - 9} + \cdots$$

$$\frac{\pi - 3}{4} = \frac{1}{2 \cdot 3 \cdot 4} - \frac{1}{4 \cdot 5 \cdot 6} + \frac{1}{6 \cdot 7 \cdot 8} - \frac{1}{8 \cdot 9 \cdot 10} + \cdots$$

The fourth verse in chapter of "Carana Paddhati" is of measure called "Indra Vajravrittam."

vagairyjámvadwigunairnnirècaih
vargìcritairvarjitayugma vargaih
vyásancha shadghnam vibhajèt phalam swam
vyásè trinighnè paridhistadásyat

i.e., "Divide six times the diameter individually by the square of twice the square diminished by unity of the even numbers 2, 4, 6, 8, 10,... and subtracting from each the square of 2, 4, 6, 8, 10,... in order to form the first term, second term, and so on. If we add three times the diameter to the sum of above terms, we obtain the circumference of the circle with the given diameter."

If the above rule is followed, we obtain the following infinite series:

$$c = 3d + \frac{6d}{(2\cdot 2^2-1)^2-2^2} + \frac{6d}{(2\cdot 4^2-1)^2-4^2} + \frac{6d}{(2\cdot 6^2-1)^2-6^2} + \cdots$$

where c is the circumference and d is the diameter of the circle.

Since

$$\frac{c}{d} = \pi$$

we get

$$\frac{\pi-3}{6} = \frac{1}{(2\cdot 2^2-1)^2-2^2} + \frac{1}{(2\cdot 4^2-1)^2-4^2} + \frac{1}{(2\cdot 6^2-1)^2-6^2} + \cdots$$

$$= \frac{1}{1\cdot 3\cdot 3\cdot 5} + \frac{1}{3\cdot 5\cdot 7\cdot 9} + \frac{1}{9\cdot 11\cdot 19\cdot 21} + \cdots$$

Charles M. Whish pointed out that the author of "Carana Paddhati" provides a verse for finding the arc of the circumference of the circle in terms of the sum obtained by dividing the product of the radius and the sine of the arc by its cosine. The following is that verse:

> vyásárdhènahatádabhìshta gunatah cotyaptamádyam phalam
> jyávar 'ena vinighnamádimaphalam tattat phalanchá harèl
> crityá còti gunasya tatratuphalèshwècatripanchá dibhi
> bhuctèshwòjayutèstyajèt samayutim jìvádhanussishtyatè

i.e., "Multiply the given sine by the radius and divide this product by the cosine to form the first quote."

Multiply the first quote by the square of the sine and divide the product by the square of the cosine to form the second quote. Proceed this way, by multiplying each previous quote by the square of the sine and dividing the product by the square of the cosine. Divide the quotes obtained this way in succession by odd numbers 1, 3, 5, 7, 9,... in order to form the first term, second term, third term, and so on. Subtract the sum of the even terms from the sum of odd terms. This difference will be the arc of the sine.

If r = radius, a = arc, S = sine, cos = cosine, the above rule gives the following infinite series:

$$a = \frac{r \times S}{\cos} - \frac{rS^3}{3\cos^3} + \frac{rS^5}{5\cos^5} - \frac{rS^7}{7\cos^7} + \frac{r\cos^9}{9\cos^9} - \cdots$$

Hence, if the cosine be less than the sine of the given angle, interchange the names of the two and proceed as laid down in the above rule. It means that the smaller of the sine and cosine chords is to be used for this calculation.

The first term

$$\frac{r \times S}{\cos} = \text{tangent of the arc}$$

We have

$$a = t - \frac{t^3}{3r^2} + \frac{t^5}{5r^4} - \frac{t^7}{7t^6} + \cdots$$

This is the series for the arc a, as laid down in the "carana paddhati."

If the arc of 45° is taken and if the radius = 1, the above series becomes

$$1 - \frac{1}{3} + \frac{1}{5} - \frac{1}{7} + \cdots = a\frac{\pi}{4}$$

Simple Continued Fraction and its Convergents

We know from modern continuous fractions, if $\{a_n\}$, $n = 0, 1, 2, 3, \ldots$ is a sequence of integers with $a_0 \geq 0$ and $a_n \geq 1$ ($n = 1, 2, 3, \ldots$), the expression

$$\alpha = a_0 + \cfrac{1}{a_1 + \cfrac{1}{a_2 + \cfrac{1}{a_3 + \cdots + \cdots + \cfrac{1}{a_n + \cdots}}}}$$

is called a single continued fraction.

If

$$c_n = a_0 + \cfrac{1}{a_1 + \cfrac{1}{a_2 + \cfrac{1}{a_3 + \cfrac{1}{\vdots + \cfrac{1}{a_n}}}}}$$

then c_n is called the nth convergent of α. It is convenient to write

$$c_n = a_0 + \frac{1}{a_1} + \frac{1}{a_2} + \frac{1}{a_3} + \cdots + \frac{1}{a_n}$$

or simply $c_n = [a_0, a_1, a_2, \ldots, a_n]$.

Now, if

$$c_0 = a_0 = \frac{p_0}{q_0}$$

$$c_1 = a_0 + \frac{1}{a_1} = \frac{a_0 a_1 + 1}{a_1} = \frac{p_1}{q_1}$$

$$c_2 = a_0 + \frac{1}{a_1 + (1/a_2)} = a_0 + \frac{1}{(a_1 a_2 + 1)/a_2} = a_0 + \frac{a_2}{a_1 a_2 + 1}$$

$$= \frac{a_0 a_1 a_2 + a_2 + a_0}{a_1 a_2 + 1} = \frac{p_2}{q_2}$$

and so on and if

$$c_n = \frac{p_n}{q_n}$$

We know the following:

i. $p_n = a_n p_{n-1} + p_{n-2}$

$q_n = a_n q_{n-1} + q_{n-2}$

ii. If α is irrational and c_n is its nth convergent, then

a. $\lim_{n\to\infty} c_n = \alpha$

b. $|\alpha - c_n| < \dfrac{1}{q_n q_{n-1}}$ for $n = 1, 2, \ldots$

Convergents of π = 3.1415926536 as Given in Carana Paddati

$$\pi = 3.1415926536 = \frac{3.415926536}{10000000000} = 3 + \frac{1415926536}{10000000000}$$

1415926536)10000000000(7
9911485752
88514248)141592653(15
88514248
530784056
442571240
88212816)88514248(1
88212816
301432)8821286(292
602864
2792641
2712888
797536
602864
194672)301432(1
194672
106760
106760)194672(1
106760
87912)106760(1
87912
18848)87912(4
12520
6328)12520(1
6320
6192)6328(1
6192
136)6192(
and so on.

Hence,

$$\pi = \frac{31415926536}{1000000000} = [a_0, a_1, a_2, a_3, a_4, a_5, a_6, a_7, \ldots]$$

$$= [3, 7, 15, 292, 1, 1, 1, 4, 1, 1, ...]$$

Hence,

$$c_0 = a_0 = [3] = 3 = \frac{p_0}{q_0} \Rightarrow p_0 = 3,\ q_0 = 1$$

Now,

$$c_1 = [a_0\ a_1] = [3 \quad 7] = 3 + \frac{1}{7} = \frac{22}{7} = \frac{p_1}{q_1} \Rightarrow p_1 = 22,\ q_1 = 7$$

Now,

$$c_2 = [a_0, a_1, a_2] = [3, 7, 15]$$

From (1) on page 666, we have

$$p_2 = a_2 p_1 + p_0 = 15(22) + 3 = 330 + 3 = 333$$

$$q_2 = a_2 q_1 + q_0 = 15(7) + 1 = 105 + 1 = 106$$

So,

$$c_2 = [3, 7, 15] = \frac{p_2}{q_2} = \frac{330}{106}$$

Now, $c_3 = [a_0,a_1,a_2,a_3] = [3,7,15,1]$.
Again from (1) on page 666, we have

$$p_3 = a_3 p_2 + p_1 = 1(333) + 22 = 333 + 22 = 355$$

$$q_3 = a_3 q_2 + q_1 = 1(106) + 7 = 106 + 7 = 113$$

So,

$$c_3 = [3, 7, 5, 1] = \frac{p_3}{q_3} = \frac{355}{113}$$

Now, $c_4 = [a_0,a_1,a_2,a_3,a_4] = [3,7,15,1,292]$.
Again, from (1) on page 666, we have

$$p_4 = a_4 p_3 + p_2 = (292)(355) + 333 = 103,660 + 333 = 103{,}993$$

$$q_4 = a_4 q_3 + q_2 = (292)(113) + 106 = 32,996 + 106 = 33{,}012$$

So,

$$c_4 = [3, 7, 15, 1, 292] = \frac{p_4}{q_4} = \frac{103,993}{33,012}$$

Now, $c_5 = [a_0,a_1,a_2,a_3,a_4,a_5] = [3,7,15,1,292,1]$.
Again from (1) on page 666, we have

$$p_5 = a_5 p_4 + p_3 = 1(103,993) + 355 = 103,993 + 355 = 104,348$$

$$q_5 = a_5 q_4 + q_3 = 1(33,012) + 113 = 33,012 + 113 = 33{,}125$$

So,

$$c_5 = [3, 7, 15, 1, 292, 1] = \frac{p_5}{q_5} = \frac{104,348}{33,125}$$

Now, $c_6 = [a_0,a_1,a_2,a_3,a_4,a_5,a_6] = [3,7,15,1,292,1,1]$.

$$p_6 = a_6p_5 + p_4 = 1(104,348) + 103,993 = 104,348 + 103,993 = 208,341$$

$$q_6 = a_6q_5 + q_4 = 1(33,125) + 33,012 = 66,137$$

So,

$$c_6 = [3, 7, 15, 1, 292, 1, 1] = \frac{p_6}{q_6} = \frac{208,341}{66,137} \text{ and so on.}$$

Estimate of the err of the convergents as successive approximations for π. From 2(b) on page 666, we have

$$|\pi = c_0| = |\pi - 3| < \frac{1}{q_0q_1} = \frac{1}{1 \cdot 7} = \frac{1}{7} < \frac{1}{10^0}$$

$$|\pi = c_1| = |\pi - \frac{22}{7}| < \frac{1}{q_1q_2} = \frac{1}{7 \cdot 106} = \frac{1}{742} < \frac{1}{100} = \frac{1}{10^2}$$

$$|\pi = c_2| = |\pi - \frac{336}{106}| < \frac{1}{q_2q_3} = \frac{1}{106 \cdot 113} = \frac{1}{11,978} < \frac{1}{10^4}$$

$$|\pi = c_3| = |\pi - \frac{355}{113}| < \frac{1}{q_3q_4} = \frac{1}{113 \cdot 33102} = \frac{1}{3,740,526} < \frac{1}{10^6}$$

$$|\pi = c_4| = |\pi - \frac{103,993}{33,102}| < \frac{1}{q_4q_5} = \frac{1}{33,102 \cdot 33,215} = \frac{1}{109,948,230} < \frac{1}{10^8}$$

$$|\pi = c_5| = |\pi - \frac{104,348}{33,215}| < \frac{1}{q_5q_6} = \frac{1}{33,215 \cdot 66,317} = \frac{1}{2,202,719,155} < \frac{1}{10^9}$$

and so on.

Thus, the convergent c_n ($n = 1,2,\ldots$) gives excellent approximations for π as n increases.

Kriyakramakari

The "Kriyakramakari" is an exhaustive commentary of "Lilavati" of Bhaskara II. It is believed that two Kerala mathematicians, Sankara Vāriyar (AD 1500–1560) and Narayana (AD 1494–1570) of Mahisha mangalam family are the authors of this work. Sankara is said to have written the first part of this commentary up to the verse 199 of "Lilavati." Narayana is believed to have finished the second part of this work by writing his commentary on the remaining verse of "Lilavati."

The significant feature of this work is a crystal clear analysis of the theories propounded by Bhaskara II in "Lilavati," as well as providing easy to understand logical rationale for some of the formulae stated by Bhaskara II in "Lilavati." This very detailed commentary is invaluable for the exhaustive references made in this work to the writings of earlier authors, some of which has been lost and still have to be unearthened.

The "Kriyakramakari" gave a remarkably close approximation to π, credited to Sangamagrama Madhava:

बिबुत्र नेत्त्रगजाहिहुताशनत्रिगुणवेदभवारणबाहबः ।
नवनिखर्वमिते वृतिविस्तरे पतिधिमानमिदं जगदुब्रुंघा ॥

(K.K. 688)

i.e., "The circumference of a circle of diameter 900,000,000 is 2,827,433,388,233."

Since $\pi = \dfrac{c}{d}$ where c is the circumference and d is the diameter of a circle, we have

$$\pi = \frac{2,827,433,388,233}{900,000,000,000} = 3.1415926359$$

which is correct to seven decimal places.

In the following verse, the "Kriyakramakari" provided a still better approximation for π:

वृत्तव्यासे हते नागवेदवह्नथब्धिखेन्दुभिः।
तिथ्यश्विविबुघैर्भक्ते सुसूक्ष्मः परिघिर्भत् ॥

$$\pi = \frac{104,348}{33,215} = 3.1415926539211$$

which is correct to nine decimal places.

Madhava's remarkable discovery of series now known as Gregory-Leibnitz series.

"Kriyakramakari" quoted the following verse assigned to Madhava who lived in the fourteenth century:

इष्टज्यात्त्रिज्ययोर्धातात् कोट्याप्तं प्रथमं फलं ।
ज्यावगं गुणकं कुत्वा कोटिवगं च हारकम् ॥

प्रथमादिफलेभ्योऽथ नेया फलततिर्मुहुः।
एकत्याद्योजसंख्याभिर्भक्तेष्वेतेष्वनुक्रमात् ।
दोःकोट्ग्योरल्पमेवेष्टं कल्पनीयमिह स्मृतम् ॥

(K.K. 692–693)

i.e., "The given sine chord is multiplied by the radius and then the product is divided by the cosine chord. The quotient obtained is the first quote. A series of quotes have to be obtained from the first quote. The succeeding quotes have to be obtained by making the square of the sine chord as the multiplier and the square of the cosine chord as the divisor. When these quotes are divided in order by the odd numbers 1, 3, 5, 7, 9,... We obtain the first term, second term, and so on. When the sum of even terms is subtracted from the sum of odd terms, we get the arc."

This means, if S is sine chord, C is the cosine chord, and r is the radius of the circle, the arc θ is given by

$$\theta = \frac{Sr}{C} - \frac{Sr}{3C} \cdot \frac{S^2}{C^2} + \frac{Sr}{5C} \cdot \frac{S^4}{C^4} - \frac{Sr}{7C}\frac{S^6}{C^6} + \frac{Sr}{9C}\frac{S^8}{C^8} - \cdots$$

i.e.,

$$\theta = \frac{S}{C}r - \frac{1}{3}\frac{S^3}{C^3}r + \frac{1}{5}\frac{S^5}{C^5} \cdot r - \frac{1}{7}\frac{S^7}{C^7}r + \frac{1}{9}\frac{S^9}{C^9}r - \cdots \quad (*)$$

But,

$$\frac{S}{C} = \text{tangent } \theta = \tan \theta$$

Let

$$\frac{S}{C} = t \text{ and } r = 1$$

Then, we have

$$\tan^{-1} t = t - \frac{t^3}{3} + \frac{t^5}{5} - \frac{t^7}{7} + \frac{t^9}{9} - \cdots \quad (**)$$

This series was rediscovered in Europe by James Gregory in AD 1671, nearly three century after Madhava had discovered this series. This series is presently known as Gregory's series.

If $t = 1$, in (**), we get

$$\tan^{-1} t = 1 - \frac{1}{3} + \frac{1}{5} - \frac{1}{7} + \cdots$$

i.e.,

$$\frac{\pi}{4} = 1 - \frac{1}{3} + \frac{1}{5} - \frac{1}{7} + \cdots$$

This is Madhava series for π.

This series occurs in both "Tantra-Sangraha" and "Caranapaddhiti." This series was rediscovered in Europe and is now known as Euler's series.

Arc in terms of sine and cosine chords

Proof of (*): Inscribe a circle in a square *EFGH* of side $2r$. Let *C* be the center of this circle. Let *P* be a point on the circumference of this circle. Join *CP* and produce it to cut *EF* at *Q*.

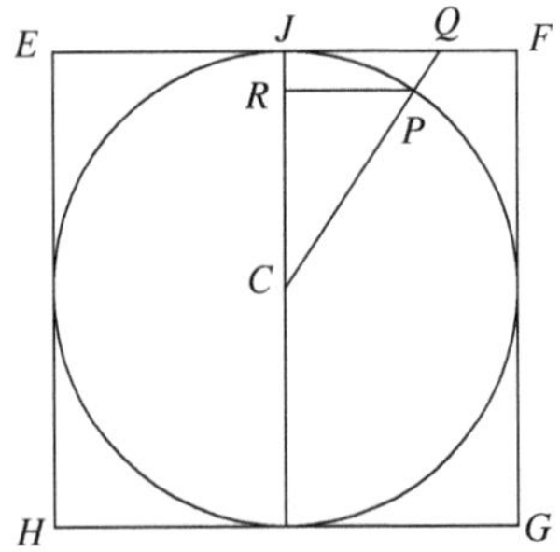

Draw *PR* perpendicular from *P* on *CJ*, where *J* is the point where the circle touches *EF*. Then, *PR* is the half chord = *S* (sine chord) and *RC* = cosine chord = *C*.

Since the number of terms that have to be summed up is equal to the number of unit divisions in *JF*, the infinite series for the arc

$$\theta = JF - \frac{JF^3}{3r^2} + \frac{JF^5}{5r^4} - \cdots \quad (***)$$

We observe that triangles *CJQ* and *CRP* are similar.

So,

$$\frac{JQ}{RP} = \frac{CJ}{CR}$$

Hence,

$$JQ = \frac{CJ \cdot RP}{CR} = \frac{rS}{C}$$

Substituting this in (***), we get

$$\text{arc } \theta = \frac{S}{C}r - \left(\frac{S \cdot r}{C}\right)^3 \cdot \frac{1}{3r^2} + \left(\frac{S \cdot r}{C}\right)^5 \cdot \frac{1}{5r^4} - \cdots$$

i.e.,

$$\text{arc } \theta = \frac{S \cdot r}{C} - \frac{S \cdot r}{3C}\frac{S^2}{C^2} + \frac{S \cdot r}{5C}\frac{S^4}{C^4} - \cdots$$

This infinite series enables to find the series for π by taking chords of known arcs.

Rational Right Triangles

"Kriyakramakari" provided the following method to determine the sides of a rational right triangle, when one side is given.

Let a, b, c be the sides of a right triangle. Let a be the given side and c the hypotenuse.

Then, $c^2 = a^2 + b^2$.

Then,

$$b = \frac{b^2}{b} = \frac{c^2 - a^2}{b} = \frac{(c+a)(c-a)}{b}$$

$$= \frac{c+a}{(b/(c-a))} = \frac{(c+a)2a}{(b/(c-a))2a}$$

$$b = \frac{(c+a)[c+a-(c-a)]}{(b/(c-a))[c+a-(c-a)]}$$

Dividing both numerator and denominator by $c + a$, we get

$$b = \frac{(c+a)-(c-a)}{(b/(c-a))[1-((c-a)/(c+a))]}$$

$$= \frac{2a}{(b/(c-a)) - (b/(c-a))((c-a)/(c+a))}$$

Let

$$m = \frac{b}{c-a}$$

Then,

$$b = \frac{2a}{m - m((c-a)/(c+a))} = \frac{2am}{m^2 - m^2((c-a)/(c+a))}$$

i.e.,

$$b = \frac{2am}{m^2 - (b^2/(c-a)^2)((c-a)/(c+a))} = \frac{2am}{m^2 - (b^2/(c^2-a^2))} = \frac{2am}{m^2 - 1}$$

Also,

$$c + a = \frac{b^2}{c - a}$$

(since $b^2 = c^2 - a^2$), we have

$$c = \frac{b^2}{c-a} - a = b \cdot \frac{b}{c-a} - a = bm - a$$

The solution Bhaskara II has given in the following sloka (verse):

इष्टो भुजोऽस्माद्द्विगुणेष्टनिध्नात् इष्टस्य कृत्यैकवियुक्तयाप्तम् ।
कोटिपुथक् सेष्टगुणा भुजोना कर्णौं भवेत् त्र्यस्रमिदं तु जात्यम् ॥

(Lil. 141)

"Kriyakramakari" has provided the following method to determine the second solution given by Bhaskara II in sloka 142 of "Lilavati" for the construction rational right triangle:

(i) Let a be the base ("*bhuja*"), b be the perpendicular side ("Koti"), and c be the hypotenuse ("Karna"). Let a be given side, then

$$c^2 = a^2 + b^2$$

Let

$$c - b = m \tag{1}$$

Then,

$$c + b = \frac{c^2 - b^2}{c - b} = \frac{a^2}{c - b} = \frac{a^2}{m} \tag{2}$$

Adding (1) and (2), we get

$$2c = m + \frac{a^2}{m} \Rightarrow c = \frac{1}{2}\left(m + \frac{a^2}{m}\right)$$

Subtracting (1) from (2), we get

$$2b = \frac{a^2}{m} - m \Rightarrow b = \frac{1}{2}\left(\frac{a^2}{m} - m\right)$$

(ii) Let the hypotenuse c be the given side, with a being the base and b, the perpendicular side.

Then,

$$c^2 = a^2 + b^2. \quad \text{So, } b^2 = c^2 - a^2$$

Let

$$m = \frac{b}{c - a}$$

Then,

$$m^2 = \frac{b^2}{(c-a)^2} = \frac{c^2 - a^2}{(c-a)^2} = \frac{c + a}{c - a}$$

So, $m^2(c - a) = c + a$.
i.e., $cm^2 - am^2 = c + a$.

$$c(m^2 - 1) = a(m^2 + 1)$$

So,

$$a = c\frac{(m^2 - 1)}{m^2 + 1}$$

Now, $b^2 = c^2 - a^2$

$$= c^2 - c^2\frac{(m^2 - 1)^2}{(m^2 + 1)^2}$$

$$b^2 = \frac{c^2}{(m^2+1)^2}\left[(m^2+1)^2 - (m^2-1)^2\right] = \frac{c^2 \cdot 4m^2}{(m^2+1)^2}$$

Thus,

$$b = \frac{2cm}{(m^2 - 1)}$$

This is the solution given by Bhaskara II in sloka 144 of "Lilavati."

Sadrantamala

The "Sadrantamala" consists of six chapters with 211 verses of different measures and is a well written treatise on the Kerala School of astronomy and mathematics.

This work was composed by prince Sankara Varma, known as Appu Tampuran. According to Charles M. Whish Sankara Varma was the younger brother of Raja of "Cadattanāda" near Tellichery in Kerala. The date of composition of this work has been indicated by the chronogram "lokāmbe siddhisevye" (1797313 Kaliyuga days), which correspond to Friday, December 10, 1819. According to Whish, this work is abundant with fluxional forms and series, nowhere found in the works of foreign country. The author Sankara varma has written very detailed commentary on his own work, except for the last 32 verses of the last chapter.

Charles M. Whish has quoted the following verses from the "Sadrantamala":

The first verse which Whish has extracted is from the chapter on sines and is of measure called "Salini-vrittam." It reads like this:

vargádwyásasyárcanighnat padamyat
tatryamshò yastècha tattannavámsháh
dwigna vyècaicadwi purvaujayugmah
chinnányaicya dwyantaram vrittanáhah

i.e., "Extract the square root of twelve times the square of the diameter. Divide this square root continually by 3 and the quotients obtained by 1, 3, 5, 7,... in order to form the first term, second term, and so on. Subtract the sum of the even terms from the sum of odd terms. If this procedure is followed and measure the diameter of great circle by 100000000000000000 equal parts, the circumference will be equal to 314159265358979324 of such parts."

If we follow the above procedure, we get the following infinite series:

$$c = \sqrt{12}d\left[1 - \frac{1}{3 \cdot 3} + \frac{1}{5 \cdot 3^2} - \frac{1}{7 \cdot 3^3} + \cdots\right]$$

where c is the circumference and d is the diameter of the circle. It is also stated that if $d = 100000000000000000$, $c = 314159265358979324$, and since

$$\frac{c}{d} = \pi$$

this gives the value of π as $\pi = 314159265358979324$.

And

$$\pi = \sqrt{12}\left[1 - \frac{1}{3 \cdot 3} + \frac{1}{5 \cdot 3^2} - \frac{1}{7 \cdot 3^3} + \cdots\right]$$

Another verse in the same chapter of a measure called "vasntitilacavrittam" is as follows:

cotihrita triguna báhu vadhè chatasmá
tattat phaláchcha bhujavarga hatáttu cotyáh
crityá critèshucha dharágni sarádi bhacte
shwojaicyatastyajatuyugmayutin dhanustat

i.e., "Divide the product of the radius and the sine by the cosine. That is the first result. Multiply the first result by the square of the sine and divide the product by the square of the cosine. That is the second result. Repeat this process of multiplying the previous result by the square of the sine and dividing the product by the square of the cosine. Divide the above found results by 1, 3 5, 7, 9,... in order to make the first term, second term, and so on. Subtract the sum of the even terms from the sum of the odd terms. This difference will give the arc whole sine is taken."

This rule is exactly the same as given in "Carana Paddhati."

Hence,

$$a = \frac{r \times S}{\cos} - \frac{rS^3}{3\cos^3} + r\frac{S^5}{5\cos^5} - \frac{rS^7}{7\cos^7} + \cdots \quad (*)$$

where a = arc, S = sine, and cos = cosine.

Suppose $a = 30°$. Then, the arc is the arc of 30°. Let $r = 1$.

Substituting in (*), we get

$$\frac{\pi}{6} = \tan 30^\circ - \frac{\tan^3 30^\circ}{3} + \frac{\tan^5 30^\circ}{5} - \frac{\tan^7 30^\circ}{7} + \cdots$$

Now,

$$\tan 30^\circ = \frac{1}{\sqrt{3}}$$

So, we have

$$\frac{\pi}{6} = \frac{1}{\sqrt{3}}\left[1 - \frac{1}{3 \cdot 3} + \frac{1}{5}\frac{1}{3^2} - \frac{1}{7}\frac{1}{3^3} + \cdots\right]$$

So,

$$2\sqrt{3}\left[1 - \frac{1}{3 \cdot 3} + \frac{1}{5}\frac{1}{3^2} - \frac{1}{7}\frac{1}{3^3} + \cdots\right] = \pi$$

i.e.,

$$\pi = \sqrt{12}\left[1 - \frac{1}{3 \cdot 3} + \frac{1}{5}\frac{1}{3^2} - \frac{1}{7}\frac{1}{3^3} + \cdots\right]$$

The rule for finding the sum of any arc in terms of tangent has been described in "Sadrantamala" in four verses of "Anustub" measure as follows:

vrittánayana márgìnabreshta jyáná dhanurnayèt
ishta jya trijyaorghátát còtyáptam prathamam phalam
jyávargom gunacam critwá cotì vargancha háracam
prathamádi phalèbhyòthá nèyáphalalatirmuhuh
ecatryadyaja sanchyábhirbhactèshwètèshwanucramát
ojánám yutitastyacte yugma yògè dhanurbhavet
dòh còtyòralpamèrèshtam calpaniyamihoditam
muhuh critèpi labd' hìnánnanyathavásitirbhavet

i.e., "According to the rule that has been laid out for the circumference of the circle, the arc of any given sine can be found in the following way: Multiply the given sine by the radius and divide this product by the cosine. This gives the first result for the given sine. Making the square of the sine as a multiplier and the square of cosine as a divisor, continue the process of multiplying the previous result by the multiplier and dividing the product by the divisor. After placing these results in order, divide them individually in succession by the odd numbers 1, 3, 5,... Then, subtract the sum of the even results from the sum of the odd results. This difference gives the arc, whose sine was taken. If the sine of the given arc is greater than its cosine, interchange them. The smaller of the sine and cosine is to be used for calculation."

So,

$$a = \frac{rS}{\cos} - \frac{rS^3}{3\cos^3} - \frac{rS^5}{5\cos^5} - \frac{rS^7}{5\cos^7} - \cdots$$

If $a = \frac{\pi}{4}$ four times this sum will be semi circumference. When the radius is 1,

$$4\left(1 - \frac{1}{3} + \frac{1}{5} - \frac{1}{7} + \cdots\right)$$

will be the semi circumference $= \pi$.

So,

$$\frac{\pi}{4} = 1 - \frac{1}{3} + \frac{1}{5} - \frac{1}{7} + \cdots$$

The same result is also given in "Tantra-Sangraha."

14 Sixteenth and Seventeenth Century Commentators of Bhaskara II

Jnanaraja

Jnanaraja is believed to be a resident of Parthapure in Maharashtra and of Bharadwaja gotra. He composed "Siddhanta Sundara" and a commentary "Bijadhyaya" on Bhaskara II's "Bijaganita" in AD 1503.

Lemma on "Vargaprakrti"

If $x=\alpha$ and $y=\beta$ be a solution of $Nx^2 \pm k = y^2$, then $x=\alpha/m$, $y=\beta/m$ is a solution of $Nx^2 \pm (k/m^2) = y^2$ and $x = n\alpha$, $y = n\beta$ is a solution of $Nx^2 \pm n^2k = y^2$, where m and n are arbitrary rational numbers.

Proof: If $x=\alpha$, $y=\beta$ be a solution of $Nx^2 \pm k = y^2$, then

$$N\alpha^2 \pm k = \beta^2 \tag{1}$$

Dividing (1) by m^2, we get

$$N\frac{\alpha^2}{m^2} \pm \frac{k}{m^2} = \frac{\beta^2}{m^2}$$

which means that

$$x = \frac{\alpha}{m}, \quad y = \frac{\beta}{m}$$

is a solution of

$$Nx^2 \pm \frac{k}{m^2} = y^2$$

Mathematical Achievements of Pre-modern Indian Mathematicians. DOI: http://dx.doi.org/10.1016/B978-0-12-397913-1.00014-4

Again multiplying (1) by n^2, we get

$$N\alpha^2 n^2 \pm kn^2 = \beta^2 n^2$$

which means that

$$x = \alpha n, \quad y = \beta n$$

is a solution of $Nx^2 \pm kn^2 = y^2$.

Example 1

Using the above lemma, find the solution of the equations:

1. $6x^2 + 12 = y^2$
2. $6x^2 + 75 = y^2$
3. $6x^2 + 300 = y^2$

We observe that $6 \cdot 1^2 + 3 = 9 = 3^2$.
So, $(1\ \ 3)$ is a solution of $6x^2 + 3 = y^2$.
Then, from the above lemma, $(n\ \ 3n)$ is a solution of $6x^2 + 3n^2 = y^2$.
If $n = 2$, we get $(2\ \ 6)$ as a solution of $6x^2 + 12 = y^2$.
If $n = 5$, we get $(5\ \ 15)$ as a solution of $6x^2 + 75 = y^2$.
If $n = 10$, we get $(10\ \ 30)$ as a solution of $6x^2 + 300 = y^2$.
If $(l\ \ m) = (2\ \ 6)$ is a solution of $6x^2 + 12 = y^2$, the general solution is given by

$$x = lq \pm mp = 2q \pm 6p$$

and

$$y = mq \pm 6lp = 6q \pm 12p$$

where $6p^2 + 1 = q^2$.

Similarly, if $(l\ \ m) = (5\ \ 15)$ is a solution of $6x^2 + 75 = y^2$, the general solution is given by $x = lq \pm mp = 5q \pm 15p$ and $y = mq \pm 6lp = 15q \pm 30p$ where $6p^2 + 1 = q^2$.

Again, if $(l\ \ m) = (10\ \ 30)$ is a solution of $6x^2 + 300 = y^2$, the general solution is given by

$$x = lq \pm mp = 10q \pm 30p$$

and

$$y = mq \pm 6lp = 30q \pm 60p$$

where $6p^2 + 1 = q^2$.

Rational solution of $Nx^2 + 1 = y^2$

$$\left(\frac{2m}{m^2 \sim N}\ \frac{m^2 + N}{m^2 \sim N}\right)$$

is a rational solution of $Nx^2 + 1 = y^2$, where m is any rational number.

Proof: Let $y = mx - 1$.
Substituting in $Nx^2 + 1 = y^2$, we get

$$Nx^2 + 1 = (mx - 1)^2 = m^2x^2 - 2mx + 1$$

We get $(m^2 - N)x^2 - 2mx = 0$.
If $x \neq 0$, we have $(m^2 - N)x = 2m$

$$x = \frac{2m}{m^2 - N}$$

$$y = mx - 1 = m \cdot \frac{2m}{m^2 - N} - 1 = \frac{2m^2 - m^2 + N}{m^2 - N} = \frac{m^2 + N}{m^2 - N}$$

So,

$$x = \frac{2m}{m^2 \sim N}$$

and

$$y = \frac{m^2 + N}{m^2 \sim N}$$

is a rational solution of $Nx^2 + 1 = y^2$.

Example 2

Jnanaraja has given an example of solving in integers of the equation

$$ax^2 + bx + c = a^1y^2 + b^1y + c$$

Find the integral solutions of $3(x^2 + 4x) = y^2 + 4y$.

(Continued)

Example 2 (cont'd)

Solution

Multiplying by 3, we get

$$9(x^2 + 4x) = 3y^2 + 12y$$

$$9x^2 + 36x = 3y^2 + 12y$$

So,

$$(3x+6)^2 = 3y^2 + 12y + 36$$

Let $3x + 6 = z$.
Then, we have $z^2 = 3 \cdot y^2 + 12y + 36$.
Multiplying by 3, we get

$$3z^2 = 9y^2 + 36y + 108$$

$$3z^2 = (3y + 6)^2 + 72$$

Let $3y + 6 = w$.
Then, we have $3z^2 = w^2 + 72$ or $3z^2 - 72 = w^2$.
We observe that $3 \cdot 1^2 + 6 = 9 = (3)^2$.
So, $(1 \;\; 3)$ is a solution of $3z^2 + 6 = w^2$.
Again, we observe that $3 \cdot 4^2 - 12 = 36 = (6)^2$.
So, $(4 \;\; 6)$ is a solution of

$$3z^2 - 12 = w^2$$

$$(1 \;\; 3 \;\; 6) \circ (4 \;\; 6 - 12) = (1.6 + 3.4 \quad 3.6 + 1.4.3 \quad (-12)\,6) = (18 \; 30 - 72)$$

which means that
$(z \;\; w) = (18 \;\; 30)$ is a solution of $3z^2 - 72 = w^2$.
Let $(l \;\; m) = (18 \;\; 30)$.
Then, the general solution of $w^2 = 3z^2 - 72$ will be

$$z = lq \pm mp = 18q \pm 30p$$

$$w = mq \pm 3lp = 30q \pm 54p$$

where

$$3p^2 + 1 = q^2$$

We observe that if $p = 0$, $q = 1$; if $p = 1$, $q = 2$; if $p = 4$, $q = 7$; and so on.
If $p = 0$, $q = 1 \Rightarrow z = 18$, $w = 30$

$$z = 3x + 6 = 18 \Rightarrow 3x = 12 \Rightarrow x = 4$$

$$w = 3y + 6 = 30 \Rightarrow 3y = 24 \Rightarrow y = 8$$

If $p = 1$, $q = 2 \Rightarrow z = 36 \pm 30 = 66$ or 6

$$w = 60 \pm 54 = 114 \text{ or } 6$$

Discarding $z = 6$ and $w = 6$, we get $z = 66$ and $w = 114$

$$z = 3x + 6 = 66 \Rightarrow 3x = 60 \Rightarrow x = 20$$

$$w = 3y + 6 = 114 \Rightarrow 3y = 108 \Rightarrow y = 36$$

If $p = 4$, $q = 7$, $z = 18(7) \pm 30(4) = 126 \pm 120 = 246$ or 6

$$w = 30(7) \pm 54(4) = 210 \pm 216 = 426 \text{ or } -6$$

Discarding $z = 6$ and $w = -6$, we get

$$z = 246 \quad \text{and} \quad w = 426$$

$$z = 3x + 6 = 246 \Rightarrow 3x = 240 \Rightarrow x = 80$$

$$w = 3y + 6 = 426 \Rightarrow 3y = 420 \Rightarrow y = 140$$

So, $(x \;\; y) = (4 \;\; 8), (20 \;\; 36), (80 \;\; 140)$, and so on.

Suryadasa

Suryadasa is the son of Jnanaraja. He composed "Ganithamrita" in AD 1538 and "Suryaprakasha" in AD 1541, which were commentaries of Bhaskara II's "Lilavati" and "Bijaganita."

The rational solutions of $y^2 - a^2x^2 = \pm c$ is given by

$$x = \frac{1}{2a}\left(\frac{\pm c}{m} - m\right) \quad \text{and} \quad y = \frac{1}{2}\left(\frac{\pm c}{m} + m\right)$$

where m is a rational number.

The above solution is given by both Suryadasa and Krsna.

Proof:

$$y^2 - a^2x^2 = \pm c$$

$$(y + ax)(y - ax) = \pm c$$

Let

$$y - ax = m \tag{1}$$

Then,

$$y + ax = \frac{\pm c}{m} \tag{2}$$

Adding (1) and (2), we get

$$2y = m \pm \frac{c}{m} \Rightarrow y = \frac{1}{2}\left(m \pm \frac{c}{m}\right)$$

Subtracting (1) from (2), we get

$$2ax = \pm \frac{c}{m} - m$$

So,

$$x = \frac{1}{2a}\left(\pm \frac{c}{m} - m\right)$$

Construction of a Rational Right Triangle Given a Side a

Let $n^2 - 1$, $2n$, $n^2 + 1$ be the sides of a rational right triangle, where n is a rational number.

If (x, y, z) are the sides of another right triangle, such that

$$\frac{x}{n^2 - 1} = \frac{y}{2n} = \frac{z}{n^2 + 1} = k \quad \text{(say)}$$

Then,

$$x = (n^2 - 1)k, \quad y = 2nk$$

and

$$z = k(n^2 + 1)$$

If $x = a$, then $a = (n^2 - 1)k$.
So,

$$k = \frac{a}{n^2 - 1}$$

$$y = 2nk = \frac{2na}{n^2 - 1}, \quad z = k(n^2 + 1) = \frac{a(n^2 + 1)}{n^2 - 1}$$

A second method of construction has been given by Suryadasa, Ganeśa, and Ranganatha as follows:

If $x = a$ is one side, y and z are the other sides, we have

$$y^2 + a^2 = z^2, \text{if } z \text{ is the hypotenuse}$$

If z is the hypotenuse

$$a^2 = (z^2 - y^2) = (z + y)(z - y)$$

Let

$$z - y = m, \text{ where } m \text{ is a rational number } 0 \tag{1}$$

then

$$z + y = \frac{a^2}{m} \tag{2}$$

Adding (1) and (2), we get

$$2z = m + \frac{a^2}{m} \Rightarrow z = \frac{1}{2}\left(m + \frac{a^2}{m}\right)$$

Subtracting (1) from (2), we get

$$2y = \frac{a^2}{m} - m \Rightarrow y = \frac{1}{2}\left(\frac{a^2}{m} - m\right)$$

Construction of a Rational Right Triangle Given the Hypotenuse

Let $n^2 - 1$, $2n$, $n^2 + 1$ be the sides of rational right triangle, when n is a rational number.

Let c be the given hypotenuse.

If (x, y, c) are sides of the right triangle, then

$$\frac{x}{n^2-1} = \frac{y}{2n} = \frac{c}{n^2+1} = k \quad \text{(say)}$$

Then,

$$x = k(n^2-1), \quad y = 2nk, \quad \text{and} \quad c = k(n^2+1) \Rightarrow k = \frac{c}{n^2+1}$$

Thus,

$$x = \frac{c}{n^2+1}(n^2-1)$$

$$y = \frac{2nc}{n^2+1}$$

Thus,

$$\frac{c(n^2-1)}{n^2+1}, \quad \frac{2nc}{n^2+1}, \quad \text{and} \quad c$$

are the sides of a rational triangle.

Krsna (Sixteenth Century) AD 1580

Krsna belongs to a family of Benares astronomers of Devarāta gotra. This family is believed to have moved to Benares from Maharashtra in the late sixteenth century.

Krsna stated that equations in general can be classified into two categories: (i) equations with single unknown and (ii) equations with several unknowns. He pointed out that the category (i) itself consists of two subcategories: (1) simple equations and (2) quadratic and higher degree equations, whereas category (ii) itself consists of three subcategories: (1) simultaneous linear equations, (2) second and higher degree equations with several variables, and (3) equations that contain product of unknowns. He further pointed out that the above five subcategories can be made to four subcategories by grouping together the subcategory (2) of both categories (i) and (ii) to form one subcategory called "*madyamaharana*."

Krsna has provided the following lemmas regarding the solution of "Kuttaka," namely,

$$by = ax \pm c \tag{1}$$

Lemma 1: Let g be the greatest common factor of a and c.

If $x = \alpha$, $y' = \beta$ be a solution of

$$by' = a'x \pm c'$$

where

$$y' = \frac{y}{g}, \quad a = a'g, \quad c = c'g$$

then $x = \alpha$, $y = g\beta$ is a solution of (1).

Proof: Now, $x = \alpha$, $y' = \beta$ is a solution of

$$by' = a'x \pm c'$$

So,

$$b\beta = a'\alpha \pm c'$$

Multiplying by g, we get

$$bg\beta = a'g\alpha \pm gc'$$

i.e.,

$$b(g\beta) = a\alpha \pm c$$

Hence, $(x \;\; y) = (\alpha \;\; g\beta)$ is a solution of $by = ax \pm c$.

Lemma 2: Let $b = g'b'$, $c = g'c''$.
If $(\alpha' \;\; \beta')$ is a solution of $b'y = ax' \pm c''$, then $(g'\alpha' \;\; \beta')$ is a solution of $by = ax \pm c$.

Proof: $(\alpha' \;\; \beta')$ is a solution of $b'y = ax' \pm c''$.
So, $b'\beta' = a\alpha' \pm c''$.
Multiplying by g', we get

$$g'b'\beta' = ag'\alpha' \pm g'c$$

$$b\beta' = a[g'(\alpha')] \pm c$$

So, $(g'\alpha' \;\; \beta')$ is a solution of $by = ax \pm c$.

Lemma 3: Let $a = a'g$, $C = C'g$, $b = b''g''$, $C' = C''g''$.

If $(\alpha \ \beta)$ is a solution of $b''y' = a'x' \pm C''$, where $x' = x/g''$, $y' = y/g$, then $(g''\alpha \ g\beta)$ is a solution of $by = ax \pm C$.

Proof: $(\alpha \ \beta)$ is a solution of $b''y' = a'x' \pm C''$.
So, $b''\beta = a'\alpha \pm C''$.
Multiplying by gg'', we get

$$gg''b''\beta = a'gg''\alpha \pm gg''C''$$

$$b(g\beta) = a(g''\alpha) \pm C$$

So, $(g''\alpha \ g\beta)$ is a solution of $by = ax \pm C$.

Lemma 4: If $(\alpha \ \beta)$ is a minimum solution of $ax + c = by$, then $(b - \alpha \ a - \beta)$ is a solution $by = ax - c$. If $c < a$ and $c < b$, this solution will be a minimum solution of $by = ax - c$.

Proof: $(\alpha \ \beta)$ is a solution of $ax + c = by$.
So,

$$a\alpha + c = b\beta$$

Adding ab to both sides of the equations, we get

$$a\alpha + ab + c = b\beta + ab$$

$$ab - b\beta + c = ab - a\alpha$$

$$b(a - \beta) + c = a(b - \alpha)$$

So,

$$a(b - \alpha) - c = b(a - \beta)$$

So,

$$(b - \alpha \ a - \beta)$$

is a solution $ax - c = by$.

Krsna gave the following proof of rational solutions of the equation $Nx^2 + 1 = y^2$.

Proof:

$$4Nm^2 = (m^2 + N)^2 - (m^2 - N)^2$$

where m is an arbitrary integer, i.e.,

$$N(2m)^2 + (m^2 - N)^2 = (m^2 + N)^2$$

Dividing by $(m^2 - N)^2$, we get

$$N\left(\frac{2m}{m^2 - N}\right)^2 + 1 = \left(\frac{m^2 + N}{m^2 - N}\right)^2$$

which means that

$$\left(\frac{2m}{m^2 - N} \quad \frac{m^2 + N}{m^2 - N}\right)$$

is a solution of $Nx^2 + 1 = y^2$.

The following is the proof of Brahmagupta's lemma 1 on page 164.

Proof: $(\alpha \quad \beta)$ is a solution of $Nx^2 + K = y^2$.
So,

$$N\alpha^2 + K = \beta^2 \tag{1}$$

Similarly, $(\alpha' \quad \beta')$ is a solution of $Nx^2 + K' = y^2$.
So,

$$N\alpha'^2 + K' = \beta'^2 \tag{2}$$

Multiplying (1) by β'^2, we get

$$N\alpha^2\beta'^2 + K\beta'^2 = \beta^2\beta'^2 \tag{3}$$

Substituting the value of β'^2 from (2), equation (3) becomes

$$N\alpha^2\beta'^2 + K(N\alpha'^2 + K') = \beta^2\beta'^2$$

$$N\alpha^2\beta'^2 + N\alpha'^2(\beta^2 - N\alpha^2) + KK' = \beta^2\beta'^2$$

by substituting the value of K from (1).
So, we have

$$N(\alpha^2\beta'^2 + \alpha'^2\beta^2) + KK' = \beta^2\beta'^2 + N^2\alpha'^2\alpha^2$$

Adding $\pm 2N\alpha\alpha'\beta\beta'$ to both sides of the equation, we get

$$N(\alpha\beta' \pm \beta\alpha')^2 + KK' = (\beta\beta' \pm N\alpha'\alpha)^2$$

which means that $(\alpha\beta' \pm \alpha'\beta \; \beta\beta' \pm N\alpha\alpha')$ is a solution of the equation $Nx^2 + KK' = y^2$.

This completes the proof of Brahmagupta's lemma 1 on page 164.

Astronomers of Maharashtra

Ganeśa Daivajna

Ganeśa Daivajna was a reputed astronomer and was born in Saka 1429 (AD 1507) at a village called Nandigrama on the Western Coast, about 40 miles south of Mumbai (Bombay) in the Konkan region of Maharashtra state. His father was a well-known astronomer, Keśava Daivajña, and his mother's name is Lakshmi. Ganeśa's paternal grandfather, Kamalakara, was also an eminent astronomer. The family belonged to Kaushika gotra. Keśava composed the astronomical treatise called "Graha Kautuka" and several other works and commentaries on astronomy and astrology. Keśava has been considered as one of the best observational astronomers of ancient India.

Ganeśa himself authored several important treatises and manuals on astronomy and astrology. Included among them are "Grahalāghava," "Laghu- and Brht-Tithi Chintamani," a commentary on Bhaskara II's "Siddharta Siromani," "Buddhi Vilasini," a commentary on Bhaskara II's "Lilavati," "Sraddha nirnaya," "Patasārani," and "Parva nimaya." It is believed that Ganeśa composed his first work, "Graha Laghava," when he was only 13 years old. "Laghu Tithi Chintamani" was composed in AD 1525 when he was 18 years old and "Buddhi Vilasini" in AD 1545.

Eminent Maharashtrian astronomers Gangadhara (AD 1586), Mallari (AD 1602), and Visvanatha (about AD 1612) have written commentaries on the "Graha Lāghava." The "Graha Laghava" is extensively used by the Panchanga makers in Maharashtra, Gujarat, Andhra Pradesh, and Karnataka. In the "Graha Lāghava," planetary positions have been given at the instant of sunrise of Monday, the new moon day of Palguna of Saka 1441, which corresponds to March 19, AD 1520.

For Panchanga makers or for persons who do not know trigonometry, "Laghu Lāghava" is easy to use. In spite of dropping off the trigonometric ratios, the accuracy of the results has not been seriously affected. Since every Karana manual using the sine tables generally give the sine value of angles at regular intervals of either 15° or $3\frac{3^\circ}{4}$, the sine values for intermediate angles by linear interpolations are only approximate.

Gangadhara's commentary on "Laghu Lāghava" is called "manorama." His father himself has composed "Muhurta Mārtanda." Gangadhara was a Vajasaneyi Brahmin belonging to Kaushika gotra.

Visvanātha and Mallari are brothers; they had another brother Vishnu who was an eminent astronomer himself. He composed not only a Karana text based on the "Surya Siddhanta" but also a commentary on Ganeśa's "Brhat Tithi Cintamani." Visvanatha has also authored "Udaharana" on his brother, Vishnu's Karana.

Kamalakara (AD 1658), the author of "siddhanta tattva viveka," and Ranganatha (AD 1602), who wrote a commentary on "Surya Siddhanta," are descendents of Visvanatha and Mallari. Kamalakara was an astronomer in the court of Mogul emperor Jahangir.

Rational Right Triangle

For the construction of right triangle, given a side a, Bhaskara II has enunciated the following rule for the sides of this right triangle as one of the solutions:

$$a,\ \frac{1}{2}\left(\frac{a^2}{n}-n\right)$$

and

$$\frac{1}{2}\left(\frac{a^2}{n}+n\right)$$

where n is an arbitrary rational number. This solution is exactly the same as that of Brahmagupta. Ganesha has provided a rationale for this rule.

Suppose the sides of the right triangle are a, x, and z, where z is the hypotenuse and a is the given side.

Then,

$$a^2+x^2=z^2$$

i.e.,

$$a^2=z^2-x^2=(z+x)(z-x)$$

Let

$$z-x=n \qquad (1)$$

Then,

$$z + x = \frac{z^2 - x^2}{z - x} = \frac{a^2}{n} \tag{2}$$

Adding (1) and (2), we get

$$2z = n + \frac{a^2}{n} \Rightarrow z = \frac{1}{2}\left(n + \frac{a^2}{n}\right)$$

Subtracting (1) from (2), we get

$$2x = \frac{a^2}{n} - n \Rightarrow x = \frac{1}{2}\left(\frac{a^2}{n} - n\right)$$

Then,

$$(a,\ x,\ z) = \left[a,\ \frac{1}{2}\left(\frac{a^2}{n} - n\right),\ \frac{1}{2}\left(\frac{a^2}{n} + n\right)\right]$$

Rational Quadrilateral

Ganesha demonstrated that the quadrilateral is formed by the juxtaposition of four right triangles formed by multiplying by the sides of each of the two right triangles by the "Koti" (upright) and the "bhuja" (base) of the other. He further stated that a quadrilateral will be divided into four triangles by the intersection of the diagonals. So, by the juxtaposition of these four triangles, a quadrilateral will be formed. For that purpose, the four triangles have to be selected in a special way. Two right triangles should be formed as Brahmagupta has suggested, namely, $(m^2 - n^2,\ 2mn,\ m^2 + n^2)$ and $(a^2 - b^2,\ 2ab,\ a^2 + b^2)$, where m, n, a, and b are arbitrary rational numbers. Recalling if the "Koti" (upright) and the "bhuja" (base) and the hypotenuse are multiplied by a rational number, a new right triangle with rational sides will be formed, by multiplying the sides of each of the above selected triangles by a rational number equal to the base of the other and again by a rational number equal to the upright of the other, four rational triangles will be formed with the following sides:

1. $(m^2 - n^2)(a^2 - b^2)$, $2mn(a^2 - b^2)$, $(m^2 + n^2)(a^2 - b^2)$;
2. $(m^2 - n^2)2ab$, $2mn(2ab)$, $(m^2 + n^2)2ab$;
3. $(a^2 - b^2)(m^2 - n^2)$, $2ab(m^2 - n^2)$, $(a^2 + b^2)(m^2 - n^2)$;
4. $2mn(a^2 - b^2)$, $2mn(2ab)$, $2mn(a^2 + b^2)$.

A rational quadrilateral will be formed with the juxtaposition of these four right triangles. He then actually demonstrated (shown in the below figures) with the specific example of starting with two right triangles (4, 3, 5) and (12, 5, 13).

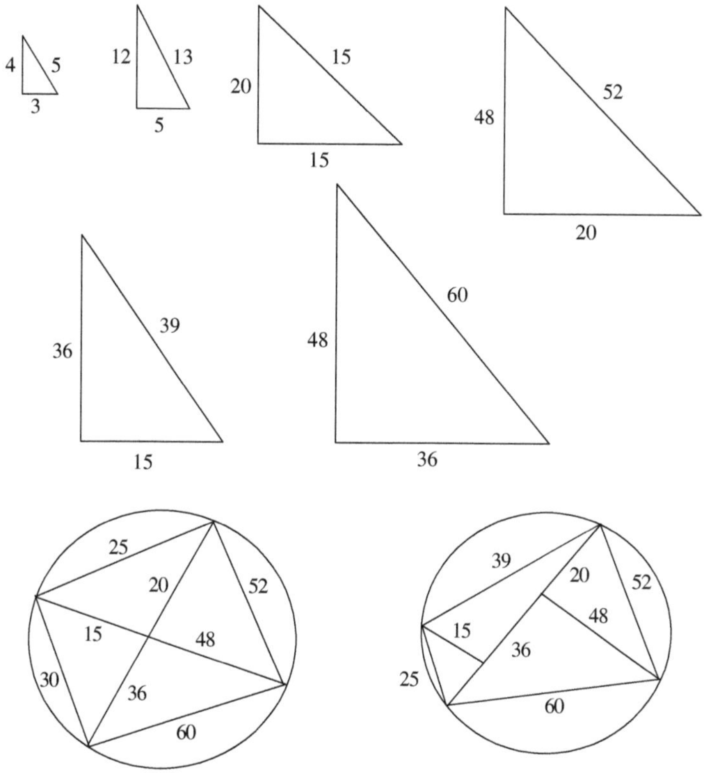

We can obtain in this way, from the set of two rational right triangles, two different variety of rational convex quadrilaterals; one in which the diagonals intersect at right angles and the other in which the diagonals intersect obliquely.

Kamalakara (AD 1658)

Kamalakara belongs to a family of astronomers of Bharadwaja gotra, who migrated from Maharashtra to Benares in the sixteenth century. He composed "Siddhanta tattva viveka," a commentary of Bhaskara II's "Bijaganita," in AD 1658.

Kamalakara has provided other methods to solve particular cases of the equation $ax^2 + by^2 + c = z^2$.

Case I: Let b and c be of opposite signs. Here two subcases arise:

1. c is negative and b positive.
2. b is negative and c positive.

Subcase 1 of (1)

Solve $ax^2 + by^2 - c = z^2$.

Find u and v from the "vargaprakrti" equations $au^2 - c = v^2$ by Brahmagupta–Bhaskara II method.

Let

$$x = \sqrt{\frac{b}{ac}}vy + u$$

Then,

$$\begin{aligned} ax^2 + by^2 - c &= a\left(\sqrt{\frac{b}{ac}}vy + u\right)^2 + by^2 - c \\ &= \frac{b}{c}v^2y^2 + 2\sqrt{\frac{ab}{c}}uvy + by^2 + au^2 - c \\ &= \frac{b}{c}(au^2 - c)y^2 + 2\sqrt{\frac{ab}{c}}uvy + by^2 + v^2 \end{aligned}$$

since $au^2 - c = v^2$.

Hence,

$$\begin{aligned} ax^2 + by^2 - c &= \frac{b}{c}au^2y^2 - by^2 + 2\sqrt{\frac{ab}{c}}uvy + by^2 + v^2 \\ &= \frac{b}{c}au^2y^2 + 2\sqrt{\frac{ab}{c}}uvy + v^2 \end{aligned}$$

Hence,

$$ax^2 + by^2 - c = \left(\sqrt{\frac{ab}{c}}uy + y\right)^2$$

But,

$$ax^2 + by^2 - c = z^2$$

Hence,

$$z = \sqrt{\frac{ab}{c}}uy + v$$

Subcase 2 of (1)

Solve $ax^2 - by^2 + c = z^2$.

Find u and v from the "*vargaprakrti*" equation $au^2 + c = v^2$.

Let

$$x = \sqrt{\frac{b}{ac}}vy + u$$

Then,

$$ax^2 - by^2 + c = a\left[\sqrt{\frac{b}{ac}}vy + u\right]^2 - by^2 + c$$

$$ax^2 - by^2 + c = \frac{b}{c}v^2y^2 + 2\sqrt{\frac{ab}{c}}vyu + au^2 - by^2 + c$$

But,

$$au^2 + c = v^2$$

Hence, we have

$$ax^2 - by^2 + c = \frac{b}{c}v^2y^2 + 2\sqrt{\frac{ab}{c}}uvy - by^2 + v^2$$

since $au^2 + c = v^2$, i.e.,

$$ax^2 - by^2 + c = by^2\left(\frac{v^2}{c} - 1\right) + 2\sqrt{\frac{ab}{c}}uvy + v^2$$

$$= by^2\frac{(v^2 - c)}{c} + 2\sqrt{\frac{ab}{c}}uvy + v^2$$

But,

$$v^2 - c = au^2$$

$$ax^2 - by^2 + c = \frac{aby^2u^2}{c} + 2\sqrt{\frac{ab}{c}}uvy + v^2$$

$$= \left(\sqrt{\frac{ab}{c}}yu + v\right)^2$$

But,

$$z^2 = ax^2 - by^2 + c^2 = \left(\sqrt{\frac{ab}{c}}yu + v\right)^2$$

So,

$$z = \sqrt{\frac{ab}{c}}yu + v$$

Kamalakara gave the following illustrative examples.

Example 1

Solve

$$5x^2 + 16y^2 - 20 = z^2 \quad (*) \quad \text{(Sitvi. xii. 279)}$$

This is an example of subcase (1) of Case I.
Here, $a = 5$, $b = 16$, $c = +20$.
Solution of (*) is

$$x = \sqrt{\frac{b}{ac}}vy + u \quad \text{and} \quad z = \sqrt{\frac{ab}{c}}uy + v$$

where $au^2 - c = v^2$, i.e., $5u^2 - 20 = v^2$.
Hence,

$$x = \sqrt{\frac{16}{5 \cdot 20}}vy + u = \frac{2}{5}vy + u$$

$$z = \sqrt{\frac{5 \cdot 16}{20}}uy + v = 2uy + v$$

where $au^2 - c = v^2 \Rightarrow 5u^2 - 20 = v^2$.
Now we have to solve "Vargaprakrti" equation $5u^2 - 20 = v^2$.
We observe that $5 \cdot 3^2 - 20 = 45 - 20 = 25 = (5)^2$.
So, $(u \;\; v) = (3 \;\; 5)$ is a solution of $5u^2 - 20 = v^2$.
Hence,

$$x = \frac{2}{5}(5y) + 3 = 2y + 3$$

$$z = 2(3)y + 5 = 6y + 5$$

are the solutions of (*).

Hence, $(y, x, z) = (1, 5, 11), (2, 7, 17)$, and so on.

Example 2

Solve

$$5x^2 - 20y^2 + 16 = z^2 \quad (*)$$

This is an example of subcase (2) of Case I.

$$a = 5, \quad b = 20, \quad c = 16$$

Solution is

$$x = \sqrt{\frac{b}{ac}}vy + u = \sqrt{\frac{20}{5 \cdot 16}}vy + u = \frac{1}{2}vy + u$$

$$z = \sqrt{\frac{ab}{c}}uy + v = \sqrt{\frac{5 \cdot 20}{16}}uy + v = \frac{5}{2}uy + v$$

where

$$au^2 + c = v^2$$

i.e.,

$$5u^2 + 16 = v^2$$

We observe that $5 \cdot 2^2 + 16 = 20 + 16 = 36 = (6)^2$.

So, (2, 6) is a solution of $5u^2 + 16 = v^2$.

Substituting $u = 2$, $v = 6$, we get

$$x = \frac{1}{2}(6)y + 2 = 3y + 2$$

(*Continued*)

Example 2 (cont'd)

$$z = \frac{5}{2}(2)y + 6 = 5y + 6$$

are solutions of (*).

Hence, $(y, x, z) = (1, 5, 11), (2, 8, 16)$, and so on.

Case II: Coefficient of either x^2 or y^2 is a perfect square, i.e., let a or b be a perfect square.

Kamalakara divided this case into two subcases:

i. c is a perfect square.
ii. c is not a perfect square.

Subcase (i)
(a) Let the coefficient of x^2 be a perfect square and c a perfect square.
Let $a = d^2$ and $c = e^2$.
Then, we have to solve $ax^2 + by^2 + c = z^2$, i.e.,

$$d^2x^2 + by^2 + e^2 = z^2$$

Let

$$x = \frac{b}{2de}y^2$$

Then,

$$d^2x^2 + by^2 + e^2 = d^2\left(\frac{by^2}{2de}\right)^2 + by^2 + e^2$$

$$= \frac{b^2y^4}{4e^2} + by^2 + e^2 = \left(\frac{by^2}{2e} + e\right)^2 = z^2$$

Hence,

$$z = \frac{by^2}{2e} + e$$

(b) Let the coefficient of y^2 be a perfect square and c a perfect square.
Let $b = d^2$, $c = e^2$.

We have to solve $ax^2 + by^2 + c = z^2$, i.e.,

$$ax^2 + d^2y^2 + e^2 = z^2$$

Let

$$y = \frac{a}{2de}x^2$$

Then,

$$ax^2 + d^2y^2 + e^2 = ax^2 + d^2\left(\frac{ax^2}{2de}\right)^2 + e^2$$

$$= ax^2 + \frac{a^2x^4}{4e^2} + e^2 = \left(\frac{ax^2}{2e} + e\right)^2 = z^2$$

So,

$$z = \frac{ax^2}{2e} + e$$

Subcase (ii)
Suppose c is not a square.
(a) $a = d^2$. We have to solve $d^2x^2 + by^2 + c = z^2$.
Find α and β from the "vargaprakrti" equation $\alpha^2 + c = \beta^2$.
Then, let

$$n = \frac{d\beta n - (b/2)}{d\alpha}$$

Then,

$$n\alpha\alpha = d\beta n - \frac{b}{2}$$

$$d\beta n - n\alpha\alpha = \frac{b}{2}$$

$$nd(\beta - \alpha) = \frac{b}{2}$$

So,

$$n = \frac{b}{2d(\beta - \alpha)}$$

Let

$$x = ny^2 + \frac{\alpha}{d}$$
$$= \frac{b}{2d(\beta - \alpha)}y^2 + \frac{\alpha}{d}$$

Then,

$$d^2x^2 + by^2 + c = d^2\left[\frac{b}{2d(\beta-\alpha)}y^2 + \frac{\alpha}{d}\right]^2 + by^2 + c$$

So,

$$d^2x^2 + by^2 + c = \frac{b^2y^4}{4(\beta-\alpha)^2} + \frac{by^2\alpha}{\beta - \alpha} + \alpha^2 + by^2 + c$$

$$d^2x^2 + by^2 + c = \frac{b^2y^4}{4(\beta-\alpha)^2} + \frac{by^2\alpha}{\beta - \alpha} + by^2 + \beta^2$$

since $\alpha^2 + c = \beta^2$.

$$d^2x^2 + by^2 + c = \frac{b^2y^4}{4(\beta-\alpha)^2} + by^2\left[1 + \frac{\alpha}{\beta - \alpha}\right] + \beta^2$$

$$d^2x^2 + by^2 + c = \frac{b^2y^4}{4(\beta-\alpha)^2} + \frac{by^2\beta}{\beta - \alpha} + \beta^2 = \left(\frac{by^2}{2(\beta-\alpha)} + \beta\right)^2 = z^2$$

So,

$$z = \frac{by^2}{2(\beta - \alpha)} + \beta, \quad b = d^2$$

(b) We have to solve $ax^2 + d^2y^2 + c = z^2$ (here, $b = d^2$).
Find α and β from the "vargaprakrti" equation

$$\beta^2 + c = \alpha^2$$

Let

$$n = \frac{d\alpha n - (a/2)}{d\beta}$$

Then,

$$nd\beta = d\alpha n - \frac{a}{2}$$

$$nd(\alpha - \beta) = \frac{a}{2}$$

$$n = \frac{a}{2d(\alpha - \beta)}$$

Let

$$y = nx^2 + \frac{\beta}{d} = \frac{a}{2d(\alpha - \beta)}x^2 + \frac{\beta}{d}$$

Then, we have

$$\begin{aligned} ax^2 + d^2y^2 + c &= ax^2 + d^2\left[\frac{a}{2d(\alpha-\beta)}x^2 + \frac{\beta}{d}\right]^2 + c \\ &= ax^2 + \frac{a^2x^4}{4(\alpha-\beta)^2} + \frac{ax^2\beta}{\alpha - \beta} + \beta^2 + c \\ &= ax^2\left[1 + \frac{\beta}{\alpha - \beta}\right] + \frac{\alpha^2x^4}{4(\alpha-\beta)^2} + \alpha^2 \end{aligned}$$

since

$$\begin{aligned} \alpha^2 &= \beta^2 + c \\ &= \frac{ax^2\alpha}{\alpha - \beta} + \frac{a^2x^4}{4(\alpha-\beta)^2} + \alpha^2 \end{aligned}$$

$$ax^2 + d^2y^2 + c = \left(\frac{ax^2}{2(\alpha-\beta)} + \alpha\right)^2 = z^2$$

So,

$$z = \frac{ax^2}{2(\alpha - \beta)} + \alpha$$

Kamalakara provided the following illustrative example.

Example 3

Solve

$$4x^2 + 48y^2 + 20 = z^2 \quad (*)$$

This is an example, where a is a perfect square and c not a perfect square. Let $a = d^2 = 4 \Rightarrow d = 2$ and $c = 20$ (not a perfect square) and $b = 48$. Find α and β from the "vargaprakrti" equation $\alpha^2 + 20 = \beta^2$.

Let

$$n = \frac{b}{2d(\beta - \alpha)} = \frac{48}{2 \cdot 2(\beta - \alpha)} = \frac{12}{\beta - \alpha}$$

where $\alpha^2 + 20 = \beta^2$, which is a "vargaprakrti" equation.

We observe that $4^2 + 20 = 16 + 20 = 36 = (6)^2$.

So, $(\alpha \ \beta) = (4 \ 6)$ is a solution of $\alpha^2 + 20 = \beta^2$.

Hence,

$$n = \frac{12}{6 - 4} = 6$$

Solutions are

$$x = ny^2 + \frac{\alpha}{d}, \quad z = \frac{by^2}{2(\beta - \alpha)} + \beta$$

So,

$$x = 6y^2 + \frac{4}{2} = 6y^2 + 2$$

$$z = \frac{48y^2}{2(6 - 4)} + 6 = 12y^2 + 6$$

are the solutions of (*).

Hence, $(y, x, z) = (1, 8, 18), (2, 26, 54)$, and so on.

Observation: It may be observed that the solution stated by Kamalakara follows easily from the solution of Bhaskara II on page 443.

Let

$$m = \beta - \alpha$$

where

$$\alpha^2 + c = \beta^2$$

$$x = \frac{1}{2a}\left[\frac{by^2 + c}{m} - m\right] = \frac{1}{2a}\left[\frac{by^2 + \beta^2 - \alpha^2}{\beta - \alpha} - (\beta - \alpha)\right]$$

$$= \frac{1}{2a}\left[\frac{by^2}{\beta - \alpha} + \beta + \alpha - (\beta - \alpha)\right] = \frac{1}{2a}\left[\frac{by^2}{\beta - \alpha} + 2\alpha\right]$$

$$= \frac{1}{2a}\frac{by^2}{\beta - \alpha} + \frac{\alpha}{a}$$

Appendix

Solution of Generalized "*Kuttaka*" Equation

Proposition 1: Solve for integral solutions of the equations

$$A_1X_1 + A_2X_2 + A_3X_3 + \cdots\cdots + A_nX_n = C(^*)$$

where A_i $(i = 1,2,\ldots, n)$ and C are integers.

Solution: Let $X_i = (a_ix + b_iy)$, where a_i and b_i for $(i = 1, 2,\ldots, n)$ are integers.
Then (x) becomes

$$(A_1a_1 + A_2a_2 + A_3a_3 + \cdots + A_na_n)x + (A_1b_1 + A_2b_2 + A_3b_3 + \cdots + A_nb_n)y = C \quad (^{**})$$

Let

$$a = A_1a_1 + A_2a_2 + A_3a_3 + \cdots + A_na_n$$

and

$$b = A_1b_1 + A_2b_2 + A_3b_3 + \cdots + A_nb_n$$

Then $(^{**})$ becomes

$$ax + by = C$$

which can be solved in integers as a "Kuttaka" equation.

Illustrative Examples

Example 1

Find the integral solutions of the equation

$$3x_1 - 5x_2 + 6x_3 - 4x_4 + 7x_5 - 11x_6 - 9x_7 + 12x_8 + 275 = 0 \quad (^*)$$

Solution

Let

$$x_1 = 3x + 2y,\ \ x_2 = 4y - 5x,\ \ x_3 = 6x + 7y,\ \ x_4 = 17x - 6y,$$
$$x_5 = 8x + 3y,\ \ x_6 = 2x + 9y,\ \ x_7 = 14x + 5y,\ \ \text{and}\ \ x_8 = 9x + 2y.$$

Then (x) becomes

$$3(3x + 2y) - 5(4y - 5x) + 6(6x + 7y) - 4(17x - 6y) + 7(8x + 3y)$$
$$- 11(2x + 9y) - 9(14x + 5y) + 12(9x + 2y) + 275 = 0$$

i.e.,

$$9x + 6y - 20y + 25x + 36x + 42y - 68x + 24y + 56x + 21y$$
$$- 22x - 99y - 126x - 45y + 108x + 24y + 275 = 0$$

i.e.,

$$18x - 47y + 275 = 0$$

or

$$18x + 275 = 47y \quad (^{**})$$

We now solve the equation (**)

$$47)18(0 = q_1$$
$$\underline{0}$$
$$18)47(2 = q_2$$
$$\underline{36}$$
$$11)18(1 = q_3$$
$$\underline{11}$$
$$7)11(1 = q_4$$
$$\underline{7}$$
$$4)7(1 = q_5$$
$$\underline{4}$$
$$3)4(1 = q_6$$
$$\underline{3}$$
$$1$$

So, we have the quotients $q_1 = 0$, $q_2 = 6$, $q_1 = 3$, $q_4 = 1$, $q_5 = 1$, and $q_6 = 1$. Hence, the number of quotients is even.

We now form the following table:

0	0	0	0	0	0	1375
2	2	2	2	2	3575	3575
1	1	1	1	1375	1375	1375
1	1	1	825	825	825	
1	1	550	550	550		
1	275	275	275			
275	275	275				
0	0					

So,

$$y = 1375 \quad \text{and} \quad x = 3575$$

Now,

$$y = 1375 \equiv 7(\text{mod } 18)$$

$$x = 3575 \equiv 3(\text{mod } 47)$$

Then, $(x\ y) = (3\ 7)$ is the least positive integral solution of the equation (**).

Hence, the general solution of (**) is given by

$$x = 3 + 47m \quad \text{and} \quad y = 7 + 18m$$

where m is any arbitrary integer.

Now,

$$\begin{aligned} x_1 &= 3x + 2y = 3(3 + 47m) + 2(7 + 18m) \\ &= 9 + 141m + 14 + 36m = 23 + 177m \end{aligned}$$

$$\begin{aligned} x_2 &= 4y - 5x = 4(7 + 18m) - 5(3 + 47m) \\ &= 28 + 72m - 15 - 235m = 13 - 163m \end{aligned}$$

$$\begin{aligned} x_3 &= 6x + 7y = 6(3 + 47m) + 7(7 + 18m) \\ &= 18 + 282m + 49 + 126m = 67 + 408m \end{aligned}$$

$$\begin{aligned} x_4 &= 17x - 6y = 17(3 + 47m) - 6(7 + 18m) \\ &= 51 + 799m - 42 - 108m = 9 + 691m \end{aligned}$$

(*Continued*)

Example 1 (cont'd)

$$\begin{aligned} X_5 &= 8x + 3y = 8(3 + 47m) + 3(7 + 18m) \\ &= 24 + 376m + 21 + 54m = 45 + 430m \end{aligned}$$

$$\begin{aligned} X_6 &= 2x + 9y = 2(3 + 47m) + 9(7 + 18m) \\ &= 6 + 94m + 63 + 162m = 69 + 256m \end{aligned}$$

$$\begin{aligned} X_7 &= 14x + 5y = 14(3 + 47m) + 5(7 + 18m) \\ &= 42 + 658m + 35 + 90m = 77 + 748m \end{aligned}$$

$$\begin{aligned} X_8 &= 9x + 2y = 9(3 + 47m) + 2(7 + 18m) \\ &= 27 + 423m + 14 + 36m = 41 + 459m \end{aligned}$$

are the general solutions of (*), where m is any arbitrary integer.

Example 2

Find the integral solutions of the equation

$$217X_1 + 521X_2 - 324X_3 + 174X_3 + 91X_5 + 186X_6 + 31X_7 = 213 \quad (*)$$

Solution

Let

$$X_1 = 3x - 4y,\ X_2 = 9y - 2x,\ X_3 = 5x + 8y,\ X_4 = 7x - 5y,$$
$$X_5 = 4x - 2y,\ X_6 = 2x - 4y,\ \text{and } X_7 = 3x + 17y$$

Then (*) becomes

$$217(3x - 4y) + 521(9y - 2x) - 324(5x + 8y) + 173(7x - 5y)$$
$$+ 91(4x - 2y) + 186(2x - 4y) + 31(3x + 17y) = 213$$

i.e.,

$$651x - 868y + 4689y - 1042x - 1620x - 2592y + 1211x - 865y$$
$$+ 364x - 182y + 372x - 744y + 93x + 527y = 213$$

i.e.,

$$29x - 35y = 213 \quad \text{or} \quad 29x - 213 = 35y \quad (**)$$

We now solve (**)

$$
\begin{array}{l}
35)29(0 = q_1 \\
\quad \underline{0} \\
\quad 29)35(1 = q_2 \\
\qquad \underline{29} \\
\qquad 6)29(4 = q_3 \\
\qquad\quad \underline{24} \\
\qquad\quad 5)6(1 = q_4 \\
\qquad\qquad \underline{5} \\
\qquad\qquad 1
\end{array}
$$

Hence, the quotients are $q_1 = 0$, $q_2 = 1$, $q_3 = 4$, and $q_4 = 1$ and the number of quotients is even.

We now form the following table:

0	0	0	0	1065
1	1	1	1278	1278
4	4	1065	1065	1065
1	213	213	213	
213	213	213		
0	0			

Now,

$$Y = 1065 \quad \text{and} \quad X = 1278$$

So,

$$Y = 1065 \equiv 21 (\text{mod } 29)$$

$$X = 1278 \equiv 18 (\text{mod } 35)$$

Hence, $(x\ y) = (18\ 21)$ is that least positive integral solution of the equation $29x + 213 = 35y$.

Then, by Bhaskara II's lemma on page 360, we have

$$x = (35 - 18) + 35m = 17 + 35m$$

$$y = (29 - 21) + 29m = 8 + 29m$$

as the general solution of (**) $29x - 213 = 35y$, where m is any arbitrary integer.

(Continued)

Example 2 (cont'd)

Then, the general solution of (*) is given by

$$X_1 = 3x - 4y = 3(17 + 35m) - 4(8 + 29m)$$
$$= 51 + 105m - 32 - 116m = 19 - 11m$$

$$X_2 = 9y - 2x = 9(8 + 29m) - 2(17 + 35m)$$
$$= 72 + 261m - 34 - 70m = 38 + 191m$$

$$X_3 = 5x + 8y = 5(17 + 35m) + 8(8 + 29m)$$
$$= 85 + 175m + 64 + 232m = 149 + 407m$$

$$X_4 = 7x - 5y = 7(17 + 35m) - 5(8 + 29m)$$
$$= 119 + 245m - 40 - 145m = 79 + 100m$$

$$X_5 = 4x - 2y = 4(17 + 35m) - 2(8 + 29m)$$
$$= 68 + 140m - 16 - 58m = 52 + 82m$$

$$X_6 = 2x - 4y = 2(17 + 35m) - 4(8 + 29m)$$
$$= 34 + 70m - 32 - 116m = 2 - 46m$$

$$X_7 = 3x + 17y = 3(17 + 35m) + 17(8 + 29m)$$
$$= 51 + 105m + 136 + 493m = 187 + 598m$$

where m is any arbitrary integer.

Example 3

Find the integral solutions of the equation

$$7x_1 + 8x_2 - 5x_3 + 4x_4 - 2x_5 + 3x_6 = 230 \quad (*)$$

Solution

Let

$$x_1 = 2x - 3y,\ x_2 = 5x + 4y,\ x_3 = 7x - 2y,\ x_4 = 4x - 5y,$$
$$x_5 = 3x + 7y,\ \text{and}\ x_6 = 11x - 9y$$

Then the equation (*) becomes

$$7(2x - 3y) + 8(5x + 4y) - 5(7x - 2y) + 4(4x - 5y) - 2(3x + 7y)$$
$$+ 3(11x - 9y) = 230$$

i.e.,

$$14x - 21y + 40x + 32y - 35x + 10y + 16x - 20y - 6x - 14y + 33x - 27y = 230$$

i.e.,

$$62x - 40y = 230 \quad \text{or} \quad 62x - 230 = 20y$$

Dividing by 2, this equation becomes

$$31x - 115 = 20y \quad (^{**})$$

We now solve this equation (**)

$$\begin{array}{l} 20)31(1 = q_1 \\ \quad \underline{20} \\ \quad 11)20(1 = q_2 \\ \qquad \underline{11} \\ \qquad 9)11(1 = q_3 \\ \qquad\quad \underline{9} \\ \qquad\quad 2)9(4 = q_4 \\ \qquad\qquad \underline{8} \\ \qquad\qquad 1 \end{array}$$

So, the quotients are $q_1 = 1$, $q_2 = 1$, $q_3 = 1$, and $q_4 = 4$ and the number of quotients is even.

We now form the following table:

1	1	1	1	1610
1	1	1	1035	1035
1	1	575	575	575
4	460	460	460	
115	115	115		
0	0			

Hence,

$$Y = 1610 \quad \text{and} \quad X = 1035$$

(*Continued*)

Example 3 (cont'd)

Now,

$$Y = 1610 \equiv 29 (\text{mod } 31)$$

$$X = 1035 \equiv 15 (\text{mod } 20)$$

Hence, $(x \quad y) = (15 \; 29)$ is the least positive integral solution of $31x + 115 = 20y$.

Then, by Bhaskara II's lemma on page 360

$$x = (20 - 15) + 20m = 5 + 20m$$

$$y = (31 - 29) + 31m = 2 + 31m$$

is the general solution of $31x - 115 = 20y$, where m is any arbitrary integer.

Then,

$$x_1 = 2x - 3y = 2(5 + 20m) - 3(2 + 31m)$$
$$= 10 + 40m - 6 - 93m = 4 - 53m$$

$$x_2 = 5x + 4y = 5(5 + 20m) + 4(2 + 31m)$$
$$= 25 + 100m + 8 + 124m = 33 + 224m$$

$$x_3 = 7x - 2y = 7(5 + 20m) - 2(2 + 31m)$$
$$= 35 + 140m - 4 - 62m = 31 + 78m$$

$$x_4 = 4x - 5y = 4(5 + 20m) - 5(2 + 31m)$$
$$= 20 + 80m - 10 - 155m = 10 - 75m$$

$$x_5 = 3x + 7y = 3(5 + 20m) + 7(2 + 31m)$$
$$= 15 + 60m + 14 + 217m = 29 + 277m$$

$$x_6 = 11x - 9y = 11(5 + 20m) - 9(2 + 31m)$$
$$= 55 + 220m - 18 - 279m = 37 - 59m$$

are the general solutions of (*), where m is any arbitrary integer.

Example 4

Solve in integers the equation

$$33X_1 - 193X_2 + 257X_3 + 200X_4 + 5500 = 0 \quad (*)$$

Solution

Let

$$X_1 = 5x - 4y, \; X_2 = 7x + 3y, \; X_3 = 3y - 2x, \; \text{and} \; X_4 = 2x + 2y$$

Then (*) becomes

$$33(5x - 4y) - 193(7x + 3y) + 257(3y - 2x) + 200(2x + 2y) + 5500 = 0$$

i.e.,

$$165x - 132y - 1351x - 579y + 771y - 514x + 400x + 400y + 5500 = 0$$

i.e.,

$$-1300x + 460y + 5500 = 0 \quad \text{or} \quad 1300x - 5500 = 460y$$

Dividing by 20, this becomes

$$65x - 275 = 23y \quad (**)$$

We now solve (**)

$$
\begin{array}{l}
23)65(2 = q_1 \\
\quad \underline{46} \\
\quad 19)23(1 = q_2 \\
\qquad \underline{19} \\
\qquad 4)19(4 = q_3 \\
\qquad\quad \underline{16} \\
\qquad\quad 3)4(1 = q_4 \\
\qquad\qquad \underline{3} \\
\qquad\qquad 1
\end{array}
$$

So, the quotients are $q_1 = 2$, $q_2 = 1$, $q_3 = 4$, and $q_4 = 1$ and the number of quotients is even.

We now form the following table:

2	2	2	2	4675
1	1	1	1650	1650
4	4	1375	1375	1375
1	275	275	275	
275	275	275		
0	0			

(*Continued*)

Example 4 (cont'd)

So,

$$Y = 4675 \quad \text{and} \quad X = 1650$$

Now,

$$Y = 4675 \equiv 60(\text{mod } 65)$$

$$X = 1650 \equiv 17(\text{mod } 23)$$

Hence, $(x \quad y) = (17\ 60)$ is the least positive integral solution of $65x + 275 = 23y$.

Then, by Bhaskara II's lemma on page 360

$$x = (23 - 17) + 23m = 6 + 23m$$

$$y = (65 - 60m) + 65m = 5 + 65m$$

is the general solution of $65x - 275 = 23y$, where m is any arbitrary integer.

Then,

$$X_1 = 5x - 4y = 5(6 + 23m) - 4(5 + 65m)$$
$$= 30 + 115m - 20 - 260m = 10 - 145m$$

$$X_2 = 7x + 3y = 7(6 + 23)m + 3(5 + 65m)$$
$$= 42 + 161m + 15 + 195m = 57 + 356m$$

$$X_3 = 3y - 2x = 3(5 + 65m) - 2(6 + 23m)$$
$$= 15 + 195m - 12 - 46m = 3 + 149m$$

$$X_4 = 2x + 2y = 2(5 + 65m) + 2(6 + 23m)$$
$$= 10 + 130m + 12 + 46m = 22 + 176m$$

are the general solutions of (*), where m is any arbitrary integer.

Solution of Some Generalized Equations Using the Solution of "*Vargaprakrti*" Equations

Proposition 2: 1. Solve for rational solutions of

$$C_1(A_1X_1+B_1)^2 + C_2(A_2X_2+B_2)^2 + \cdots + C_n(A_nX_n+B_n)^2 + K^* = C_{n+1}(A_{n+1}Y+B_{n+1})^2 \quad (^{**})$$

where A_k, B_k, C_k ($k = 1, 2, \ldots, n+1$) and K^* are constants and X_k ($k = 1, 2, \ldots, n$) and Y are variables.

Proof: Let $A_kX_k + B_k = U_k$ for $k = 1, 2, \ldots, n$ and $A_{n+1}Y + B_{n+1} = V$.
Then, we have

$$C_1U_1^2 + C_2U_2^2 + \cdots + C_nU_n^2 + K^* = C_{n+1}V^2$$

Let $U_k = a_kZ$ for $k = 1, 2, \ldots, n$.
Then, we have

$$(C_1a_1^2 + C_2a_2^2 + \cdots + C_na_n^2)Z^2 + K^* = C_{n+1}V^2$$

We have, (*) $AZ^2 + K^* = C_{n+1}V^2$, where

$$A = C_1a_1^2 + C_2a_2^2 + \cdots + C_na_n^2$$

Then two cases arise.

1. C_{n+1} is a perfect square.
2. C_{n+1} is not a perfect square.

Case 1: C_{n+1} is a perfect square.
Let $C_{n+1} = B^2$.
Then, we have

$$AZ^2 + K^* = B^2V^2$$

i.e.,
$AZ^2 + K^* = W^2$, where $W = BV$.
If (l, m) be a solution of $AZ^2 + K^* = W^2$, another solution of this equation is

$$Z = ql \pm pm \quad \text{and} \quad W = qm \pm Apl$$

where $Ap^2 + 1 = q^2$.

Now,

$$A_kX_k + B_k = U_k = a_kZ = a_k(ql \pm pm) \quad \text{for } k = 1, 2, \ldots, n$$

$$A_kX_k = a_k(ql \pm pm) - B_k \quad \text{for } k = 1, 2, ..., n$$

$$X_k = \frac{1}{A_k}[a_k(ql \pm pm) - B_k] \quad \text{for } k = 1, 2, ..., n$$

Also,

$$B(A_{n+1}Y + B_{n+1}) = BV = W = qm \pm Apl$$

where

$$B^2 = C_{n+1}$$

$$BA_{n+1}Y + BB_{n+1} = qm \pm Apl$$

So,

$$Y = \frac{1}{BA_{n+1}}[qm \pm Apl - BB_{n+1}]$$

where $B^2 = C_{n+1}$.

Hence,

$$X_k = \frac{1}{A_k}[a_k(ql \pm pm) - B_k] \quad \text{for } k = 1, 2, \ldots, n$$

and

$$Y = \frac{1}{BA_{n+1}}[qm \pm Apl - BB_{n-1}]$$

where

$$B^2 = C_{n+1}$$

$$AP^2 + 1 = q^2$$

are the solutions of (**).

Case 2: C_{n+1} is not a perfect square.

Then, multiplying the equation (*) on page 715 by C_{n+1}, we get

$$AC_{n+1}Z^2 + K^*C_{n+1} = C_{n+1}^2V^2 = W^2$$

where $W = C_{n+1}V$.

So, we have

$$A^*Z^2 + K^{**} = W^2 \quad (***)$$

where $A^* = AC_{n+1}$ and $K^{**} = K^*C_{n+1}$.

If $(l\ m)$ is a solution of (***), then another solution of (***) is

$$Z = q_* \cdot l \pm p_* \cdot m$$

and

$$W = q_* \cdot m \pm p_* \cdot lA^* \text{where}$$

$$A^*p_*^2 + 1 = q_*^2$$

Now,

$$A_kX_k + B_k = a_kZ = a_k(q \cdot l \pm p \cdot m) \quad \text{for } k = 1, 2, ..., n$$

where

$$A^*P_*^2 + 1 = q_*^2$$

$$X_k = \frac{1}{A_k}[a_k(q_* \cdot l \pm p_* \cdot m) - B_k] \quad \text{for } k = 1, 2, \ldots, n$$

$$C_{n+1}(A_{n+1}y + B_{n+1}) = C_{n+1}V = W = q \cdot m \pm p \cdot lA^*$$

So,

$$y = \frac{1}{A_{n+1}}\left[\frac{q \cdot m \pm p \cdot lA^*}{C_{n+1}} - B_{n+1}\right]$$

where $A^*p_*^2 + 1 = q_*^2$ is the solution of (**) on page 715.

We give the following example of Case 2.

Example 5

Find the rational solutions of the equation

$$72x_1^2 - 240x_1 + 20x_2^2 + 60x_2 + 240x_3^2 + 120x_3 + 250x_4^2 - 200x_4$$
$$+ 324x_5^2 - 756x_5 + 2650 = 180y^2 - 300y \quad (*)$$

(*Continued*)

Example 5 (cont'd)

The given equation (*) can be written as

$$8(9x_1^2 - 30x_1) + 5(4x_2^2 + 12x_2) + 15(16x_3^2 + 8x_3) + 10(25x_4^2 - 20x_4) + 9(36x_5^2 - 84x_5) + 2650 = 5(36y^2 - 60y)$$

i.e.,

$$8[(3x_1 - 5)^2 - 25] + 5[(2x_2 + 3)^2 - 9] + 15[(4x_3 + 1)^2 - 1] + 10[(5x_4 - 2)^2 - 4] + 9[(6x_5 - 7)^2 - 49] + 2650 = 5[(6y - 5)^2 - 25]$$

Let

$$3x_1 - 5 = u_1,\ 2x_2 + 3 = u_2,\ 4x_3 + 1 = u_3,\ 5x_4 - 2 = u_4$$
$$6x_5 - 7 = u_5,\ \text{and}\ 6y - 5 = v$$

Then, we have

$$8u_1^2 - 200 + 5u_2^2 - 45 + 15u_3^2 - 15 + 10u_4^2 - 40 + 9u_5^2 - 441 + 2650 = 5v^2 - 125$$

i.e.,

$$8u_1^2 + 5u_2^2 + 15u_3^2 + 10u_4^2 + 9u_5^2 + 2034 = 5v^2$$

Let $u_1 = 3z$, $u_2 = 2z$, $u_3 = 6z$, $u_4 = 5z$, and $u_5 = 9z$.
Then, we get

$$72z^2 + 20z^2 + 540z^2 + 250z^2 + 729z^2 + 2034 = 5v^2$$

Thus, we have $1611z^2 + 2034 = 5v^2$.
Multiplying by 5, we get

$$8055z^2 + 10,170 = w^2 \quad (**)$$

which is a "vargaprakrti" equation, where $w = 5v$.
Now, we have to solve this equation (**).
We observe that $8055 \cdot 1^2 + 45 = 8100 = (90)^2$.
So, (1, 90) is a solution of $8055z^2 + 45 = w^2$.
Again, we note that $8055 \cdot 1^2 + 226 = 8281 = (91)^2$.
Hence, (1, 91) is a solution of $8055z^2 + 226 = w^2$.

So,

$$\begin{aligned}&(1\ 90\ 45)\circ(1\ 91\ 226)\\&=[1\cdot(91)+(90)\cdot 1\quad (90)\,(91)+1\cdot 1\cdot 8055\quad (45)(226)]\\&=(181\ 8190+8055\ 10{,}170)\\&=(181\ 16{,}245\ 10{,}170)\end{aligned}$$

So, $(l, m) = (181\ 16{,}245)$ is a solution of $8055z^2 + 10{,}170 = w^2$ (**). Another solution of (**) is

$$z = 181q \pm 16,245p$$

$$w = 16,245q \pm (181p)(8055) = 16{,}245q \pm 1{,}457{,}955p$$

where $8055p^2 + 1 = q^2$.

Now,

$$3x_1 - 5 = u_1 = 3z = 3(181q \pm 16{,}245p)$$

$$2x_2 + 3 = u_2 = 2z = 2(181q \pm 16{,}245p)$$

$$4x_3 + 1 = u_3 = 6z = 6(181q \pm 16{,}245p)$$

$$5x_4 - 2 = u_4 = 5z = 5(181q \pm 16{,}245p)$$

$$6x_5 - 7 = u_5 = 9z = 9(181q \pm 16{,}245p)$$

and

$$5(6y - 5) = 5v = w = 16{,}245q \pm 1,457{,}955p$$

$$6y - 5 = 3249q \pm 291{,}591p$$

So,

$$x_1 = 181q \pm 16{,}245p + \frac{5}{3}$$

$$x_2 = 181q \pm 16{,}245p - \frac{3}{2}$$

$$x_3 = \frac{3}{2}\left(181q \pm 16{,}245p\right) - \frac{1}{4}$$

(*Continued*)

Example 5 (cont'd)

$$x_4 = \left(181q \pm 16{,}245p\right) + \frac{2}{5} \quad (***)$$

$$x_5 = \frac{3}{2}\left(181q \pm 16{,}245p\right) + \frac{7}{6}$$

and

$$y = \frac{1}{6}(3249q \pm 291{,}591p) + \frac{5}{6}$$

where $8055p^2 + 1 = q^2$ is the solution of (*).

If $p = 0$, $q = 1$.

Then, we have

$$x_1 = 181 + \frac{5}{3} = \frac{548}{3}$$

$$x_2 = 181 - \frac{3}{2} = \frac{359}{2}$$

$$x_3 = \frac{3}{2}(181) - \frac{1}{4} = \frac{543}{2} - \frac{1}{4} = \frac{1085}{4}$$

$$x_4 = 181 + \frac{2}{5} = \frac{907}{5}$$

$$x_5 = \frac{543}{2} + \frac{7}{6} = \frac{818}{3}$$

$$y = \frac{3249}{6} + \frac{5}{6} = \frac{1627}{3}$$

Other solutions are obtained from (***), where $8055p^2 + 1 = q^2$, for example, if $p = 4$, $q = 359$, and so on.

Proposition 3: Find the rational solutions of the equation

$$C_1(A_1X_1 + B_1)^2 + C_2(A_2X_2 + B_2)^2 + \cdots + C_n(A_nX_n + B_n)^2 + K^*$$
$$= d_1(D_1y_1 + E_1)^2 + d_2(D_2y_2 + E_2)^2 + \cdots + d_m(D_my_m + E_m)^2 \quad (*)$$

where A_i, B_i, C_i ($i = 1, 2, \ldots, n$) and d_j, D_j, E_j ($j = 1, 2, \ldots, m$) and K^* are constants.

Proof: Let

$$A_kX_k + B_k = u_k \quad \text{for } k = 1, 2, \ldots, n$$

and

$$D_j y_j + E_j = V_j \quad \text{for } j = 1, 2, ..., m$$

Then, we have

$$C_1 u_1^2 + C_2 u_2^2 + \cdots + C_n u_n^2 + K^* = d_1 V_1^2 + d_2 V_2^2 + \cdots + d_m V_m^2$$

Let

$$u_k = a_k z, \quad \text{for } k = 1, 2, ..., n$$

$$V_j = b_j W, \quad \text{for } j = 1, 2, ..., m$$

Then, we have

$$(C_1 a_1^2 + C_2 a_2^2 + \cdots + C_n a_n^2) z^2 + k^* = (d_1 b_1^2 + d_2 b_2^2 + \cdots + d_m b_m^2) W^2$$

Let

$$A = C_1 a_1^2 + C_2 a_2^2 + \cdots + C_n a_n^2$$

and

$$B = (d_1 b_1^2 + d_2 b_2^2 + \cdots + d_m b_m^2)$$

Thus, we get (**) $Az^2 + K^* = BW^2$, which is a "vargaprakrti" equation.
As before, two cases arise:

1. B is a perfect square.
2. B is not a perfect square.

Case 1: B is a perfect square $= D^2$.
Then (**) becomes

$$Az^2 + K^* = D^2 W^2 = W_*^2 \quad (***)$$

where $W_* = DW$.
If $(l\ m)$ is a solution of

$$Az^2 + K^* = W_*^2 \quad (***)$$

then another solution of (***) is $z = ql \pm pm$ and $W_* = qm \pm Apl$, where $q^2 = Ap^2 + 1$.

$$A_k X_k + B_k = u_k = a_k z = a_k(ql \pm pm) \quad \text{for } k = 1, 2, ..., n$$

So,

$$X_k = \frac{1}{A_k}[a_k(ql \pm pm) - B_k] \quad \text{for } k = 1, 2, ..., n$$

$$D_jY_j + E_j = V_j = b_jW = \frac{b_jW_*}{D} = \frac{b_j}{D}[qm \pm Apl] \quad \text{for } j = 1, 2, ..., m$$

So,

$$y_j = \frac{b_j}{DD_j}[qm \pm Apl] - \frac{E_j}{D_j}$$

where $q^2 = Ap^2 + 1$.

Hence,

$$X_k = \frac{1}{A_k}[a_k(ql \pm pm) - B_k] \quad \text{for } k = 1, 2, ..., n$$

and

$$y_j = \frac{b_j}{DD_j}[qm \pm Apl] = \frac{E_j}{D_j} \quad \text{are solutions of (*)}$$

where $Ap^2 + 1 = q^2$.

Here,

$$A = C_1a_1^2 + C_2a_2^2 + \cdots + C_na_n^2$$

$$D^2 = B = [d_1b_1^2 + d_2b_2^2 + \cdots + d_mb_m^2]$$

Case 2: B is not a perfect square.

Then, multiplying (**) on page 721 by B, we get

$$ABz^2 + BK^* = B^2W^2$$

Let $AB = A^*$, $BK^* = K^{**}$, and $BW = W_*$.

Then, the equation becomes

$$A^*z^2 + K^{**} = W_*^2 \quad (***)$$

If $(l\ m)$ is a solution of (***), then another solution of (***) is

$$z = q \cdot l + p \cdot m \quad \text{and} \quad W_* = q \cdot m \pm A^*pl$$

where $q^2 = A^*p^2 + 1$.

Now,

$$A_kX_k + B_k = u_k = a_kz = a_k(ql \pm pm)$$

$$X_k = \frac{1}{A_k}[a_k(ql \pm pm)] - \frac{B_k}{A_k} \quad \text{for } k = 1, 2, ..., n$$

$$D_jY_j + E_j = b_jW = \frac{b_jW_*}{B} = \frac{b_j}{B}\left[qm \pm A^*pl\right]$$

$$y_j = \frac{b_j}{BD_j}\left[qm \pm A^*pl\right] - \frac{E_J}{D_J} \quad \text{for } J = 1, 2, ..., m$$

where $q^2 = A^*p^2 + 1$.

Here

$$A^* = AB = [C_1a_1^2 + C_2a_2^2 + \cdots C_na_n^2][d_1b_1^2 + d_2b_2^2 + \cdots + d_mb_m^2]$$

We will now give example of Proposition 3.

Example 6

Find the rational solutions of the equation

$$\begin{aligned} &45x_1^2 + 60x_1 + 100x_2^2 - 120x_2 + 98x_3^2 - 140x_3 + 144x_4^2 \\ &- 336x_4 - 32x_5^2 - 96x_5 + 48x_6^2 - 120x_6 + 601 \\ &= 12y_1^2 - 12y_2 + 196y_2^2 - 168y_2 + 45y_3^2 + 60y_3 - 32y_4^2 - 80y_4 \quad (*) \end{aligned}$$

Solution: (*) can be written as

$$\begin{aligned} &5(9x_1^2 + 12x_1) + 4(25x_2^2 - 30x_2) - 2(49x_3^2 - 70x_3) + 4(36x_4^2 - 84x_4) \\ &- 8(4x_5^2 + 12x_5) + 3(16x_6^2 - 40x_6) + 601 \\ &= 3(4y_1^2 - 4y_1) + 4(49y_2^2 - 42y_2) + 5(9y_3^2 + 12y_3) - 2(16y_4^2 + 40y_4) \end{aligned}$$

i.e.,

$$\begin{aligned} &5[(3x_1+2)^2 - 4] + 4[(5x_2 - 3^2) - 9] - 2[(7x_3-5)^2 - 25] + 4[(6x_4-7)^2 - 49] \\ &- 8[(2x_5 + 3)^2 - 9] + 3[(4x_6-5)^2 - 25] + 601 \\ &= 3[(2y_1-1)^2 - 1] + 4[(7y_2 - 3)^2 - 9)] + 5[(3y_2+2)^2 - 4] - 2[(4y_4+5)^2 - 25] \end{aligned}$$

Let

$$\begin{aligned} &3x_1 + 2 = u_1,\ 5x_2 - 3 = u_2,\ 7x_3 - 5 = u_3,\ 6x_4 - 7 = u_4,\ 2x_5 + 3 = u_5, \\ &4x_6 - 5 = u_6,\ 2y - 1 = v_1,\ 7y_2 - 3 = v_2,\ 3y_3 + 2 = v_3,\ \text{and } 4y_4 + 5 = v_4 \end{aligned}$$

(*Continued*)

Example 6 (cont'd)

Then, we get

$$5u_1^2 - 20 + 4u_2^2 - 36 - 2u_3^2 + 50 + 4u_4^2 - 196 - 8u_5^2 + 72 + 3u_6^2 - 75 + 601$$
$$= 3v_1^2 - 3 + 4v_2^2 - 36 + 5v_3^2 - 20 - 2v_4^2 + 50$$

Thus, we have

$$5u_1^2 + 4u_2^2 - 2u_3^2 + 4u_4^2 - 8u_5^2 + 3u_6^2 + 405 = 3v_1^2 + 4v_2^2 + 5v_3^2 - 2v_4^2$$

Let

$$u_1 = 2z, \quad u_2 = 5z, \quad u_3 = 4z, \quad u_4 = 7z, \quad u_5 = 3z, \quad u_6 = 6z$$

and

$$v_1 = 5w, \quad v_2 = 3w, \quad v_3 = 4w, \quad \text{and} \quad v_4 = 9w$$

We get

$$5(4z^2) + 4(25z^2) - 2(16z^2) + 4(49z^2) - 8(9z^2) + 3(36z^2) + 405$$
$$= 3(25w^2) + 4(9w^2) + 5(16w^2) - 2(81w^2)$$

Hence, we have

$$20z^2 + 100z^2 - 32z^2 + 196z^2 - 72z^2 + 108z^2 + 405 = 75w^2 + 36w^2 + 80w^2 - 162w^2$$

This reduces to

$$320z^2 + 405 = 29w^2 \quad (^{**})$$

This is a "vargaprakrti" equation.
Multiplying (**) by 29, we get

$$9280z^2 + 11{,}745 = 841w^2 = (29w)^2 = w_*^2$$

where $w_* = 29w$.

So, we have $9280z^2 + 11{,}745 = w_*^2$, which is a "vargaprakrti" equation.
We observe that $9280 \cdot 1^2 + 11{,}745 = 21{,}025 = (145)^2$.
So, $(l\ m) = (1\ 145)$ is a solution of

$$9280z^2 + 11{,}745 = w_*^2 \quad (^{***})$$

Another solution of (***) is given by

$$z = ql \pm pm \quad \text{and} \quad w_* = qm \pm 9280pl$$

i.e.,

$$z = q \pm 145p \quad \text{and} \quad w_* = 145q \pm 9280p$$

where $9280p^2 + 1 = q^2$.

Now,

$$3x_1 + 2 = u_1 = 2z = 2(q \pm 145p)$$

$$5x_2 - 3 = u_2 = 5z = 5(q \pm 145p)$$

$$7x_3 - 5 = u_3 = 4z = 4(q \pm 145p)$$

$$6x_4 - 7 = u_4 = 7z = 7(q \pm 145p)$$

$$2x_5 + 3 = u_5 = 3z = 3(q \pm 145p)$$

$$4x_6 - 5 = u_6 = 6z = 6(q \pm 145p)$$

and

$$2y_1 - 1 = v_1 = 5w = \frac{5w_*}{29} = \frac{5}{29}(145q \pm 9280p)$$

$$7y_2 - 3 = v_2 = 3w = \frac{3}{29}w_* = \frac{3}{29}(145q \pm 9280p)$$

$$3y_3 + 2 = v_3 = 4w = \frac{4}{29}w_* = \frac{4}{29}(145q \pm 9280p)$$

$$4y_4 + 5 = v_4 = 9w = \frac{2}{29}w_* = \frac{9}{29}(145q \pm 9280p)$$

where

$$9280p^2 + 1 = q^2$$

So,

$$x_1 = \frac{1}{3}[2(q \pm 145p) - 2]$$

$$x_2 = \frac{1}{5}[5(q \pm 145p) + 3]$$

$$x_3 = \frac{1}{7}[4(q \pm 145p) + 5]$$

(*Continued*)

Example 6 (cont'd)

$$x_4 = \frac{1}{6}\left[7(q \pm 145p) + 7\right]$$

$$x_5 = \frac{1}{2}\left[3(q \pm 145p) - 3\right]$$

$$x_6 = \frac{1}{4}\left[6(q \pm 145p) + 5\right]$$

$$y_1 = \frac{1}{2}\left[\frac{5}{29}(145q \pm 9280p) + 1\right]$$

$$y_2 = \frac{1}{7}\left[\frac{3}{29}(145q \pm 9280p) + 3\right]$$

$$y_3 = \frac{1}{3}\left[\frac{4}{29}(145q \pm 9280p) - 2\right]$$

$$y_4 = \frac{1}{4}\left[\frac{9}{29}(145q \pm 9280p) - 5\right]$$

where

$$9280p^2 + 1 = q^2$$

We observe that if $p = 0$, $q = 1$.
Then,

$$x_1 = 0,\ x_2 = \frac{8}{5},\ x_3 = \frac{9}{7},\ x_4 = \frac{7}{3},\ x_5 = 0,\ x_6 = \frac{11}{4}$$

and

$$y_1 = 13,\ y_2 = \frac{18}{7},\ y_3 = 6,\ y_4 = 10$$

Other solutions can be found, using $9280p^2 + 1 = q^2$, like if $p = 3$, $q = 289$, and so on. Obviously, there will be infinitely many solutions.

Proposition 4: Find the rational solutions of the equation

$$c_1(A_1X_1 + A_2X_2 + \cdots + A_nX_n + d_1)^2 + K = c_2(By + d_2)^2 \quad (*)$$

where c_1, c_2, d_1, d_2, B, and A_j ($j = 1, 2, \ldots, n$) are constants.

Solution: Let

$$X_J = a_J u \text{ for } j = 1,\ 2,\ \ldots,\ n$$

and

$$By + d_2 = V$$

Then, we get

$$c_1[(A_1a_1 + A_2a_2 + \cdots + A_na_n)u + d_1]^2 + K = c_2V^2$$

Let

$$(A_1a_1 + A_2a_2 + \cdots + A_na_n)u + d_1 = u^*$$

Then, we have

$$c_1u^{*2} + K = c_2v^2 \quad (^{**})$$

which is a "vargaprakrti" equation.

As before, two cases arise:

1. c_2 is a perfect square.
2. c_2 is not a perfect square.

Case 1: c_2 is a perfect square.

Let $c_2 = g^2$.

From (**), then, we have $c_1u^{*2} + K = g^2V^2 = w^2$, where $w = gV$.

So, if $(l\ m)$ is a solution of

$$c_1u^{*2} + K = w^2 \quad (^{***})$$

another solution of (***) is given by

$$u^* = lq \pm mp \quad \text{and} \quad w = mq \pm c_1lp$$

where $q^2 = c_1p^2 + 1$.

Now,

$$X_j = a_ju = a_j\frac{u^* - d_1}{\sum_{j=1}^{n} A_ja_j}$$

$$= a_j\frac{(lq \pm mp) - d_1}{\sum_{j=1}^{n} A_ja_j} \quad j = 1, 2, \ldots, n$$

and

$$By + d_2 = V = \frac{1}{g}w = \frac{1}{g}[mq \pm clp] = \frac{1}{\sqrt{c_2}}[mq \pm c_1 lp]$$

i.e.,

$$y = \frac{1}{B\sqrt{c_2}}[mq \pm c_1 lp] - \frac{d_2}{B}$$

are the solutions of (*), where $q^2 = c_1 p^2 + 1$.

Case 2: c_2 is not a square.

Then, multiplying (**) by c_2, we get

$$c_1 c_2 u^{*2} + c_2 K = c_2^2 V^2$$

Let $c_1 c_2 = c^*$, $c_2 K = K^*$, and $c_2 V = W_*$.

Then, we have

$$c^* u^{*2} + K^* = W_*^2 \quad (***)$$

which is a "vargaprakrti" equation.

If (l m) is a solution of (***), another solution of (***) is

$$u^* = lq \pm mp \quad \text{and} \quad W_* = mq \pm c^* pl$$

where $c^* p^2 + 1 = q^2$.

Now,

$$X_j = a_j u = \frac{u^* - d_1}{\sum_{j=1}^{n} A_J a_J} = a_J \frac{(lq \pm mp) - d_1}{\sum_{j=1}^{n} A_j a_j} \quad \text{for } j = 1, 2, \ldots, n$$

and

$$By + d_2 = V = \frac{1}{c_2} W_* = \frac{1}{c_2}(mq \pm c^* pl)$$

where

$$q^2 = c^* p^2 + 1 = c_1 c_2 p^2 + 1$$

i.e.,

$$y = \frac{1}{B}\left[\frac{1}{c_2}(mq \pm c^* pl) - d_2\right]$$

We will now give an example to illustrate the above method.

Example 7

(1) Find the rational solutions of the equation

$$7(7x_1+9x_2-5x_3+6x_4-8x_5+2x_6-18)^2 - 16 = 3(17y+7)^2$$

Let $x_1 = 2u$, $x_2 = 5u$, $x_3 = 3u$, $x_4 = 7u$, $x_5 = 6u$, and $x_6 = 9u$ and $17y + 7 = v$, then we have

$$7(14u+45u-15u+42u-48u+18u-18)^2 - 16 = 3v^2$$

i.e., $7(56u - 18)^2 - 16 = 3v^2$.

Let $56u - 18 = u_*$.

Thus, we have $7u_*^2 - 16 = 3v^2$, which is a "vargaprakrti" equation.

Multiplying by 3, we get

$$21u_*^2 - 48 = 9v^2 = (3v)^2 = w^2$$

where $w = 3v$.

We have

$$21u_*^2 - 48 = w^2 \quad (**)$$

We observe that $21 \cdot 2^2 - 3 = 84 - 3 = 81 = (9)^2$.

So, (2 9) is a solution of $21u_*^2 - 3 = w^2$.

We also observe that $21 \cdot (2)^2 + 16 = 84 + 16 = 100 = (10)^2$.

So, (2, 10) is also a solution of $21u_*^2 + 16 = w^2$.

Now, $(2\ 9\ -3) \cdot (2\ 10\ 16)$

$$= [2 \cdot (10) + 9 \cdot (2) \quad 9 \cdot (10) + 2 \cdot 2 \cdot (21) \quad (-3)(16)] = (38 \quad 174 \quad -48)$$

which means that $(l\ m) = (38\ 174)$ is a solution of $21u_*^2 - 48 = w^2$.

Another solution of (**), namely, $21u_*^2 - 48 = w^2$ is given by

$$u_* = 38q \pm 174p \quad \text{and} \quad w = 174q \pm (38)(21)p$$

$$= 174q \pm 798p$$

where $q^2 = 21p^2 + 1$.

(*Continued*)

Example 7 (cont'd)

Now,

$$56u - 18 = u_* = 38q \pm 174p \Rightarrow u = \frac{1}{56}(38q \pm 174p) + \frac{9}{28}$$

$$17y + 7 = v = \frac{w}{3} = \frac{1}{3}[174q \pm 798p]$$

$$\Rightarrow y = \frac{1}{51}[174q \pm 798p] - \frac{7}{17}$$

is the solution of (**), where $q^2 = 21p^2 + 1$.

Now,

$$x_1 = 2u = \frac{1}{28}(38q \pm 174p) + \frac{9}{14}$$

$$x_2 = 5u = \frac{5}{56}(38q \pm 174p) + \frac{45}{28}$$

$$x_3 = 3u = \frac{3}{56}(38q \pm 174p) + \frac{27}{28}$$

$$x_4 = 7u = \frac{1}{8}(38q \pm 174p) + \frac{9}{4}$$

$$x_5 = 6u = \frac{3}{28}(38q \pm 174p) + \frac{27}{14}$$

$$x_6 = 9u = \frac{9}{56}(38q \pm 174p) + \frac{81}{28}$$

where $q^2 = 21p^2 + 1$.

If $p = 0$, $q = 1$.

So,

$$y = \frac{174}{51} - \frac{7}{17} = \frac{174 - 21}{51} = \frac{153}{51} = 3$$

$$x_1 = \frac{38}{28} + \frac{9}{14} = \frac{38 + 18}{28} = \frac{56}{28} = 2$$

$$x_2 = \frac{190}{56} + \frac{45}{28} = \frac{190 + 90}{56} = \frac{280}{56} = 5$$

$$x_3 = \frac{114}{56} + \frac{27}{28} = \frac{114+54}{56} = \frac{168}{56} = 3$$

$$x_4 = \frac{38}{8} + \frac{9}{4} = \frac{38+18}{8} = \frac{56}{8} = 7$$

$$x_5 = \frac{114}{28} + \frac{27}{14} = \frac{114+54}{28} = \frac{168}{28} = 6$$

$$x_6 = \frac{342}{56} + \frac{81}{28} = \frac{342+162}{56} = \frac{504}{56} = 9$$

Other solutions are obtained from the equation $q^2 = 21p^2 + 1$ like if $p = 12$, $q = 25$, and so on. Obviously, there are infinitely many solutions.

Proposition 5: Find the rational solutions of the equation

$$C_1(A_1X_1 + A_2X_2 + \cdots + A_nX_n + d_1)^2 + K$$

$$= C_2(B_1Y_1 + B_2Y_2 + \cdots + B_mY_m + d_2)^2 \quad (*)$$

where C_1, C_2, d_1, d_2, A_j ($j = 1, 2, \ldots, n$), and B_{j*} ($j_* = 1, 2, \ldots, m$) are constants.

Let

$X_j = a_j u_*$ for $j = 1, 2, 3, \ldots, n$

$Y_j = b_{j_*} V_*$ for $j_* = 1, 2, 3, \ldots, m$

Then, we have

$$C_1[(A_1a_1 + A_2a_2 + \cdots + A_nu_n)u_* + d_1]^2 + K = C_2[(B_1b_1 + B_2b_2 + \cdots + B_mb_m)V_* + d_2]^2$$

Let

$$(A_1a_1 + A_2a_2 + \cdots + A_na_n)u_* + d_1 = u$$

and

$$(B_1b_1 + B_2b_2 + \cdots + B_mb_m)V_* + d_2 = v$$

Then, we have

$$C_1u^2 + K = C_2v^2 \quad (**)$$

which is a "vargaprakrti" equation.

As before, two cases arise:

1. C_2 is a perfect square.
2. C_2 is not a perfect square.

Case 1: C_2 is a perfect square.

Let $C_2 = d^2$.

Then from (**), we have $C_1u^2 + K = d^2v^2 = w^2$, where $w = dv$, so we have the equation

$$C_1u^2 + K = w^2 \quad (***)$$

If $(u\ w) = (l\ m)$ is a solution of (***), then another solution is given by

$$u = ql \pm pm \text{ and } w = qm \pm C_1pl$$

where $q^2 = C_1p^2 + 1$.

$$X_j = a_ju_* = a_j\frac{u - d_1}{\sum_{K=1}^{n} A_Ja_J} = a_j\frac{(ql \pm pm)}{\sum_{j=1}^{n} A_ja_j}, \quad j = 1, 2, 3, ..., n$$

and

$$Y_{j*} = b_{j*}V_* = b_{j*}\frac{v - d_2}{\sum_{j*=1}^{m} B_{j*}b_{j*}} = b_{j*}\frac{1}{d}\frac{(qm \pm c_1pl) - d_2d}{\sum_{j=1}^{m} B_{j*}b_{j*}}, \quad j = 1, 2, 3, ..., n$$

are the solutions of (*), where $q^2 = C_1p^2 + 1$ where $d = \sqrt{C_2}$.

Case 2: C_2 is not a perfect square.

Multiplying (**) by C_2, we get

$$C_1C_2u^2 + KC_2 = C_2^2v^2$$

Let $C_1C_2 = C_*$ and $KC_2 = K_*$ and $C_2v = W_*$.

Then, we have

$$C_*u^2 + K_* = W_*^2 \quad (***)$$

If $(l\ m)$ is a solution of (***), another solution of (***) is given by

$$u = q_*l \pm p_*m \quad \text{and} \quad W_* = q_*m \pm C_*p_*l$$

where $q_*^2 = C_* p_*^2 + 1$.

$$X_j = a_j u_* = a_J \frac{u - d_1}{\sum_{j=1}^{a_j} A_j a_j} = a_J \frac{q_* l \pm p_* m - d_1}{\sum_{j=1}^{n} A_j a_j} \quad \text{for } j = 1, 2, 3, ..., n$$

$$Y_j = b_{j^*} V_* = b_{J^*} \frac{v - d_2}{\sum_{j_*=1}^{m} B_{j_*} b_j} = b_{J_*} \frac{1}{C_2} \frac{(q_* m \pm C_* p_* l) - d_2 C_2}{\sum_{j_*=1}^{m} B_{j_*} b_{j_*}} \quad \text{for } j = 1, 2, 3, ..., m$$

where

$$q_*^2 = C_* p_*^2 + 1 = C_1 C_2 p_*^2 + 1$$

We give an example to illustrate the above method.

Example 8

Find the rational solutions of the equations

$$2(3x_1 - 4x_2 + 2x_3 - 5x_4 + 7x_5 - 7)^2 + 19 = 3(2y_1 - 5y_2 - 3y_3 + 4y_4 - 5)^2$$

Let

$$x_1 = 2u^*, \ x_2 = 5u^*, \ x_3 = 3u^*, \ x_4 = u^* \ x_5 = 4u^*$$

and

$$y_1 = 3v^*, \ y_2 = 2v^*, \ y_3 = 4v^*, \ y_4 = 5v^*$$

$$2(6u^* - 20u^* + 6u^* - 5u^* + 28u^* - 7)^2 + 19 = 3(6v^* - 10v^* - 12v^* + 20v^* - 5)$$

i.e.,

$$2(15u^* - 7)^2 + 19 = 3(4v^* - 5)^2$$

Let $u = 15u^* - 7$ and $v = 4v^* - 5$.
Then, we have

$$2u^2 + 19 = 3v^2$$

which is a "vargaprakrti" equation.
Multiplying by 3, we get

$$6u^2 + 57 = 9v^2 = w^2$$

(Continued)

Example 8 (cont'd)

where $w = 3v$.

So, we have the equation

$$6u^2 + 57 = w^2 \quad (^{**})$$

We have to solve this equation.

We note that $6 \cdot 1^2 + 19 = 25 = (5)^2$.

So (1 5) is a solution of $6u^2 + 19 = w^2$ (**).

Again, we observe that $6 \cdot 1^2 + 3 = 9 = (3)^2$.

So (1 3) is a solution of the equation $6u^2 + 3 = w^2$.

Then,

$$(1\ 5\ 19) \circ (1\ 3\ 3) = [1 \cdot 3 + 5 \cdot 1 \quad 5 \cdot 3 + 1 \cdot 1 \cdot 6 \quad (19) \cdot 3] = (8\ 21\ 57)$$

which means that

$$(u\ w) = (l\ m) = (8\ 21)$$

is a solution of $6u^2 + 57 = w^2$ (**).

Another solution of (**) is

$$u = ql \pm pm \quad \text{and} \quad w = qm \pm 6pl$$

where $6p^2 + 1 = q^2$, i.e.,

$$u = 8q \pm 21p \Rightarrow 15u^* = 8q \pm 21p + 7 \quad \text{and} \quad w = 21q + 48p$$

So,

$$u^* = \frac{1}{15}(8q \pm 21p + 7)$$

$$x_1 = 2u^* = \frac{2}{15}(8q \pm 21p + 7)$$

$$x_2 = 5u^* = \frac{1}{3}(8q \pm 21p + 7)$$

$$x_3 = 3u^* = \frac{1}{5}(8q \pm 21p + 7)$$

$$x_4 = u^* = \frac{1}{15}(8q \pm 21p + 7)$$

$$x_5 = 4u^* = \frac{4}{15}(8q \pm 21p + 7)$$

$$4v^* - 5 = v = \frac{w}{3} = \frac{1}{3}(21q \pm 48p) = 7m \pm 16p$$

$$v^* = \frac{1}{4}[7q \pm 16p + 5]$$

$$y_1 = 3v^* = \frac{3}{4}[7q \pm 16p + 5]$$

$$y_2 = 2v^* = \frac{1}{2}[7q \pm 16p + 5]$$

$$y_3 = 4v^* = [7q \pm 16p + 5]$$

$$y_4 = 5v^* = \frac{5}{4}[7q \pm 16p + 5]$$

where $6p^2 + 1 = q^2$.

If $p = 0$, $q = 1$.

Then,

$$x_1 = \frac{2}{15}(8+7) = 2 \quad x_2 = \frac{1}{3}(8+7) = 5$$

$$x_3 = \frac{1}{5}(8+7) = 3 \quad x_4 = \frac{1}{15}(8+7) = 1$$

$$x_5 = \frac{4}{15}(8+7) = 4 \quad y_1 = \frac{3}{4}(7+5) = 9$$

$$y_2 = \frac{1}{2}(7+5) = 6 \quad y_3 = (7+5) = 12$$

$$y_4 = \frac{5}{4}(7+5) = 15$$

Other solutions are obtained from the solutions of $6p^2 + 1 = q^2$. There are infinitely many solutions of $6p^2 + 1 = q^2$, which will give infinitely many solutions of (*), like $p = 2$, $q = 5$; $p = 20$, $q = 49$; and so on. Thus, (*) will have infinitely many solutions.

Bibliography

[1] Apte DV, editor. Titi-Cintāmaṇi of Ganeśa Daivajña with commentary of Viswanatha. Poona: Anandasrama Sanskrit Series; 1942.

[2] Bag AK. Madhava's sine and cosine series. Indian J Hist Sci 1976;11:54–57.

[3] Bag AK. Mathematics in ancient and medieval India. Varanasi: Chaukhambha Orientalia; 1979.

[4] Bag AK, Sarma SR, editors. The concept of Sunya. New Delhi: Indian National Science Academy; 2003.

[5] Balachandra Rao S. Indian mathematics and astronomy. Bangalore: Jnana Deepa Publications; 1994.

[6] Ball WWR. A short account of history of mathematics. 3rd ed. London: Macmillan Company; 1901.

[7] Balagangadharan K. A consolidated list of Hindu mathematical works. Math Student 1947;15:59–69.

[8] Bannerjee HC. Colebrooke's translation of Lilāvati. Calcutta: The Book Company Limited; 1927.

[9] Bapudeva S, Chandradeva P, editors. Siddhanta Siromani of Bhaskara II with Vasana Bhasya, edited by Bapudeva Sastry and revised by Pandit Chandradeva, Benares; 1891.

[10] Bell ET. Development of mathematics. New York, NY: McGraw-Hill; 1945.

[11] Bell ET. Men of mathematics. New York, NY: Simon and Schuster; 1960.

[12] Bernal JD. 4 volumes Science in history. London: Penguin; 1969.

[13] Bhanu Murthy TS. A modern introduction to ancient indian mathematics. New Delhi: Wiley Eastern Ltd; 1992.

[14] Bose DM, Sen SN, Subbarayappa BV. A concise history of science in India. New Delhi: Indian National Science Academy; 1971.

[15] Boyer CB. A history of mathematics. New York, NY: John Wiley; 1978.

[16] Burgess E. Surya Siddhanta [Translation]. New Delhi: Indological Book House; 1977. Reprinted from the first edition of 1860.

[17] David B. Was calculus invented in India? College Math J (Math Assoc Am) 2002;33 (11):2–13.

[18] Burk A. Das Apasthamba Sulvasutra. Zeit d Deutschen Morgenlandischen Ges 1902;55(1901):56.

[19] Cajori F. History of mathematics. New York, NY: Macmillan Company; 1958.

[20] Cantor M. Vorlesungen uber Geschichte der Mathematik, vol. 4. Leipzig: Abhandlungen Zur Geschichte der Mathematik, Germany; 1907.

[21] Chandra Das A. Rigvedic India. New Delhi: Cosmo Publications; 1980.

[22] Sarasvati C. The vedas. Bombay: Bharatiya Vidya Bhavan; 1993.

[23] Channabasappa MN. A note on Colebrooke's translation of a stanza from Bhaskara's Lilavati. Indian J Hist Sci 1974;9(2):221–3.

[24] Channabasappa MN. On the square root formula in Bakshāli manuscript. Indian J Hist Sci 1976;2(2):112–24.

[25] Clark WE, editor. The Āryabhaṭiya, translated with notes. Chicago, IL: University of Chicago Press; 1930.

[26] Colebrooke HT. Algebra with arithmetic and mensuration from the sanskrit of Brahmagupta and Bhāskara II. London: John Murray; 1817.

[27] "Leelavati" of Bhaskara II, English translation by H. T. Colebrooke, edited with notes by Haran Chandra Banerjee, 2nd edition, Calcutta, 1927 (Reprint, Kitab Mahal, Allahabad, 1967).

[28] Datta DB. On Mahavira's solution of rational triangles and quadrilaterals. Bull Calcutta Math Soc 1928–1929;20:267−94.

[29] Datta BB. The scope and limitations of Hindu Ganita. Indian Hist Q 1929;5:479−512.

[30] Datta BB. The Jaina School of mathematics. Bull Calcutta Math Soc 1929;21:115−45.

[31] Datta BB. The science of sulvas, vol. 38. Calcutta: Calcutta University Press; 1931. p. 371–6

[32] Datta BB. On the origin of Hindu terms for root. Am Math Mon 1931;38:371−6.

[33] Datta BB. Early literary evidence for the use of zero in India. Am Math Mon 1931;38:569.

[34] Datta BB. The Bakshāli mathematics. Bull Calcutta Math Soc 1929;21:1−60.

[35] Datta BB, Singh AV. History of hindu mathematics, vols. 1 and 2. Bombay: Asian Publishing House; 1962.

[36] Delire JM. Quadrature, circulature and the approximation of in Indian "Sulbasutras". Centaurus 2005;47:60−71.

[37] Divakaran PP. The first text book of calculus: Yuktibhasa. J Indian Philos 2007;35:417−43.

[38] Siddanta Siromani of Bhaskara II With Bhaskara's commentary Vasana, by S, Dvevedi, Kashi Sanskrit Series No. 72, Benares, 1929. [This reference you can find in the Encyclopedia of exact sciences from antiquity to 1947.]

[39] Dvivedi S, editor. Trisatika of Sridharacārya. Benares: Chowkhamba Sanskrit Book Depot; 1899.

[40] Emch GG, Sridharan R, Srinivas MD, editors. Contributions to the history of indian mathematics (culture and history of mathematics). New Delhi: Hindustan Book Agency; 2005.

[41] Eves H. An introduction to the history of science. New York, NY: Holt, Reinhart and Winston; 1964.

[42] Feuerstein G, Kak S, Frawley D. In search of the cradle of civilization. Wheaton, IL: Quest Books; 1955.

[43] Frawley D. The myth of the Aryan invasion of India. New Delhi: Voice of India; 1994.

[44] Ganapathi S. Tantrasamuccaya of Narayana with commentary, Vimarsini of Sankara, Trivandrum Series, 67, 1921.

[45] Ganguli S. Notes on Indian mathematics: a criticism of G.R. Kaye's interpretation. Isis 1929;12:132−45.

[46] Ganguli SK. Bhāskara Cārya and simultaneous indeterminate equations of first degree. Bull Calcutta Math Soc 1916;18:89−98.

[47] Ganguli SK. Notes of Āryabhata I. J Bihar Orissa Res Soc 1926;12:89−99.

[48] Ganguli SK. Indian's contribution to the theory of indeterminate equations of first degree. J Indian Math Soc 1931;19:110−21 and 129–42

[49] Gidwani BS. Return of the Aryans. Canada and India: Penguin Books; 1994.

[50] Gupta RC. Circumference of the Jambudvipa in Jain cosmography. Indian J Hist Sci 1975;10(1):38−46.

[51] Gupta RC. Hindu method of solving quadratic equations. J Birla Inst Technol 1966–1967;26–28.

[52] Gupta RC. Bhaskara I's approximation to sine. Indian J Hist Sci 1967;2(2):121–36.

[53] Gupta RC. Second order interpolation in Indian mathematics up to the fifteenth century AD. Indian J Hist Sci 1969;4(1–2):86–98.

[54] Gupta RC, Danben JW, Scriba CJ, editors. Writing the history of mathematics: its historical development. Basel: Birkhäuser; 2002.

[55] Gupta RC. Vedic circle–square conversions: new texts and rules. Ganita Bharati 2004;26:27–39.

[56] Gupta SP. Longer chronology of Indus–Sarasvati civilization. Puratatva 1992–1993;23:21 29

[57] Halstead GB. On the foundation and techniques of arithmetic. Chicago, IL: Open Court; 1912.

[58] Halstead GB. Evidence of early use of zero in India. Am Math Mon 1926;33:449–54.

[59] Hayashi TA, Kasuba T, Yano M. Indian values for derived from Aryabhata's value. Hist Sci 1989;37:1–16.

[60] Hayashi TA, Kasuba T, Yano M. The correction of the Madhava series for the circumference of a circle. Centaurus 1990;33:149–74.

[61] Hayashi T. Indian mathematics. In: Helen Dunmore and Ivor Grattan-Guiness, editors. Companion encyclopedia of the history and philosophy of the mathematical sciences, vol. 1. Baltimore, MD: JHV Press; 2003. p. 118–80.

[62] Hayashi T. Indian mathematics. In: Flood G, editor. The Blackwell companion to Hinduism. Oxford: Basil Blackwell; 2005978-1-4051-3251-0. p. 360–75.

[63] Hayashi T. The Bakshali manuscript: an ancient Indian mathematical treatise. Redwood City, California: John Benjamin Publishing Company; 1995.

[64] Hoerule AFR. The Bakshali manuscript. Indian Antiq 1888;17:33–48 and 275–9

[65] Hunter GR. The script of Harappa and Mohenjodaro and its connection with other scripts. London: Kegan Paul, Trench, Trubner; 1934.

[66] Ikeyama S. The Brahma-Sphuta-Siddhanta Chapter 21 with Commentary of Pruthudakasvamin. Ph.D. dissertation. Providence, RI: Brown University; 2002.

[67] Jain PK. The Suryaprakaśa of Sūryadāsa, vol. 1. Vadodara: Oriental Institute; 2001.

[68] Joseph GG. The crest of peacock. Princeton, NJ: Princeton University Press; 2000.

[69] Victor K, editor. A source book in the mathematics of Egypt, Mesopotamia, China, India and Islam. Princeton, NJ: Princeton University Press; 2007.

[70] Kaye GR. Indian mathematics. Calcutta and Simla: Thacker, Spink; 1915.

[71] Kaye GR. The Bakshali manuscript: a study of medieval mathematics. Archaeological Study of India, New Imperial Series 43; 1933.

[72] Agathe K. Expounding the mathematical seed: a translation of Bhaskara I on the mathematical chapter of Aryabhatiya, vol. 2. Basel, Boston, Berlin: Birkhäuser Verlag; 2006.

[73] Agathe K. Making diagrams speak, in Bhaskara I's commentary on the Aryabhatiya. Historia Math 2005;32(3):275–302.

[74] Kichenassamy S. Brahmagupta's derivation of the area of a cyclic quadrilateral. Historia Math 2010;37(1):28–61.

[75] Knudsen T. The Siddhanta-Sundara of Jnanaraja. Ph.D. dissertation. Providence, RI: Brown University; 2008.

[76] Krishna Swamy Iyengar AA. Bhāskara II and Samslista Kuttaka. J Indian Math 1929;18:232–45.

[77] Kunjunni Raja K. Astronomy and mathematics in Kerala. Adyar Library Bull 1963;27:117−67.

[78] Sastri K, editor. Mahābhāskariya of Bhāskara I with Bhāsya of Govindaswami. Madras Government Oriental Manuscript Library; 1957.

[79] Lakshminatham V, Leela S. The origin of Mathematics. Lanham, MD: United Press of America; 2000.

[80] Lakshminatham V, Leela S, Vasundara Devi J. The origin and history of mathematics. Cambridge: Cambridge Scientific Publishers; 2005.

[81] Lakshminatham V, Vatsala AS. The origin of mathematics. Nonlinear Stud 1988;5:255−7.

[82] Lancelot H. Mathematics for the millions. London: George Allen and Unwin Limited; 1957.

[83] Leach L. Aryan invasions over four millennia. In: Ohnuki-Tierney E, editor. Culture through time. Stanford, CA: Stanford University Press; 1990.

[84] Mazumdar NK. The Manava Sulva Sutra by Manu English translation by N. K. Mazumdar, Journal of Department of Letters, Calcutta University 1922;8:372−442.

[85] Mazumder NK. Āryabhata's rule in relation to indeterminate equations of first degree. Bull Calcutta Math Soc 1912;3:2−9.

[86] Christopher M. On Suryadasa and the invention of bidirectional poetry. J Am Oriental Soc 2004;124:325−33.

[87] David M. Review of the book, Mathematics in India by Kim Plofker. Notices of the American Mathematical Society; March 2010. p. 385–90.

[88] Nayar S.K., "Karanapaddhati", edited with two Malayalam commentaries, Madras Govt. Oriental Service, No. 98, 1956, Madras.

[89] O'connor JJ, Robertson EF. Paramesvara, Mac Tutor History of Mathematics Archive; 2000.

[90] Parameswaran S. The golden age of Indian mathematics. Kerala: Swadeshi Science Movement; 1998.

[91] Shankara Patawardan, Somashekara Amrita Naimpally and Shyam Lal Singh. First edition. Lilavati of Bhaskaracharya: a treatise of mathematics of vedic tradition. New Delhi: Motilal Banarsidass; 2001, ISBN: 81-208-1420-7.

[92] Patte F. Le Siddhantasiromani: l'oeuvre Mathematique et Astronomique de Bhaskaracharya. Geneva: Librarie, Floz; 2004.

[93] Pillai SK, editor. Tantrasangraha of Nilakantha, with commentary of T. Sankara Variyar. Trivandrum Sanskrit Series 188; 1958.

[94] Pingree D. The logic of non-western science: mathematics discoveries in medieval India. Daedalus 2003;132(4):45−54.

[95] Pingree D. On the date of Mahasiddhanta of second Aryabhata. Ganita Bharati 1992;14:55−56.

[96] Plofker K. Mathematics in India. Princeton, New Jersey: Princeton University Press; 2009.

[97] Puttaswamy TK. Mathematical accomplishments of ancient Indian mathematics. In: Selin H, editor. Mathematics across cultures: the history of non-western mathematics. Great Britain: Kluwer Academic Publishers; 2000. p. 409−22.

[98] Rajagopal CT, Murar KM. Hindu quadrature of a circle. J R Asiat Soc (Bombay Branch) 1944;20:66−82.

[99] Rajagopal CT, Venkataraman R. The sine and cosine power series in Hindu mathematics and with an addendum by K.M. George. J R Asiat Soc (Bengal Branch) 1949;25 (1):1−13.

[100] Rajagopal CT, Vedamurthi Iyer TV. On the hindu proof of Gregory's series. Scr Math 1952;18:65–74.
[101] Rajaram NS, Frawley D. Vedic Aryans and the origins of civilization. Quebec: W.H. Press; 1995.
[102] Raju CK. Cultural foundations of mathematics: the nature of mathematics proof and the transmission of calculus from India to Europe in the 16th Century C.E. (History of science, philosophy and culture in India Civilization, vol. X, part 4. New Delhi: Pearson Longman; 2007.
[103] Ramakrishna Bhatt M, trans. Varamihira's Brihat Samhita. New Delhi: Motilal Banarsidass; 1981.
[104] Rangacharya M, editor. Ganita Sara Samgraha of Mahavira, with translation and notes. Madras: Madras Government Publications; 1912.
[105] Rangacharya M, Pre-historic India, vol. 1; 1937.
[106] Rao SR. Dawn and devolution of the Indus Civilization. New Delhi: Aditya Prakashan; 1991.
[107] Sachu EC, trans. Al-Biruni's India, vols. I and II. New Delhi: Low Price Publications; 1989 [edited with notes and indices].
[108] Sambasiva Sastry K, editor. Āryabhatiya with Bhāsya of Nilakantha, Trivandrum. Trivandrum Sanskrit Series 68; 1930.
[109] Sarasvatiamma T. Development of mathematical series in India. Bull Natl Inst Sci 1963;21:320–43.
[110] Sarma KV. Ganita-Yuktibhasa of Jyesthedeva, 2 volumes. New Delhi: Hindustan Book Agency; 2008 [notary notes in English by Ramasubramanian K, Srinivas MD, Sriram MS].
[111] Sarma KV. Science texts in sanskrit in mansucripts repositories of Kerala and Tamilnadu. New Delhi: Rashtriya Sanskrit Sansthan; 2002.
[112] Sarma KV, editor. A history of Kerala school of hindu astronomy. Hoshiarpur: Vishveshvaranand Institute; 1972.
[113] Sarma KV. Goladipikā of Parameśvara. Madras; 1956–1957 [edited with introduction notes and translations]. Research in Sanskrit, Kuppaswami Sastry Institute, Adarshe Sanskrit shodha Samasths, Madras 600004.
[114] Sarma KV, editor. Tantra Sangraha of Nilakantha Somayaji. Hoshiarpur: V.V.B.I.S. and I.S.; 1977.
[115] Sarma KV, editor. Sphuta-Candraptih of Madhava of Sangamagrama with notes and translation. Hoshiarpur: Vishveshvaranand Vedic Research Institute; 1973.
[116] Sarma KV. Some highlights of astronomy and mathematics in medieval India. Sanskrit World Culture 1986;18:595–605.
[117] Sarma KV, editor. Dṛgganitam of Parameśvara. Hoshiarpur: Vishveshvaranand Vedic Research Institute; 1963.
[118] Sarasvatiamma T. Geometry in ancient and medieval India. New Delhi: Motilal Banarsidass; 1979.
[119] Sathe S. Aryans: who are they? Mysore: Bharatiya Itihasa Sankalana Samiti; 1991.
[120] Selin H, editor. Mathematics across cultures, non-western mathematics. Princeton, New Jersey: Kluwer Academic Publications; 2000.
[121] Sen SN, Bag AK. Sulvasutra of Baudhayana, Āpasthmba and Kātyayana. New Delhi: I.N.S.A.; 1983.
[122] Sengupta PC. Brahmagupta on interpolation. Bull Calcutta Math Soc 1931;23:123–8.
[123] Sethna KD. The problem of Aryan origins. New Delhi: Aditya Prakashan; 1992.
[124] Sethna KD. Ancient India in a new light. New Delhi: Aditya Prakashan; 1989.

[125] Sharma V, editor. Sulvasutra of Katyayana. Benares: Achuta Grauthamla; 1928.

[126] Shukla KS, editor. Patiganita of Sridharacarya, astronomical and mathematical text series, vol. 2. Lucknow: Lucknow University; 1959.

[127] Shukla KS, editorial translator, Bhaskara I and his works, "Mahabhaskariya" in Sanskrit, Lucknow, 1960.

[128] Shukla KS. A critical study of the laghu-manasa of Manjula. New Delhi: Indian National Science Academy; 1990.

[129] Smith DE. History of Mathematics. New York, NY: Dover Publications; 1958.

[130] Somyaji DA. A critical study of ancient hindu astronomy. Dharwar: Karnataka University; 1971.

[131] Srinivasachar D, Narasimhachar S, editors. Āpasthamba Sulbasutras, vol. 73. Mysore: Mysore University Oriental Library Publications; 1931.

[132] Srinivasiengar CN. History of ancient indian mathematics. Calcutta: The World Press Private Limited; 1967.

[133] Staal F. Greek and vedic geometry. J Indian Philos 1999;27:105–27.

[134] Stedall Jacqueline, Robson Eleanor, editors. The Oxford handbook of the history of mathematics. Oxford: Oxford University Press; 2008.

[135] Stillwell J. Mathematics and its history. 2nd ed. New York: Springer; 2004, ISBN: 0-37-95336-1.

[136] Struik DJ. A concise history of mathematics. New York, NY: Dover Publications; 1967.

[137] Subbarayappa BV, Sarma KV. Indian astronomy—a science book. Bombay: Nehru Center; 1985.

[138] Swarup E, Bag AK, Shukla KS. History of oriental astronomy. Cambridge: Cambridge University Press; 1987.

[139] Swarup R. Sharma (editor and trans.). Brahmasphuta Siddhanta of Brahmagupta. New Delhi: Indian Institute of Astronomical and Sanskrit Research; 1981.

[140] Talegeri SG. The Aryan invasion theory: a reappraisal. New Delhi: Aditya Prakashan; 1993.

[141] Thampuran R, Akileswara Iyer AR. Yuktibhasa—part I with notes. Trichur: Mangalodayam Limited; 1952.

[142] Thibaut G. On the Sulvasutras. Journal of Asiatic Society of Bengal 1875;44 (3):227–75.

[143] "Baudhayana Sulba Sutra" by Baudhayana, edited by G. Thibaut, with English translation, critical notes and extracts with commentary of Dwarakanath Yajavau, the Pandit Series IX and X, Benares, 1874–75.

[144] Thibaut G, Dvivedi S. Panchasiddhantika of Varamihira. New Delhi: Motilal Banarsidass; 1930.

[145] Tirthaji BK. Vedic mathematics. New Delhi: Motilal Banarsidass; 1998.

[146] Upadye AN, Jain H, editors. Jambudiva Prjnapti Samgraha. Sholapur: Jain Samkriti Samraksak Sangh Press; 1958.

[147] Van der Waerden BL. Science awakening. 2nd ed. Groningen: P. Noordhoff Limited; 1961.

[148] Varma Raja Raja V. History of sanskrit literature in Kerala. Trivandrum: Kamalalaya Book Depot; 1949.

[149] Whish CM. On the Hindu quadrature of a circle. Trans R Asiat Soc GB Irel 1835;3:509–23.

[150] Dickson LE. Theory of numbers, Vol. 2. American Mathematical Society; 1999.

[151] Dutta BB. The present mode of expressing numbers, Indian Hist. Quar. 3, 530–540 (rule for the solution).
[152] Dutta BB. Elder Aryabhata's rule for the solution of indeterminate equation first degree. Bull. Calcutta Math. Soc. 1932;24:35−53.
[153] Drummond JH. Am. Math. Monthly 1902;IX:232.
[154] Genocchi MA. Genocchi's solution. Nouvelles Annal. Math., Series 10 1910;16:225−7.
[155] Gupta RC. Functional parts of Govindasvamin Bhasya on the Mahabhaskariya. Indian J. Hist. Sci. 1971;6:51−59.
[156] Hankel H. Zur Geschichte bder Mathematik, Leipzig, 1874.
[157] Jinsberg J. New light on our numerals. Bull. Am. Math. Soc. 1917;25:366−7.
[158] Kamalamma KN. Bull. Calcutta Math. Soc. 1948;40:140−4.
[159] Keith AB. History of Sanskrit Literature. Delhi: Motilal Banarsidas Publication; 1948.
[160] Krishnaswamy Iyyangar AA. Hindu sine table. J. Indian Math. Soc. 1924–25;15:121−6, First Part.
[161] Naraharayya SN. Notes and questions. J. Indian Math. Soc. 1924–25;15:113, Second Part.
[162] Nau MF. Notes d'asronome syrienne. J. Asiatique Series 10. 1910;16:225−7.
[163] K.V. Sharma, Jyesthedeva and the identification as the author of Yuktibhasa, Adyar Library Bull. XXVI, 35–40.

www.ingramcontent.com/pod-product-compliance
Ingram Content Group UK Ltd.
Pitfield, Milton Keynes, MK11 3LW, UK
UKHW020129290926
15471UKWH00003B/364

9 780323 282642